GREAT EVENTS FROM HISTORY II

GREAT EVENTS FROM HISTORY II

Science and Technology Series

Volume 1
1888-1910

Edited by
FRANK N. MAGILL

SALEM PRESS
Pasadena, California Englewood Cliffs, New Jersey

∞ The paper used in these volumes conforms to the American National Standard for Permanence of Paper for Printed Library Materials, Z39.48-1984.

Library of Congress Cataloging-in-Publication Data
Great events from history II. Science and technology series / edited by Frank N. Magill.
p. cm.
Includes bibliographical references and index.
1. Science—History—20th century. 2. Technology—History—20th century. I. Magill, Frank Northen, 1907-

Q125.G825 1991
509′.04—dc20
ISBN 0-89356-637-3 (set) 91-23313
ISBN 0-89356-638-1 (volume 1) CIP

PRINTED IN THE UNITED STATES OF AMERICA

PUBLISHER'S NOTE

Great Events from History II: Science and Technology Series inaugurates a new series in the Magill family of reference books. This series joins the original, twelve-volume *Great Events from History: Ancient and Medieval* (three volumes, 1972), *Modern European* (three volumes, 1973), *American* (three volumes, 1975), and *Worldwide Twentieth Century* (three volumes, 1980), which surveyed, in chronological order, major historical events throughout the world from 4000 B.C. to 1979.

The new series, *Great Events from History II* (GEFH II), will present entirely original articles, treating twentieth century events never before covered. The current five volumes of *Science and Technology* address 458 modern scientific and technological breakthroughs, thereby supplementing and extending the scope of the original series. Future volumes in GEFH II will concentrate on other major areas of contemporary culture.

Like the original *Great Events*, the articles in GEFH II are arranged chronologically by date of event, starting at the beginning of the twentieth century and ending in 1991. A broad range of scientific and technological categories are addressed here: anthropology, applied science, archaeology, astronomy, biology, chemistry, earth science, mathematics, medicine, physics, space and aviation. From the patenting of the vacuum tube by Fleming (1904) through the invention of Xerography by Carlson (1934-1938) to the introduction of optical disks for the storage of computer data (1984), the technological milestones of our technological age are represented by 66 articles in applied science. Discoveries concerning humankind and our evolutionary and historical development are treated in 28 articles on anthropology and archaeology. The major strides made during the twentieth century in the understanding of the universe and its phenomena appear in 65 articles on astronomy, and—combined with 40 articles on earth scientists' discoveries concerning our own planet and the 67 articles treating the events that have revealed the physical principles underlying these phenomena—present the broad mosaic of twentieth century advances in these areas. Biology, particularly at the molecular and genetic levels, and its "applied" side, medicine, have witnessed dramatic advances in this century; it was difficult to limit these areas to the 130 events covered in as many articles. Chemistry is represented by 24 and mathematics by 16 events. Finally, the twentieth century is the age of aviation and space exploration, and 44 articles address the major technological achievements by both Americans and Soviets.

The articles covering these events run somewhat longer (2,500 words) than the briefer entries of the original *Great Events*, allowing more in-depth coverage of events that—characteristic of scientific and technological developments—may have been preceded by months or years of research. Accordingly, the format of these articles, while retaining the basic outline of the articles in the original series, has been somewhat revised to reflect the needs of the new subject orientation.

Each article still begins with the ready-reference listings: "Category of event" (astronomy, medicine, and so on), "Time" (year and, where applicable, specific date of event), and "Locale," followed by a brief (two- or three-line) summary of the event and its significance, and then by a listing of "Principal personages" who were key players in the event. The text begins with "Summary of Event," a rehearsal of the background and circumstances leading up to and forming the event, followed by "Impact of Event." This latter section (which did not appear in the original *Great Events* series) devotes itself to an assessment of the immediate and long-range significance of the invention, discovery, or other scientific/technological development. Next, an annotated bibliography lists and describes between five and ten sources for further study, enabling the reader to choose among books and articles that will cast more light on the topic and its context. Finally, the "Cross-References" appearing at the end of each article lead the reader to other articles of interest which appear in the current five volumes.

In a reference work of this type, in which articles are arranged chronologically to provide a temporal perspective on the individual events, the editors thought it useful to add a complete chronological listing of all 458 articles at the end of each volume. In addition, several indexes were thought to be not only desirable but also necessary to provide maximum flexibility in retrieving information. At the end of volume 5, therefore, the reader will find four indexes: The "Alphabetical List of Events" indexes events alphabetically by the article title; the "Key Word Index" cross-references events by two or three key words per title; the "Category Index" lists all articles by scientific/technological discipline (such as physics); and finally, "Principal Personages" lists all key figures appearing in the corresponding subsection of each article. In all four of these useful indexes, both the volume number and the page number appear following each index entry.

Great Events from History II: Science and Technology Series, unlike the original series, lists complete names of the academicians and scholars who wrote these articles, both at the end of the article, in a byline, and in a listing of contributors to be found in the front matter to volume 1. We wish to acknowledge the efforts of these specialists and thank them for making their expert knowledge accessible to the general reader.

CONTRIBUTING REVIEWERS

Richard Adler
University of Michigan—Dearborn

Margaret I. Aguwa
Michigan State University

Michael S. Ameigh
State University of New York College at Oswego

James H. Anderson
Somerset Community College

David W. Appenbrink
Chicago State University

Richard W. Arnseth
Science Application International Corp.

Iona C. Baldridge
Lubbock Christian University

Grace A. Banks
Chestnut Hill College

Russell J. Barber
California State University, San Bernardino

Charles A. Bartocci
Dabney S. Lancaster Community College

Christopher J. Biermann
Oregon State University

Michael S. Bisesi
Medical College of Ohio

Daniel P. Boehlke
Gustavus Adolphus College

Paul R. Boehlke
Dr. Martin Luther College

Nathaniel Boggs
Alabama State University

Lucy Jayne Botscharow
Northeastern Illinois University

Kenneth H. Brown
Northwestern Oklahoma State University

Michael L. Broyles
Collin County Community College

John T. Burns
Bethany College

Byron Cannon
University of Utah

James A. Carroll
University of Nebraska at Omaha

P. John Carter
St. Cloud State University

E. L. Cerroni-Long
Eastern Michigan University

Dennis Chamberland
Science Writer

Victor W. Chen
Chabot College

Eric Howard Christianson
University of Kentucky

Jaime S. Colome
California Polytechnic State University, San Luis Obispo

Albert B. Costa
Duquesne University

Jennifer L. Cruise
College of St. Thomas

Scott A. Davis
Mansfield University of Pennsylvania

Ronald W. Davis
Western Michigan University

Dennis R. Dean
University of Wisconsin—Parkside

Jeffrey R. DiLeo
Indiana University, Bloomington

Matthias Dörries
University of California, Berkeley

Steven I. Dutch
University of Wisconsin—Green Bay

George R. Ehrhardt
Duke University

Harry J. Eisenman
University of Missouri—Rolla

Suzanne Knudson Engler
University of Southern California

K. Thomas Finley
State University of New York College at Brockport

David G. Fisher
Lycoming College

George J. Flynn
State University of New York College at Plattsburgh

Robert G. Font
Font Geosciences Consulting

Michael J. Fontenot
Southern University, Baton Rouge

John M. Foran
Independent Scholar

Robert J. Forman
St. John's University

Thomas P. Gariepy
Stonehill College

Roberto Garza
San Antonio College

Judith R. Gibber
New York University

Karl Giberson
Eastern Nazarene College

Harold Goldwhite
California State University, Los Angeles

Douglas Gomery
University of Maryland

Hans G. Graetzer
South Dakota State University

Harold J. Grau
University of the Virgin Islands

Noreen A. Grice
Charles Hayden Planetarium

Gershon B. Grunfeld
SIU School of Medicine

Ronald B. Guenther
Oregon State University

Lonnie J. Guralnick
Western Oregon State College

William J. Hagan, Jr.
Saint Anselm College

Robert M. Hawthorne, Jr.
Unity College in Miami

Judith E. Heady
University of Michigan—Dearborn

John A. Heitmann
University of Dayton

Jean S. Helgeson
Collin County Community College

Arthur W. Helweg
Western Michigan University

Richard A. Hendry
Westminster College

Paul Hodge
University of Washington

David Wason Hollar, Jr.
Rockingham Community College

Earl G. Hoover
Science Writer

Ruth H. Howes
Ball State University

Mary Hrovat
Indiana University, Bloomington

David Hsu
Boston University

Richard C. Jones
Rollins College

Richard N. Jones
Independent Scholar

Thomas W. Judd
State University of New York College at Oswego

John P. Kenny
Bradley University

Robert Klose
University of Maine

Richard S. Knapp
Belhaven College

Jeffrey A. Knight
Mount Holyoke College

Kevin B. Korb
Indiana University, Bloomington

Ludwik Kowalski
Montclair State College

Craig B. Lagrone
Birmingham Southern College

Leon Lewis
Appalachian State University

CONTRIBUTING REVIEWERS

Ronald W. Long
West Virginia Institute of Technology

Robert Lovely
University of Wisconsin—Madison

Michael J. McAsey
Bradley University

David F. MacInnes, Jr.
Guilford College

William J. McKinney
Indiana University, Bloomington

Paul Madden
Hardin-Simmons University

V.L. Madhyastha
Fairleigh Dickinson University

David W. Maguire
Mott Community College

Laura Gray Malloy
Bates College

Joseph T. Malloy
Hamilton College

Nancy Farm Mannikko
Virginia Polytechnic Institute and State University

Susan M. Maskel
Western Connecticut State University

Joseph M. Maturo III
C. W. Post College of Long Island University

Grace Dominic Matzen
Molloy College

Maureen S. May
Rochester Institute of Technology

Andre Millard
University of Alabama at Birmingham

Gordon L. Miller
Independent Scholar

Randall L. Milstein
Oregon State University

Ellen F. Mitchum
Space Center

Susan J. Mole
Siena Heights College

Rodney C. Mowbray
University of Wisconsin-La Crosse

Turhon A. Murad
California State University, Chico

Charles Murphy
Independent Scholar

Taha Mzoughi
Governor's School of Sciences and Math

Indira Nair
Carnegie Mellon University

Peter Neushul
University of California, Santa Barbara

Anthony J. Nicastro
West Chester University of Pennsylvania

William D. Niemi
Russell Sage College

Edward B. Nuhfer
University of Wisconsin—Platteville

Paul G. Nyce
Eastern Nazarene College

Marilyn Bailey Ogilvie
Oklahoma Baptist University

Robert J. Paradowski
Rochester Institute of Technology

Gordon A. Parker
University of Toledo

Keith Krom Parker
Western Montana College

Joseph G. Pelliccia
Bates College

Robert T. Pennock
University of Pittsburgh

John R. Phillips
Purdue University-Calumet

Bernard Possidente, Jr.
Skidmore College

Wen-yuan Qian
Blackburn College

Marc D. Rayman
Jet Propulsion Laboratory

Robert Reeves
Science Writer

N.A. Renzetti
Jet Propulsion Laboratory
California Institute of Technology

Brian L. Roberts
Northeast Louisiana University

Charles W. Rogers
Southwestern Oklahoma State University

Daniel W. Rogers
Somerset Community College
University of Kentucky

Deborah L. Rogers
Somerset Community College

René R. Roth
University of Western Ontario, Canada

Emanuel D. Rudolph
Ohio State University

Virginia L. Salmon
Virginia Polytechnic Institute and State University

Rosemary Scheirer
Chestnut Hill College

Nancy Schiller
State University of New York at Buffalo

Joseph A. Schufle
New Mexico Highlands University

Robert W. Seidel
Los Alamos National Laboratory

Nancy J. Sell
University of Wisconsin-Green Bay

Roger Sensenbaugh
Indiana University, Bloomington

John M. Shaw
Education Systems Incorporated

Martha Sherwood-Pike
University of Oregon

Stephen J. Shulik
Science Writer

R. Baird Shuman
University of Illinois at Urbana-Champaign

Patricia J. Siegel
State University of New York College at Brockport

Sanford S. Singer
University of Dayton

Paul P. Sipiera
Harper College

Genevieve Slomski
Independent Scholar

Clyde J. Smith
South Carolina Governor's School for Science and Mathematics

Roger Smith
Willamette University

Kenneth S. Spector
Saint Louis University School of Medicine

Joseph L. Spradley
Wheaton College

Grace Marmor Spruch
Rutgers, State University of New Jersey, Newark

Michael A. Steele
Wilkes University

William F. Steirer, Jr.
Clemson University

Joan C. Stevenson
Western Washington University

Anthony N. Stranges
Texas A&M University

Patricia Summers
Virginia Polytechnic Institute and State University

Charles E. Sutphen
Blackburn College

Robert M. Swerdlow
New York University

Larry N. Sypolt
West Virginia University

Joyce Tang
University of California, Berkeley

Gerardo G. Tango
Consulting Geophysicist

Eric R. Taylor
University of Southwestern Louisiana

Wilfred R. Theisen
St. John's University

Russell R. Tobias
Spaceflight Editor
Ærospace Educator

Robin S. Treichel
Oberlin College

CONTRIBUTING REVIEWERS

Mary S. Tyler
University of Maine

Charles F. Urbanowicz
California State University, Chico

Peter J. Walsh
Fairleigh Dickinson University

Richard L. Warms
Southwest Texas State University

Nan White
Maharishi International University

Donald H. Williams
Hope College

Bradley R. A. Wilson
University of Cincinnati

John Wilson
Independent Scholar

Shawn Vincent Wilson
Independent Scholar

David Wu
University of California, Berkeley

Jay R. Yett
Orange Coast College

Ivan L. Zabilka
Independent Scholar

Debra Zehner
Wilkes University

LIST OF EVENTS IN VOLUME I

GREAT EVENTS FROM HISTORY II

RAMÓN Y CAJAL ESTABLISHES THE NEURON AS THE FUNCTIONAL UNIT OF THE NERVOUS SYSTEM

Categories of event: Biology and medicine
Time: 1888-1906
Locale: Madrid, Spain

Ramón y Cajal showed that nerve cells operate as the discrete entities that transmit impulses unidirectionally in the nervous system through specific points of contact

Principal personages:

SANTIAGO RAMÓN Y CAJAL (1852-1934), a Spanish histologist and physician who was a cowinner of the 1906 Nobel Prize in Physiology or Medicine

CAMILLO GOLGI (1843-1926), an Italian anatomist and physician who was the developer of the Golgi staining technique and cowinner of the 1906 Nobel Prize in Physiology or Medicine

RUDOLF ALBERT VON KÖLLIKER (1817-1905), a Swiss embryologist and histologist who wrote the first formal treatise on histology and who helped disseminate the findings and theories of Ramón y Cajal

PAUL EHRLICH (1854-1915), a German chemist who developed a new methylene blue stain for living tissues

Summary of Event

In the late 1800's, a controversy existed among brain scientists as to the nature of impulse transmission through the nervous system. The more popular school of thought began in 1872, when the gray matter of the cerebrum (brain) was described as a diffuse nerve net with fusion of dendrites (fine processes). Such notables as Theodor Meynert and Camillo Golgi agreed with this "reticular theory." According to this theory, the proper impulses were directed somehow out from this network of fused fibers to the appropriate muscles and organs, much like the streams flowing out from a lake being directed to specific locations. A. H. Forel showed that retrograde degeneration was confined to the damaged cells, and Wilhelm His showed that in embryos the nerve cells behave as centers giving origin to fiber outgrowths. At the time, the physical evidence was inadequate for determining which theory was more correct. No one had been able to see, with any clarity, the nerve fiber endings. Yet, Santiago Ramón y Cajal was able to provide the irrefutable evidence that resolved the issue eventually.

Ramón y Cajal's legacy was born of the microscope, an instrument he first became familiar with in 1877 while studying in Madrid. He soon became expert in histology, the field of biology devoted to the study of tissues. In the process, he made innovations in the staining techniques of the time. Tissue staining is a technique of applying dyes or other chemicals to the material studied so that particular structures or features are seen more easily. In 1888, while at the University of Barcelona, Ramón y Cajal made a major finding. He worked with bird and small mammal embryos because embryonic nerve cells have fewer interconnections and are not covered yet with myelin, a fatty layer of insulation that covers the axons, or long processes, of most adult neurons, allowing for individual cells to stand out. Ramón y Cajal also modified the chrome silver staining technique invented by Golgi. Focusing on the cerebellum of the brain, Ramón y Cajal discovered basket cells and mossy fibers. He found that the long processes (axons) of the nerve cells terminate in proximity, not continuity, with other cells or dendrites. This suggested that the nervous system works by contact, and thus Ramón y Cajal developed his "law of transmission by contact." For the first time, it was proposed that the cells were the important components of the nervous system, as opposed to the fibers, which previously were believed to be a continuous network that transmitted impulses in the system. The popular notion was that the cells played a relatively minor, supportive role. Ramón y Cajal's evidence of definitely limited conduction paths in the gray matter would be substantiated by later investigations of the retina, spinal cord, and other brain regions. Yet, the reticular theory proponents would not be defeated so easily.

After giving a presentation to the German Anatomical Society in Berlin in 1889, Ramón y Cajal won the support of Rudolf Albert von Kölliker, then editor of a scientific journal. Kölliker helped to promote the theories of Ramón y Cajal throughout Europe.

In 1890, Ramón y Cajal turned to a different set of studies that would support his theory. He demonstrated that developing nerve cells send out axon "growth cones" that later sprout dendrites and make connections, supporting the theory of His and Kölliker. There was an alternate theory favored by the "reticular" advocates stating that developing nerves arise from the fusion of a row of cells.

In 1891, Ramón y Cajal returned to studying the cerebrum, a subject of some investigations several years past. He published a well-received book of his results. Waldeyer coined the term "neuron"; Ramón y Cajal's ideas became known as the "neuron theory." In 1892, Ramón y Cajal proposed his neurotropic theory to explain how the growth cone of the developing neuron is directed to its proper target. For the next few years, he would work on the retinas of various animals and on the hippocampus. The consistency of the results, with respect to his neuron theory, was unequivocal.

In 1896, Ramón y Cajal learned of Paul Ehrlich's methylene blue technique for staining tissues. This was a nontoxic substance and thus could be applied to living animals. Ramón y Cajal repeated many of his findings on living specimens using this technique to refute the criticism from such "reticularists" as Golgi that his results were an anomaly of his previous methods on dead tissues. Again, the results were irrefutable: Nerve cells existed as distinct units, making only the barest contact with other nerve cells in the system. Detractors of the neuron theory persisted. They objected on the grounds that Ramón y Cajal had not shown that neuron fibrils (internal structures) were not continuous, as they had claimed in support of the reticular theory.

The beginning of the twentieth century found Ramón y Cajal in Paris to receive the Moscow Prize of the 13th Medical Congress. Not only did he receive the prestigious award, with its monetary bonus, but also the Congress awarded Madrid the venue for the next Congress to be held in 1903. This brought great praise and adulation to him at home. He was appointed director of the new National Institute of Hygiene, where he promptly convinced the authorities to establish a laboratory of biological research. The government finally was supporting science in Spain.

By 1903, Ramón y Cajal had perfected yet another staining technique—reduced silver nitrate—which made the tissue transparent, allowing him to discover details about the internal structures of the nerve cells, including fibrils. He published twenty-two papers that year on fibrillar discontinuity, consistent with his neuron doctrine. Dozens of other scientists confirmed his work. For all intents, the debate over reticular versus neuron was won, but some die-hard reticular advocates persisted, even into the 1950's.

The recognition of Ramón y Cajal's theories reached a pinnacle in 1906, when he was awarded the Nobel Prize in Physiology or Medicine. Somewhat ironically, he shared the prize with Golgi, whose staining technique had made many of Ramón y Cajal's early findings possible, but who was still an advocate of the reticular hypothesis and a severe critic of Ramón y Cajal, so much so that he embarrassingly used his Nobel lecture to attempt a critique of Ramón y Cajal's methods and results. History would forgive Golgi his myopia and would establish Ramón y Cajal as a scientist of tremendous impact on modern neuroscience.

Impact of Event

The neuron doctrine of Ramón y Cajal had an impact on many different fields within biology and medicine. The knowledge of the actual functional structure of the brain—being made of discrete, contacting units that transmit impulses in one direction—gave a better physical basis for understanding many nervous or mental disorders. Treatment now could be approached with a more accurate perspective on the possible deficiencies in the disorders. A modern understanding of impulse transmission through the nervous system, from sensory neurons to central nervous system and then to muscle, is directly reflective of Ramón y Cajal's theory. The "black box" that had existed between the reception of a stimulus event and the control of a motor response, while still not completely revealed, was at least partly illuminated by this work. Possible mechanisms of learning and memory were developed with the framework established by the neuron doctrine. The foundation for the concept of the final common pathway of Sir Charles Scott Sherrington, the principle of reflex activity that is adhered to today, was laid by Ramón y Cajal and his microscope.

Ramón y Cajal's neuron doctrine influenced histologists, physiologists, surgeons, pathologists, neurosurgeons, psychiatrists, and psychologists. The concepts of neural inhibition, summation, and facilitation that are accepted now are possible only within the context of the neuron doctrine. The English physiologists were led to their work on the synapse, the site of neuron-to-neuron communication, by Ramón y Cajal's studies, and this knowledge of the synapse has had a great impact on medicine, such as in drug therapy and in the treatment of neurological disorders. Ivan Petrovich Pavlov's famous treatise on conditioning was molded, at least in part, by the neuron doctrine, as was the work of Walter Bradford Cannon on the physiological aspects of emotion. Ramón y Cajal's work could be said to have had an impact on education because his neuron doctrine has influenced subsequent theories on mechanisms for learning.

In the course of developing his neuron doctrine, Ramón y Cajal made a number of advances and innovations in histological staining techniques, many of which are still the techniques of choice. The work conducted on regeneration in the nervous system has played a seminal role in nerve damage therapy.

It is possible that had it not been Ramón y Cajal, someone else would eventually have discovered the truth about the structure of the nervous system and the nature of impulse transmission. Nevertheless, this man of meager means and modest beginnings had the inspiration and fortitude to pioneer the murky waters of the time, to bring science out of the ignorance of complacency. His work had a domino, rippling effect on the direction of science, such that the rate of progress in the field of neuroscience has been phenomenal. If Ramón y Cajal had not paved the way, scientists may have been stumbling along into untold dead-ends for many years.

Bibliography

Bullock, Theodore Holmes, Richard Orkand, and Alan Grinnell. *Introduction to Nervous Systems*. San Francisco: W. H. Freeman, 1977. Although this book is somewhat advanced—written for college and medical students—portions of it are accessible to the general reader. In particular, the introduction gives a good overview of the field of neuroscience, and there are numerous illustrations throughout. The work of Ramón y Cajal is evidenced in nearly every chapter. A special section of chapter 3 gives a chronology of the debate over the structure of the brain. Supplemented by a glossary, a reference section, and an index.

Cannon, Dorothy F. *Explorer of the Human Brain: The Life of Santiago Ramón y Cajal (1852-1934)*. New York: Henry Schuman, 1949. This biography is written in a wonderful literary style. The narrative often includes the history of Spain during the periods discussed, putting the reader in a better perspective to appreciate his life. Numerous references and footnotes introduce the reader to the many other scientists who influenced and were influenced by Ramón y Cajal. The epilogue serves as a fine testament to his influence on the world. Includes a bibliography and an index.

Carola, Robert, John P. Harley, and Charles R. Noback. "The Action of Nerve Cells." In *Human Anatomy and Physiology*. New York: McGraw-Hill, 1990. An introductory college-level book. Includes many colorful diagrams and illustrations that will help the reader learn more about the structures and concepts of the nervous system and nerve cells. Full-color photomicrographs also are included. Fully indexed, with an extensive glossary and several appendices.

Fincher, Jack. *The Brain: Mystery of Matter and Mind*. Washington, D.C.: U.S. News Books, 1981. One volume in The Human Body series, written especially for the general public in a very nontechnical style. Several references to Ramón y Cajal are given, including a brief description of the reticular versus neuron debate. Many colorful illustrations and photographs are included, with a section in chapter 3 that covers the details about the neuron. Other chapters cover memory and learning, emotional aspects of the brain, language processing, and medical advances. Glossary and index.

Williams, Harley. *Don Quixote of the Microscope: An Interpretation of the Spanish Savant, Santiago Ramón y Cajal (1852-1934)*. London: Jonathan Cape, 1954. A biography of Ramón y Cajal written in a very literary style, at times a bit overly dramatic and prosaic. Provides a different perspective. A brief reference list is included.

Harold J. Grau

Cross-References

Pavlov Develops the Concept of Reinforcement (1902), p. 163; Sherrington Delivers *The Integrative Action of the Nervous System* (1904), p. 243; Moniz Develops Prefrontal Lobotomy (1935), p. 1060; Cerletti and Bini Develop Electroconvulsive Therapy for Treating Schizophrenia (1937), p. 1086.

BEHRING DISCOVERS THE DIPHTHERIA ANTITOXIN

Category of event: Medicine
Time: 1890-1901
Locale: Berlin, Germany

Behring discovered that a toxin produced by the causative agent of diphtheria could be destroyed by blood serum derived from immunized animals, thus leading to the development of a vaccine

Principal personages:

EDWARD JENNER (1749-1823), an English physician who developed the concept of "vaccination" to induce immunization against smallpox in 1796

EDWIN KLEBS (1834-1913), a German bacteriologist who isolated the diphtheria bacillus in 1883 from humans with the disease

FRIEDRICH LÖFFLER (1852-1915), a German bacteriologist who in 1884 confirmed that the bacillus isolated by Klebs was the causative agent of diphtheria and observed that some animals exhibited immunity to the disease

PIERRE-PAUL-ÉMILE ROUX (1853-1933), a French bacteriologist who demonstrated in 1888 that a toxin produced by the diphtheria bacillus was the true causative factor of diphtheria

ALEXANDRE YERSIN (1863-1943), a Swiss bacteriologist who worked with Roux and was the codiscoverer of the diphtheria toxin

HENRY SEWALL (1855-1936), an American physiologist who discovered that animal models could develop an immunity to toxic snake venom via vaccination

EMIL ADOLF VON BEHRING (1854-1917), a German bacteriologist who discovered antitoxins for tetanus and diphtheria in 1890 and won the 1901 Nobel Prize in Physiology or Medicine

PAUL EHRLICH (1854-1915), a German bacteriologist who worked with Behring to determine the best dosage and technique for the diphtheria antitoxin

SHIBASABURO KITASATO (1852-1931), a Japanese bacteriologist who isolated the causative agents of tetanus and anthrax in 1889 and worked with Behring during development of tetanus and diphtheria antitoxins

Summary of Event

During the nineteenth century, there was an enormous growth of knowledge in the field of bacteriology. Much of the expansion resulted from discoveries by German bacteriologist Robert Koch and French chemist Louis Pasteur. The two scientists are considered the founders of modern medical bacteriology and were instrumental in demonstrating the relationship between exposure to bacteria and specific human and animal diseases. These diseases involved causative agents, or pathogens, such as bacteria, which would interact with human and animal hosts and induce illness caused by infection or toxicity. The illnesses were collectively classified as "communicable diseases" since they could be transmitted from one host to another.

Several scientists and physicians focused on prevention and eradication of communicable diseases. Diphtheria was a major communicable disease studied during this period since it was responsible for many deaths, especially among young children. It is transmissible mainly via direct contact with an infected host and ingestion of contaminated raw milk, and possibly via contact with contaminated articles. The disease is caused by a bacterium called *Corynebacterium diphtheriae*. Certain strains of these bacteria are susceptible to genetic alteration by a virus that causes the organisms to produce a toxin which is responsible for the onset of diphtheria and the related symptoms. Infection of humans with these bacteria and absorption of their toxin can result in the formation of lesions in the nasal, pharyngeal, and laryngeal regions of the upper respiratory system. The toxin can also negatively impact other organs and systems, such as the nerves, heart, and kidneys. If the disease is extensive, diphtheria is fatal; the probability of contracting the disease is increased under crowded conditions.

The prevalence of the disease increased concomitantly as population densities increased in cities around the world. The causative agent of the disease was not discovered until 1883, when a German bacteriologist, Edwin Klebs, isolated the bacteria from people with diphtheria. The discovery was not confirmed, however, until 1884, when a German bacteriologist, Friedrich Löffler, demonstrated that pure cultures of these organisms would induce diphtheria in experimental animals. Indeed, the original name of the causative agent of diphtheria was Klebs-Löffler bacillus, later renamed to *Corynebacterium diphtheriae*.

Five years later, two other scientists, French bacteriologist Pierre-Paul-Émile Roux and Swiss bacteriologist Alexandre Yersin, were able to separate a chemical toxin from *C. diphtheriae* and demonstrate that the chemical was the actual factor that caused diphtheria. Thus, the foundation was established for development of a means for rendering the toxin innocuous in order to prevent the onset or eradicate the symptoms of the disease in people who were exposed to the toxin-producing bacteria.

The German bacteriologist Emil Adolf von Behring and his assistant, a Japanese bacteriologist, Shibasaburo Kitasato, focused on immunization of animals via vaccination. The scope of their research was influenced by concepts established by several other scientists, including an English physician named Edward Jenner, Pasteur, and the American physiologist Henry Sewall. Jenner developed the concept of vaccination during the late 1790's. Jenner knew that people who had acquired cowpox and survived were immune against future outbreaks and a very fatal disease, called smallpox. Based on this premise, Jenner demonstrated that smallpox could be prevented in humans if they were injected with a small dose of fluid from an active cowpox lesion. He named this process "vaccination" from the Latin *vaccinia*, which means cowpox.

Pasteur applied Jenner's concept to other diseases and developed vaccinations consisting of attenuated bacteria for the prevention of anthrax and rabies during the 1880's. In turn, based on Pasteur's success, Sewall applied the concept to develop a vaccine that would induce immunity against toxic snake venoms. In 1887, he was successful in demonstrating that an animal could be protected from the toxic venom if previously vaccinated with sublethal doses of the toxin.

Behring and Kitasato attempted to extend the already proven concept of immunization via vaccination and apply the technique for control of diphtheria. Thus, combined with data generated by Klebs, Löffler, Roux, and Yersin regarding the toxin-producing *C. diphtheriae*, Behring and Kitasato initiated a series of their own experiments. In 1889, Kitasato had discovered the causative agent of tetanus, which was also found to be a toxin-producing bacterium.

Behring's experimental design involved preparing a pure culture of a live, toxin-producing strain of *C. diphtheriae* in a nutrient broth, separating the toxin generated by the bacteria in the broth from the organisms via filtration, injecting graduated sublethal doses of the toxin under the skin of healthy rabbits and mice, and several days later, injecting the inoculated animals with live, active *C. diphtheriae* bacteria. Behring's experiment was a success. On December 11, 1890, Behring reported in a journal article that the animals vaccinated with *C. diphtheriae* toxin prior to injection with active *C. diphtheriae* bacteria did not develop diphtheria. Control animals which were not vaccinated, however, developed the disease subsequent to injection with active organisms. Thus, Behring demonstrated that the experimental animals were able to develop an induced immunity to the *C. diphtheriae* toxin via vaccination because of the formation of a protective toxin-destroying agent produced within their blood serum. (One week earlier, Behring and Kitasato had coauthored a journal article which reported similar findings for experiments using toxin produced by tetanus bacilli.) The two scientists referred to the protective toxin-destroying agent within the blood sera of immunized animals as an "antitoxin."

Impact of Event

As a result of Behring's discovery of diphtheria antitoxin, the foundation was established to develop an efficient vaccine and to determine an optimal dose for human use. Progress was demonstrated within a year, because of experiments conducted by German bacteriologist Paul Ehrlich, whose work involved determining if serum derived from animals and humans known to contain the antitoxin could be injected in others to induce immunization. This concept became the foundation of what is called "serotherapy" to induce "passive immunity." A person is considered to have been passively immunized when they become immune to toxin because of injection with serum containing antitoxin from another immunized person or animal. In other words, passive immunity implies the transfer of immunity from one host to another via vaccination with antitoxin instead of active toxin.

Ehrlich's assistance to Behring was also instrumental in establishing some insight into the administration of safe and effective doses of vaccine for clinical use. Within a year of Behring's discovery of the diphtheria antitoxin, clinical trials were established with humans to determine if diphtheria could be prevented and possibly cured. The clinical trials were successful; thus, the era of vaccinating humans, especially children, with diphtheria antitoxin had begun. The process, however, was not totally efficient, and scientific research continued. Nevertheless, immunization to prevent and cure diphtheria via vaccination was gaining widespread use, and a significant decline in the disease was apparent by the beginning of the twentieth century.

Behring's discovery of the diphtheria antitoxin influenced several major advances in the area of medical science. The concept of serotherapy as a form of vaccination was developed to induce passive immunity against the *C. diphtheriae* toxin. The process was later applied by other scientists to control the impact of other bacterial and viral agents found to be pathogenic to humans and animals. Concomitantly, a greater understanding of the human immune system was gained, especially relative to the concept of antibody (for example, antitoxic protein in blood serum) response to antigen (for example, *C. diphtheriae* toxin). Finally, as a result of the vaccine, countless lives of people who were afflicted with the dreaded disease of diphtheria were saved, while even more people were spared the experience of contracting the illness. In acknowledgment of Behring's discovery and its positive impact realized at that time and perceived for the future, he was awarded the first Nobel Prize in Physiology or Medicine in 1901.

Bibliography

Asimov, Isaac. *Asimov's Biographical Encyclopedia of Science and Technology*. New rev. ed. Garden City, N.Y.: Doubleday, 1972. This reference provides biographical summaries of 1,195 great scientists, including those who established the foundation and influenced the discovery of the diphtheria antitoxin.

Chase, Allan. *Magic Shots*. New York: William Morrow, 1982. This book provides a historical perspective regarding discoveries of various vaccines that had a positive influence in medicine. The book refers the reader to several of the original journal articles that reported the discoveries.

Walker, M. E. M. *Pioneers of Public Health*. New York: Macmillan, 1930. This book contains a chapter about Hermann Biggs, an American physician instrumental in encouraging the use of the diphtheria antitoxin to prevent diphtheria. The chapter provides a historical account of the discovery of the antitoxin and development of vaccine.

Walsh, James Joseph. *Makers of Modern Medicine*. Reprint. Freeport, N.Y.: Books for Libraries Press, 1970. The book contains a chapter about Joseph O'Dwyer, an American physician who relentlessly studied the efficacy of using diphtheria antitoxin and eventually succeeded in demonstrating its true value.

Wood, William Barry. *From Miasmas to Molecules*. New York: Columbia University

Press, 1961. The book provides a historical perspective as well as expanded scientific explanations of the diphtheria bacilli: antitoxin and immunity.

Michael S. Bisesi

Cross-References

Ehrlich Introduces Salvarsan as a Cure for Syphilis (1910), p. 476; Schick Introduces the Schick Test for Diphtheria (1913), p. 567; Calmette and Guérin Develop the Tuberculosis Vaccine BCG (1921), p. 705; Zinsser Develops an Immunization Against Typhus (1930), p. 921; Theiler Introduces a Vaccine Against Yellow Fever (1937), p. 1091; Salk Develops a Polio Vaccine (1952), p. 1444; Sabin Develops an Oral Polio Vaccine (1957), p. 1522; A Vaccine Is Developed for German Measles (1960), p. 1655.

STROWGER INVENTS THE AUTOMATIC DIAL TELEPHONE

Category of event: Applied science
Time: 1891-1905
Locale: Kansas City, Missouri

Strowger's automatic dialer provided one-wire signaling that ultimately proved to be a practical way to provide automatic switching on the Bell telephone network

Principal personages:

ALMON BROWN STROWGER, a Kansas City, Missouri, undertaker who developed a way for telephone subscribers to bypass manual operators at central office switches

A. E. KEITH, an associate of Strowger who helped refine the original Strowger concept

JOHN ERICKSON and CHARLES J. ERICKSON, the members of a technical team who, with Keith, developed, manufactured, and deployed equipment based on the Strowger patent

Summary of Event

Following the invention of the telephone, research and development efforts by pioneers in telephony were related directly to the need for switching equipment that would provide reliable and convenient access for an ever-increasing number of subscribers to a growing telephone network. During the decade and a half that followed Alexander Graham Bell's first dramatic demonstration of that invention in 1876, all telephone calls between two points had to be manually connected by operators, a procedure that resulted in problems during times of peak demand, when callers were forced to wait in line for operators. Subscribers also realized that, during periods of high demand, operators were in a position to control who would or would not be given priority access. Additionally, they were concerned about privacy and fairness; operators could listen in on conversations and were in a position to control whether calls went through. Some subscribers expressed concern that operators might be tempted to become involved in schemes concocted by third parties to interfere with or even sabotage competition among commercial establishments, a concern that was expressed by Almon Brown Strowger, a Kansas City, Missouri, undertaker who, despite a lack of technical training and expertise, set out to invent a way for callers to bypass operators automatically to ensure that calls would go through and that they would be private.

Strowger had become convinced that his funeral business was being victimized by unscrupulous telephone operators who would deliberately give incorrect phone numbers to callers attempting to reach him, or, when the correct number was dialed, engage a busy signal rather than put the call through. Recognizing that the only recourse was to find a way to bypass those operators, he became determined to invent a technical system capable of doing just that.

All previous attempts to develop an automatic switching system had been unsuccessful. In 1879, the Connolly-McTighe patent, issued on December 9 to M. D. Connolly, T. A. Connolly, and T. McTighe, had provided specifications for the first automatic switch. Unfortunately, the Connolly-McTighe switch ultimately proved a failure. In 1884, Ezra Gilliland developed a primitive system that proved reliable but inefficient. It was workable but capable of serving only fifteen telephones. In 1891, Strowger learned from those ideas when he proposed and patented a system that could handle up to ninety-nine telephones.

Each Strowger telephone had two buttons that could be pressed up to nine times each to achieve the required digits for any two-digit number from zero to ninety-nine. In the telephone central office, each electric pulse created when the buttons were pushed moved a mechanical arm so that the contact points were placed directly in line with those of the exact numbers requested. This innovation was quickly accepted as the key to automatic switching, and in 1891 Strowger set up the "Strowger Automatic Telephone Exchange" to exploit his patent by manufacturing the switches designed according to the specifications set forth in his patent. That company was absorbed into the Automatic Electric Company in 1908.

In the beginning, the Strowger automatic switch was not a complete success, although the underlying principle upon which it was based was sound. Shortly after acquiring the patent, Strowger became associated with A. E. Keith, John Erickson, and Charles J. Erickson. All were talented engineers who helped develop the automatic switch concept invented by Strowger into a workable, dependable system. In November, 1892, the first Strowger system was installed in LaPorte, Indiana, making it the first automatic central switching facility in the world. For the next fifteen years, the Strowger automatic dialing system was repeatedly tested, modified, and remodified.

Meanwhile, the American Telephone and Telegraph Company continued to look at and experiment with automatic switching systems that had been proposed by other companies during the last decade of the nineteenth century and the first decade of the twentieth century. The Automatic Electric Company pressed on with development and deployment of its Strowger system, installing systems of up to six thousand lines each in several cities including Grand Rapids, Michigan, and Dayton, Ohio.

Still, the future of automatic switching was seen as uncertain by the Bell company, which had set up its own automatic systems principally in small towns and rural areas. These systems had proved somewhat unreliable and more costly than manual switching because they required substantial technical maintenance and support, and they did not eliminate completely the need for manual operators. To early telephone designers, this meant that automatic switching was not a true substitute for manual switching. Instead, it was perceived to be a costly enhancement of the existing system that, over the years, had become increasingly efficient and less costly per subscriber as the number of subscribers continued to increase. Also, company officials were concerned that automatic switching put the responsibility for making call connections on the shoulders of the customer instead of the operator and that, they reasoned, might result in the perception by the public that the quality of service had been lowered.

From the beginning, American Telephone and Telegraph officials had insisted that automatic switching must be cost effective and reliable and, for the customer, an improvement in the overall quality of telephone service. Despite the concerns about privacy and access, manual switching centers had been evolving also. They were streamlined gradually for efficiency and cost effectiveness, often handling ten thousand lines or more. Telephone company officials remained unconvinced regarding the practicality of automatic switches until 1914. By that time, the automatic Electric Company had refined and enhanced the Strowger switch to the point where it could do most of the functions of a manual operator. The breakthrough came in 1905, when a new electronic pulsing scheme was incorporated that required only one wire. Until that time, two wires were required to send electronic signal pulses back and forth from each station to the central switch. A timing scheme was introduced that eliminated the need for the second line, and a major technical efficiency was suddenly at hand.

Strowger died several years before his automatic switching concept achieved its full potential as the primary automatic switching technology employed by the Bell system. In May, 1916, the Automatic Electric Company and Western Electric, the telephone equipment manufacturing branch of American Telephone and Telegraph, signed an agreement that gave Western Electric the right to manufacture Strowger automatic dialing equipment, although Automatic Electric would continue to supply equipment to the parent company for many years. In 1917, the Strowger switch was ordered for a new Bell central office in Norfolk, Virginia, and from that point on was the accepted switching technology in all Bell system installations. Automatic Electric continued to supply equipment to Bell until 1936.

Impact of Event

The Strowger automatic dialing system represented a change from the way telephone service had been provided during the early years. Bell officials had assumed that manual operators would always be involved in connecting calls because the system was too technically complicated for most subscribers. During those years, a subscriber simply picked up the telephone receiver and gave a quick pull on a hand crank to signal the operator to come on the line. The caller then gave the operator the number—sometimes only the name—of the desired party, and the operator placed the call. Station dialing, on the other hand, required the caller to do it all: to punch in the number of the desired party accurately and then wait for the connection to go through. This process was complicated when a long-distance toll call was placed, and company officials worried that the public would perceive automatic dialing to be a degradation in the quality of the service to which they had become accustomed.

By 1914, Bell operators were discovering that a significant portion of the public had already been exposed to automatic dialing with little or no complaint about the added work of station dialing. For almost fifteen years, the Automatic Electric Company had been selling ever-improving Strowger equipment to independent telephone companies around the country that could not afford to employ full-time operators. That equipment had gained a good reputation among the independent companies for reliability and economy, and was particularly valuable in systems where peak demand could not be handled efficiently by manual operators. During the same period, Bell was buying up many of these independents, in the process acquiring the installed equipment. Telephone subscribership was continuing to skyrocket, and, in desperate need of a way to accommodate that growth, the company began to run its own tests and use the Bell Research Laboratory to refine the Strowger concept for use on the nationwide system. That the concept ultimately became the basis for the design of the automatic rotary dial telephone bears witness to the significance of the unique and innovative ideas proposed by Strowger.

Bibliography

Brooks, John. *Telephone: The First Hundred Years*. New York: Harper & Row, 1976. An excellent corporate history of the Bell system that includes many anecdotes and colorful stories about the early years of telephony, giving life and context to what could otherwise be described as a highly technical description of the birth and development of one of the world's most remarkable companies. Includes pictures of principal telephone pioneers and early facilities and equipment, including the Strowger dial-up telephone.

Casson, Herbert. *The History of the Telephone*. Chicago: A. C. McClurg & Co., 1910. A rare glimpse into the early days of telephony as described by many of the innovators who participated in the development of the nationwide telephone network.

Danielian, Noorbar R. *AT&T: The Story of Industrial Conquest*. New York: Vanguard Press, 1939. A good look at some of the personalities involved in the development of the world's largest telephone network. Provides a strong backdrop for gaining an understanding of how and why decisions were made regarding the adoption of technical innovations at American Telephone and Telegraph Company through the years.

Fagen, M. D., ed. *A History of Engineering and Science in the Bell System: The Early Years (1875-1925)*. New York: Bell Telephone Laboratories, 1975. A large work prepared by members of the technical staff at the Bell Telephone Laboratories as part of a multivolume set tracing the corporate and technical history of the company. Contains detailed descriptions of the Strowger dialing apparatus from the very first prototype through its evolution to the rotary dialing mechanisms still in use. Includes pictures and graphs that describe this equipment in detail.

Hill, R. B. "Early Work on Dial Telephone Systems." *Bell Laboratories Record* 31 (March, 1953): 95.

____________. "The Early Years of the Strowger System." *Bell Laboratories Record* 31 (March, 1953): 95. The above two articles represent a good overview of the technical developments in dial switching during the earliest years.

Rhodes, Frederick Leland. *Beginnings of Telephony.* New York: Harper & Bros., 1929. Another early look at behind-the-scenes research and development that shaped telephone technology during its first quarter century.

Michael S. Ameigh

Cross-References

The First Transcontinental Telephone Call Is Made (1915), p. 595; Transatlantic Radiotelephony Is First Demonstrated (1915), p. 615; The First Transatlantic Telephone Cable Is Put into Operation (1956), p. 1502; Direct Transoceanic Dialing Begins (1971), p. 1934; The First Commercial Test of Fiber Optic Telecommunications Is Conducted (1977), p. 2078.

ABEL AND TAKAMINE INDEPENDENTLY ISOLATE ADRENALINE

Category of event: Biology
Time: 1897-1901
Locale: Johns Hopkins University, Baltimore, Maryland (Abel); New York (Takamine)

Abel and Takamine simultaneously isolated adrenaline, the main hormone of the adrenal medulla, which controls blood pressure and the "fight or flee" mechanism in higher organisms

Principal personages:

JOHN JACOB ABEL (1857-1938), an American physician, biochemist, and pharmacologist who isolated a form of adrenaline

CARL FRIEDRICH WILHELM LUDWIG (1816-1895), a famous German physiologist and physician with whom Abel trained

JOKICHI TAKAMINE (1854-1922), a Japanese-born American chemist-industrialist who isolated the enzyme Taka-diastase and pure adrenaline from bovine adrenal glands

Summary of Event

Adrenaline, or epinephrine, is a member of the very important group of biological messengers called hormones. These messengers control many biochemical processes necessary for the life and well-being of the whole body. They are produced in tiny amounts by organs called endocrine glands. Endocrinology, the study of the hormones, is named after these glands, which secrete hormones into the blood. The blood then takes hormones to other, "target" organs where their effects occur. The messages embodied in hormones integrate body functions, allowing them to occur with great flexibility.

Some hormones cause their effects by a mechanism that involves direct stimulation of the expression of genes in target organs. Adrenaline exemplifies another group of hormones that cause their effects by a process called a second messenger mechanism. Here, a hormone causes the production of a second chemical, the second messenger, that actually produces the effect.

Adrenaline is made by the central portion (medulla) of the adrenal glands, twin endocrine glands located atop the kidneys. It is produced from the amino acids phenylalanine and tyrosine. Adrenaline secretion triggers a series of body processes often called the "fight or flee" responses. These responses include accelerated heart rate and output, accompanied by increased energy and strength. They may explain, for example, the unexpected strength that sometimes enables a 150-pound man to lift a telephone pole off a loved one in an emergency situation. A very prominent biochemical event related to fight or flee responses is the greatly increased breakdown of the main energy reserve, glycogen (a complex carbohydrate), in muscle and liver. This prepares the body to use the energy reserve and leads to the physical responses required to fight or flee. The understanding of this process was elucidated in the 1970's. Yet, it might not have occurred if John Jacob Abel and Jokichi Takamine had not isolated adrenaline between 1897 and 1901.

Abel's contribution is a consequence of his lifelong effort to carry the use of the precepts and methodology of chemistry to the practice of medicine and biology. This commitment developed after he received his doctoral degree from the University of Michigan in 1883. At that time, Abel engaged in six years of postgraduate work in Germany, where he worked with famous biologists, including Carl Friedrich Wilhelm Ludwig, at Leipzig. For the next two years, Abel engaged in additional training in biochemistry and pharmacology, fine-tuning his interests in those fields. He returned to the United States and accepted a position at the University of Michigan, where Abel offered courses on various aspects of biochemistry and pharmacology, including "the influence of drugs in metabolism of tissues." By 1893, Abel had become the first professor of pharmacology at Baltimore's Johns Hopkins University, a position he occupied until retirement. Abel steadfastly worked on many aspects of drug and hormone research throughout his life and trained numerous biomedical scientists in the intelligent use of chemistry in their endeavors. He warned developing researchers constantly that sensible investigators must work in areas where "molecules and atoms rather than multicellular tissues or unicellular organisms are the units of study."

The work with epinephrine that Abel carried out, from 1895 to 1905, was only one of his major efforts. It is one for which he became well known. In 1897, Abel published on the isolation of this substance from adrenal glands. Then, in 1899, he named it epinephrine (the preferred name for the hormone today). The chemical Abel actually isolated was not the free hormone, but rather a modified substance, called monobenzyl adrenaline. It was not until 1901 that Takamine isolated the pure, "free" hormone from the adrenal glands of beef cattle. The chemical structure of the hormone was not proven, however, until it was chemically synthesized by a third researcher in 1904.

Takamine received undergraduate and graduate degrees at the Universities of Tokyo and Glasgow. In 1894, he moved to the United States, where he founded a private research laboratory, where he isolated bovine adrenaline. In 1901, Takamine described his endeavor at a medical convention at Johns Hopkins University and patented his method for production of the hormone.

Takamine and Abel each recognized the other's contribution to the discovery and isolation of adrenaline, which was the first hormone ever isolated in a pure and stable form. Both became prominent for their efforts, and the most immediate benefit of their dual discovery was the fact that medical practitioners found out quickly that adrenaline could be utilized to stop bleeding during surgery.

Abel is credited for work in many other areas, including efforts that led to the later development of the artificial kidney; endeavors of use in the evolution of modern blood banking; contributions to the isolation of insulin; and very important discoveries about tetanus toxin. He also played an essential role in the foundation of the American Society of Biological Chemists and the American Society for Pharmacology and Experimental Therapeutics. Moreover, Abel shaped the empirical field of science called *materia medica* into modern, experimental pharmacology.

Abel received many honors for his endeavors, including presidency of the American Society of Biological Chemists (1908) and the American Society for the Advancement of Science (1932); memberships in the National Academy of Sciences and the Royal Society of London; the Gibbs Medal (1926), the Conne Medal (1932), and the Kober Medal (1934). Also, he was awarded many honorary degrees. Takamine was the founder and president of Takamine Laboratory Inc., a member of the Royal Chemical Society of England and the Japanese Academy of Science, and a recipient of the Japanese Order of the Rising Sun.

Impact of Event

The pioneering efforts of Abel and Takamine to isolate adrenaline are viewed as being among the most important fundamental discoveries in life science. This results from the fact that they produced pure samples of the first hormone ever isolated. Study of the pure adrenaline—in their laboratories and by numerous other researchers—laid many ground rules for the methodology utilized in study of other types of hormones. Furthermore, it led to many important discoveries, including the identification of several hormones related structurally to epinephrine (the catecho lamines); explanation of the endocrine actions of the catecholamines; and development of modern explanations for aspects of nervous transmission in mental health and disease.

The widespread impact of the discovery of adrenaline on medical science arose from development of understanding of its actions in preparing living organisms to flee from danger or to fight—the so-called fight or flee responses. This was caused by the biological effects of the hormone including increased alertness, elevated heart rate, blood pressure, and increased blood levels of glucose (the source of most of the body's energy). For example, it was examination of the basis for adrenaline effects on glucose levels that led Eugene Sutherland to identify the second messenger mechanism for adrenaline action. Once the second messenger process in adrenaline action was understood, other researchers showed that the mechanism could be applied to explanation of the action of many other hormones.

The observation that adrenaline caused increased mental alertness led to the development of many stimulants, once its chemical structure had been identified as 3,4-dihydroxyphenyl-2-methyl-aminoethanol. The first such chemicals were the structurally related drugs known as the amphetamines (for example, benzedrine and methedrine). The earliest amphetamines were made by German organic chemists of the 1930's to mimic adrenaline action in a manner that would enable soldiers to fight better in situations where they needed to stay awake for long periods of time.

Eventually, abuse of catecholamine-related drugs (such as the amphetamines) was useful also; understanding of such abuse helped other scientists identify the catecholamine involvement in nervous transmission. In fact, the current explanations of mania, depression, and their effective treatment began with the catecholamine hypothesis of affective disorders, which was proposed in 1965. This hypothesis supposed that depression arises from suboptimum catecholamine production or utilization, while mania is caused by its excess. Spinoffs of successful application of the hypothesis include choice of new drugs for therapeutic use on the basis of their effect on catecholamine levels and implication of other "biogenic amines" (for example, serotonin), related structurally to catecholamines, in nerve transmission and its diseases.

Bibliography

Geiling, E. M. K. "John Jacob Abel." In *Dictionary of Scientific Biography*, edited by Charles Coulston Gillispie. Vol. 1. New York: Charles Scribner's Sons, 1971. This brief biographical sketch is a useful source of basic information on Abel. It focuses on his career and provides additional sources of biographical information. Insight into Abel's personality and wide impact on biochemistry and pharmacology is given also.

Harrow, Benjamin. *Textbook of Biochemistry.* 5th ed. Philadelphia: W. B. Saunders, 1950. This older textbook is interesting both because it credits Abel and Takamine as the discoverers of adrenaline and because it describes briefly the original methodology for isolation and for chemical synthesis of the hormone. Also, several important early references on the hormone are given.

Lamson, Paul D. "John Jacob Abel—A Portrait." *Bulletin of The Johns Hopkins Hospital* 68 (1941): 119-157. This very pleasant and entertaining article is a detailed personal account of Abel's life and career. It provides very interesting and engaging reading and describes Abel both as a man and as a scientist.

Lehninger, Albert L. *Principles of Biochemistry.* New York: Worth, 1982. Chapter 25 of this excellent college text gives a concise description of the chemistry, biochemistry, and mechanism of epinephrine. It is especially valuable to readers who wish to have a relatively simple, technical explanation of the hormone and its actions and to understand second messenger mechanisms of hormone action with adrenaline and other hormones.

Murnaghan, Jane H., and Paul Talalay. "John Jacob Abel and the Crystallization of Insulin." *Perspectives in Biology and Medicine* 10 (1967): 334-380. This review of Abel's work on insulin emphasizes its importance and describes the skepticism with which his contemporaries met the concept of a protein hormone. It also places the impact of Abel's research in the broad context of biochemistry of the first half of the twentieth century and cites references on Abel's life and endeavors.

Sutherland, E. W. "Studies on the Mechanism of Hormone Action." *Science* 177 (1972): 401-408. This article explains the important aspects of the Sutherland group's development of understanding of cyclic AMP function in epinephrine action. The evolution of the second messenger concept is well developed by its originator.

Voegtlin, Carl. "John Jacob Abel, 1857-1938." *Journal of Pharmacology and Experimental Therapeutics* 67 (1939): 373-406. This article describes Abel's scientific work, giving detailed explanation on important issues, including his role in the discovery and study of adrenaline. Abel's excellent scientific ability and aspects of his personality are made clear. Many valuable literature references to Abel's work are included.

White, James T. "Jokichi Takamine." In *The National Cyclopedia of American Biography.* Vol 40. New York: James T. White, 1955. This brief biographical sketch is one of the only readily available references on Takamine's life and work. It contains some aspects of his early life, his education, his career, and his accomplishments. Takamine is portrayed as a very solid scientist and a successful entrepreneur.

Sanford S. Singer

Cross-References

Bayliss and Starling Discover Secretin and Establish the Role of Hormones (1902), p. 179; Abel Develops the First Artificial Kidney (1912), p. 512; Banting and Macleod Win the Nobel Prize for the Discovery of Insulin (1921), p. 720; Li Isolates the Human Growth Hormone (1950's), p. 1358; Du Vigneaud Synthesizes Oxytocin, the First Peptide Hormone (1953), p. 1459; Sanger Wins the Nobel Prize for the Discovery of the Structure of Insulin (1958), p. 1567; Janowsky Publishes a Cholinergic-Adrenergic Hypothesis of Mania and Depression (1972), p. 1976; The First Commercial Genetic Engineering Product, Humulin, Is Marketed by Eli Lilly (1982), p. 2221.

BJERKNES PUBLISHES THE FIRST WEATHER FORECAST USING COMPUTATIONAL HYDRODYNAMICS

Categories of event: Earth science and physics
Time: July, 1897-July, 1904
Locale: Theoretical Physics Institute, Stockholm, Sweden

Bjerknes developed and tested the first hydrothermodynamic mathematical model capable of making weather predictions

Principal personages:

VILHELM BJERKNES (1862-1951), a Norwegian theoretical physicist and quantitative meteorologist who was the founder of the Bergen School of dynamic meteorology

NILS EKHOLM (1848-1923), a Swedish aeronomist and experimental meteorologist responsible for pioneering experimental studies of the formation of extratropical atmospheric cyclones (low-pressure systems)

J. W. SANDSTRØM (1874-1947), a Swedish postdoctoral student in fluid dynamics and atmospheric geophysics who undertook mathematical calculations of Bjerknes' circulation theory of cyclones and made the first successful predictions of a low-pressure system

Summary of Event

The gradual development of modern predictive meteorology, from a purely empirical and numerical methodology, was the result of a complex interaction of technological, scientific, and military-economic factors. During the nineteenth century, earlier efforts at studying weather processes predominantly used historical data statistics (the rudiments of so-called synoptic meteorology) and personal experience. The would-be forecaster learned to infer roughly how a weather system—then defined by crudely mapping field measurements of barometric pressure near the earth's surface—would move and/or change character, with generally poor accuracy. Especially following the near destruction of both French and English fleets in the Black Sea during the Crimean War, use of telegraphy determined how much and how swiftly geographically sparse atmospheric information was disseminated. Even after the International Meteorological Committee's efforts, beginning in 1873, to standardize and synchronize as well as increase the number of observational stations for international meteorological data exchange, predictions remained neither detailed nor specific as to time and location, and typically amounted only to general forecasts (for example, dry, changeable, wet) and coastal gale warnings for broad areas at most only eighteen hours in advance.

After three decades of empirically studying the formation and progression of large-scale weather systems, by about 1900, meteorologists had reaped only a meager harvest in terms of organized theory and improved predictive capability. Although several separate hydro- and thermodynamic theories for idealized atmospheric conditions had been advanced, these theoretical efforts were divorced almost entirely both from practical forecasting requirements and from detailed physical understanding of the complex processes actually responsible for change and motion of weather phenomena. As detailed by the historical studies of Gisela Kutzbach on the theory of cyclones, some meteorologists even abandoned the possibility of improving rational weather prediction.

Several technological advances and political events, however, brought a new impetus to a new generation of physical meteorologists and oceanographers. These advances included a wider-spread network of greater economic and military requirements, as well as opportunities for meteorological observations and predictions of greater geographic extent and accuracy. Other motivations for developing meteorology included the associated aerodynamic stimuli to enhance theories of aerodynamic fluid flow for improved aircraft construction and, independently, attempts at theoretical mechanization of electromagnetic wave fields via quantitative analogs with hydrodynamics (the latter as developed by the physicists Heinrich Hertz, Philipp Lenard, Carl Anton Bjerknes, and his son, Vilhelm).

Despite general neglect by the physics community at large, the Bjerkneses concentrated on the problem of formulating a complete and common set of equations for electromagnetic and hydrodynamic force and flux fields. One of the chief analogs was the mathematical similarity between fluid and electromagnetic "solenoids of circulation." When, in 1897, Vilhelm Bjerknes first presented his two extensions of Lord Kelvin's circulation theorems in a lecture to the Stockholm Physics Society, he mentioned no applications whatsoever. Soon after, Bjerknes was approached by his colleagues Nils Ekholm and Svante August Arrhenius, experts in aeronomy and meteorology, respectively, concerning the potential applications of Bjerknes' fluid mechanical analogs to the quantitative study of meso-scale atmospheric motions. As evidenced by Bjerknes' subsequent 1898 lecture and paper "On a Hydrodynamic Circulation Theorem and Its Application to the Mechanics of the Atmosphere and Global Oceans," Bjerknes had begun to consider circulation-theorem applications to polar and continental atmospheric and ocean-flow phenomena.

From 1850 to 1905, it was widely believed that so-called extratropical cyclones were initiated solely by local thermal motion and maintained by the liberated thermal heat of convection. In 1891, Ekholm had shown empirically that the atmosphere frequently has characteristics similar to those of the fluid, later postulated by Bjerknes for his electrohydrodynamic circulation theorems. Most noticed by Ekholm in 1898 was the incongruity between lines of equal value of pressure and density in the vicinity of cyclonic (low pressure) systems. In response to these suggestions (and the receipt of unique upper-air data recording the passage of a cyclone and anticyclone over the Blue Hill Observatory near Weston, Massachusetts), Bjerknes, together with his assistant J. W. Sandstrøm, returned to the circulation theorems, now directly analyzing them from a geophysical applications perspective rather than electromagnetic analogy. During his initial studies, Bjerknes generalized the previous propositions of Kelvin and Hermann von Helmholtz on the velocity of circulation and conservation of vorticity. Bjerknes' generalization was based on introducing a broader interpretation of the definition of fluids. Whereas earlier views assumed a unique (hydrodynamic) relationship between pressure and volume, in his 1900 and 1901 publications ("The Dynamic Principles of Circulation and Motion in the Atmosphere" and "Circulation Relative to the Earth"), Bjerknes expanded his circulation theorem further to its now-classic form by including terms for Coriolis forces arising from the earth's rotation, and approximations for viscous-thermal losses caused by friction from the atmospheric fluid. Using their equations, Bjerknes and Sandstrøm were theoretically able to reconstruct successfully the changes in direction and intensity of the Massachusetts low-pressure system, believed to be the first "pre"-diction of weather phenomena.

Except for acoustic waves, all atmospheric motions may be characterized as flow circulations along closed streamlines. The area distribution of horizontal atmospheric velocity may be represented in terms of two scalar quantities: relative vertical vorticity and horizontal divergence. The former is twice the angular velocity of an air particle around a vertical axis relative to the earth. The latter is the relative expansion rate of an infinitesimal horizontal area moving with the air. The horizontal velocity field can be decomposed mathematically into one horizontal component containing all the vorticity and no divergence, and likewise into another all-divergence/no-vorticity-component. These two components of total atmospheric air mass motion have different behaviors. For example, whereas small-scale convection and internal-gravity waves are associated with the first component, large-scale atmospheric motions of interest to international forecasting are predominantly controlled by the latter, revealed by isobaric contour maps having predominantly horizontal circulations and, hence, vorticity-component dominance. In Bjerknes' initial geometrical interpretation of his circulation theorems as a series of solenoids, because of the tendencies toward rotation in the crisscross lattice of intersecting surfaces, occurrence of skewed distributions of pressure and density should result in overall spatial circulation. Bjerknes' circulation theorem also can explain smaller-scale reciprocating air circulations, such as land-sea breezes and mountain-valley winds. In both cases, circulation is maintained to satisfy the basic vorticity continuity and conservation conditions. Mountain slope air temperature increases more in the day than at night because of radiative heating than air at equal pressure away from the mountain. Consequently, there arises a day wind blowing up the slope, and a night wind blowing downward, as first rigorously demonstrated by Julius Wagner von Jauregg in 1932 using Bjerknes' continuity equation.

In his 1903 *Lehrbuch der Rosmischen Physik* (treatise of cosmic physics), Arrhenius was the first to include independently Bjerknes' circulation theorem as the basis for a chapter on the thermomechanics of the atmosphere and oceans. Later in 1903, Bjerknes explicitly formulated and later published a major proposal for "A Rational Method of Weather Prediction," more fully explained in his publication "The Problem of Weather Prediction Considered from the Standpoint of Mechanics and Physics" (1904). These latter publications outline two basic conditions for an improved predictive meteorology based on his prognostic circulation equations: sufficient knowledge of the state of the atmosphere at a given place and time based on a sufficient number of accurate measurements, and sufficiently accurate knowledge of the quantitative physical laws by which one atmospheric state at a given place and time evolves into another. Since manual solution of equations over many separate observation (grid) points and times was practically impossible (until the advent of computers), Bjerknes and Sandstrøm suggested employing a physical-graphical approximation method. To evaluate the resulting circulation integrals, approximate two-dimensional surfaces of equal pressure (isobars) and of equal specific volume (1/density = isosters) were drawn at specified regular intervals. These surfaces subdivide the three-dimensional atmospheric space into tubes of isobaric-isosteric solenoids. It can be shown, through application of Stokes's theorem of integral calculus, that the integration value is equal to the number of solenoids enclosed by the curve around which the circulation integral is taken.

Impact of Event

Although a number of English and German observational meteorologists in the pre-World War I period had varyingly suggested that cyclones can be associated with regional motion of large-scale circulation air bodies, these anticipations were formalized and subsumed in Bjerknes' work. Because it not only included rigorous derivations and predictions but also tied these to the best available observational data, as underscored by Hans Ertel, Bjerknes' initial publications had a greater impact on reconceiving the atmosphere as multiple dynamically related air masses instead of a unified global static airmass than the earlier (1903) but purely theoretical hydrodynamic studies of French physicist Jacques Hadamard. In addition to becoming the object of further theoretical efforts to reprove and improve its formulation, Bjerknes' circulation theorems not only became the basis of his later Bergen School of dynamic meteorology but also had the greatest impact on German, French, English, and American quantitative meteorology. In addition to Vilhelm and his son Jacob's 1910 and 1918 publications, respectively, on atmospheric flow-line convergence and divergence in cloud/precipitation formation and cyclonic squall-lines, in 1921, Jacob Bjerknes and H. Solberg extended this work in a paper "Meteorological Conditions for the Formation of Rain," introducing the concepts of warm and cold fronts into predictive meteorology. A year later, they published "The Life Cycle of Cyclones and the Polar-Front Theory of Atmospheric Circulation," in which their prior descriptive model of cyclones was shown to be a special case, or rather a single stage, in the genesis and development of cyclones, from incipient unstable atmospheric waves through occluded/stalled fronts and dying frontal vortices. Notwithstanding the fact that polar fronts do not account for all low-pressure systems, much of the conceptual and theoretical apparatus as well as nomenclature remains intact in contemporary pedagogic and predictive meteorology. Further confirmatory and developmental studies by Tor Harold Percival Bergeron in 1926, and later by Carl-

Gustaf Arvid Rossby, Erik Herbert Palmén, and others in the 1930's, more closely defined additional frontal and cyclonic phenomena and locally predictive models by developing the kinematical principles of air-mass analysis.

Bibliography

Bates, Charles C., and John F. Fuller. *America's Weather Warriors, 1814-1985*. College Station: Texas A&M Press, 1986. Discusses the development and application of the Bergen School's efforts of fifty years of military weather predictions.

Bjerknes, Vilhelm, and Johann Sandstrøm. *Dynamic Meteorology and Hydrography*. Vol. 1. Washington, D.C.: Carnegie Institution, 1910. Discusses the theoretical derivations, experimental confirmations, and potential applications of the circulation theorems to predicting atmospheric conditions. Illustrated with numerous maps and diagrams. Written at a higher technical level requiring a solid undergraduate background in partial differentiation equations and/or hydrodynamics.

Friedman, Robert Marc. *Appropriating the Weather: Vilhelm Bjerknes and the Construction of a Modern Meteorology*. Ithaca, N.Y.: Cornell University Press, 1989. Friedman traces in nonmathematical detail the historical-conceptual details of the revolution in meteorology initiated by Bjerknes. Friedman's philosophical thesis is that Bjerknes "appropriated" the tools of hydrodynamics and the problems of synoptic meteorology to construct a new predictive meteorology.

Holmboe, Jorgen, George E. Forsythe, and William Gustin. *Dynamic Meteorology*. New York: John Wiley, 1945. Basic textbook used to instruct flyers and meteorologists. Develops Bjerknes' basic thermohydrodynamic principles directly from physical concepts.

Kutzbach, Gisela. *The Thermal Theory of Cyclones: A History of Meteorological Thought in the Nineteenth Century*. Boston: American Meteorological Society, 1979. Presents a thematic analysis of the historical evolution of Euro-American meteorology. Discusses Bjerknes' numerical-observational predictive program from 1903 to 1905.

Gerardo G. Tango

Cross-References

Teisserenc de Bort Discovers the Stratosphere and the Troposphere (1898), p. 26; Zeppelin Constructs the First Dirigible That Flies (1900), p. 78; Fabry Quantifies Ozone in the Upper Atmosphere (1913), p. 579; Tiros 1 Becomes the First Experimental Weather Reconnaissance Satellite (1960), p. 1667; Rowland and Molina Theorize That Ozone Depletion Is Caused by Freon (1973), p. 2009; The British Antarctic Survey Confirms the First Known Hole in the Ozone Layer (1985), p. 2285.

TEISSERENC DE BORT DISCOVERS THE STRATOSPHERE AND THE TROPOSPHERE

Category of event: Earth science
Time: 1898-1902
Locale: Trappes (Paris), France

Based on experimental balloon measurements of atmospheric temperature versus height, Teisserenc de Bort discovered the stratosphere's and troposphere's vertical layering on the basis of thermal inversion

Principal personages:

LÉON TEISSERENC DE BORT (1855-1913), a French physicist and meteorologist

RICHARD ASSMANN (1845-1918), a German physicist and meteorologist

Summary of Event

The details of the rate of change of atmospheric temperature versus height have been of basic importance for many years in trying to determine and predict the processes governing weather. For example, the variation of wind with height also depends upon vertical temperature variation.

Until the violent eruption of the volcano Krakatau in the Java Sea in 1883, which produced abnormally high atmospheric concentrations of dust, implying the existence of higher-level global temperature and wind patterns, the body of air above the earth's surface was considered generally a uniform body. William Morris Davis' 1894 text, *Elementary Meteorology*, is representative of knowledge of the upper atmosphere before large-scale kite and balloon sondings. Davis simply divides the earth into geosphere (rock), hydrosphere (water), and atmosphere (air). An empirical formula for atmospheric temperature gradient was developed by Austrian meteorologist Julius Ferdinand von Hann in 1874, based on indirect atmospheric measures such as astronomical observations of the duration of twilight and of meteor burns. Davis proposed that successive isobaric (equipressure) surfaces are separated by greater and greater distances indefinitely out into space. Here, the general distribution of temperature with elevation is simply illustrated as a nearly linear decreasing function.

Manned balloon ascents to measure upper air temperature were first undertaken by John Jeffries and François Blanchard in 1784 and subsequently by Jean-Baptiste Biot and Joseph-Louis Gay-Lussac in 1804, and continued in England in 1852. Factors influencing balloon performance included the excess of buoyancy forces over balloon gross weight (including human observers) and the maximum size to which the balloon's silk or India rubber envelope would expand in response to decreasing atmospheric pressure. These factors control both maximum ascent ceiling and ascent rate. The need for light gases, such as hydrogen or helium, is to keep the balloon's envelope sufficiently distended. The buoyancy force, which arises from Archimedes' principle, is equal to the air mass displaced by the balloon. As the balloon rises, the air density falls by a factor of about ten for every 10 kilometers of ascent; therefore, the balloon's envelope expands in exact proportion to falling density.

Prior to 1890, balloon observations were, for the most part, limited to heights of only a few kilometers by human oxygen consumption, recording mainly local rather than regional or global temperature behaviors. The first attempts at global isothermal charts were published by Hann in Vienna and Alexander Buchan in Edinburgh in 1887 and 1889, respectively. To overcome the human limitation, kites were first employed by Cleveland Abbe in studying winds under a thunder cloud at the Blue Hill Observatory in Massachusetts. Nevertheless, for technical reasons, the maximum heights attained by kites were only about 8 kilometers.

Because of proven dangers to human life in high ascents, small free rubber balloons carrying recently developed self-recording temperature and pressure recorders were first deployed in 1893 by French aeronomist Georges Besançon and were rapidly adopted elsewhere for meteorological observations. When atmospheric visibility is sufficiently good, larger meteorological balloons could be followed visually by theodolites to obtain supplementary wind direction data. Theodolites are grid-mounted survey telescopes permitting measurement of height and angular motion. These various observations demonstrated that to at least about 9,000 meters, temperature decreased in a fairly uniform fashion at a rate of about 1 degree Celsius per every 180 meters rise.

After extensive work in Europe and North Africa with the French government undertaking barometric and other weather observations, in 1897, Léon Teisserenc de Bort founded his own private aeronomic observatory at Trappes near Paris. Earlier, Teisserenc de Bort had pioneered self-recording temperature and barometric pressure sensors; the Austrian physicist Richard Assmann developed the first self-recording hygrometer to measure atmospheric humidity. Using hydrogen-filled balloons specially designed for rapid and near vertical ascents, Teisserenc de Bort named his surveys "soundings" or "sondings," in analogy to bathymetric depth soundings by sonde-line or acoustic sound at sea. A critical factor was sufficient protection of thermometers from direct solar radiation, as well as recorders that could respond to changing temperature faster than the balloon would rise.

In April, 1898, Teisserenc de Bort, using his improved apparatus, began a long series of regular balloon sondings from Trappes, France. Among other details, he soon discovered unusual temperature records, first believed to be instrument errors, of constant or even increasing temperature conditions from the extreme upper limits of his balloon's ascents. After precluding instrument error and repeating many measurements, in 1899, he published a report indicating that temperatures at heights above 0.1 atmospheric pressure (100 millibars) cease to decline with altitude but remain constant over a specific height interval, thereafter slowly increasing.

In his papers of 1904 in the noted French journal *Comptes rendus physique* and his own *Travaux scientifiques de l'observatoire de météorologie de Trappes*, Teisserenc de Bort gave mean temperatures versus height measured at Trappes between

1899 and 1903. Out of 581 balloon ascents, 141 attained temperature "isothermal" and "inverted" measurements at height records of 14 kilometers or more. His data showed that there is a slow temperature decrease up to about 2 kilometers above sea level. This is followed by a more rapid decrease up to about 10 kilometers. A very slow or total lack of decrease was measured between 11 and 14 kilometers (with an ambient temperature of about −55 degrees Celsius). He called this the "thermal" zone or boundary.

Teisserenc de Bort's observations were almost concurrently confirmed by Assmann's independent series of ascents from Berlin. Assmann and Artur Berson, beginning in 1887, undertook a more extensive series of upper atmospheric soundings, under the aegis of the Prussian Meteorological Office and Aeronautical Section of the German Army, and later as an independent scientific station at Lindenberg. The details of their seventy ascents between 1887 to 1889 were the first published aeronometric measurements of temperature for several locations, in 1900, and thereafter published regularly in the German journal *Das Wetter*. From a particularly long series of kite soundings from Berlin between October, 1902, and December, 1903, Assmann showed that atmospheric temperature is much more variable at 6 or 7 kilometers height than at ground surface. The effects of diurnal and seasonal changes on upper-level temperatures were also measured. Following the systematic planned simultaneous ascents from many European cities between 1895 and 1899, Assmann assembled a data base of more than one thousand of his own observations, with 581 of Teisserenc de Bort, and others from England, Holland, and the Soviet Union, enabling him to compute monthly and annual temperature and wind velocity averages of many altitudes between 0 and 11 kilometers over central Europe. Assmann also argued that at about 12 kilometers, the upper limit for cirrus clouds, temperature remains constant and later increases slowly. The atmospheric region above these heights of constant temperature was called the stratosphere, the lower region nearest the ground was called the troposphere, and the transition zone was called the tropopause. The mesosphere and thermosphere are above the stratosphere.

Impact of Event

Meteorologic sonding heights of more than 25 kilometers were achieved in France and Belgium between 1905 and 1907. The Fifth Conference of the International Committee on Scientific Aeronautics at Milan in 1906 saw an increasing number of measurements confirming the temperature results of Teisserenc de Bort and Assmann, notably kite ascents from 1904 to 1905 from the Soviet Union. These data established that above a height that geographically varied from about 18 kilometers near the equator to about 11 kilometers at 50 degrees north latitude to only about 6 kilometers at the poles, atmospheric temperature remained approximately constant over a certain level. (The English meteorologist W. Dines subsequently showed that the stratosphere is high and cold over high pressure and low and warm over low pressure.)

As soon as diverse independent observations had established the troposphere/

tropopause/stratosphere, many efforts were made to explain rigorously the occurrence of stationary upper-level discontinuities on the basis of the rapidly developing hydrothermodynamics of Vilhelm Bjerknes, Ludwig Prandtl, and others—initially, however, with only very limited success. Finally, W. Humphreys in the United States (*Astrophysical Journal*, vol. 29, 1909) and F. Gold in England (*Proceedings of the Royal Society*, vol. 82, 1909) published what became essentially the generally accepted explanation. In both approaches, it was recognized that it is necessary to consider the thermodynamic balance between absorbed and reemitted solar radiation. Humphreys' account is less mathematical but equivalent to Gold's. Briefly, since the average annual temperature in the atmosphere at any location had been shown experimentally not to vary greatly, Humphreys concluded that the absorption of solar radiation is equal basically to the net outgoing reradiation by Earth (discovered previously by S. Langley), using a simple thermodynamic "black body" model. Humphreys concluded that the isothermal/tropopause zone marks the limit of vertical thermal convection and, from this, correctly deduced that the above-lying layers are warmed almost entirely by direct solar radiation (later shown to be dependent upon atmospheric ozone). The increasing temperature trend was shown later to be caused directly by the heat released during the interaction between incoming ultraviolet radiation and atmospheric ozone molecules.

Further direct and indirect studies of the stratosphere and troposphere continued by a variety of means. In studies of ground versus air waves from earthquakes by Emil Wiechert in 1904, and later during World War I, it was noted that loud noises could be heard occasionally at distances ranging from 150 kilometers to more than 400 kilometers from their source, even when observers near the source could barely hear the sounds. Between 1928 and 1931, H. Benndorf, P. Duckert, and O. Meissner made recordings of seismic-acoustic wave propagation in atmospheric temperature inversions associated with the troposphere. Sound waves are bent gradually or refracted resulting from the increased velocity of sound in air resulting from a gradient of rising temperature at about 35 to 60 kilometers height. These observations provided another method of estimating temperature then inaccessible to aircraft, balloon, and kite soundings. In 1926, G. Dobson and F. Lindemann employed data from hundreds of meteor burn observations to extrapolation temperature, pressure, and chemical observations to heights of up to 160 kilometers, confirmed by V-2 flights during and after World War II.

Subsequent studies of the stratosphere by Earth-orbiting satellites include the mapping of the (polar) jet streams and the twenty-six-month quasi-biennial cycle. The original motivation and basis for these and other studies, however, remain the methods and results of Teisserenc de Bort and Assmann.

Bibliography

Anthes, Richard A., et al. *The Atmosphere*. Columbus, Ohio: Charles Merrill, 1975.
A good general-reader text incorporating almost all meteorological techniques and findings up to the late 1970's.

Davis, William Morris. *Elementary Meteorology.* Boston: Ginn, 1894. Representative of atmospheric science prior to the experimental results of Teisserenc de Bort and Assmann and the dynamic meteorological theory of Bjerknes. Widely available.

Goody, Richard M. *The Physics of the Stratosphere*. London: Cambridge University Press, 1954. A technical account devoted to stratospheric processes. Recommended.

Humphreys, W. J. *The Physics of the Air.* New York: Dover Books, 1964. Historical-technical account of upper atmospheric science.

__________. "Vertical Temperature Gradient of the Atmosphere, Especially in the Region of the Upper Inversion." *Astrophysical Journal* 29 (1909): 14-26. The first detailed study to incorporate and explain the stratosphere and tropopause in the context of physical theories of atmospheric heating and thermodynamics.

Massey, Harrie Stewart Wilson, Sir. *The Middle Atmosphere as Observed by Balloons, Rockets, and Satellites*. London: Royal Society, 1980. General descriptions of many remote sensing methods and typical self-recording instruments.

Gerardo G. Tango

Cross-References

Bjerknes Publishes the First Weather Forecast Using Computational Hydrodynamics (1897), p. 21; Wiechert Invents the Inverted Pendulum Seismograph (1900), p. 51; Fabry Quantifies Ozone in the Upper Atmosphere (1913), p. 579; The Germans Use the V-1 Flying Bomb and the V-2 Goes into Production (1944), p. 1235; Tiros 1 Becomes the First Experimental Weather Reconnaissance Satellite (1960), p. 1667; Rowland and Molina Theorize That Ozone Depletion Is Caused by Freon (1973), p. 2009; The British Antarctic Survey Confirms the First Known Hole in the Ozone Layer (1985), p. 2285.

HILBERT DEVELOPS A MODEL FOR EUCLIDEAN GEOMETRY IN ARITHMETIC

Category of event: Mathematics
Time: September, 1898-July, 1900
Locale: University of Göttingen, Göttingen, Germany

Hilbert's The Foundations of Geometry *established the basic axiomatic-formalist approach to systematizing mathematics, initiated by compactly deriving a formal axiomatic model for Euclid's geometry*

Principal personages:
DAVID HILBERT (1862-1943), a German mathematician and logician
FELIX KLEIN (1849-1925), a German mathematician and educator
MORITZ PASCH (1843-1930), a German mathematician
GIUSEPPE PEANO (1858-1932), an Italian mathematician and logician

Summary of Event

The modern hypothetico-deductive method in mathematics and the concurrent drive toward abstraction, formalization, and establishing universally applicable foundations may be traced to two principal and near-contemporary sources: the development of diverse non-Euclidean geometries and the "Erlanger Programme" efforts at their reconciliation, and the paradoxes in formal logic and set theory following the invention of quantification theory by Gottlob Frege and the axiomatization of arithmetic by Richard Dedekind and Giuseppe Peano. Carl Friedrich Gauss, Nikolay Ivanovich Lobatchevsky, and János Bolyai in the first three decades of the nineteenth century developed alternate non-Euclidean geometries following the realization that logical negation of Euclid's parallel postulate need not lead to contradiction. Subsequently, they and others found that although the theorems resulting from the new geometric axioms were at odds with observational results of everyday experience, or Immanuel Kant's intuitions, none of the expected logical contradictions appeared in these new geometries.

In the late 1860's, new attention was drawn to non-Euclidean geometry by publications of Hermann Helmholtz and Ernesto Beltrami. In 1870, Felix Klein discovered a more general model or interpretation of non-Euclidean geometries in the 1859 work of Alexander Cayley, by means of which Klein was able to identify systematically all the primitive objects and relations of the new geometries with corresponding primitives in Euclidean geometry. In 1872, Klein published an important paper, "Equivalency Considerations of Recent Geometric Research," which expounded his so-called Erlanger Programme that strongly influenced several generations of mathematicians. This program for the "algorithmic systematization" of geometry and other areas of mathematics expounded the thesis that diverse non-Euclidean geometries could be unified and classifiably related by reconsidering geometry as the more general study of the particular "forms," or formal algebraic properties of spatial configurations of a "manifold" that are left unchanged (invariant) by an underlying group of transformations (such as rotation, translative motion, and the like). The new group-theoretical viewpoint of Klein, varyingly adopted by Sophus Lie, Henri Poincaré, Friedrich Schur, Eli Cartan, and others, was subsequently applied to their work on the theory of equations, automorphic functions, and complex function theory, all considered by Klein as "higher geometry." This liberal and novel use of the term "geometry" reflected not only disdain for ancient Euclidean axiomatics employing verbal definitions but also broader visions of the group-theoretic concept as a unifying principle for all mathematics. This view found even greater cultivation in Klein's later work at the University of Göttingen during the 1890's.

Concurrent and partially linked to the Erlanger Programme was reorganization of Euclidean projective geometry by the German mathematician Moritz Pasch in 1882. In the light of the variety of non-Euclidean geometries and the new symbolic logic, concurrent with but independent of Giuseppe Peano, Pasch underscored the distinctions between "explicit" and "implicit" definitions in geometry. An explicit definition expresses a new term by means of terms already accepted in the technical vocabulary at hand. By contrast, implicit definitions broadly define a new term from the total context in which it occurs, recognizing that it is logically impossible to define all terms explicitly without infinite regress or vicious circularity. Whereas Euclid attempted explicit definitions for all his basic terms such as point, line, and plane, Pasch accepted these terms as primitive or "nuclearly irreducible" implicit definitions. Although the origin of these nuclear propositions in geometry might be based directly on assertions of physical or psychological origins, Pasch emphasized that these propositions, insofar as they are stated and used in mathematics, are stated totally without any regard to extra-geometrical aspects. Following Pasch, Italian mathematician Peano in 1889 gave a new and rigorous reinterpretation of Euclidean geometry using his new symbolic logic. Like Pasch, Peano based his treatment on specific primitive terms and their relations to the intuitive notion of "betweenness." Effectively translating Pasch's treatise into more compact symbolic notation, Peano's geometry remained a purely formal calculus of relations between variables, which remained without any ready applications or continuations by other mathematicians.

As a student at Königsberg between 1884 and 1890, and later as a professor at Halle University in 1891, David Hilbert was exposed to the theory of invariants and notably a wide variety of abstract formalistic approaches to the new axiomatizations of (non) Euclidean geometry. Some of Hilbert's earliest university lectures, between 1889 and 1891, concerned algebraic and projective geometry. In the fall of 1891, Hilbert attended a lecture by German mathematician Hermann Wiener, where he first learned about the more general validity and scope of the axiomatic method. According to Constance Reid's biography of Hilbert, as early as 1894 in his summer lectures on *Grundlagen der Geometrie* (the foundations of geometry), Hilbert intended to produce the purest possible algebraic system of exact axiomatic non-Euclidean geometry, with Euclid's geometry as a special case. After extensive studies at the

University of Göttingen of number theory and algebra, Hilbert was reinspired to continue his axiomatic foundations of geometry in early 1898, as noted through communications with Schur and Klein. In the summer of 1898, Hilbert gave a lecture series on elements of Euclidean geometry. Hilbert tried not only to synthesize more effectively other efforts to reorganize geometry axiomatically but also to reverse partially the trends toward purely abstract symbolization in geometry, by returning to Euclid's points, lines, and planes and the basic relations of incidence, order, and congruence. Yet, instead of considering only Euclidean geometry, Hilbert began his lectures by explaining that the specific content of Euclid's definitions, interpretation of which had proven so difficult in the case of non-Euclidean geometries, was irrelevant for mathematics and not for the purposes of philosophy, psychology, or physics. For Hilbert, proper definitional meanings or interpretations of geometrical entities emerge only via their interconnections with whatever basic axioms are selected to define a given system of geometry. As Hilbert and others emphasized, all meanings in geometry are implicit and context-dependent. The "objects" are not to be determined a priori by an individual's psychological or historical perceptions of geometrical shapes in the real world. Hilbert explicitly states that the intuitive basis of fundamental geometrical concepts is mathematically insignificant and that their interconnections come into consideration only through the axioms. In his lectures, Hilbert subsequently proposed to set up on this foundation a simple and (unlike Euclid) complete set of independent axioms, by means of which it would be deductively possible to prove systematically all the long-held theorems of Euclid's geometry, and by implication to do the same for any other non-Euclidean geometry. By employing an algebra utilizing a minimum of new abstract symbols and by keeping his examples in Euclid's axioms, Hilbert was able to formulate and present his (non-Archimidean) conception of the new axiomatic method more clearly and convincingly to a wider audience than did his predecessors.

In-depth analyses of Hilbert's published lecture notes, in *Grundlagen der Geometrie* (1899; *The Foundations of Geometry*, 1902), have been given by many authors. A number of key themes and methodological conclusions common throughout Hilbert's later efforts can be identified readily from Hilbert's work. For the axiomatic reconstruction of any mathematical theory, Hilbert asserted three main requirements to be met by the system of axioms: algebraic independence, set-theoretic completeness, and logical consistency. The requirement of independence asserts that it must be possible to prove any one of the axioms from any others alone or in combination. Hilbert called a geometrical system of axioms complete if it suffices for the verification of all geometric theorems therein. The specific problematics surrounding the completeness requirement are related directly to the subsequent work by Kurt Gödel and Alan Mathison Turing in foundational studies, as well as A. N. Kolmorgorov's axiomatization of Andrey Andreyevich Markov's probability theory. For Hilbert, the consistency of an axiomatic system like geometry is directly derivable from that of naïve arithmetic. As well known since René Descartes, analytical geometry simply assigns pairs of real number spatial coordinates to points in plane geometry, with two and three variable linear equations defining lines and planes. In 1900, Hilbert stated, as a more general methodological conclusion of his axiomatization of geometry, that it would be possible ultimately to prove the consistency of, for example, Georg Cantor's continuum hypothesis, subsequent arithmetic axioms, and those of mathematical physics, by establishing the correctness of the solutions through a finite number of steps, based on a finite number of hypotheses that must be exactly formulated. A formal axiomatic system is construed not only as a system of specific statements about a given subject matter but also as a system of general conditions for what has been called a general "relational structure," such that the infinite number of formulas in mathematics can be defined completely and consistently by a finite number of formal axioms. Further application of such relational structures to a specific domain of natural science is thus taken to be made by means of a further intuitive or other interpretation of the formal objects and relations of the axioms.

Impact of Event

Within months of its original German publication, *The Foundations of Geometry* was translated into French, Italian, and English, with a number of important shorter and longer effects on the reformulation and pedagogy of geometry and number theory, which Hilbert next sought to axiomatize. Hilbert and others later showed that other geometries of higher than three dimensions, as well as several other areas of mathematics, were both consistent and complete in Hilbert's sense. Notwithstanding the ingenuity, clarity, and brevity of his system, Hilbert received a number of criticisms, namely, that his axioms were semantically empty denatured symbols dealing with totally abstract things in a rarefied formal system to which no known kind of certain reality or truth could apparently be attached. A related question was in what sense the logical (nongeometric specific) meanings of key terms such as "and," "is," "not," and "when" are to be defined. As a response, after a decade of mainly applications-oriented developments in mathematical physics, over the next two decades, Hilbert and his students redoubled further development of his Formalist program by inquiring directly into the logical and philosophical structure of the new foundations. Hilbert subdivided mathematics into three levels: Level 1 is ordinary operational mathematics considered as mathematics; level 2 is a system of formal symbols employed by level 1, and defined by level 3, which is an informal metamathematical theory about level 2.

Equally serious and longer-lasting criticisms of Hilbert's program, by L. E. J. Brouwer and others, was that the question of the absolute consistency of arithmetical axioms employed in Hilbert's relatively consistent foundations of geometry was left unanswered. Under the impact of the antinomies in Cantor's, Bertrand Russell's, and Frege's systems, several critics demonstrated how far, despite all claims, mathematics actually went beyond levels 1 and 2 meaning, expunged formal axioms purportedly divorced from linguistic and psychologistic considerations of evidence or truth. Although the results of Gödel and others suggest that Hilbert's program may never be carried out fully, the practical advantages of Hilbert's abstract approach to mathematics were adopted by numerous physicists as well as mathematicians through his publication *Methoden der mathematischen Physik* (1931-1937; *Methods of Mathematical Physics*, 1953; with Richard Courant).

Bibliography

Blanché, Robert. *Axiomatics*. Translated by G. B. Kleene. London: Routledge & Kegan Paul, 1962. One of the most cited texts embodying, as well as describing, Hilbert's axiomatic method as comparatively embodied by Euclid, Hilbert, and others.

Hilbert, David. *The Foundations of Geometry.* Translated by Leo Unger. 2d ed. La Salle, Ill.: Open Court, 1971. The high technical level and brevity of Hilbert's book makes most of its discussion inaccessible to all but those with graduate-level background in abstract algebra and geometry. Primarily of historical interest.

Hilbert, David, and S. Cohn-Vossen. *Geometry and the Imagination*. Translated by P. Nemenyi. New York: Chelsea, 1952. Discusses methodological ideas in a popular presentation.

Poincaré, Henri. *The Foundations of Science*. Translated by George B. Halsted. Lancaster, Pa.: Science Press, 1946. Although written somewhat at odds with Hilbert's *Foundations* (cited above), Poincaré gives an intuition-based introduction to non-Euclidean geometries as well as some of the background to the debates between the formalist (Hilbert), logicist (Russell), and intuitionist (Brouwer) camps.

Reid, Constance. *Hilbert*. New York: Springer-Verlag, 1970. The most informative and readily obtainable English-language account of Hilbert's education, developments, debates, and publications. Discusses Hilbert's controversies with Frege on Hilbert's semantic relativism, as well as an expert summary of Hilbert's main contributions by his one-time student, Hermann Weyl.

Wilder, Raymond L. *The Foundations of Mathematics*. 2d ed. New York: J. Wiley, 1965. An extensive and accessible treatment. Contains a comprehensive bibliography.

Gerardo G. Tango

Cross-References

Levi Recognizes the Axiom of Choice in Set Theory (1902), p. 143; Russell Discovers the "Great Paradox" Concerning the Set of All Sets (1902), p. 184; Brouwer Develops Intuitionist Foundations of Mathematics (1904), p. 228; Zermelo Undertakes the First Comprehensive Axiomatization of Set Theory (1904), p. 233; Markov Discovers the Theory of Linked Probabilities (1906), p. 335; Russell and Whitehead's *Principia Mathematica* Develops the Logistic Movement in Mathematics (1910), p. 465; Gödel Proves Incompleteness-Inconsistency for Formal Systems, Including Arithmetic (1929), p. 900; Turing Invents the Universal Turing Machine (1935), p. 1045; The Bourbaki Group Publishes *Eléments de mathématique* (1939), p. 1140; Cohen Shows That Cantor's Continuum Hypothesis Is Independent of the Axioms of Set Theory (1963), p. 1751.

LEBESGUE DEVELOPS A NEW INTEGRATION THEORY

Category of event: Mathematics
Time: 1899-1902
Locale: University of Rennes, France

Lebesgue developed a new theory for integrating discontinuous functions, based on a more general set-theoretic concept of measure

Principal personages:
HENRI-LÉON LEBESGUE (1875-1941), a French mathematician
ÉMILE BOREL (1871-1956), a French mathematician and politician
WILLIAM HENRY YOUNG (1863-1943), an English mathematician

Summary of Event

Since Euclid's *Stoicheia (Elements)*, the theory of measurements in mathematics generally was thought to encompass little more than systematically comparing the points, lines, or planes to be measured to a standard reference. With Pythagoras' discovery of geometric incommensurables (irrational numbers), it was realized gradually that the question of mathematical measurement, in general, requires more precise and comprehensive consideration of seemingly infinite processes and collections. Development of the differential and integral calculus and limit theory by Sir Isaac Newton and Gottfried Wilhelm Leibniz brought with it the realization that, for most geometric figures, true mathematical measures do not exist a priori, but rather depend on the existence and computability of strictly defined associated limits. In 1822, French physicist and mathematician Joseph Fourier discovered that computation of sets of harmonic (trigonometric) series used to approximate a given function depended upon appropriate existence and calculations using integrals. Integration, or integral theory, concerns the techniques of finding a function g(x) the first derivatives of which are equal to a given function f(x). These, in turn, depend on how discontinuous the function is. A function is a mathematical expression defining the relation between one (independent) and another (dependent) variable. Although it was known that every continuous function has an integral summation, it was not clear then whether or how an integral could be defined for the many different classes of discontinuous functions; that is, those functions that are not definable at one or more specific points. A function is continuous if for every positive number, it is possible to select a corresponding positive number.

In 1854, the German mathematician Georg Friedrich Riemann offered the first partial answer to the question of how to integrate discontinuous functions, based on approximating integrands having only a finite number of definitely known discontinuous points by a sum of step-functions, instead of the curve-tangent sums of earlier calculus. The sum or measure of the Riemann integral is equal to the area of the region bounded by the curve f(x). Yet, one of the many classic functions of importance to mathematics and physics, for which Riemann integration cannot be defined, is the "salt and pepper" function of Peter Dirichelet, where f(x) = 1 if x is rational, and 0, if x is irrational. Earlier, in 1834, Austrian mathematician-philosopher Bernhard Bolzano gave examples of mathematically continuous functions that are nowhere differentiable and, thus, unintegrable by Riemann's definition, as did Karl Theodor Weierstrass in 1875. Further motivations for clarifying the notions of continuity and integration arose in 1885, when German mathematician Adolf Harnack paradoxically showed that any countable subset of the real number system could be covered by a collection of intervals of arbitrarily small total length.

As a reaction to these difficulties, between 1880 and 1885, French geometrician Jean Darboux gave a novel definition of continuity, for the first time as a locally definable (versus global) mathematical property for discontinuous functions. Likewise, Camille Jordan, in 1892, first defined analogously the more general notion of "mathematical measure," using finite unions of mathematical intervals to approximate sparse and dense subsets of real numbers. Nevertheless, the opinions of many leading mathematicians such as Henri Poincaré and Charles Hermite differed as to whether discontinuous functions and functions without derivatives were legitimate mathematical objects, as well as how to define the concept of a normal versus a "pathological" function.

The first mathematician to infer from the above results that countable unions of intervals should be used to measure the more general entity of real number subsets was Émile Borel. In 1898, in his *Leçons sur la théorie des fonctions* (lectures on the theory of functions), Borel advocated an abstract axiomatics of constructivistic definitions. Constructivistic, in this case, meant that all proposed definitions should permit explicit construction of actual examples of the mathematical entities referred to. Borel redefined the "measure" of any countable union of real number intervals to be its total length and thereby extended the notion of abstract measurability to progressively more complex sets. Borel sought to generalize Georg Cantor's set theory, as well as to explicitly study "pathological" functions definable in terms of point sets. For Borel, the main problem was how to assign consistently to each pathologic point or singularity an appropriate numerical measure, meaning a nonnegative real number precisely analogous to length, area, and volume. Starting with elementary geometrical figures, Borel sought to define constructively measures to these sets so that formal measures of a line segment, or polygon, is always the same as its Euclidean measure and that the measure of a finite or countably infinite union of nonoverlapping sets is equal to the sum of the measures of all individual sets.

Cantor's set theory had expanded the definition of continuity to include not only geometric smoothness or (non)variability of a curve but also its pointwise mapping, or set theoretic correspondence. One of the results of Borel's studies was the well-known Heine-Borel theorem, which states that if a closed set of points on a line can be covered by a set of intervals, such that every point of this set is an interior point of at least one of the intervals, then there exists a finite number of intervals with this "covering" property. For Borel, any such set obtainable by the basic mathematical properties and operations of union and intersection of sets in principle has a measure. With these ideas as background, in 1900, Henri-Léon Lebesgue sought to enlarge Borel's notion of measureable sets in order to apply it explicitly to the problem of integrating a wider class of pathological functions than those permitted by Riemann's integral. In the preface of his doctoral dissertation, Lebesgue outlined his motivations and methods. In contrast to Borel, Lebesgue employed a nonaxiomatic descriptive approach, one of the key results of which was to solve the problem of defining integral measure for discontinuous functions in general, insofar as it is necessary here that an infinite but bounded set have finite measure. Lebesgue generalized Riemann's definition of the integral by applying this new definition of measure. In the second chapter of his dissertation, Lebesgue proposed five criteria necessary for sufficiently widening integration theory, including the need to contain Riemann's definition as a special case to incorporate only assumptions and results in this extension that are natural, necessary, and computationally useful. Another key insight of Lebesgue's integration theory is that every function with bounded measure is also integrable. Perhaps the most critical property of Borel-Lebesgue measure is its property of summability, or countable additivity. In particular, Lebesque showed that this property, of term-by-term integrability, gives a definition of the integral much wider and with more stable computational properties in the limit than Riemann's integral. For example, if the approximation to a discontinuous function f(s) approaches f(x), as the number of terms of the approximations approaches infinity, then the integral of this series approximates the integral of the function in the limit; in general, this property of uniform-convergence is not true for Riemann's integration.

As noted in Thomas Hawkins' *Lebesgue's Theory of Integration* (1970), much of the power of Lebesgue's integration results from judicious use of the techniques of monotonic sequences and bracketing. Substituting equivalent monotonic sequences for complicated functions simplifies convergence in the limit. The bracketing technique consists of using the integrals of two well-behaved ("tame") functions to bracket as upper and lower bounds the integral of the pathological function. Instead of subdividing the domain of the independent variable x (abscissa axis), Lebesgue subdivided the range (ordinate axis) of the corresponding function f(x) into subintervals. Therefore, Lebesgue's integration replaces Riemann's integral sums in the limit as sampling intervals approach 0. Lebesgue's theory of the integral also yields other important results, such as extending the fundamental theorem of calculus.

Impact of Event

Initially, Lebesgue's work met with strong and lasting controversy from Borel. The main point of contention was not so much the mathematical results as the metamathematical methods used by each. Lebesgue subsequently developed the ideas of his dissertation further, and soon published these in his two classic texts, *Leçons sur les séries trigonométriques* (1906; lessons on the trigonometric series) and *Leçons sur l'intégration et la recherche des fonctions primitives* (1904; lessons on integration and analysis of primitive functions). Despite the fact that it was recognized early by some as an important innovation, Lebesgue's integration was comparatively slow to be adopted by the mathematical community at large. In 1906, Lebesgue's contemporary, the English mathematician William Henry Young, independently arrived at a somewhat more general but operationally equivalent definition of Lebesgue-type integration, using the method of monotone sequences. Most textbook discussions of Lebesgue integration have incorporated a combination of Young's notation and formalism for Lebesgue's arguments and examples.

Lebesgue's integral, despite its major advantages, did not generalize completely the concept of integration for all discontinuous functions. For example, Lebesgue integration did not treat the case of unbounded functions and intervals. Subsequently, in 1912 Arnaud Denjoy, Thomas Stieltjes in 1913, and Johann Radon and Maurice-René Fréchet in 1915 created other more encompassing definitions of the definite integral over complicated functions. Fréchet, in particular, showed how to generalize Lebesgue's integral to treat functions defined on an arbitrary set without any reference to topological or metric concepts of measure, later leading to Hausdorff dimensional or (Mandelbrot) fractal measures. As further reformulated by Beppo Levi, Lebesgue integrable functions are those that almost always equal the sum of a series of step functions.

Many of the complicated functions of aero- and fluid-dynamics, electromagnetic theory, and the theory of probability were for the first time analytically integrable using Lebesgue's method. In his book on the axiomatic foundations of Andrey Markov's probability theory, A. N. Kolmogorov defined a number of operational analogues between the Borel-Lebesgue measure of a set and the probability of an event.

Bibliography

Craven, B. O. *The Lebesgue Measure and Integral*. Boston: Pitman Press, 1981. An intermediate treatment, but is more comprehensive. Includes a modern presentation of Borel's set measure.

Kestelman, Hyman. *Modern Theories of Integration*. Oxford: Clarendon Press, 1937. Gives a complete but rather abstract synopsis of contemporary integration theory.

Kline, Morris. *Mathematical Thought from Ancient to Modern Times*. New York: Oxford University Press, 1990. The standard reference for the history of limit and function theory.

Monna, A. F. "The Integral from Riemann to Bourbaki." In *Sets and Integration*, edited by Dirk Van Dalen and A. F. Monna. Gröningen: Wolters-Noordhoff, 1972. Provides a more technical discussion.

Temple, George F. J. *The Structure of Lebesgue Integration Theory*. Oxford: Clarendon Press, 1971. The most detailed, step-by-step treatment of the modern integration theory.

Young, W. H., and G. C. Young. *The Theory of Sets of Points*. Cambridge, England: Cambridge University Press, 1906. Contains Young's alternative independent development of what is essentially Lebesgue measure-based integration theory.

Gerardo G. Tango

Cross-References

Levi Recognizes the Axiom of Choice in Set Theory (1902), p. 143; Russell Discovers the "Great Paradox" Concerning the Set of All Sets (1902), p. 184; Cohen Shows That Cantor's Continuum Hypothesis Is Independent of the Axioms of Set Theory (1963), p. 1751; Mandelbrot Develops Non-Euclidean Fractal Measures (1967), p. 1845.

EINTHOVEN DEVELOPS THE FORERUNNER OF THE ELECTROCARDIOGRAM

Category of event: Medicine
Time: The early 1900's
Locale: Leiden, The Netherlands

Einthoven led the study of the electrical currents of the heart by inventing the string galvanometer, which evolved into the modern-day electrocardiogram

Principal personages:

WILLEM EINTHOVEN (1860-1927), a Dutch physiologist and winner of the 1924 Nobel Prize in Physiology or Medicine

AUGUSTUS D. WALLER (1856-1922), a German physician and researcher who conducted some of the initial research on the electrical events of the heart

CLÉMENT ADER (1841-1926), a French electrical engineer who developed a similar string galvanometer for use in telegraph recorders

SIR THOMAS LEWIS (1881-1945), an English physiologist who is recognized for increasing the general interest in electrocardiography

Summary of Event

In the late 1800's, there was substantial research interest in the electrical events that took place in the human body. Researchers studied many organs and systems in the body, including the nerves, eyes, lungs, muscles, and heart. As a result of lack of technology available to the researchers, this research was very tedious and frequently inaccurate. Therefore, the development of the appropriate instrumentation was as important as the research itself.

The initial work on the electrical activity of the heart (detected from the surface of the body) was conducted by Augustus D. Waller and published in 1887. Many credit him with the development of the first electrocardiogram. Waller used a Lippmann's capillary electrometer to determine the electrical changes in the heart and called his recording a "cardiograph." The recording was made by placing a series of small tubes on the surface of the body. The tubes contained mercury and sulfuric acid. As an electrical current passed through the tubes, the mercury would expand and contract. The images were projected onto photographic paper to produce the first cardiograph. Despite the contributions Waller made to science, he failed to accomplish any of the clinical applications. As a result, he did not pursue this area of study.

In the early 1890's, Willem Einthoven, who became a good friend of Waller, began using the same type of capillary tube to study the electrical currents of the heart. Einthoven also had a difficult time working with the instrument. His laboratory was located in an old wooden building near a cobblestone street. Teams of horses pulling heavy wagons would pass by and cause his laboratory to vibrate. This vibration affected the capillary tube, causing the cardiograph to be unclear. In his frustration, Einthoven began to modify his laboratory. He removed the floorboards and dug a hole ten to fifteen feet deep. He lined the walls with large rocks to stabilize his instrument. When this failed to solve the problem, Einthoven, too, abandoned the Lippmann's capillary tube. Yet, Einthoven did not abandon the idea and began to experiment with other instruments.

In order to continue his research on the electrical currents of the heart, Einthoven began to work with a new device, the d'Arsonval galvanometer. This instrument had a heavy coil of wire suspended between the poles of a horseshoe magnet. Changes in electrical activity would cause the coil to move; however, Einthoven found that the coil was too heavy to record the small electrical changes found in the heart. Therefore, he modified the instrument by replacing the coil with a silver-coated quartz thread (string). The movements could be recorded by transmitting the deflections through a microscope and projecting them on photographic film. Einthoven called the new instrument the "string galvanometer."

Einthoven's development of the string galvanometer was very important for science. Some have questioned whether the string galvanometer was Einthoven's original idea. In 1897, Clément Ader, a French electrical engineer, reported his development of a device similar to the string galvanometer. Ader was working with telegraph recorders; he also found a need to replace the slow coil in the galvanometer with a lighter component: a wire. Both scientists needed to make their respective instruments more sensitive, and both chose to use a wire in place of the coil. Yet, there were some fundamental differences between the two instruments. The major difference was the size of the wire. Ader used a wire that was ten times the diameter of Einthoven's. It is believed that Ader's instrument would not have been sensitive enough to detect the electrical currents in the heart. Although Ader developed the concept first, he was not given credit for developing the forerunner to the electrocardiogram because his instrument could not be used for studying the heart.

After developing the string galvanometer, Einthoven began work validating the string galvanometer with the instrument accepted at the time: the capillary tube. Using both instruments on the same subjects, he made comparisons between the two outputs. He was able to identify the similarities in the wave changes between the two instruments. The waves on the tracing from the capillary electrometer had been labeled "A," "B," "C," and "D." Einthoven labeled the waves on his galvanometer "P," "Q," "R," "S," and "T." He changed the labels on the waves because, in geometry, points on curved lines begin with "P" and points on straight lines begin with "A." Since the output was obviously a curve, he chose to use a new system. Einthoven's labels became universally known and continue to be used in electrocardiography.

By 1905, Einthoven had improved the string galvanometer to the point that he could begin using it for clinical studies. In 1906, he had his laboratory connected to the hospital in Leiden by a telephone wire. With this arrangement, Einthoven was able to study electrocardiograms in his laboratory, which were derived from patients in the hospital located one mile away. With this source of subjects, Einthoven was able to use his galvanometer to study many heart problems. As a result of these studies, Einthoven identified the following heart problems with his galvanometer: blocks in the electrical conduction system of the heart; premature beats of the heart, including two premature beats in a row; and enlargements of the various chambers of the heart. He was also able to study the heart during the administration of cardiac drugs. These studies and studies conducted by other researchers with the string galvanometer added to the knowledge that has made the electrocardiogram an important diagnostic tool.

A major researcher who communicated with Einthoven about the electrocardiogram was Sir Thomas Lewis. Lewis is credited with developing the electrocardiogram into a useful, clinical tool. One of his important credits was his identification of atrial fibrillation, the overactive state of the upper chambers of the heart. During World War I, Lewis was involved with studying soldiers' hearts. He designed a series of graded exercises, which he used to test the soldiers' ability to perform work. From this study, Lewis was able to use similar tests to diagnose heart disease and screen recruits who had some heart problems.

Throughout his career, Einthoven was influenced by one of his professors, Johannes Bosscha. Bosscha was also a professor in Leiden. In the 1850's Bosscha published a study showing the function of the forces involved in measuring small amounts of electricity. He proposed the idea that a galvanometer, modified with a needle hanging from a silk thread, would be more sensitive when determining the minute currents in the heart. His idea was studied by Einthoven and Ader and was responsible, in part, for the development of their instruments. Also, it was Bosscha's suggestion to link the hospital with Einthoven's laboratory. Therefore, Bosscha assisted Einthoven by providing both theoretical and practical ideas for his work, which led to the development of the electrocardiogram.

The electrocardiogram was not created in a short period of time. It took many years and the work of many people to produce a workable instrument. Then, it took additional years of study to determine the clinical significance of the instrument's output. Although the forerunner to the electrocardiogram was first created in 1901, it took several years for it to be used clinically.

Impact of Event

As Einthoven published additional studies on the string galvanometer in 1903, 1906, and 1908, greater interest in his instrument was generated around the world. In 1910, the instrument, now called the "electrocardiograph," was installed in the United States. It was the foundation of a new laboratory for the study of heart disease at The Johns Hopkins University Hospital. The electrocardiograph was used to diagnose various types of heart problems, which previously were very difficult to diagnose.

In 1924, Einthoven visited the United States to discuss the electrocardiograph with researchers and clinicians. His lecture tour was part of a lecture series at Harvard University founded in memory of Dr. Edward K. Dunham. During his visit to the United States, he was awarded the Nobel Prize in Physiology or Medicine. Since he was on tour, he was unable to attend the award ceremony. Yet, one year later, the award was made to Einthoven, and he had the opportunity to present the Nobel lecture. In his lecture, he discussed the string galvanometer and its use in electrocardiography. This presentation helped to promote the instrument for clinical purposes.

Through the 1920's, Einthoven's instrument included four electrodes, one placed one each limb. One of the electrodes acted as a ground and the other three could be combined several ways to make three different leads. These three leads were used to diagnose many heart problems. In the 1930's, researchers at the University of Michigan, led by Dr. Frank N. Wilson, developed a system that combined the same four electrodes into three new leads. They also added six new electrodes, which created six new leads to make a total of twelve leads. The addition of the nine leads to Einthoven's three leads significantly improved the diagnostic capabilities of the instrument.

As time passed, the use of the electrocardiogram—"ECG," as it is now known—increased substantially. The major advantage of the ECG is that it can be used to diagnose problems in the heart without incisions or the use of needles. It is relatively painless for the patient. Also, in comparison to other diagnostic techniques, it is relatively inexpensive. Therefore, Einthoven's invention has resulted in an important diagnostic tool that can be widely used.

Recent developments in the use of the ECG have been in the area of stress testing. Since many heart problems are not shown on the ECG while idle, the ECG is used on a patient who is exercising. This is generally done on a treadmill. The clinician gradually increases the intensity of work the patient is doing while monitoring the patient's heart with the ECG. Since many heart problems are more evident during exercise when the heart is working much harder, the use of stress testing makes the ECG a valuable diagnostic tool.

The electrocardiogram is a very important diagnostic tool for heart problems. It is continually being improved and new applications are being studied. Although the electrocardiogram was initially developed at the turn of the century, it is still an integral part of the fight against heart disease and other heart problems.

Bibliography

Barron, S. L. "The Development of the Electrocardiogram in Great Britain." *British Medical Journal* 1 (1950): 720-725. This article focuses on how the ECG was developed in Great Britain. The works of Sir Thomas Lewis are an important part of this piece, which is directed at an audience with an appropriate science background.

Burch, George E., and Nicholas P. De Pasquale. *A History of Electrocardiography.* Chicago: Year Book Medical Publishers, 1964. This book traces the history of the development of the electrocardiogram through the 1960's. It includes the works of the major contributors, including Einthoven. It is a lengthy book, with a wealth of references, and should be understandable to a cross-section of readers.

Cooper, James K. "Electrocardiography One Hundred Years Ago." *The New England Journal of Medicine* 315 (August, 1986): 461-464. The history of electrocardiography is discussed beginning with the contributions of Augustus Waller and concluding with the works of Willem Einthoven. The article celebrates the one hundredth anniversary of electrocardiography, which occurred in 1987. This article is concise, understandable, and includes references.

Ershler, Irving. "Willem Einthoven: The Man." *Archives of Internal Medicine* 148 (February, 1988): 453-455. This article features the experiences of Dr. George E. Fahr during the years he worked with Einthoven beginning in 1909. The author obtained much of this information from Dr. Fahr while they worked together from 1935 to 1938. It is light, easy reading and includes nine references.

Snellen, H. A. *Two Pioneers of Electrocardiography: The Correspondence Between Einthoven and Lewis from 1908-1926*. Rotterdam: Donker Academic Publications, 1983. This book highlights the correspondence between two of the major researchers in electrocardiography. It includes 140 pages and a limited reference list. Although it is interesting, it is technical and may be difficult for some to understand.

Bradley R. A. Wilson

Cross-References

Wilkins Introduces Reserpine for the Treatment of High Blood Pressure (1950's), p. 1429; Favaloro Develops the Coronary Artery Bypass Operation (1967), p. 1835; Barnard Performs the First Human Heart Transplant (1967), p. 1866; Gruentzig Uses Percutaneous Transluminal Angioplasty, via a Balloon Catheter, to Unclog Diseased Arteries (1977), p. 2088.

HOPKINS DISCOVERS TRYPTOPHAN, AN ESSENTIAL AMINO ACID

Category of event: Biology
Time: 1900
Locale: Cambridge, England

Hopkins discovered tryptophan, an essential amino acid in human diets, leading to a better understanding of diet and nutrition in humans

Principal personages:

SIR FREDERICK GOWLAND HOPKINS (1861-1947), an English biochemist who was a cowinner of the 1929 Nobel Prize in Physiology or Medicine

SYDNEY W. COLE (1877-1952), a student of Hopkins who aided in the discovery and isolation of tryptophan

WILLIAM CUMMING ROSE (1887-1985), an American biochemist who completed the pioneering work of Hopkins in the essentiality of amino acids

Summary of Event

In the late nineteenth and early twentieth centuries, the field of biochemistry and biochemical thought was a relatively new area. Sir Frederick Gowland Hopkins brought a new insight into the field. He was able to visualize the cell as a biochemical machine and piece together the significance of isolated chemical processes on the working of the cell. This research was a result of the background with which he entered the biochemical field. Hopkins had originally been trained in the field of analytical chemistry and soon became an assistant to Sir Thomas Stevenson, the Medical Jurist at Guy's Hospital. Hopkins became involved in several murder cases, and his analytical chemistry skill helped to gain convictions. Hopkins worked in the medical school by day and in a clinical research laboratory by night. At the invitation of Michael Foster, Hopkins came to The University of Cambridge in September, 1898, to teach chemical physiology. Hopkins had no special training in biochemistry.

Hopkins' background in the area of biochemistry and the cell started with protein chemistry. Proteins are long chains of amino acids that are linked together. There are many types of amino acids, but only twenty of these are used to build proteins in humans. Of the twenty amino acids, ten are essential to the diet of humans because humans are unable to synthesize them. Therefore, they must be obtained from the diet. The amino acids required in the diet are phenylalanine, isoleucine, leucine, lysine, threonine, tryptophan, cysteine, arginine, valine, and histidine (in infants).

Hopkins' initial work with proteins dealt with more of the pure chemical properties of the proteins. His initial research was concerned with the isolation and crystallization of proteins. His keen analytical background enabled him to devise a new technique to crystallize pure albumin protein from egg white. This was his first published work from The University of Cambridge. This led to his work with Sydney W. Cole, which led to the isolation of tryptophan. A paper published in 1901 dealt with purely chemical techniques that show the Adamkiewicz "acetic acid" reaction resulted from an impurity in the acetic acid. This was proposed because Hopkins had obtained a sample of acetic acid that gave no Adamkiewicz reaction. Hopkins obtained many samples of acetic acid and compared them to determine if they would react with the protein. Hopkins and Cole concluded eventually that glyoxylic acid was the impurity in the acetic acid and that it was responsible for the Adamkiewicz reaction.

A second paper published in 1901 discussed an undescribed product of protein digestion, which was responsible for the Adamkiewicz reaction. Again, Hopkins' analytical chemistry skill played a large role in the isolation of this product. After the isolation of this product, it was tested and found to give the tryptophane reaction; thus, it was the product that was giving the Adamkiewicz reaction. Cole and Hopkins concluded, then, that the product of the protein digestion was, indeed, tryptophan. The final paper in the series, published in 1903, described the structure of the tryptophan molecule. Hopkins and Cole concluded that the structure was a skatole-amido-acetic acid (this was later revised in 1907 to be indole-amido-acetic acid structure by later researchers). This paper was important because it was the first to employ bacterial decomposition as an aid in determining the structure of a chemical molecule. This, in turn, led to the start of bacterial biochemistry, which was pursued by Majorie Stephenson, a student of Hopkins.

Hopkins then published several papers on proteins, but then his interest turned from a purely chemical nature of proteins to nutritional studies of proteins. In the nineteenth century, it was known that gelatin, a protein, would not support life if it was the only source of protein in a rat's diet. It was found later that gelatin was deficient in the amino acids tyrosine and tryptophan.

Hopkins also published two papers in 1906 on the dietary role of individual amino acids. The first paper proposed the role of tryptophan as a precursor (starter molecule) for the synthesis of adrenalin. The second paper alluded to the concept of the "essential amino acid" and was entitled "The Importance of Individual Amino-Acids in Metabolism: Observations on the Effect of Adding Tryptophane to a Dietary in Which Zein is the Sole Nitrogenous Constituent," published in the *Journal of Physiology.* This work used zein—a protein from corn—which was made deficient in the amino acid tryptophan. Young mice fed a diet of this protein were unable to maintain growth. Addition of tryptophan enabled the mice to prolong their survival, but they were unable still to maintain growth. In comparison, the addition of tyrosine, an amino acid present in zein, did not prolong survival. Thus, tryptophan was concluded to be an important dietary constituent. It was published in the relatively new *Biochemical Journal*.

The scope of Hopkins' research was shown by the following passage: "A deficiency in a nitrogenous dietary need not necessarily be one of quantity; the form in which the nitrogen is supplied may determine its efficiency." Thus, Hopkins was able to realize that not only was the quantity of protein important but also the quality of the protein was important.

Further research with Harold Ackroyd in 1916 showed that other amino acids were essential also in the diet. Work with the amino acids arginine and histidine indicated that they were important in nucleic acid metabolism (the genetic material). Exclusion from the diet resulted in nutritive failure. When either one or the other was restored to the diet, then normal bodily maintenance and growth of the organism occurred.

The body of this research was reviewed by Hopkins in a lecture presented to the chemical society. As the father of biochemistry in England, Hopkins was gratified to deliver a lecture on the chemistry of animals. He discussed the fate of protein in the body.

Hopkins remained active in his career. In 1914, he became the first professor in the new department of biochemistry at The University of Cambridge. After World War I, biochemistry began to be taught at the entry level in college throughout the English university system. In 1925, Hopkins was knighted and awarded the Copley Medal of the Royal Society. The most impressive distinction came in 1929 when he shared the Nobel Prize in Physiology or Medicine with Christiann Eijkman for what became known as the vitamin concept.

Hopkins performed research studies in many other fields of biochemistry and physiology. Research with Walter Fletcher showed that muscle activity and lactic acid production are correlated with each other. Hopkins later worked to discover sulfur-containing vitamins and discovered a tripeptide (three amino acids linked together) called glutathione. He then published a series of papers on the structure and function of the molecule. The impact of Hopkins on the field of biochemistry was immense; at the time of his death in 1947, seventy-five of his students occupied professorial positions around the world.

Impact of Event

The "essentiality of amino acids" was important in the history of diet and nutrition throughout the world. Hopkins' work was at the forefront of this area of research. Hopkins had realized that proteins differ, and the difference lay in the amino acid building blocks. By 1935, the last of the essential amino acids (threonine) was isolated by an American biochemist, William Cumming Rose.

Rose continued the research that Hopkins had begun. Rose found that of the twenty amino acids found in proteins, ten were essential to the diet of rats. Rose then utilized graduate students in his nutritional experiments. He fed them isolated amino acids and found only eight amino acids to be essential. Rose then calculated the minimum daily requirement for each amino acid. Thus, Rose firmly established what Hopkins had worked on in the essentiality of amino acids.

The pioneering work of Hopkins has been refined greatly since the beginning of the twentieth century. It is known now that human adults require nine essential amino acids and infants require ten essential amino acids in their diet. The nutritional quality of a protein depends on the content of the amino acids found in the protein (as Hopkins had suggested) and how well it is digested.

Most animal proteins have a high nutritional quality in that they are easily digestible and they contain all the essential amino acids. On the other hand, plant proteins may lack either an essential amino acid or they may be only partially digestible. Proteins of wheat and other grains may be partly digested only because they are surrounded by husks that cannot be digested. Thus, an adult male would have to consume approximately seventy-three slices of wheat bread to meet the daily allowance of protein recommended for the diet.

Though plant proteins alone may not provide an adequate supply of amino acids, such as corn protein, some combinations of plant proteins will provide an adequate and balanced mixture of amino acids. Beans (which are low in tryptophan but high in lysine) and corn (which is low in lysine but contains enough tryptophan), which is a mixture called succotash used by New World Indians, would provide a nutritionally adequate supply of amino acids. Another combination used by Asian populations would be soybeans and rice. These combinations would provide the essential amino acids required in a diet and can be equivalent almost to milk protein.

Nutrition is a worldwide problem because in most parts of the world animal protein is not readily available, while plants contain low amounts of protein and are usually lacking in essential amino acids. In areas of the world where population growth is outstripping food production, chronic protein deficiencies (called kwashiorkor, or "weaning disease" in Africa) are occurring. In Central and South America, protein defiency is common; an international nutrition board has introduced caparina, a mixture of corn, sorghum, and cottonseed meal. Separately, these plant proteins are poor, but together they have the protein value of milk. Much research in the area of nutrition is still continuing. The pioneering work of Hopkins was only one step toward a better understanding of diet and nutrition.

Bibliography

Asimov, Isaac. "Sir Frederick Gowland Hopkins." In *Asimov's Biographical Encyclopedia of Science and Technology.* 2d ed. Garden City, N.Y.: Doubleday, 1982. This book contains the biography of 1,510 great scientists from the ancient times to the early 1980's.in a chronological fashion. It is an easy-to-read book, and the biographies are short. Some of the biographies are cross-referenced for major advances in the field.

Baldwin, Ernest. "Frederick Gowland Hopkins." In *Dictionary of Scientific Biography,* edited by Charles Coulston Gillispie. Vol. 6. New York: Charles Scribner's Sons, 1972. A nice reference set on famous scientists. The biographies are well written and detailed. References at the end of each biography. For a wide audience.

Lehninger, Albert L. "Human Nutrition." In *Principles of Biochemistry.* New York:

Worth, 1982. A very nicely written chapter in an elementary biochemistry college textbook. It is easy to understand, even for the beginner who has little understanding of biochemistry. References at the end of the chapter.

Needham, Joseph, and Ernest Baldwin, eds. *Hopkins and Biochemistry.* Cambridge, England: W. Heffer and Sons, 1949. A book containing Hopkins' uncompleted autobiography, excerpts from papers with commentaries, and selected addresses of Hopkins. A nice collection. The autobiography is quite interesting. The book gives one a feel for the accomplishments of Hopkins. For the general reader.

Williams, Trevor I., ed. "Sir Frederick Gowland Hopkins." In *A Biographical Dictionary of Scientists*. 3d ed. New York: John Wiley & Sons, 1982. The biographies are short and easy to read, but they do not contain cross-references to other biographies. References at the end of each biography. For interested readers; nontechnical.

Lonnie J. Guralnick

Cross-References

Abel and Takamine Independently Isolate Adrenaline (1897), p. 16; Hopkins Suggests That Food Contains Vitamins Essential to Life (1906), p. 330; Steenbock Discovers That Sunlight Increases Vitamin D in Food (1924), p. 771; Krebs Describes the Citric Acid Cycle (1937), p. 1107; Miller Reports the Synthesis of Amino Acids (1953), p. 1465.

WIECHERT INVENTS THE INVERTED PENDULUM SEISMOGRAPH

Category of event: Earth science
Time: 1900
Locale: Göttingen, Germany

Wiechert introduced a damping mechanism that restrained the seismograph pendulum and greatly increased its accuracy

Principal personages:

EMIL WIECHERT (1861-1928), a German seismologist and geophysicist who invented the inverted pendulum seismograph

HUGO BENIOFF (1899-1968), an American seismologist who developed a seismograph that permitted the accurate recording and analysis of the tiniest movements of the ground, both vertical and horizontal

BORIS GOLITSYN (1862-1916), a Soviet physicist and seismologist who invented the electromagnetic seismograph

JOHN MILNE (1850-1913), an English seismologist, geologist, and engineer, who invented the first seismograph in 1893

Summary of Event

Prior to the development of the science of seismology, the study of earthquakes and measurement of the elastic properties of the earth, the natural philosophers of the late nineteenth century held varying views on the composition inside the earth. One view was that if the inside were liquid, then the surface of the earth would rise and fall almost like the tides of the oceans. Another view was that the geological hypothesis of a fluid interior is untenable, and the overall rigidity of the earth's interior is considerable. The invention of the seismograph, an instrument for recording the phenomena of earthquakes, at the end of the nineteenth century was to open the twentieth century with an explosion of discoveries about the inner earth.

The first seismographs were delicate horizontal pendulums that registered singular waves. Such an instrument was used by the German seismologist E. von Rebeur Paschwitz on April 18, 1889, when he correlated horizontal pendulum recordings at Potsdam and Wilhemshaven with a great earthquake in Tokyo. Four years later, John Milne, an English seismologist, geologist, and mining engineer, invented the first clockwork-powered seismograph. As a result of this invention, when an earthquake occurred, it became possible to record continuously the seismic waves, earth vibrations, produced during an earthquake. Seismic waves carry information to the surface on the structure through which they have passed. The Milne seismograph was the first to record the movements of the earth in all three of their components: up and down, back and forth, and side to side.

An important fact in the study of earthquakes is that there are two types of waves that can be transmitted through a homogeneous, isotropic, elastic solid such as the earth. Isotropic means having the same properties in like degree in all directions. These waves are: dilatational (compressional) waves, such as sound waves, which involve particle motions parallel to the direction of the energy; and transverse (shear) waves, which involve motion at right angles to this direction. Emil Wiechert, in 1899, independent of similar work by Richard D. Oldham, an Irish seismologist, determined that the p waves were dilatational and the s waves shear waves.

The Milne seismograph, and others developed at the end of the nineteenth century, fell short of meeting the demands of a science that was asking ever more complex questions. These early machines measured only a portion of the broad band of wave size and frequencies. Another shortcoming was the tendency of the seismograph's pendulum to keep swinging indefinitely once a strong motion had started it. Without a way to control the pendulum's motion, the seismograph was unable to record accurately the other kinds of waves that arrived later. Wiechert, a German seismologist, introduced a damping mechanism that restrained the seismograph pendulum and greatly increased its accuracy. It had one major shortcoming in that it was bulky, since it depended upon weights large enough to remain at rest despite the energy transmitted both by the shaking of the instrument's frame and by the mechanical linkage that inscribed the seismic waves on paper.

The Wiechert seismograph, an inverted pendulum, permitted the detection of the vertical component of long earthquake waves. The amplification of the boom movement was achieved by using a system of mechanical levers and recorded by a stylus scratching on smoked paper. It had a mass of 17 tons supported on a vertical column, which acted as the boom. Its period was about a second and its magnification was two thousand. In his experimentation, he found that the only way to make a seismometer (an instrument for measuring the direction, intensity, and duration of earthquakes) react to the vertical component above was to counteract the force of gravity by a spring. In the inverted pendulum, the mass is held on the right and left by spiral springs. It rests and oscillates upon a sharp horizontal edge turned at right angles to the direction of the movements.

Prior to Wiechert's invention of the inverted pendulum seismograph, he suggested the existence of a central core within the earth. He speculated that this central core was appreciably different from the outer shell. In 1901, a year after he invented his seismograph, Wiechert founded the Geophysical Institute in Göttingen, Germany. The institute quickly became a center for the study and compilation of earthquake data from observatories worldwide. One of the important seismologists affiliated with the Geophysical Institute was Beno Gutenberg, a German seismologist, who in 1914, after studying seismic data collected by the Institute from worldwide sources, estimated the depth of the boundary between the inner and outer cores at 2,900 kilometers from the earth's surface. Later, Harold Jeffreys, an English seismologist, precisely measured this depth at 2,898 kilometers, plus or minus 4 kilometers.

Wiechert observed that seismograms at many observatories revealed the presence of additional small earth movements, called microseisms. These small movements complicated the problem of the accurate recording of ordinary earthquakes; their form is likely to be related to features that are similar on records traced at observatories distributed over a wide area. One of the features is the approximately simultaneous occurrence of maximum amplitudes at all the observatories involved. These microseisms may persist for many hours at a time and may have more or less regular periods of from 2 to 10 or more seconds. Wiechert suggested that microseisms are generated by the action of rough surf against an extended steep coast; this hypothesis was followed up by Gutenberg. Norwegian and Japanese seismologists have concluded a similar relationship, as suggested by Wiechert.

Shortly after Wiechert's invention, in 1906, Boris Golitsyn, a Soviet physicist and seismologist, invented the first electromagnetic seismograph, which did away with the need for mechanical linkage between the pendulum that revealed the earth's movement and the record that transcribed it. With slight modifications, the state of the art in seismographs after Golitsyn was established until 1932, when Hugo Benioff, an American seismologist, perfected a completely different kind of seismograph. It was based on the relative movement of two points on the ground, drawing near or separating during the passage of elastic waves of an earthquake, and not the inertia of pendulum as in earlier seismographs.

Impact of Event

With the development of the seismograph in the late nineteenth century came an explosion of knowledge about the inner earth. Very early in the twentieth century, Oldham and Wiechert independently postulated the presence in the center of the earth of a large, dense, and at least partially molten core. In 1909, Andrija Mohorovičić, a Yugoslav meteorologist and seismologist, discovered the discontinuity between seismograph recordings of earthquake waves, which led to his discovery of the boundary between the earth's crust and upper mantle. This boundary, called the Mohorovičić Discontinuity, varies from about 10 kilometers beneath the basaltic ocean floor to between 32 and 64 kilometers beneath the granitic continents. In 1914, Gutenberg, later affiliated with California Institute of Technology, discovered the boundary between the mantle and the outer core. In 1936, Inge Lehmann, a Danish seismologist, after a number of years of observing waves through the core from Pacific earthquakes and a mathematical model, discovered the boundary between the outer and inner core. These discoveries established the existence of boundaries for the inner and outer core, the mantle and crust.

Around 1924, portable seismographs were being introduced to record seismic waves from quarry blasts and other relatively small explosions. Initially, records were taken at distances of about 20 to 30 kilometers from the source and later at greater distances. Wiechert and others inferred p wave speeds of 4.8, 6.0, and 6.7 kilometers per second from quarry blasts near Göttingen, Germany, between 1923 and 1929. Seismic monitoring of rock blasting in quarries and other methods of mining are now regulated by local laws or are common practice in practically all mining operations in the United States and Europe.

A worldwide network of recording stations in the 1960's, improved seismographs, and computers have all contributed to a greater understanding of the earth in recent times. Some of these discoveries include that the Earth's core is not a single molten mass, but consists of an inner and outer core with a transition zone dividing a solid inner core from its surrounding molten rock. The mantle has been found to consist of an upper and lower section. There are concentric envelopes of elusive discontinuities with new ones being detected frequently. P waves rebound sometimes from mysterious interfaces at depths of 257 kilometers, 402 kilometers, and also at 644 kilometers. At one layer in the upper mantle, just below the crust, the p and s waves suddenly lose velocity, indicating that the rocks become less rigid and begin to ooze. It is the region of the upper mantle that commands the special attention of seismologists because it is at the juncture of the mantle and the crust that most earthquakes have their source. Understanding this region is also important for understanding the movement of continental plates.

Wiechert was an important contributor to both the evolution of the modern seismograph and the understanding of the inner earth. His discovery of the inverted pendulum was a major step forward.

Bibliography

Bolt, Bruce A. *Inside the Earth: Evidence from Earthquakes*. New York: W. H. Freeman, 1982. An introductory text of a nonmathematical nature for earth science. The principal focus is on earthquake waves, main shells of the earth, structural detail, earth vibrations, and physical properties of the earth. Illustrated with simplified diagrams and boxed excerpts of major discoveries in seismicity.

Eiby, G. A. *Earthquakes*. Exeter, N.H.: Heinemann, 1980. This book is intended for the nontechnical reader who wants elemental knowledge about earthquakes; it is a good reference for the high school student and college undergraduate. It is well written and easy to read since it contains no mathematical formulas. Of particular interest is the chapter on recording an earthquake.

Halacy, D. S., Jr. *Earthquakes: A Natural History*. Indianapolis: Bobbs-Merrill, 1974. This is a well-written elementary book designed for the high school and lower-level college student as well as the general reader. Illustrations and photographs complement the text and are well prepared. It is a short book covering the basics on earthquakes and is highly recommended.

Stacey, Frank D. *Physics of the Earth*. New York: John Wiley & Sons, 1969. The principal audience is the graduate and advanced undergraduate student of physics. Topics covered include the solar system, rotation and figure of the earth, the gravity field, seismology and the internal structure of the earth, and the geomagnetic field.

Tazieff, Haroun. *When the Earth Trembles*. New York: Harcourt, Brace & World, 1964. An excellent book for the general reader and high school student. The book is composed of three sections: The first covers the great Chilean earthquake of May, 1960; the second section describes the geography of earthquakes in the

Mediterranean Belt, Asiatic Belt, Pacific Belt, and Midoceanic Ridge; and the third section is on instruments that record earthquakes. The text is well written, illustrated, and contains a brief bibliography.

Walker, Bryce. *Earthquake*. Alexandria, Va.: Time-Life Books, 1982. This book is structured for the average reader, high school student, and undergraduate college student. It has excellent photographs, many in color and full page, and the text is written in narrative style. Areas covered include the Alaska earthquake of 1964, history of earthquakes, and seismology from earliest times to 1982.

Wood, Robert Muir. *Earthquakes and Volcanoes*. New York: Weidenfeld & Nicolson, 1987. This is an excellent book for the nontechnical reader and is copiously illustrated with color photographs. It is written in a very understandable format on the causes and prediction of earthquakes. The second part of the book discusses the effects of volcanoes.

Earl G. Hoover

Cross-References

Oldham and Mohorovičić Determine the Structure of the Earth's Interior (1906), p. 340; Gutenberg Discovers the Earth's Mantle-Outer Core Boundary (1913), p. 552; Richter Develops a Scale for Measuring Earthquake Strength (1935), p. 1050; Lehmann Discovers the Earth's Inner Core (1936), p. 1065; Hess Concludes the Debate on Continental Drift (1960), p. 1650.

LANDSTEINER DISCOVERS HUMAN BLOOD GROUPS

Category of event: Medicine
Time: 1900-1901
Locale: Vienna, Austria

Landsteiner investigated the chemistry of the immune response and discovered the A-B-O blood groups, the most significant advance toward safe blood transfusions

Principal personages:

KARL LANDSTEINER (1868-1943), an Austrian pathologist, immunologist, and winner of the 1930 Nobel Prize in Physiology or Medicine

PHILIP LEVINE (1900-1987), a physician, immunoserologist, and Landsteiner's student, who revived interest in blood groups with his extensive clinical research findings

ALEXANDER S. WEINER (1907-1976), an immunoserologist who wrote on blood grouping and blood transfusion and who, with Landsteiner, discovered the rhesus blood system

Summary of Event

In the late 1800's, immunology was developing rapidly as scientists examined the various physiological changes associated with bacterial infection. Some pathologists studied the way cells helped the body fight disease; others favored a central role for noncellular, or humoral, factors. In 1886, George Nuttall showed how serum (the fluid remaining after a clot is removed from blood) can be toxic to microorganisms. Similar results were obtained by others, and although debate between advocates of cellular versus humoral immunity continued, most researchers focused on the humoral response to disease. It was during this period that the term "antigen" was used to refer to any substance inducing a reaction against itself by the host organism. "Antibody" referred to the factor in the serum which could react with the foreign substance.

In 1894, Richard Pfeiffer, of the Institute for Infectious Diseases in Berlin, injected cholera into guinea pigs and observed a series of changes in the cholera organisms: loss of mobility, clumping, becoming unstainable, and eventual disappearance. His work inspired others. Max von Gruber, a bacteriologist at the Hygiene Institute in Vienna, was particularly interested in the clumping, or agglutination, of foreign cells. He and his student, Herbert Edward Durham, discovered that antibodies cause the agglutination of disease organisms and that particular antibodies react only with like or closely related microorganisms. Also influenced by Pfeiffer, Jules Bordet of the Pasteur Institute demonstrated in the late 1890's that agglutination and hemolysis occurred after red blood cells were injected from one species into another. He pointed out that this response was similar to the body's reaction to foreign microorganisms.

Paul Ehrlich, a German physician and brilliant biologist, along with his coworker,

Julius Morgenroth, injected blood from goats into other goats and discovered that the goats produced "isoantibodies," a term coined by Bordet to describe antibodies to cells of the same species. Ehrlich could discern no pattern and barely missed discovering goat blood groups, but he interpreted his results to signify biochemical differences between individuals. He inferred from the artificial nature of the experiment that the reaction may not have any obvious function for the animal. Rather, it might be an unrelated effect of another physiological process and not part of the body's response to disease.

Samuel Shattock, an English pathologist, almost discovered the human blood groups. In 1899 and 1900, he described the clumping of red cells by serum from patients with acute pneumonia and certain other diseases. He could not find the clumping in the serums of normal persons because of his small sample size, and he concluded that his results reflected a disease process.

Karl Landsteiner synthesized the results of all the above experiments and provided a simple but correct explanation. He joined the University Hospital in Vienna and began advanced study in organic chemistry. At the age of twenty-eight, he joined Gruber at the Hygiene Institute in Vienna and soon began a lifelong interest in immunology. Almost two years later, he transferred to the Institute of Pathological Anatomy of Anton Weichselbaum. Landsteiner recognized early the chemical nature of the serological reactions that make immune processes visible. He used Bordet's work as his model. He made antibodies by injecting harmless forms of bacteria into a variety of animals, and, like Bordet, he had independently produced antibodies to another species' blood cells by injecting guinea pig blood into rabbits. Landsteiner withdrew his manuscript that had been submitted already for publication when Bordet's article was published in 1898. In 1900, before Ehrlich and Morgenroth had published their work on goats, he commented in a footnote that serums from healthy humans agglutinated animal red cells and cells from other humans (the first mention of human isoantibodies). He did not believe there was sufficient evidence to tell whether agglutination was solely a function of disease or a result of individual differences; however, by late 1901, he had elaborated on his footnote and provided the supporting experimental evidence.

Pauline Mazumdar considers the 1901 paper a logical extension of Ehrlich's goat studies. He took blood samples from his colleagues, separated the cells from the serum, and suspended the red blood cells in a saline solution. He then mixed individually each person's serum with a sample from every cell suspension. Agglutination occured in some cases; there was no reaction in others. From the pattern he observed, he hypothesized that there were two types of red blood cell, A and B, whose serum agglutinated the other type of red cell. There was another group, C (in later papers, group O), whose serum agglutinated red blood cells of both types A and B, but whose red cells were not agglutinated by serum from individuals with either blood type A or B. He concluded that there were two types of antibodies, now called anti-A and anti-B, found in persons of blood types B and A, respectively, and together in persons with blood type C. The absence of antibodies to one's own antigens was referred to later as Landsteiner's rule.

In 1902, two students of Landsteiner, Alfred von Decastello and Adriano Sturli, working at Medical Clinic II in Vienna with more subjects, tested blood samples with the three kinds of cells. Four out of 155 individuals had no antibodies in their serums (2.5 percent), but their cells were clumped by the other types of serums. This fourth rare kind of blood was type AB, because both A and B substances are present on red cells. Decastello and Sturli proved also that the red cell substances were not part of a disease process when they found the markers equally distributed in 121 patients and 34 healthy subjects.

In Landsteiner's 1901 paper, he anticipated forensic uses of the blood by observing that serum extracted from fourteen-day-old blood that had dried on cloth would still cause agglutination. He suggested that the reaction could be used to identify blood. He noted also that his results could explain the devastating reactions that occurred after some blood transfusions. Human-to-human transfusions had replaced animal-to-human transfusions, but cell agglutination and hemolysis still resulted after some transfusions using human donors. In a brief paper, Landsteiner interpreted agglutination as a physiological and not a pathological process. He laid the basis for safe transfusions, forensic serology, and became the "father of blood groups."

Louis Diamond points out that Landsteiner's experiments were performed at room temperature in dilute saline suspensions which made possible the agglutination reaction of anti-A and anti-B antibodies to antigens on red cells but hid the reaction of warm "incomplete antibodies" (small antibodies that coat the antigen but require a third substance before agglutination occurs) to other, yet undetected antigens such as the rhesus antigens, which are so important for understanding hemolytic disease of the newborn. Further developments in blood grouping would not take place for another thirty years.

Landsteiner's discovery of the A-B-O blood groups was made early in his career. His later contributions to immunology and medicine are numerous. He was particularly interested in how an antibody reacts with a particular antigen—and only that antigen—and demonstrated that the specificity was dependent on the chemical structure of blood factors (usually more than one) present on an antigen. His extensive investigations in this area are presented in his monograph, *Die Spezifität der serologischen Reaktionen* (1933; *The Specificity of Serological Reactions*, 1936).

Impact of Event

Landsteiner's blood group work was initially ignored. Others rediscovered the A-B-O blood groups and provided new classifications which led to confusion over nomenclature and credit for the discovery. In 1908, Albert Epstein and Reuben Ottenberg suggested that the blood groups were inherited, but the exact mode of inheritance remained controversial until the work of Felix Bernstein in 1924. Results of tests for the A-B-O groups were soon admitted as evidence in paternity cases, and state legislation followed which permitted courts to order such tests.

Ludwik Hirszfeld and Hanna Hirszfeld, army doctors in the Balkans during World

War I, tested many soldiers from different countries and showed that racial and ethnic groups differed in their A-B-O blood type frequencies. Much later, William Boyd, an immunochemist, convinced anthropologists to use blood marker distributions in order to infer population history (which is now routine).

The most important practical outcome of the discovery of blood groups was the increased safety of blood transfusions which, initially, few were able to appreciate. In 1907, Ottenberg was the first to apply Landsteiner's discovery by matching blood types for a transfusion. Citrate had been discovered to be an anticoagulent in 1890, but it was not until 1914 that it was used in a blood transfusion. A New York pathologist, Richard Weil, discussed this breakthrough in an influential article published in the *Journal of the American Medical Association* in 1915. He was also familiar with Landsteiner's work and argued for testing blood to ensure compatibility. Anticoagulents and "cross-matching" were soon part of standard practice.

Subgroups of blood type A were discovered in 1911, but it was not until 1927 that Landsteiner, now working at the Rockefeller Institute in New York, and his student, Philip Levine, discovered additional blood group systems. They injected different human red cells into rabbits and eventually obtained antibody that could distinguish human blood independently from A-B-O differences. The new M, N, and P factors were not important for blood transfusion but were used for resolving cases of disputed parentage. More scientists were aware of the multiple applications of Landsteiner's blood group research, and in 1930, he was awarded the Nobel Prize in Physiology or Medicine.

Transfusion reactions between A-B-O compatible people still occurred; finally, in 1940, Landsteiner and Alexander S. Weiner reported an antibody to another antigen, the rhesus factor, responsible for most of the adverse transfusion reactions and hemolytic disease of the newborn. This and Levine's work reawakened interest in blood groups, and new methods developed in the 1940's that led to discoveries of many more antigens.

The addition of new antigens has facilitated population analysis and individual identification in paternity and criminal cases, and two world wars and additional methodological advances have hastened the acceptance of the use of blood transfusions for the treatment of injuries and sickness. Today, Diamond estimates that more than 10 million blood transfusions are given yearly in the United States.

Bibliography

Diamond, Louis Klein. "The Story of Our Blood Groups." In *Blood, Pure and Eloquent*, edited by Maxwell M. Wintrobe. New York: McGraw-Hill, 1980. An excellent, very readable discussion (by a knowledgeable insider) of people, their ideas, and methodological advances. See also the preceding chapter, entitled "A History of Blood Transfusion," by Diamond.

Erskine, Addine G., and Wladyslaw Socha. *The Principles and Practice of Blood Grouping*. 2d ed. St. Louis: C. V. Mosby, 1978. The first chapter, entitled "History of Blood Transfusion and Blood Grouping," is recommended. Weiner's work is discussed more than Levine's. The text also includes a selective bibliography after each chapter and a helpful glossary.

Lechevalier, Hubert A., and Morris Solotorovsky. *Three Centuries of Microbiology.* New York: McGraw-Hill, 1965. The various related discoveries that preceded and followed Landsteiner's work are detailed. The connectedness of the scientific ideas are broken up by minibiographies of all the major personalities. Limited bibliography.

Mazumdar, Pauline M. H. "The Purpose of Immunity: Landsteiner's Interpretation of the Human Isoantibodies." *Journal of the History of Biology* 8 (Spring, 1975): 115-133. A detailed discussion of the relevant ideas and experiments that set the stage for Landsteiner's work. The elegant simplicity of his explanation is emphasized.

Race, R. R., and R. Sanger. *Blood Groups in Man*. Oxford, England: Blackwell Scientific Publications, 1975. Known as the "blood groupers bible," this sixth and last edition (first published in 1950) describes all the blood group antigens.

Speiser, Paul, and Ferdinand Smekal. *Karl Landsteiner.* Translated by Richard Rickett. Vienna, Austria: Hollinek Brothers, 1975. A revision and translation of the 1961 German edition written solely by Speiser. A comprehensive and fascinating description of his personal life and each of his many scientific achievements. Complete bibliography included.

Joan C. Stevenson

Cross-References

Crile Performs the First Direct Blood Transfusion (1905), p. 275; Ehrlich and Metchnikoff Conduct Pioneering Research in Immunology (1908), p. 422; McLean Discovers the Natural Anticoagulant Heparin (1915), p. 610; Boyd Defines Human Races by Blood Groups (1950), p. 1373.

DE VRIES AND ASSOCIATES DISCOVER MENDEL'S IGNORED STUDIES OF INHERITANCE

Category of event: Biology
Time: March-June, 1900
Locale: Amsterdam, Tübingen, and Vienna

Three scientists independently discovered in 1900 that in 1865, Mendel had reported that characteristics of plants were inherited as though they were particles

Principal personages:

HUGO DE VRIES (1848-1935), a Dutch botanist and horticulturalist, who became famous for his mutation theory for the mechanism of evolution

CARL ERICH CORRENS (1864-1933), a German botanist, who made significant contributions to the modern theory of Mendelian genetics in its early years

ERICH TSCHERMAK VON SEYSENEGG (1871-1962), an Austrian botanist and plant breeder who contributed to the improvement of various crop plants and ornamental flowers

WILLIAM BATESON (1861-1926), an English geneticist and evolutionist, who after 1900 was the major champion and popularizer of the new Mendelian genetics

GREGOR JOHANN MENDEL (1822-1884), a Czechoslovakian teaching monk and naturalist who experimented with garden peas and published the results

Summary of Event

Before 1900, some investigators of heredity were concluding that its mechanism was not a blending of characteristics, as was generally thought, but rather that the characteristics were particulate and unchanged by mixing with others, reappearing unchanged in future generations. Thus, inheritance was similar to combining various colored balls, rather than to the mixing of two different colored liquids. In 1900, three of these scientists who were studying plant crosses, Hugo de Vries, Carl Erich Correns, and Erich Tschermak von Seysenegg, independently discovered that in 1866 Gregor Johann Mendel had published a paper giving this particulate explanation with the regular ratios for the appearance of contrasting characteristics in garden peas over several generations. In the first generation of such a cross, only one trait that was termed dominant, of a pair of contrasting ones, was found in all of the offspring. This is usually known as Mendel's law of dominance. In the second, and succeeding, generations the lost trait, termed recessive, would appear in some individuals. In the second inbred generation, it would show up in one-fourth of the offspring of self-pollinated plants. After the second generation, only two-thirds of the dominant-appearing plants carried the recessive traits, the other third were pure dominants. Those that carried the recessive traits would have them show up in one-fourth of their offspring when selfed. Those that were pure dominants would breed true and show only that trait in future selfed generations. Further, in crosses in which several traits were followed at the same time, each one of them was inherited independently of the others and each showed these same ratios. This result is called Mendel's law of independent assortment.

A number of plant hybridizers, as early as the eighteenth century, had noticed the uniformity of the first-generation offspring of a hybrid cross and the lack of uniformity in succeeding generations. None of them noticed that the individual traits were inherited in a systematic way that could be described as percentages or ratios. Thus, Mendel was unique in keeping careful records of each trait in each generation and in using mathematical ratios to characterize them. He also noted that it did not matter which trait was examined; they all behaved in the same manner. In the initial crossing of the two types, it did not matter from which parent the pollen or ovules came.

De Vries, who in 1900 was a professor at Amsterdam University in The Netherlands, realized as early as 1889 when he published a book entitled *Intracelluläre Pangenesis* (translated into English in 1910 as *Intracellular Pangenesis*), that the key to understanding plant heredity and evolution could best be discovered by hybridizing different types of plants and studying the characteristics of their offspring over several generations. In the following years, he made many such crosses using a wide variety of flowering plants in order to study his hypothesized "pangenes," or particles that he thought carried the hereditary information. Also, he was interested in sudden changes in plants from one generation to the next, called saltation or mutation. He thought that new species could originate in that manner. In his crosses, de Vries found that the first generation exhibited only one of a pair of contrasting traits and that the second generation had about a fourth with the lost trait. Sometime before March, 1900, he discovered that Mendel had explained this for peas in 1866; thus, he rediscovered Mendel's ratios. In March, de Vries submitted a paper for publication in which he reported his ratios for many different kinds of plants and in a footnote referred to Mendel's original publication. Because of this, he was the first person to be credited with the rediscovery.

Correns, who in 1900 was teaching at Tübingen University in Germany, had been hybridizing different races of maize and peas. He was particularly interested in a characteristic in maize heredity called xenia, in which the color of the endosperm of the grain is caused by the kind of pollen the plant received from the male parent. Correns, too, found that the "lost" traits in both types of plants reappeared in one-fourth of the offspring in the second inbred generation. Like de Vries, he independently discovered Mendel's paper sometime before March, 1900, but only submitted his paper for publication in April, after seeing de Vries' paper. He also gave Mendel full credit for the discovery of the ratios and their mechanism. Correns, however, reported cases where the first generation lacked dominance and showed some uniform intermediate state for a particular characteristic. He had questions about the universality of Mendel's intrepretation for all cases. Correns was a highly respected researcher, as was de Vries, and his papers and those of de Vries were read widely, encouraging others to do research to test their explanations.

Tschermak von Seysenegg, who had been breeding garden peas and studying their various traits, also noticed the regularity of their appearance in different generations. He, too, independently found Mendel's paper when preparing to present a lecture on his breeding work as part of his new teaching position at the Land Cultivation University in Vienna (Hochschule für Bodenkultur). He credited Mendel in a paper submitted for publication in April, 1900. He was a less well-known scientist than either de Vries or Correns, and thus his research was less influential.

William Bateson, in England, also had been studying actively discontinuous inheritance of characters of plants and animals; he published a book in 1894, *Materials for the Study of Variation Treated with Especial Regard to Discontinuity in the Origin of Species*. When he received a copy of Correns' paper with reference to Mendel's work, Bateson searched out the paper immediately and realized the great value of Mendel's explanations for explaining discontinuous variation. Bateson quickly became the promoter of Mendelianism by having Mendel's paper translated into English and publishing it together with an explanation of its importance for understanding heredity in a book, *Mendel's Principles of Heredity, a Defense*, in 1902. The reason for needing a defense of Mendel resulted from the fact that, in England, the study of inheritance was dominated by the biometric school founded by Sir Francis Galton. His explanation of heredity proposed in 1897 was based upon a continuous blending hypothesis giving each ancestor a certain proportion of the mix. Each parent contributed half; each grandparent contributed one-fourth, and so on backward by generations. The Mendelian explanation did not fit the biometricians' laws of ancestral inheritance. This disagreement led to heated controversy. When one of the biometricians, the zoologist Raphael Weldon, published an unfavorable review of Mendel's work and those who followed, Bateson was compelled to respond. By doing so, he split the English biology community into two feuding factions. Bateson had difficulty in getting his research published and in finding a university appointment. Nevertheless, his zeal for the new genetics—genetics was the name he gave the field in a public lecture in 1906—led others, particularly young botanists and zoologists, to join him in experimental research on heredity.

Mendel's paper was cited in 1881 in a major analytical reference book, W. O. Focke's *Die Pflanzen-Mischlinge* (the plant hybrids), that tells that Mendel had found constant ratios for characters in second and succeeding generations of peas. Both Correns and Seysenegg found Mendel's paper through this reference. De Vries may have first learned about Mendel's work in the bibliography of a paper by the American horticulturalist Liberty Hyde Bailey, who did not see the paper, but included the reference from Focke's book in 1892. Or, de Vries may have first seen Mendel's paper when a colleague sent him an offprint of it from his library because he thought it would be useful to him in his work. De Vries was unclear about this when he later tried to recall that event. The literature about the rediscovery of Mendel's experiments contains some controversy as to whether de Vries gave Mendel enough credit. Correns believed that de Vries never gave Mendel the credit he deserved. In any case, it is very clear that each of the rediscoverers came to an understanding similar to that of Mendel for their experiments before finding the Mendel paper. Both de Vries and Correns continued to do significant genetic research after 1900. Bateson became the strong champion of the Mendelian explanation for all types of plant and animal inheritance.

Impact of Event

The modern science of genetics began with the rediscovery of Mendel's explanation for inheritance. After 1900, de Vries, Correns, Tschermak von Seysenegg, and many other scientists began to explain their experimental results in terms of Mendelian ratios, or exceptions to those ratios. Particulate inheritance with genes on chromosomes replaced the notion of blending inheritance. The appearance of offspring having characteristics of both parents now could be explained by the large number of characteristics involved, each one of which operated independently in a particulate manner except when they were linked together on the same chromosome. Within a few years, cytological data connected contrasting traits with sections of chromosomes. This connection with the use of the term gene to describe the controlling unit provided a physical basis for the particulate explanation of heredity. One early finding was that sexuality was controlled in a Mendelian fashion. Experimental work with various types of organisms showed that the three-to-one dominant to recessive Mendelian ratio for the second generation of inbred crosses was the normal situation. When these ratios were not found, further explanations were necessary and were produced: for example, lack of dominance; more than two possible contrasting traits for the same chromosomal location, only two of which were present in normal cells; more than one pair of genes controlling the expression of a particular trait; linkage of genes on the same chromosome; and modifier genes that changed the expression of a trait. Quickly, cytological studies began to supplement breeding experiments. Genetics became a leading field of biological study. In applied areas, genetics became an important and useful tool. Agriculturists used their understanding of Mendelian genetics to develop better plants and animals. This resulted in the discovery of hybrid vigor in corn and other crops, which significantly increased agricultural production. In medicine, it led to a better understanding of heredity diseases such as sickle-cell anemia and color blindness. Genetics was to play a major role in interpreting data in other fields of biology; for example, morphology, embryology, taxonomy, physiology, and evolutionary biology. The search for the nature of the gene led to the development of molecular biology. The twentieth century truly can be said to be the century of genetics in biological sciences.

Bibliography

Bowler, Peter J. *The Mendelian Revolution: The Emergence of Hereditarian Concepts in Modern Science and Society.* Baltimore: The Johns Hopkins University

Press, 1989. The importance of Mendelian genetics after 1900 is well documented, as is the social and scientific importance of the rediscovery. Provides good background information for better appreciating Mendelian genetics.

Mendel, Gregor J. *Experiments in Plant Hybridisation*, translated by William Bateson. Cambridge, Mass.: Harvard University Press, 1948. The English translation of Mendel's paper supervised by Bateson and published in his book *Mendel's Principles of Heredity, a Defense* in 1902. The paper is a clear description of Mendel's methods, results, and his interpretation, as well as providing background on breeding experiments of other scientists. It is basic to understanding why those who discovered it thirty-four years later were so impressed with its insights.

__________, et al. "The Birth of Genetics: Mendel, de Vries-Correns-Tschermak in English Translation." *Genetics*, supp. 35 (September, 1950): 1-47. The original first papers of the three scientists who rediscovered Mendel's work are given in English translations. Also in translation are nine letters written by Mendel between 1866 and 1873 to Carl Nägeli, professor of botany at Vienna University, explaining his pea experiments and his later experiments with crossing hawkweeds.

__________, et al. *Fundamenta Genetica: The Revised Edition of Mendel's Classic Paper with a Collection of Twenty-seven Original Papers Published During the Rediscovery Era*, edited by Jaroslav Kříženecky. Oosterhout, The Netherlands: Anthropological Publications, 1965. The twenty-eight papers in their original languages published between 1899 and 1904 give the flavor of excitement of the early Mendelian geneticists. A good source to find Bateson's papers. An introductory essay in English about concepts before Mendel is useful in understanding why Mendel's work was so revolutionary and was therefore ignored. Published to celebrate the centenary of the publication of Mendel's paper.

Olby, Robert C. *Origins of Mendelism*. New York: Schocken Books, 1966. A well-balanced discussion of the origins of Mendelism, considerations of other interpretations of heredity before and during Mendel's life, and detailed explanation of the rediscovery of Mendel's discovery. Clearly written without extensive quotes from original writings, which are given in a large appendix.

__________. "William Bateson's Introduction of Mendelism to England: a Reappraisal." *British Journal of the History of Science* 20 (1987): 399-420. Documents the rediscovery of Mendel's work, but mostly elaborates on Bateson's role as an early defender of Mendelism. Claims that Bateson took several years to conclude that Mendelian principles were broadly applicable to inheritance in different kinds of organisms.

Roberts, H. F. *Plant Hybridization Before Mendel*. Princeton, N.J.: Princeton University Press, 1929. Reprint. New York: Hafner Publishing Co., 1965. The classical presentation of the background of Mendels' hybridizing experiments, Mendel's work, and its rediscovery in 1900. Roberts was able to obtain written comments quoted from de Vries, Correns, and Seysenegg about their recollections of how they found Mendel's paper. Includes a chapter on Bateson's contributions. Essential for any serious study of the origins of Mendelism.

Stern, Curt, and Eva R. Sherwood, eds. *The Origin of Genetics: A Mendel Source Book*. San Francisco: W. H. Freeman, 1966. Includes new English translations of Mendel's papers on plant hybrids and those of de Vries and Correns. Contains a reprint of a paper by the English geneticist R. A. Fisher on the statistics in Mendel's 1866 paper and why they seem too perfect. Sewall Wright, an American geneticist, responds to Fisher. Stern gives a good summary of Mendel's work and its rediscovery.

Emanuel D. Rudolph

Cross-References

McClung Plays a Role in the Discovery of the Sex Chromosome (1902), p. 148; Sutton States That Chromosomes Are Paired and Could Be Carriers of Hereditary Traits (1902), p. 153; Punnett's *Mendelism* Contains a Diagram for Showing Heredity (1905), p. 270; Bateson and Punnett Observe Gene Linkage (1906), p. 314; Hardy and Weinberg Present a Model of Population Genetics (1908), p. 390; Morgan Develops the Gene-Chromosome Theory (1908), p. 407; Johannsen Coins the Terms "Gene," "Genotype," and "Phenotype" (1909), p. 433.

EVANS DISCOVERS THE MINOAN CIVILIZATION ON CRETE

Category of event: Archaeology
Time: March 23, 1900
Locale: Kephála (Knossos), Crete

Evans altered the chronology assigned to the development of Hellenic civilization through his excavations at Knossos and initiated a scholarly controversy on the dominance of Minoan culture on the Greek Peloponnisos

Principal personages:

SIR ARTHUR EVANS (1851-1941), an English archaeologist who was the excavator of Knossos and curator of the Ashmolean Museum, Oxford, England

DUNCAN MACKENZIE (1859-1935), a Scottish archaeologist who directed field operations at Knossos and kept the excavation's "day books"

JOHN PENDLEBURY (1904-1941), an English archaeologist and successor to Mackenzie

MICHAEL VENTRIS (1922-1956), an Engish classicist and cryptographer who deciphered the script known as "Linear B"

Summary of Event

Well into his middle age, a period at which lesser individuals settle into careers and comfortably contemplate retirement, Sir Arthur Evans, eldest child of Sir John Evans and Harriet Ann Dickinson, discovered a civilization to which he gave the name "Minoan." Though the Minos of legend is familiar to even the most casual reader of myth, no one until Evans' excavations were under way could appreciate its distinctive character. By Evans' calculations (restated by most contemporary archaeologists to begin only about 2600 B.C.), Cretan origins, dated from 3500 B.C., reached their zenith around 2200 B.C., and ended suddenly about 1100 B.C., probably because of a devastating earthquake which necessitated the relocation of government from Crete to the eastern Peloponnisos.

A fortuitous combination of circumstances made Evans' discovery possible. He had intended as early as 1871 to write an archaeological history of the Balkans. Indeed, he planned permanent residence in the town of Ragusa (Dubrovnik, Yugoslavia) and emigrated there after his marriage to Margaret Freeman in 1878. The couple shared an enthusiasm for the Balkans, in general, and Ragusa, in particular. They lived in Ragusa for almost four years, through a combination of support from Evans' wealthy father and regular articles Evans wrote for *The Manchester Guardian* on the tense political situation in the Balkans under Austrian occupation. Ultimately, it was these articles and Evans' increasingly outspoken position on Balkan independence which led to his arrest, imprisonment, and expulsion in April, 1882. It was only through the concerted efforts of his wife, brother, and sister that he avoided trial.

Once returned to the family home in Hemel Hempstead, Hertfordshire, England, Evans' family secured for him the post of Keeper of the Ashmolean Museum, Oxford, in 1884. The Ashmolean did not resemble the prestigious institution it would become under Evans' leadership. Essentially, it was like a disorganized warehouse, a polyglot collection of archaeological, geological, and historical artifacts. Evans organized and refined the collections, placed the institution on a firm financial basis, and supervised the construction of its handsome building in 1894.

Evans traveled often during his tenure at the Ashmolean and incurred frequent criticism for his long absences. The sudden death of his wife in 1893 was partly responsible for Evans' extensive forays to possible archaeological sites in the Mediterranean, but it is certain that having met Heinrich Schliemann in 1873, only three years after the latter had astounded the archaeological world by unearthing Troy, had excited Evans' imagination to the possibilities of even earlier sites awaiting discovery.

What led Evans to Crete, however, was his having come upon in 1894 several three- and four-sided sealstones of Cretan origin that were offered for sale in an Athens antique shop. They were engraved with the distinctive picture writing Evans would eventually call "Linear A." He recalled similar markings on two vases found by Greek archaeologist Chrestos Tsountas at Mycenae on the Greek Pelponnisos. Evans was convinced that this picture writing indicated a distinct Cretan culture, one that antedated that of the mainland and one that was truly pre-Hellenic. He was determined to prove his hypothesis through excavations on the great mound at Kephála, the modern name for the region that surrounded Knossos, considered the site of the palace of Minos. Evans coined the word "Minoan" to distinguish this culture from that of Mycenae. He used it to identify the culture not only with the king associated with the Daedalean labyrinth and the Minotaur of myth but also to delineate the palace period that preceded the Trojan War.

Evans' friend Federico Halbherr, an Italian archaeologist, convinced him that important discoveries were likely to be found at Knossos. In 1878, Halbherr had unearthed several *pithoi* (large storage jars) on the site, about 6.5 kilometers from the town of Candia. Schliemann, still fresh from his triumphs at Troy and Mycenae, had expressed an interest in excavating there as early as 1878, and might have done so were it not for his inability to reach financial agreement with the owner of the land. The Turkish authorities then administering Crete repeatedly posed obstacles, but Evans eventually surmounted these by purchasing the land outright. He combined substantial family resources with contributions from the England-based Cretan Exploration Fund he created and, thus, became one of the few archaeologists ever to own the site upon which he excavated.

Evans' unflagging success at Crete lay in his ability to marshal shrewd managerial talent with an uncanny ability to dig in the right place. Immediate and continuing success ensured constant financial support, and though there were interruptions because of the onset of World Wars I and II, the Knossos excavations never lacked the substantial funding they required.

Never had any excavation paid such immediate, continuing, and spectacular dividends. The first day of digging (March 23, 1900) yielded walls and pottery fragments a mere 33 centimeters beneath the topsoil. The second day produced an ancient house and fresco fragments. The third day revealed smoke-blackened walls and broken pottery, including the rims of *pithoi* similar to those discovered by Halbherr. Exactly one week after excavations began, Evans unearthed baked clay bars inscribed with the intriguing linear script, which had first interested Evans in 1894. Evans would discover two kinds of linear script at Knossos. The first, designated by him as "Linear A," is hieroglyphic in genre and virtually indecipherable, because no inscription bilingual with some known language exists to function as a starting point for translation. "Linear B," however, is different; Evans had hoped until his death that he would be able to decipher it, and even withheld unpublished several thousand tablets at his home. It was not until 1953 that the English classicist Michael Ventris, a cryptographer during World War II, established that the script was ideographic (syllabic) pre-Greek. Working from the earlier hypothesis of American Alice Kober that the upright strokes marked word divisions and by noting repetitions, Ventris proved the script's syllabic character and prepared a grid with the ideographs' syllabic equivalents. In effect, Ventris broke the code, and revealed, thereby, that the Linear B tablets were largely shorthand inventory lists. The mud-brick upon which they were written was ideal to keep a changing inventory of larder stores and exports. Subsequent discoveries of large tablets at Plyos, Mycenae, and other locations on the Peloponnisos are evidence that Evans' hypothesis regarding the distinctive and independent character of Cretan civilization is questionable.

Evans defended Minoan civilization as dominant and discrete until his death, relegating Mycenaean Greece to the role of its successor, important because successive earthquakes at Knossos necessitated transference of power to the mainland. Archaeologists have subsequently advanced Evans' chronology of Minoan palace settlement and argued that Knossos more likely fell by Mycenaean invasion than by earthquake, probably as early as 1500 B.C.

While he was still actively involved in excavating, Evans received severe criticism for the comprehensive restorations of the great palace and its environs. His architect Theodore Fyfe in effect rebuilt large portions of it, often with considerably less evidence than would satisfy contemporary archaeologists. Similarly, the bright colors Émile Gilliéron used in restoring the palace frescos left Evans open to the charge of commercializing the excavation and catering to popular imagination. Still, Evans' daring and flair were responsible for the dramatic unveiling of a brilliant civilization whose discovery might otherwise have been postponed indefinitely.

Impact of Event

For more than thirty years of his active involvement in the Knossos excavations, Evans repeatedly captured scholarly attention and popular imagination. In part, his colorful personality sustained progress at Knossos, for Evans was the stereotype of an English archaeologist: immaculately dressed in tropical white linen suit, tie, and always photographed with his walking stick (named "Prodger"), which he used as a magic wand to uncover astonishing finds. Evans carefully nurtured his public image as a wizard of archaeology. He used it and his facile writing ability to keep the Knossos excavations on a firm financial footing. His radical restorations also satisfied the general public's notion of what archaeology should be. "Minoan red," the distinctive color used extensively in Gilliéron's restorations, became familiar to the world of art. Wealthy patrons of the Cretan Exploration Fund, as well as others who could afford them, purchased reproductions of Minoan palace art. The homes of European art lovers in the early twentieth century often displayed a bull fresco, a painting of a wasp-waisted serving boy, or the like.

Though such popular attention assured financial stability, the professionals connected with the excavation often found it vaguely embarrassing. Duncan Mackenzie, the rough-edged Scottish archaeologist who supervised daily operations, often disagreed heatedly with Evans' flamboyant methods, and this caused his temporary dismissal on several occasions. Nevertheless, Mackenzie's "day books," painstaking accounts of daily progress, remain the primary source for scholars concerned with the early progress of the excavations. Mackenzie's scientific approach, his careful measuring of soil strata, and his drawings of principal objects found to scale and in context provided a method used by subsequent archaeologists worldwide. After Mackenzie's death, John Pendlebury filled this post, and wrote *The Archaeology of Crete: An Introduction* (1939), the comprehensive single-volume account of Minoan archaeology.

The most significant effect of Evans' work was, however, the restructuring of the pre-Greek chronology of settlement on Crete and the eastern Peloponnisos. Evans posited a Neolithic period, which ended about 3500 B.C. Likewise, he maintained until his death the preeminence of Minoan over Mycenaean civilization, rejecting the theory of its fall by Mycenaean invasion. Many contemporary archaeologists reject this position and have adjusted Evans' datings; but without Evans' work, the debate would never have begun.

Bibliography

Evans, Sir Arthur. *The Palace of Minos*. Reprint. 4 vols. London: Macmillan, 1921-1936. New York: Biblo & Tannen, 1964. This profusely illustrated account of the excavations gives its entire history with numerous color plates, many prepared especially for the set, and numerous painstakingly executed drawings and plans, many of these from Mackenzie's day books. Evans explicates every major object found, discusses the frescos at length, and provides an index, which is in itself a masterly enumeration of activity at Knossos. The sections on chronology and scripts should be considered against other works cited herein.

Evans, Joan. *Time and Chance: The Story of Arthur Evans and His Forebears*. London: Longmans, Green, 1943. Reprint. Westport, Conn.: Greenwood Press, 1974. This family memoir, prepared by Evans' half-sister, focuses on the famed archaeologist but also discusses in considerable detail the backgrounds of his progenitors. The Evans and Dickinson families made their fortune in the manufacture of paper. This volume serves not only as a biography but also as a social history of England from the late seventeenth century to the beginning of World War II.

Horwitz, Sylvia. *The Find of a Lifetime: Sir Arthur Evans and the Discovery of Knossos*. New York: Viking Press, 1981. Popularly written, this volume provides an account of Evans' early life, his experiences in the Balkans, and the circumstances that led Evans to excavate at Knossos. It is a portrait of the man, and focuses on his colorful personality and uncanny flair for making astonishing discoveries. There is no sustained discussion of the excavations or what was found at Knossos, but Horwitz shows effectively how the site perfectly suited the personality of its archaeologist. A good selected bibliography suitable for general readers concludes the volume.

MacKendrick, Paul. *The Greek Stones Speak: The Story of Archaeology in Greek Lands*. New York: St. Martin's Press, 1962. Paperback reprint. New York: Mentor Books, 1966. Contains a good account for general readers on the important particulars of the Knossos excavations as a nontechnical explanation of Linear B and how Ventris deciphered it. Includes photographs of the palace excavations, principal finds, and a reproduction of the syllable grid Ventris published in connection with the decipherment of Linear B. MacKendrick, like most modern archaeologists, holds that Mycenaean civilization gained ascendancy over the Minoan world by force; he also provides discussion of the major Mycenaean sites.

Palmer, Leonard R. *A New Guide to the Palace of Knossos*. London: Macmillan, 1969. Ideal for general readers and travelers, this volume provides plates, figures, and plans adapted from Evans' *magnum opus* on the site, as well as an easy-to-follow text written to incorporate subsequent scholarship on the excavations.

____________. *On the Knossos Tablets: The Find Place of the Knossos Tablets*. Oxford, England: Clarendon Press, 1963. A companion volume to the book cited above, this work discusses the locations and contexts of the tablet hoards. Provides a less technical discussion of linear scripts than the Ventris and Chadwick work, cited below.

Pendlebury, J. D. S. *The Archaeology of Crete: An Introduction*. London: Methuen, 1939. Reprint. New York: Biblo & Tannen, 1963. A classic study of Minoan civilization by Evans' site supervisor at Knossos. Discusses the island's geography, principal sites, including those other than Knossos, and provides a concluding discussion of post-Minoan Crete.

Ventris, Michael, and John Chadwick. *Documents in Mycenaean Greek: Three Hundred Selected Tablets from Knossos, Plyos, and Mycenae*. Cambridge, England: Cambridge University Press, 1956. 2d ed., 1973. Though parts of this volume are more detailed than general readers might desire, it offers a complete account of discovery, the linear script writing system, the language, and several hundred of the most representative tablets. The second edition includes commentary on additional tablets not available at the time the first edition was prepared for publication.

Willetts, R. F. *The Civilization of Ancient Crete*. Berkeley: University of California

REED ESTABLISHES THAT YELLOW FEVER IS TRANSMITTED BY MOSQUITOES

Category of event: Medicine
Time: June, 1900-February, 1901
Locale: Quemados, near Havana, Cuba

Reed established that yellow fever was caused by an unknown infectious agent being transmitted by the Aëdes *mosquito, thereby laying the basis for experimental medicine in the twentieth century*

Principal personages:

WALTER REED (1851-1902), a physician, bacteriologist, and major in the United States Army Medical Corps, who headed the United States Army Yellow Fever Board

JESSE WILLIAM LAZEAR (1866-1900), a physician and bacteriologist who served as an entomologist on the Yellow Fever Board

JAMES CARROLL (1854-1907), a physician who served as a bacteriologist on the Yellow Fever Board

ARISTIDES AGRAMONTE (1869-1931), a physician who served as a pathologist on the Yellow Fever Board

WILLIAM CRAWFORD GORGAS (1854-1920), a physician and colonel, United States Army Medical Corps, who, as chief sanitary officer of Havana, used the findings of the Yellow Fever Board to develop methods by which to control yellow fever

Summary of Event

During the last thirty years of the nineteenth century, a new understanding of the role of microorganisms in disease, along with improvements in personal hygiene and public sanitation, led to a dramatic decrease in the incidence of diseases such as typhoid and cholera. The disease yellow fever, however, had foiled all attempts at control. Yellow fever had been known for centuries along the west coast of Africa. There it showed little effect in the native population, probably because generations of exposure to the disease had resulted in a high degree of inherent resistance. Europeans, however, had virtually no resistance to yellow fever and thousands died of the disease during efforts to develop the natural resources of Africa. So deadly was yellow fever that the continent of Africa became known to Europeans as the white man's grave.

The Western Hemisphere remained free of yellow fever until the seventeenth century, when it was probably introduced as a result of the slave trade from Western Africa. The first documented epidemics in the New World struck Barbados in 1647 and Yucatán in 1649. The first United States epidemic was in New York City in 1668. Severe epidemics continued periodically with 135 major epidemics striking American port cities between 1668 and 1893.

Yellow fever was a particularly frightening disease. It attacked suddenly, causing a high fever, headache, and nausea. The eyes became inflamed and the skin took on a yellow pallor. Many victims hemorrhaged internally, producing a black vomit. The mortality rate ranged from 30 percent to 70 percent, with most deaths occurring on the sixth or seventh day. During the Philadelphia epidemic of 1793, there were more than seventeen thousand cases and more than five thousand deaths. In 1802, an epidemic in Santo Domingo killed twenty-nine thousand of the thirty-three thousand troops sent there by Emperor Napoleon Bonaparte for a planned invasion up the Mississippi River. As a result of these devastating losses, Napoleon changed his plans and the next year negotiated the Louisiana Purchase with President Thomas Jefferson. An epidemic in the Mississippi Valley in 1878 caused the deaths of about thirteen thousand people. More than fifty-two hundred of these deaths were in Memphis. The financial impact resulting from the disease caused Memphis to lose temporarily its city charter in 1879.

Two theories developed as to the cause and spread of yellow fever. Some considered it to be a contagious disease that could be spread directly from one person to another. Others believed it to be caused by a miasma, filth, or rotting vegetables. The Philadelphia epidemic of 1793 was attributed to a shipment of spoiled coffee beans. In 1881, Juan Carlos Finlay, a Cuban physician, became a strong advocate of the mosquito transmission theory that had been proposed in 1848 by Josiah Nott. Dr. Finlay went as far as to state that it was probably the *Culex* mosquito (the name was later changed to *Aëdes*) that was spreading an unknown infectious agent. He performed experiments in which some individuals were bitten by mosquitoes and came down with yellow fever. His unsophisticated experiments, however, were easily discounted by his detractors. In 1898, at the end of the Spanish-American War, William Crawford Gorgas, United States Army, was appointed chief sanitary officer for Havana. Gorgas set out to rid Havana of the filth and disease left by the war. As a result of his campaign, typhoid and dysentery were significantly reduced. Yellow fever temporarily declined but struck again in 1900, this time bypassing the filthiest areas, where there was a high level of immunity from previous epidemics, and causing devastation in the cleaner sections of town, including the United States Army headquarters.

It was at this time that Walter Reed was appointed to head a commission to study yellow fever in Cuba. The United States Army Yellow Fever Board was made up of Reed, James Carroll, Jesse William Lazear, and Aristides Agramonte, a native Cuban. With the help of Dr. Gorgas, Dr. Finlay, and Dr. Henry R. Carter of the U. S. Public Health Service, the group was able to, during a brief six months, remove the veil of ignorance about yellow fever and provide an understanding of the disease to realize its eventual control. The initial work of the Yellow Fever Board was to attempt to identify the cause of the disease. The Surgeon General had discouraged pursuit of the mosquito transmission concept, so the investigators went to work to isolate a bacterial agent. They thoroughly studied eighteen severe cases of yellow fever but were unable to isolate any microbial cause for the disease. The team quickly reached an impasse and was forced to take a new approach. Reed knew of an outbreak of yellow fever in a military prison where there had been no possible direct contamination. It was also known that a ship leaving an infected port might be free of the disease for two to three weeks before an outbreak would occur at sea. The cases in Cuba would jump from one house to another where no contact had occurred. Carter also pointed out that a study he did in Mississippi showed an average of twelve days between the first case of yellow fever in an area and a larger second wave. All these facts forced Reed to reconsider Finlay's mosquito hypothesis.

At the request of the Yellow Fever Board, on August 1, 1900, Finlay supplied mosquito eggs for Reed's studies. Since no animals were known to acquire yellow fever, the board knew they must rely upon human experimentation. It was agreed that the first humans involved in the experiments must include the members of the board. That same week, Reed was recalled to Washington to report on some earlier work he had done in Cuba on typhoid fever. The board agreed to carry on in his absence. Dr. Lazear took charge of the mosquitoes, which were fed on yellow fever patients and then on nine volunteers. None of these first volunteers became ill, probably because the initial patients were too advanced in their illness to have the yellow fever virus in the bloodstream. On August 27, a mosquito that had twelve days earlier bitten a patient in his second day of illness was allowed to feed upon Carroll. Within three days, Dr. Carroll became seriously ill with yellow fever. Although very near death for a time, he survived. A subsequent experiment on another volunteer led to a second experimental case; this volunteer also recovered.

On September 13, Lazear was feeding one of the experimental mosquitoes on a patient when another mosquito came through a window and landed upon his hand. Rather than chance disrupting his experiment, he allowed this wild mosquito to feed upon his blood. Five days later, Dr. Lazear developed symptoms of yellow fever. His condition deteriorated rapidly, and on September 25, 1900, Dr. Lazear died. Reed returned to Cuba in October. He quickly realized that the experimental cases, while certainly important, did not completely settle the yellow fever issue, since either could have been exposed by some means other than by a mosquito. He requested approval to build an experiment station in an open field 1.6 kilometers from the town of Quemados, Cuba. This site was to be named by the board Camp Lazear, after their lamented colleague.

Camp Lazear was established November 20, 1900, and consisted of seven floored hospital tents and support facilities. Each of the tents was made mosquito-proof and was inhabited by one to three nonimmune volunteers who were paid one hundred dollars for participation and another one hundred dollars if they developed yellow fever as a result of the experiments. From November 20 through December 30, six volunteers were bitten by infected mosquitoes. Five developed yellow fever, and all survived. Even though these carefully controlled experimental cases provided strong evidence that mosquitoes could carry yellow fever, Reed set out to determine if there were other ways the disease could be transmitted. In his next series of experiments, three volunteers came down with yellow fever after being injected with blood from patients in the early stages of the disease. A third set of experiments was carried out to determine if contaminated clothing and bedding could transmit yellow fever. On November 30, Reed began an experiment during which seven men slept each night with soiled sheets, pillow cases, and blankets taken from the fellow fever ward of the hospital. All seven men remained completely healthy.

As his final experiment, Reed had constructed a mosquito-proof frame building with a center wire screen mesh partition. Each side of the building was occupied by a nonimmune volunteer. Yet, on one side of the partition mosquitoes were released which had fed on yellow fever patients. As expected, the volunteer bitten by the mosquito became ill, while the one on the other side of the partition remained healthy. The Yellow Fever Board was now satisfied with its experiments. On February 4, 1901, Dr. Reed reported the results of his findings to the Pan-American Medical Congress held in Havana. This report was then published in the February 16, 1901, edition of the *Journal of the American Medical Association*. The mosquito was clearly established as the vector in the transmission of some unknown agent which produced yellow fever. (The discovery of the yellow fever virus was almost thirty years away.)

Impact of Event

It was clear that the way to control yellow fever was to eliminate exposure to mosquitoes. Gorgas quickly set out to do just that. The approach was to protect infected patients from mosquito bites and to eliminate mosquito-breeding places. In March, 1901, Dr. Gorgas set out with great enthusiasm by launching a house-to-house attack on the mosquito. All water containers were to be emptied, covered, or layered with kerosene. Failure to comply meant a ten-dollar fine. Gorgas' tactics were first met with derision by the populace; however, during the summer of 1901, the number of yellow fever cases in Havana was drastically reduced and by October, for the first time in decades, no cases were being reported. Within a few years, the entire Western Hemisphere was for the most part rid of yellow fever. One of the most significant developments from the elimination of yellow fever was that it again became possible to consider building a canal across the Isthmus of Panama. This project had been started by the French in 1881, but was abandoned because of the deaths of more than twenty thousand workers from yellow fever. Gorgas was promoted to colonel and made the chief sanitary officer of the Canal Zone. In the spring of 1904, the project was begun. The disease that had caused the defeat of the French was no longer a major problem. Dr. Gorgas was there to see the first ship go through the canal in 1914.

The work of Reed and the Yellow Fever Board had an impact far beyond the reduction of morbidity and mortality from yellow fever. Yellow fever was the first human viral disease extensively studied, and the *Aëdes* mosquito was the first insect determined to transmit a virus to other organisms. Additionally, for the first time medical experimentation was performed following a sound scientific method. Human experimentation was carried out by the board only after careful consideration of other alternatives. All experiments were carefully controlled and meticulous records were kept. Not only was the work of Walter Reed used as a basis for subsequent studies on yellow fever, but also his methods served as a model for medical experimentation throughout the early part of the twentieth century.

Bibliography

Andrewes, C. H. "Yellow Jack." In *Natural History of Viruses*. New York: W. W. Norton, 1967. This exhaustive book presents basic facts of the chemical makeup and nature of viruses in general and explains how viruses spread, how they perpetuate themselves in nature, and how they survive in hard times. The chapter on yellow fever includes the natural history of the virus in the jungle environment.

Bean, William B. *Walter Reed: A Biography.* Charlottesville: University Press of Virginia, 1982. Bean provides a very complete biography, including discussions of Reed's childhood and work in the Indian wars. The information is taken from approximately fourteen thousand pages of letters, papers, and writings of Walter Reed, as well as other sources.

De Kruif, Paul. "Walter Reed: In the Interest of Science—and for Humanity." In *Microbe Hunters*. New York: Harcourt, Brace and Co., 1926. De Kruif was a prolific writer on the historical background of numerous developments in medicine. Although short on details and full of fictional narrative, this volume presents the work of Walter Reed in a very readable, human manner.

Kelly, Howard A. *Walter Reed and Yellow Fever.* 3d rev. ed. Baltimore: Norman, Remington, 1923. Kelly was the first biographer of Reed and provides some interesting insight into his early years. The major portion of the book deals with the work on yellow fever and its implications.

Reed, Walter, James Carroll, and Aristides Agramonte. "The Etiology of Yellow Fever." *Journal of the American Medical Association* 36 (February, 1901): 431-440. This landmark article was presented by Reed at the Pan-American Medical Congress in Havana on February 4-7, 1901. It provides some background, as well as detailed information on all phases of the Yellow Fever Board's experiments. The article was reprinted in the August 5, 1983, issue of *JAMA* (pages 649 to 658).

Williams, Greer. "Part Three: Costly Victory—Yellow Fever." In *The Plague Killers*. New York: Charles Scribner's Sons, 1969. Greer has drawn upon previously unpublished material from The Rockefeller Foundation archives for this volume. He has caught the personalities of the scientists involved and imparted a sense of the massive human implications of their work.

Daniel W. Rogers

Cross-References

Gorgas Develops Effective Methods for Controlling Mosquitoes (1904), p. 223; Rous Discovers That Some Cancers Are Caused by Viruses (1910), p. 459; Theiler Introduces a Vaccine Against Yellow Fever (1937), p. 1091.

ZEPPELIN CONSTRUCTS THE FIRST DIRIGIBLE THAT FLIES

Category of event: Space and aviation
Time: July 2, 1900
Locale: Manzell, Lake Constance, Germany

Zeppelin devised the first prototype of the rigid airship, which played a major role in World War I and in international air traffic until 1938

Principal personages:

FERDINAND VON ZEPPELIN (1838-1917), a retired German general and fervent promoter of the rigid airship

THEODOR KOBER (1865-1930), Zeppelin's private engineer, who worked out the basic design of the rigid airship in the following years

Summary of Event

The history of rigid airships is directed largely by attempts to extend the military potential of lighter-than-air crafts by propulsion and directed flight. After the Montgolfier brothers' balloon launch in 1783, engineers—especially in France—began to focus on how the direction of balloon flight could be influenced by machines. Ideas ranged from rowing through the air with silk-covered oars or movable wings to using a rotating fan, an airscrew, or a propeller powered by a steam engine (1852) or an electric motor (1882). The internal combustion engine, introduced at the end of the nineteenth century and promising higher speeds and more power, was another major step toward the realization of dirigible balloons. These, however, were not rigid yet.

Rigidity had the advantage of permitting larger airships with a wider range. Around 1890, the Austro-Hungarian War Ministry turned down a design for a rigid airship devised by the Dalmatian David Schwarz; however, the Soviet government accepted it in 1892. During trials in St. Petersburg in 1893, flaws that became apparent during inflation kept the airship on the ground. Therefore, Schwarz took his plans to a third sponsor and succeeded, in 1894, in convincing a Prussian aeronautical commission to support his research. The second test flight in 1897 in Berlin, however, was troubled by leaks in the metal hull and ended in a crash.

Whereas Schwarz's airship consisted of an entirely rigid aluminum cylinder, Ferdinand von Zeppelin's design was based on a rigid frame. Zeppelin was familiar with balloons, having been in two wars in which they were used: the American Civil War and the Franco-Prussian War of 1870-1871. The first "thoughts about an airship" to be recorded in his diary were dated March 25, 1874, inspired by an article about international post and aviation. Zeppelin soon lost interest in this idea of civil uses for an airship and came to favor the notion that dirigible balloons might become central to modern warfare. He submitted a request and asked for assistance to build an airship for Germany. In particular, he pointed to the apparent superiority of French military airships. Nationalism was the driving force behind Zeppelin's persistence in his endeavor, overshadowing enormous technical and financial obstacles.

In 1893, in order to procure money for his project, Zeppelin struggled to convince the German military and engineering experts of the utility of his invention. Although a governmental committee judged his work worth minimal funding, the army was skeptical, arguing that the ratio of cost to utility of Zeppelin's machines was too unfavorable. Ultimately, the committee chose Schwarz's design. In 1896, however, the indefatigable Zeppelin won the support of the powerful Union of German Engineers, which helped him to form a stock company, the Association for the Promotion of Airship Flights, in May, 1898. Zeppelin invested almost half of the 800,000 marks' capital in the association. In 1899, he began construction in Manzell at Lake Constance, and in July, 1900, the airship was finished and ready for its first trial.

Zeppelin, together with engineer Theodor Kober, had worked on the design of the ship since May, 1892, shortly after his retirement from the army. By 1894, they had established the basic design, which, though some of its details were later modified, was to remain the recognized form of the Zeppelin. An improved version was patented in December, 1897.

In the final prototype—the LZ 1—the engineers aimed especially to reduce the weight of the airship as far as possible. They used a light internal combustion motor and devised a latticed container from aluminum, which had become commercially available because of a new electrolytic process. The airship was 128 meters long and 11.7 meters in maximum diameter and consisted of a huge zinc-aluminum alloy framework with twenty-four longitudinal girders running from nose to tail and drawn together at each end. Sixteen traverse frame rings held the body together. Over the framework, the engineers had stretched an envelope of smooth cotton cloth to reduce friction as the ship moved through the air and to protect the gas bags from the direct rays of the sun. Seventeen separate gas cells made of rubberized cloth were placed inside the framework, all equipped with safety valves. Several were provided with maneuvering valves. Together they contained 121,615 cubic meters of hydrogen gas, which would lift 11,090 kilograms. A bridgelike construction was fixed to the side of the body, attached to which were two motor gondolas, each with a 16-horsepower gasoline motor, driving four propellers on the sides.

The trials were unsuccessful. The two main questions—whether the construction was stable enough and whether its speed was sufficient for practical requirements—could not be answered definitively because small details, such as the breaking of a crankshaft and the jamming of a lateral rudder, prevented normal flying. The first flight lasted no more than eighteen minutes; the ship attained a maximum speed of 13.7 kilometers per hour. During all three flights, the airship was in the air a total of only two hours, and its speed did not exceed 28.2 kilometers per hour. Lack of money forced Zeppelin to abandon his projects for some years, and his company was liquidated. The LZ 1 was wrecked in the spring of 1901.

Trials with a second airship, which had been financed by industry, the military, and Zeppelin (with the additional help of a lottery), took place in November, 1905, and January, 1906. Both were unsuccessful and ended in the destruction of the ship during a storm. By 1906, however, the German government was convinced of the military utility of the airship, although it demanded that in order to back the airship financially, it required future airships to fly nonstop for at least twenty-four hours. The third Zeppelin failed to fulfill these conditions in the autumn of 1907. The LZ 4, however, was a breakthrough: Not only did it prove itself to the military but also it attracted enormous public attention. In the summer of 1908, it flew for more than twenty-four hours and attained the considerable speed of 64 kilometers per hour. When, after the forced landing at the end of this flight, the ship was caught in a storm and exploded, spontaneous donations from throughout Germany amounted to more than 6 million marks. The "flying pencil" (a derogatory expression for the rigid airship coined by the German engineer of nonrigid airships, August von Parseval) had become a German national affair. Its advanced technology guaranteed a lasting dominance in the field of airship design.

Impact of Event

Rigid airship development and operation remained chiefly in German hands, particularly in those of the Zeppelin company and its operating associates. During the time of their construction between 1900 and 1938, 139 of 161 rigid airships were constructed in Germany, and 119 of these were of the Zeppelin type (the remaining 20 were the innovative Schtte-Lanz airships, the construction of which was abandoned at the end of World War I).

More than 80 percent of airships were built for military purposes. Whereas the British Royal Navy had only four rigid airships at the end of the war, the German army and navy bought their first airship in 1909 and used more than one hundred improved and enlarged Zeppelins for long-range reconnaissance and bombing during the war. Starting in May, 1915, bombing attacks were flown on the eastern front on Warsaw, Bucharest, and Salonika and on the western front predominantly on England, especially London. The airships had a considerable psychological effect on the English population, although the physical damage they caused did not entail a turning point in the war. Furthermore, the English antiaircraft defense, which used artillery and airplanes, improved rapidly. By 1916, the loss of airships had increased to such an extent that the German army abandoned them. The navy, however, continued to employ them until the end of the war, primarily for reconnaissance over water but also for bombing.

Airships were first used for civilian passenger traffic in 1910. By 1914, the Delag (German Aeronautic Stock Company) had acquired seven passenger airships to make recreational circular flights from German cities. Insufficient engine power, ignorance of meteorology, and difficulties in maneuvering the airship on the ground still posed serious problems for regular use. After World War I, the remaining Zeppelins became part of reparation payments, and airship construction for German use was forbidden until 1925.

As airship service over short distances was economically infeasible, partly because of the airplane, long-distance flights became the new task for the airship in the 1920's and 1930's. Linking the empire by air was the aim of the British airship program between 1924 and 1930. A British airship had succeeded in the first transatlantic flight in 1919. The intended commercial flights, however, especially toward the Far East, never materialized, as a result of the crash of the R-101 in 1930, in which most of the leading English aeronauts were killed.

The American rigid airship program of 1928 to 1935 intended to provide long-range naval reconnaissance and led to the construction of the *Akron* (1931) and the *Macon* (1933); however, the United States government stopped the program in 1935 after both airships crashed. Commercial overseas travel by airship was entirely a German concern and as such was exploited for nationalist and, later, Fascist propaganda. The world tour of the *Graf Zeppelin* in 1929 was a worldwide success; exploratory and promotional flights followed. Air connections between Germany and South America started in 1932 on a regular basis, and German airships bearing the Nazi swastikas flew to Lakehurst, New Jersey, in 1936. The explosion of the hydrogen-filled *Hindenburg* in 1937, however, meant the end of the rigid airship. As the U.S. secretary of the interior vetoed the sale of nonflammable helium, fearing its use for military purposes by the Nazi regime, the German government had to stop transatlantic flights for safety reasons. In 1940, the last two remaining rigid airships were wrecked.

Bibliography

Beaubois, Henry. *Airships*. Translated and adapted by Michael Kelly and Angela Kelly. New York: Two Continents Publishing Group, 1976. A comprehensive collection of pictures and drawings of all airships ever built, starting with the first balloons and ending with the radar-equipped Goodyear ZPG-3W of the U.S. Navy and blimps used for publicity.

Brooks, Peter W. *Historic Airships*. London: H. Evelyn, 1973. A dense and clear account of the history of the rigid airship. Brooks also provides useful tables and reliable statistics concerning the technical aspects and operation of airships.

Kirschner, Edwin J. *The Zeppelin in the Atomic Age*. Urbana: University of Illinois Press, 1957. This book appeared shortly before the end of the United States military airship program. It discusses the possibility of commercial transport using atomic-powered airships, with a strong emphasis on economic considerations. The bibliography includes publications of governmental organizations.

Meyer, Henry Cord. "France Perceives the Zeppelins, 1924-1937." *South Atlantic Quarterly* 78 (1979): 107-121. The article describes the French perception of Zeppelins as a symbol of new German political and technological strength after World War I. Recapitulating the struggles concerning permission for German airships to fly over France, it reflects on the political implications of modern technology.

Robinson, Douglas H. *Giants in the Sky.* Seattle: University of Washington Press, 1973. An accurate and insightful survey of the history of the airship. Robinson is the leading expert on rigid airships and has written about airships in World War I, including the *Graf Zeppelin* and the *Hindenburg*.

Robinson, Douglas H., and Charles L. Keller. *"Up Ship!" A History of the U.S. Navy's Rigid Airships, 1919-1935*. Annapolis: Naval Institute Press, 1982. Based on documents of the National Archives. Presents a detailed and precise history of the U.S. Navy's rigid airships, but does not avoid romanticism in describing military weaponry.

Matthias Dörries

Cross-References

Tsiolkovsky Proposes That Liquid Oxygen Be Used for Space Travel (1903), p. 189; The Wright Brothers Launch the First Successful Airplane (1903), p. 203; The Fokker Aircraft Are the First Airplanes Equipped with Machine Guns (1915), p. 600; Lindbergh Makes the First Nonstop Solo Flight Across the Atlantic Ocean (1927), p. 841; Piccard Travels to the Stratosphere by Balloon (1931), p. 963.

PLANCK ANNOUNCES HIS QUANTUM THEORY

Category of event: Physics
Time: December 14, 1900
Locale: Berlin, Germany

Planck assumed that the energy of light was quantized proportional to its frequency, not the intensity, spawning the field of quantum mechanics

Principal personages:

MAX PLANCK (1858-1947), a German physicist who was director of the Institute for Theoretical Physics of the University of Berlin and winner of 1918 Nobel Prize in Physics

ALBERT EINSTEIN (1879-1955), a German-born American physicist who was awarded the 1921 Nobel Prize in Physics

NIELS BOHR (1885-1962), a Danish physicist who was director of the Institute of Theoretical Physics of the University of Copenhagen and winner of the 1922 Nobel Prize in Physics

Summary of Event

Toward the end of the nineteenth century, many physicists believed that all the principles of physics had been discovered and little remained to be done, except to improve experimental methods to determine known values to a greater degree of accuracy. This attitude was somewhat justified by the great advances in physics that had been made up to that time. Advances in theory and practice had been made in electricity, hydrodynamics, thermodynamics, statistical mechanics, optics, and electromagnetic radiation.

These classical theories were thought to be complete, self-contained, and sufficient to explain the physical world. Yet, several experimental oddities remained to be explained. One of these was called "black body radiation," the radiation given off by material bodies when they are heated.

It is well known that when a piece of metal is heated, it turns a dull red and gets progressively redder as its temperature increases. As that body is heated even further, the color becomes white and eventually becomes blue as the temperature becomes higher and higher. There is a continual shift of the color of a heated object from the red through the white into the blue as it is heated to higher and higher temperatures. In terms of the frequency (the number of waves that pass a point per unit time), the radiation emitted goes from lower to a higher frequency as the temperature increases, because red is in a lower frequency region range of the spectrum than is blue. These observed colors are the frequencies that are being emitted in the greatest proportion. Any heated body will exhibit a frequency spectrum (a range of different intensities for each frequency). An ideal body, which emits and absorbs all frequencies, is called a black body; its radiation when heated is called black body radiation.

The experimental black body radiation spectrum is bell-shaped, where the highest intensity—the top of the bell—occurs at a characteristic frequency for the material. The frequency at which this maximum occurs is dependent upon the temperature and increases as the temperature increases, as is the case for any heated object. Many theoretical physicists had attempted to derive expressions consistent with these experimental observations, but all failed. The expression that most closely resembles the experimental curve was that derived by John William Strutt, third Baron of Rayleigh, and Sir James Hopwood Jeans. Like the experimental curve, it predicts low intensities for small frequencies; however, it never resumes the bell shape at high frequencies. Instead, it diverges as the frequency increases. Because the values of the theoretical expression diverge in the ultraviolet (high-frequency) region, it was termed the ultraviolet catastrophe. In other words, the Rayleigh-Jeans expression held that a body at any temperature would have its maximum frequency in the ultraviolet region. This was a clear contradiction to the experimental evidence.

On December 14, 1900, a soft-spoken, articulate professor presented a solution to the problem of black body radiation: Max Planck offered a mathematical exercise that averted the ultraviolet catastrophe. He explained why the heat energy added was not all converted to (invisible) ultraviolet light. This explanation was, to Planck, merely an experiment, a sanding down of a rough theoretical edge. It was that theoretical edge, however, that brought Planck before the august body of the German Physics Society. Six weeks earlier, he had done what he described as a "lucky guess." His discovery did not take place in a laboratory; it took place in his mind. Planck had introduced a mathematical construct into the Rayleigh-Jeans formula. Upon its use in the formula, however, Planck realized the significance of his mathematics. As Planck described, "After a few weeks of the most strenuous work of my life, the darkness lifted and an unexpected vista appeared."

Planck had discovered that matter absorbed heat energy and emitted light energy discontinuously. (Discontinuously means in discrete amounts of quantities, called quanta.) Planck had determined from his observations that the energy of all electromagnetic radiation was determined by the frequency of that radiation. This was a direct contradiction to the accepted physical laws of the time; in fact, Planck also had difficulty believing it.

Planck's theory—or quantum theory—was a fundamental departure from the theory of light of his day. That theory stated that light waves behaved as mechanical waves. Mechanical waves, much like waves in a pond, are a collection of all possible waves at all frequencies, with a preponderance for higher frequency. In the mechanical wave theory, all waves appear when energy is introduced, and high-frequency waves fit more easily and therefore should be present in greater amounts. After empirical observation of black body radiation, Planck showed this to be incorrect. He stated that light waves did not behave like mechanical waves. He postulated that the reason for the discrepancy lay in a new understanding of the relationship between energy and wave frequency. The energy either absorbed or emitted as light depended in some fashion on the frequency of light emitted. Somehow, the heat energy supplied to the glowing material failed to excite higher-frequency light waves unless the temperature of that body was very high. The high-frequency waves simply cost too much energy to be produced. Therefore, Planck created a formula that reflected this dependence of the energy upon the frequency of the waves. His formula said that the energy and frequency were directly proportional, related by a proportionality contact, now called Planck's contact. Higher frequency meant higher energy. Consequently, unless the energy of the heated body was high enough, the higher-frequency light was not seen. In other words, the available energy was a fixed amount at a given temperature. The release of that energy could be made only in exact amounts, by dividing up the energy exactly. Small divisions of the energy, resulting in large numbers of units, were favored over large divisions of small numbers of units. Lower frequencies (small units of energy) were favored over higher frequencies, and the black body radiation was explained.

Planck was quite reluctant to accept fully the discontinuous behavior of matter when it was involved with the emission of light or the absorption of heat energy. He was convinced that his guess would eventually be changed to a statement of real physical significance, for there was no way to see it, visualize it, or even connect it with any other formula. Yet, Planck's new mathematical idea had forced the appearance of a new and somewhat paradoxical physical picture.

Impact of Event

Planck's simple formula started a furor in the world of physics, although he did not accept fully its conclusions. In fact, it was not accepted by most physicists at the time and was considered to be an ad hoc derivation. It was felt that in time a satisfactory classical derivation would be found. A few years later, however, the very same idea would be used in three different applications that would establish the quantum theory. Albert Einstein would use Planck's quantum theory to explain several other experimental oddities of the late nineteenth century. In 1905, Einstein explained the photoelectric effect using Planck's ideas. The phenomenon of the photoelectric effect is that light striking the surface of metals causes electrons to be ejected from that surface. Yet, it was also found that the phenomenon was frequency-dependent (not intensity-dependent). For example, a threshold frequency was needed to allow an electron to be ejected, not a threshold amount of energy caused by a high-intensity light source. Einstein showed that it was the frequency of the incident light that determined whether electrons would be ejected. He had used Planck's theory to explain that the threshold energy required to eject electrons was frequency-dependent. His correct theoretical explanation of the photoelectric effect won for him the 1921 Nobel Prize.

Two years later, in 1907, Einstein again utilized the quantum theory to explain one more "oddity." This explanation dealt with the capacity of objects to accept heat, their heat capacity. At the time, it was accepted that the heat capacity for any object at room-temperature conditions was constant. As the temperature of the object was decreased, however, the heat capacity was no longer at the classical value. In fact, these low-temperature heat capacities are quite contrary to classical theory. The classical result was for atoms vibrating about their equilibrium lattice positions. Einstein assumed that the oscillation of the atoms about their equilibrium lattice positions were quantized. In this instance, the mechanical vibrations of the atoms are subject to quantization. Therefore, in addition to electron oscillations and radiation itself, the motion of particles was found to be quantized.

In 1913, Niels Bohr used the ideas of the quantum theory to explain atomic structure. Bohr reasoned that the emission of radiation from excited atoms could only come about from a change in the energy of the electrons of those atoms. Yet, that radiation was found to be of discrete frequencies. Bohr determined that this implied that only certain quantized energy states were available to atoms and that the only means for change from one state to another was to have an exact (quantized) energy be emitted or absorbed. His theoretical explanation of the atom won for him the 1922 Nobel Prize.

Since these early successes of the quantum theory, the explanation of the microscopic has continued in quantum mechanics. The Planck quantum theory has become the basis for understanding of the fundamental theories of physics. In fact, all the "known" physics of the late nineteenth century has been shown to have a theoretical background based on the ideas of quantized light that converge to the classical theories at high temperatures and large numbers. Yet, the most basic understanding of matter and radiation physics has derived from the quantum theory.

Bibliography

Guillemin, Victor. *The Story of Quantum Mechanics*. New York: Charles Scribner's Sons, 1968. A nontechnical book that conveys factual information and insight into the ways professional scientists think and work. In addition to presenting the historical background of quantum mechanics, the text continues into the theories and models of atomic and subatomic particles. The book concludes by considering the philosophical implications of quantum physics.

Hoffmann, Banesh. *The Strange Story of the Quantum*. 2d ed. New York: Dover, 1959. A nontechnical book that serves as a guide to the development of quantum mechanics. Written for the general reader, this small book is a treasure of insight into the persons, places, and pitfalls encountered in the early history of quantum physics. Another "must read" for the nontechnical reader interested in the development of quantum physics.

Jammer, Max. *The Philosophy of Quantum Mechanics*. New York: John Wiley & Sons, 1974. A guide to the analysis of the concepts, philosophy, and interrelation of the mechanics and ideas of the new physics. In addition, a guide to the literature of the subject, containing many general and technical references. Self-contained in its scope and material, it provides the reader with information about references in quantum mechanics.

McQuarrie, Donald. *Quantum Chemistry.* Mill Valley, Calif.: University Science Books, 1983. A textbook that traces the beginnings of quantum mechanics and utilizes the results in the study of molecules. Excellent text for background and conclusions based upon the earliest works in quantum physics. Short biographies of the principal characters in the development of the field add to the interest. Excellent treatment of the historical beginnings of quantum mechanics on a technical level.

Wolf, Fred Alan. *Taking the Quantum Leap*. San Francisco: Harper & Row, 1981. An outstanding book for the nonscientist about quantum physics. Traces the earliest debates and experiments concerning the philosophy and practice of physics. Presented in a humanized view of science that is accurate, historical, and conceptual, leading the reader to a solid, nontechnical grasp and beyond. A "must read" for anyone interested in quantum physics.

Scott A. Davis

Cross-References

Einstein Develops His Theory of the Photoelectric Effect (1905), p. 260; Bohr Writes a Trilogy on Atomic and Molecular Structure (1912), p. 507; Broglie Introduces the Theory of Wave-Particle Duality (1923), p. 741.

BOOTH INVENTS THE VACUUM CLEANER

Category of event: Applied science
Time: 1901
Locale: United States

The vacuum cleaner, a device used in the home and industry to clean surfaces, uses the principle of suction to carry dust and dirt away from carpeting and other surfaces

Principal personages:

HERBERT CECIL BOOTH, the holder of the earliest patent on a power-driven vacuum cleaner in 1901

MURRAY SPENGLER, the developer of the first practical household vacuum cleaner in 1908

WILLIAM H. HOOVER (1849-1932), the founder, with Spengler, of the Hoover company, a well-known manufacturer and distributor of vacuum cleaners

HIRAM H. HERRICK, the patentee of the first mechanical carpet sweeper in 1858

MELVILLE R. BISSELL (1843-1889), the inventor and marketer of the Bissell carpet sweeper in 1876

Summary of Event

Electricity was becoming widely available in the United States by the end of the nineteenth century, and with it came many new gadgets that utilized this clean and relatively inexpensive source of power to run motors, to create heat or cold, to provide light, and to perform many other tasks. Some were used in industry, others in the home. Among the many inventions that found application in both places was the vacuum cleaner.

The vacuum cleaner is based on a simple principle of physics. As air is forced from one place to another through the blades of a fan, a partial vacuum is created. The vacuum causes more air to rush in, filling the void, as demonstrated when an electric fan is turned on, causing air to move through it. The fan blades push air out and in the process, draw in more air. The movement of air is continuous until the fan stops. The same principle is applied in a vacuum cleaner. A fan inside the machine forces air through an output valve, at the same time drawing more air through an input valve, the nozzle or business end of the machine. The suction thus created brings with it loose dust and other particles that are caught up in the rush of air through the system. Beater bars, brushes, and other devices loosen dirt that is then drawn into the cleaner, where it is trapped in a disposable bag as the air rushes through the output valve. Once the bag is filled, it is replaced.

Vacuum cleaners are of two types: stand-up models that are pushed along, sucking up debris through a wide but thin suction opening at the front, and canister models that are held in one hand as the other is used to manipulate a small suction brush or other attachment that is connected to the canister by a long, flexible hose. Both are used in the home, and each is fairly lightweight.

The first home vacuum cleaner appeared in 1901. It was patented by Herbert Cecil Booth, who described it as a suction dust-removal machine. Booth's vacuum cleaner bore more resemblance to a lawn mower than to the electric vacuum cleaner of the 1990's. It was large and awkward to use, a five-horsepower piston-driven motor mounted atop the front. Also, it was extremely heavy, an upright model without attachments.

It should be noted that large, mobile industrial vacuum cleaners were in use as early as the 1890's. These large suction cleaners were mounted on wagons pulled by horses and connected to the vacuum nozzle by long, flexible hoses that were strung through doorways and windows to reach the surfaces to be cleaned. These machines were used primarily to provide vacuum-cleaning services for schools, hotels, theaters, and other buildings that were sites for activities that involved high levels of foot traffic. There were also smaller, nonelectric piston-driven consumer models available before 1900 that created more work than they saved for the homemaker. As a result, they were never accepted. Others were hand-operated, with bellows for pumping air. There were vacuum-based airblowers and some cleaners that ran on compressed air.

Booth's vacuum cleaner, although cumbersome and difficult to manage, provided the basis for advancements in suction technology that were to come. Ten years after the Booth patent was filed, newer, lighter-weight and more powerful electric motors began to appear. This meant that vacuum cleaners could be powered by motors small enough to allow a greater degree of portability. In 1908, Murray Spengler patented a much smaller and more popular version of the vaccum cleaner, later forming a partnership with a cousin, William H. Hoover. The new enterprise was called the Hoover Company. Almost a century later, the Hoover upright vacuum cleaner was still an industry leader and distributed worldwide.

In 1924, the first hand luggable vacuum cleaner appeared, originating in Sweden. This early portable came with a flexible hose that could be attached for cleaning hard-to-reach places, furniture, draperies, and wall hangings. Other attachments came along over the years to suit the needs of the times, some using pulsating air to loosen dust and dirt and others employing special rake brushes to clean deep-pile shag rugs.

Unlike many home appliances that have come and gone or evolved into contraptions that scarcely resemble their earliest ancestors, today's vaccum cleaners are very similar in design to the Booth vacuum of 1901. They operate on the same principle of suction and are used for the same purpose: to eliminate dust and small particles from living spaces. Over the years, vacuum cleaners have increased in popularity as more and more floors have been covered with wall-to-wall carpeting, including those in many commercial and institutional settings. Such surfaces enhance the aesthetic quality of living space, in addition to providing at least the illusion of greater warmth and a high degree of sound absorption.

As the technology employed to clean floors in the modern home and office continues to evolve, it is doubtful that Booth's invention will be pushed aside by some newer, better way of extracting dust and dirt from carpeting. Indeed, the evolution of wall-mounted carpets of centuries past to area rugs, which were, in turn, superseded by wall-to-wall carpeting, has encouraged the development of new materials that can be installed outside, on patios, pool decks, walkways, and other areas where they are subject to the merciless elements, in addition to grinding foot traffic. These new exterior surfaces and the increasingly popular indoor wall-to-wall carpeting found in millions of homes are here to stay. Herbert Cecil Booth will remain high on the list of pioneers who provided the inspiration for the development of the most useful modern-day home appliance.

Impact of Event

During earlier centuries, carpets were hung on walls as decorations long before they were used as floor coverings. Most floors were of uncovered wood, stone, or dirt, and the best way to clean was to sweep with a broom. During the nineteenth century, industrialization brought the invention of textile looms that were able to mass-produce heavy fabric that could be used to manufacture carpeting materials. Soon, the cost of mass-produced carpets began to drop as availability increased, and the occasional rug was replaced by wide area rugs that were increasingly installed in every room of the house. These early carpets were cleaned using small, hand-held whisk brooms or by taking them outside, placing them over a clothesline or similar device, and beating them with a broom, wooden paddle, or large loop of wood, cane, or wire attached to a handle. This technique was considered pure drudgery, and over time, it damaged the carpet.

Efforts to find mechanical means of cleaning floors can be traced back to the early nineteenth century. In 1811, English inventor Jane Hume received a patent for a device that was designed to make sweeping hard floors easier. It was a box on wheels attached to a handle that contained a pulley for operating a brush in the box that was supposed to pick up dirt and sweep it into the box. Unfortunately, the device did not work as well as the broom it was designed to replace, and it was never marketed. The basic concept was revisited often, however, as carpets came off the walls and onto floors.

The first patent for a device actually called a mechanical carpet sweeper was issued to Hiram H. Herrick of Massachusetts in 1858. Eighteen years later, Melville B. Bissell of Grand Rapids, Michigan, came up with a design that worked on floors and had the added advantage of working equally well on carpets. The Bissell carpet sweeper became extremely popular as a forerunner to the vacuum cleaner, in part the result of Bissell's marketing acumen as well as its excellent design. In 1896, four Bissell models were listed for sale in the Sears, Roebuck catalog. The Bissell carpet sweeper continued to enjoy a niche in the floor cleaner appliance marketplace long after the vacuum cleaner became ubiquitous in homes across the country and is still available worldwide. The advantage of the Bissell carpet sweeper was its simplicity. There were few moving parts and it could be used by anyone. It emerged in an era when carpets were used as much for ornamentation as for practical purposes; therefore, the demand for more efficiency—the mechanism's ability to pick up more debris over wider areas more often—did not emerge until later. With electrification came the electric motor, a tool that could be used to enhance the efficiency of the carpet sweeper by rotating a fan to create suction, thus drawing off dust and dirt more effectively.

As electricity reached more and more homes during the early twentieth century, the vacuum cleaner became the carpet sweeper of choice. Modern home construction often included built-in vacuum systems that allowed the operator to plug a hose into receptacles placed strategically throughout the house. Because the electric motor and fan design can be configured to operate at a wide range of suction levels, applications were found quickly for the vacuum cleaner in industry.

As with most home appliances, continuous evolution of the technology has brought new uses for the vacuum cleaner. Today, it often takes the form of rechargeable hand-held appliances that mount on walls and are used to clean drawers, kitchen cabinets, automobile interiors, and any number of other out-of-the-way places. Mini vacuum cleaners are used in high-tech industrial applications for cleaning dust and particles from micro-circuitry, and very large vacuum fans extract and collect particles from the air in and around industrial manufacturing areas to protect the health of workers. Booth's vacuum cleaner was the forerunner of a long line of vacuum innovation that has made homes and workplaces cleaner and safer.

Bibliography

Galvin, Vanessa, et al., eds. *How It Works: The Illustrated Encyclopedia of Science and Technology.* Vol. 19. London: New Caxton Library Service, 1977. Contains a good description of the vacuum principle as applied in vacuum cleaner technology. Includes a series of early photographs showing horsedrawn vacuum machines and the first vacuum cleaner salesman. Heavily illustrated.

Grossinger, Tania. *The Book of Gadgets*. New York: David McKay, 1974. Bills itself as "anything you ever wanted to know about gadgets—what they are, where they are, how they work, and how to collect them." A useful narrative that looks at a multitude of devices used in the house and office for various purposes. Contains a good description of the evolution of floor cleaning technology. Index.

Lifshey, Earl. *The Housewares Story: A History of the American Housewares Industry.* Chicago: National Housewares Manufacturers Association, 1973. Contains a detailed look at the evolution of the vacuum cleaner, history, and its applications. Contains photographs, index.

Newhouse, Elizabeth L., ed. *Inventors and Discoverers: Changing Our World*. Washington, D.C.: National Geographic Society, 1988. A heavily illustrated survey of the evolution of technology from the steam age in the early nineteenth century to

ELSTER AND GEITEL DEMONSTRATE RADIOACTIVITY IN ROCKS, SPRINGS, AND AIR

Category of event: Earth science
Time: 1901
Locale: Wolfenbüttel, Germany

Elster and Geitel pioneered research in ion conduction in gases, atmospheric electricity, photoelectric effects, and radioactivity

Principal personages:

JULIUS ELSTER (1854-1920), a German physicist who, with Geitel, studied ion conduction in gases, atmospheric electricity, photoelectric effects, and radioactivity

HANS FRIEDRICH GEITEL (1855-1923), a German physicist who built the first cathode tube, discovered selective photoelectric effect, built the photometer, invented the photocell, and formulated the law of radioactive fallout

ANTOINE-HENRI BECQUEREL (1852-1908), a French physicist who discovered radioactivity from his experiments with uranium ore in 1896

ERNEST RUTHERFORD (1871-1937), an English physicist who noted that rays from uranium were of three types: alpha, beta, and gamma

MARIE CURIE (1867-1934), a Polish physicist who, with her husband, Pierre, discovered radium in 1898

BERTRAM BORDEN BOLTWOOD (1870-1927), an American physicist who established the steps whereby uranium and thorium lose electrons and alpha particles to end up as lead

JOHANNES HANS GEIGER (1882-1945), a German physicist who worked with Rutherford to perfect a counter for beta or cosmic ray particles (Geiger-Müller counter)

Summary of Event

The late 1890's and early 1900's were a fertile period for discoveries in radioactivity, which is the emission of energetic particles and radiant energy by certain atomic nuclei. These early discoveries were the building blocks for nuclear physics and the nuclear age, which followed in the mid-1940's.

Radioactivity was first observed in 1896 by the French physicist Antoine-Henri Becquerel. He found that uranium ore emitted radiation strong enough to blacken covered photographic plates and to discharge a charged electroscope, a device for detecting the presence of electricity and whether it is positive or negative by means of electric attraction and repulsion. In 1898, Pierre and Marie Curie, French physicists, announced the discovery of two new radiation-emitting elements: polonium and radium. Ernest Rutherford, an English physicist born in New Zealand, made the discovery in 1903 that radiation from such elements was composed of three different kinds of energetic rays: alpha rays with a positive charge, which he showed were ionized helium atoms; beta particles with a negative charge, which were shown to be high-energy electrons; and gamma rays with no charge, which were found to be high-energy photons. Electrons are extremely small, negatively charged particles, whereas photons are a quantum of light that is proportional to the frequency of the radiation.

Natural radioactivity is the property possessed by roughly fifty elements—such as radium, thorium, and uranium—of spontaneously emitting alpha or beta rays and sometimes gamma rays by the disintegration of the nuclei of atoms. These elements are called radioelements. During the disintegration process, alpha or beta particles are emitted and atoms of a new element are formed. This new element is lighter than the predecessor and possesses chemical and physical properties quite different from those of the parent. The disintegration proceeds from stage to stage with measurable velocities in each case.

The existence of radioactivity cannot be discerned without the aid of instruments. The most useful procedures for the detection and measurement of alpha and beta particles and gamma-ray photons are based on the fact that gases become electrical conductors as the result of exposure to radiations from radioactive substances. Because there is a strong electrical field in its immediate vicinity, a rapidly moving charged particle ejects orbital electrons from the atoms or molecules of a gas through which it passes, thus converting them into positive ions. The expelled electrons usually remain free for some time, though a few may attach themselves to other atoms or molecules to form negative ions. The passage of a charged particle through a gas results in the formation of a number of ion pairs and free electrons. The study of radioactivity and the successful use of radiation as a research tool or for other purposes depends on its quantitative detection and measurement. The quantities most often needed are the number of particles (electrons, photons, beta) arriving at a detector per unit time and their energies. When a charged particle passes through matter, it causes excitation and ionization of the molecules of the material. This ionization is the basis of nearly all the instruments used for the detection of such particles and the measurement of their energies.

Julius Elster and Hans Friedrich Geitel discovered a method of counting alpha particles from the visible scintillations they produced on a zinc sulfide screen. Radiation of alpha particles can produce luminescence in zinc sulfide. This luminescence is not uniform but consists of a large number of individual flashes, which can be seem under a magnifying glass. Each alpha particle produces one scintillation, so the number of alpha particles that fall on a detecting screen per unit time is given directly by the number of scintillations counted per unit time. A screen is prepared by dusting small crystals of zinc sulfide containing a very small amount of copper impurity on a slip of glass. The counting is done by microscope with a magnification of about thirty, with precision but with difficulty. This method works well for counting alpha particles in the presence of other radiations because the zinc sulfide screen is comparatively insensitive to beta and gamma rays.

During this period of major discovery Elster and Geitel were actively conducting research on radioactivity in rocks, springs, and air. Elster was on the faculty of the Gymnasium, Wolfenbüttel, Germany, from 1881 to 1919. Geitel, also a physicist, was a teacher at Strosse Schule in Wolfenbüttel. Their joint research consisted of ion conduction in gases, atmospheric electricity, photoelectric effects, and radioactivity. Elster, in addition to his work with Geitel, built the first photoelectric cell, the first photometer, and the Tesla transformer. Also, he was the first to determine the electrical charge on falling raindrops in 1899. Elster demonstrated that lead is not radioactive of itself. Another important discovery was the presence of radioactive substances in the atmosphere that easily break down into unstable elements, which are responsible for atmospheric conductivity. Geitel built the first cathode tube and discovered selective photoelectric effect free energy left in an atom after the transformation of radioactive elements. He built a photometer, invented the photocell, formulated a law of radioactive fallout, and originated the concept of atomic energy.

One of Elster and Geitel's first collaborations was in 1880, when they carried out a systematic study of the electrification of hot bodies. From this early work and their perfection of instruments for detection and measurement, Elster and Geitel moved to a determination of other sources of radiation. A basic fact is that the earth is heated from within by the energy released when uranium, thorium, and other radioactive elements naturally undergo nuclear disintegration. As the disintegration takes place, radioactive elements find their way into rocks, soil, water, and air. The total energy released through nuclear disintegration over the earth's history is more than 100 calories per gram of earth material.

Elster and Geitel discovered in 1901 that it is possible to produce excited radioactivity from the atmosphere, without further agency, simply by exposing a highly charged wire to a negative potential in the atmosphere for many hours. They found that the radioactivity may dissolve with exposure to acids and that the wire would be left unchanged. This discovery, according to Elster and Geitel, had an important bearing on the theory of atmospheric electricity. In 1903, they discovered a property of alpha rays that proved of great importance in radioactive measurement. If a screen coated with small crystals of phosphorescent zinc sulfide is exposed to alpha rays, a brilliant luminosity is observed. Further, the study of penetrating radiation had its origin in their observations that there was a definite transport of charge to the insulated system even after all possible precautions had been taken to reduce electrical leakage over the insulators. The order of magnitude of this residual ionization corresponded to the production of about twenty pairs of ions per cubic centimeter.

Elster and Geitel's discoveries—though less heralded than those of their contemporaries—were a major contribution to nuclear physics and particularly to the understanding and detection of the omnipresence of radioactive elements in the environment.

Impact of Event

From their discoveries, it was found that not all nuclei are stable. Radioactive nuclei disintegrate spontaneously, releasing energy in the process. Understanding why this takes place was one of the great advances in physics during the first half of the twentieth century. Significantly affected was the field of geology. In natural radioactivity, the unstable nucleus emits several types of high-energy particles and also releases energy in the form of electromagnetic waves similar to light energy. This process results in radioactive decay, whereby an atom is changed to an atom of another element. This means that the number of atoms of radioactive element decreases with the passage of time. If the rate of disintegration is known, then measuring the amounts of the parent and daughter elements will reveal the age of the mineral containing the parent. This is the principle of radioactive dating.

One of the more important early discoveries was that of Bertram Borden Boltwood, an American physicist who, in 1907, determined that the age of a mineral crystal containing uranium could be determined chemically by ascertaining the ratio between the number of uranium atoms and the number of uranium atoms plus lead. If the crystal is part of an igneous intrusion, its date could be determined as the time of solidification of the magma. The date of a crystal in metamorphosed sediments was found to be the time of metamorphism, not the time of deposition of the sediment. Only a few of the natural radioactive isotopes are of geologic importance. Some are useful in determining the age of objects; others are sources of radioactive heating of the earth. Those that have proved most useful are carbon 14, potassium 87, uranium 235, uranium 238, and thorium 232. The requirements for dating are a reasonable rate of decay (half-life), the retention of daughter isotopes, and the existence of common minerals containing the parent element.

Radiation detection also made it possible to locate uranium ores. The early work by Elster, Geitel, Becquerel, the Curies, and Rutherford resulted in the most widely used detection instrument, the Geiger-Müller counter. The Geiger-Müller counter consists of a metal cylinder enclosed in a glass tube filled with a gas at low pressure. The entrance of a charged particle ionizes the gas enough to cause a current flow, but the current is quenched immediately by the high resistance placed in the circuit. As in the ionization chamber, an entering particle causes a momentary pulse of voltage. Geiger counters are equipped with special cylinders for counting alpha, beta, and gamma particles. The counter invented by Johannes Hans Geiger in 1913 was perfected in 1926 by Walther Müller and is still widely used.

The early work of Elster and Geitel in the second decade of the twentieth century led to the discovery that the atomic nucleus is a source of large quantities of energy. Humans have learned to make nuclear energy available in many different ways: for medical therapy, for power in industry, for energy to propel submarines and ships, for research in biological sciences, and as weapons of great destruction. The increasing interest and research in the field of artificial transmutation made it evident that the energies of particles emitted by natural radioactivity were far too small to produce the multiplicity of possible reactions. This conclusion brought about bullets of high energy for their bombardment of nuclei, which is accomplished in a linear accelerator, a machine that accelerates particles to high velocities in a straight-line path.

Bibliography

Curie, Marie Skłodowska. *Radioactive Substances*. Reprint. Westport, Conn.: Greenwood Press, 1971. This is a translation from the French of the classical thesis presented to the Faculty of Science in Paris. The text is not illustrated; however, it can be understood by advanced high school students with a background in science. A very informative reference on the early days in the study of detection and discovery of radioactive bodies by one of the researchers of the time.

Mann, Wilfred B., and S. B. Garfinkel. *Radioactivity and Its Measurement*. Princeton, N.J.: D. Van Nostrand, 1966. This compact reference is a chronicle on the discoveries of radioactivity and is an ideal source for the advanced high school and lower-level college student, as well as the interested public. It is well illustrated, and presents mathematical computations in an easy-to-follow format. Brief reference list.

Romer, Alfred. *The Discovery of Radioactivity and Transmutation*. Vol. 2. In *Classics of Science*. New York: Dover, 1964. A general collection of essays and original articles in various areas of radioactivity taken from scientific writings by Becquerel, Rutherford, Sir William Crookes, Frederick Soddy, and Pierre and Marie Curie. The articles follow the original texts verbatim. Should appeal to the interested public, historians, and scientists. Illustrations, photographs, and footnotes are included.

Rutherford, Ernest, James Chadwick, and C. D. Ellis. *Radiations from Radioactive Substances*. Reprint. London: Cambridge University Press, 1951. This scholarly text is more suited for the college student and historian. Illustrated, with references.

Taton, Rene. *History of Science in the Twentieth Century*. New York: Basic Books, 1966. A very good reference for the lower- and upper-level college student. Sections are devoted to mathematics, physical science, earth science, the universe, biology, and medicine. Copiously illustrated, and contains many figures for easier understanding of the text. Includes additional references for each section.

Earl G. Hoover

Cross-References

Röntgen Wins the Nobel Prize for the Discovery of X Rays (1901), p. 118; Becquerel Wins the Nobel Prize for the Discovery of Natural Radioactivity (1903), p. 199; Elster and Geitel Devise the First Practical Photoelectric Cell (1904), p. 208; Boltwood Uses Radioactivity to Obtain the Age of Rocks (1905), p. 285; Thomson Wins the Nobel Prize for the Discovery of the Electron (1906) p. 356; Rutherford Presents His Theory of the Atom (1912), p. 527; Rutherford Discovers the Proton (1914), p. 590.

THE FIRST SYNTHETIC VAT DYE, INDANTHRENE BLUE, IS SYNTHESIZED

Categories of event: Chemistry and applied science
Time: 1901
Locale: Ludwigshafen, Germany

The efforts of centuries to obtain the colored brilliance of nature, combined with permanence and functionality, found its birth in the dawn of the age of synthesis

Principal personages:

RENÉ BOHN (1862-1922), a synthetic organic chemist and the holder of many patents for the synthesis of organic molecules noted for their color and durability as dyes

KARL HEUMANN (1850-1894), a German chemist who taught Bohn and was responsible for several important developments in dye chemistry

ROLAND SCHOLL (1865-1945), a Swiss chemist whose detailed work on Bohn's dye established its correct structure

AUGUST WILHELM VON HOFMANN (1818-1892), a giant of nineteenth century organic chemistry

SIR WILLIAM HENRY PERKIN (1838-1907), an English student in Hofmann's laboratory who discovered the first synthetic dye and built it into a substantial industry

Summary of Event

The presence of color in living spaces and daily utensils is largely taken for granted. The precise date when the search began for attractiveness in human surroundings through the use of color is unknown. By contrast, people's search for the means to obtain and improve on nature's colors has been recorded and analyzed in great detail. One of many possible critical turning points in these researches was the introduction of the first vat dye by a chemist at the German firm Badische Anilin- und Soda-Fabrik (BASF) in 1901.

The term "vat dye" is used to describe a method of applying the dye, but it also serves to characterize the structure of the dye, because all currently useful vat dyes share a common unit. One fundamental problem in dyeing relates to the extent to which the dye is water-soluble. A beautifully colored molecule that is easily soluble in water might seem attractive given the ease with which it binds with fiber; however, this same solubility will lead to the dye's rapid loss in daily use.

Vat dyes are designed to solve this concern with molecules that can be made water-soluble, but only during the dyeing or vatting process. The structural unit that can be changed chemically and then reformed involves two groups of a carbon and an oxygen atom connected by two pairs of electrons. Such a carbon-oxygen double bond, or carbonyl group, is easily reduced by the addition of a molecule of hydrogen. After the reduced or *leuco*-dye crystals are safely trapped within the fibers, they are reoxidized, usually with air.

The excellent water-fastness of vat dyes is complemented by the fact that many of them have a very simple chemical structure. This structure results in great chemical stability and particularly in light-fastness. One modern textbook of dye chemistry asserts that "the fastness properties of vat dyes are surpassed by no other class of dyestuffs."

From prehistoric times until the mid-nineteenth century, all dyes were derived from natural sources. With a few important exceptions, these coloring materials came from plants. Over thousands of years, only about a dozen dyes proved to be of lasting practical importance. Among the most lasting dyes in nature are the red and blue dyes such as found in alizarin and indigo. Both of these substances have the carbonyl groups characteristic of vat dyes and played a central role in the earliest commercially significant synthesis of a vat dye.

The development of modern chemistry and that of dye chemistry are closely linked. The early years of the nineteenth century found chemists beginning to study the natural world in a scientific way. In France, Antoine-Laurent Lavoisier made chemistry a science by insisting that accurate nomenclature and analysis were essential to progress. In England, John Dalton stated the basic foundation by bringing the Greek *atomos*, or atoms, into modern form. In Germany, Justus von Liebig promoted intensive laboratory education for analysis and scholarly journals for results. There was great growth, but fundamental problems remained.

Organic chemistry, dealing with the compounds of the element carbon and associated with living matter, hardly existed. Synthesis of carbon compounds simply was not attempted. Considerable data had accumulated showing that organic or living matter was basically different from the compounds of the mineral world. In general, the material was fragile and difficult to work with. Carbon compounds did not fit easily into the now well-established electrical picture of rocks, metals, and salts. The problem was neatly, if unproductively, solved by stating that these molecules contained carbon, hydrogen, nitrogen, and "vital force." It was widely believed that although one could work with various organic matter in physical ways and even analyze their composition, they could be produced only in a living organism.

In 1828, Friedrich Wöhler found that it was possible to synthesize the clearly organic compound urea from clearly mineral compounds. The concept of "vitalism" began to fade as more chemists reported the successful preparation of compounds previously only isolated from plants or animals.

One especially rich field ripe for exploration was that of coal tar, and August Wilhelm von Hofmann was an active worker. He and his students made careful studies of this complex mixture. The high-quality stills they designed allowed the isolation of pure samples of important compounds for further study. Of greater importance was the collection of able students he attracted. Among them was Sir William Henry Perkin, who is regarded as the founder of the dyestuffs industry.

In 1856, Perkin undertook the task of synthesizing quinine from a nitrogen-containing coal tar material called toluidine. Luck played a decisive role in the outcome of his experiment. The sticky compound Perkin obtained contained no quinine, so he turned to investigate the simpler related compound, aniline. A small amount of the impurity toluidine in his aniline gave Perkin the first synthetic dye, Mauveine.

From this beginning, the great dye industries of Europe, particularly Germany, grew. The trial-and-error methods gave way to more systematic searches as the structural theory of organic chemistry was formulated. The academic and the industrial, the theoretical and the synthetic advanced in large measure through mutual stimulation. As the twentieth century began, great progress had been made, and German firms dominated the industry. Badische Anilin-und Soda-Fabrik was incorporated at Ludwigshafen in 1865 and undertook extensive explorations of both alizarin and indigo. A chemist, René Bohn, had made important discoveries in 1888, which helped the company recover lost ground in the alizarin field. Bohn was well trained, having earned his doctorate at the University of Zurich in studies with Karl Heumann, who had developed two successful syntheses of indigo. In 1901, he undertook the synthesis of a dye he hoped would combine the desirable attributes of both alizarin and indigo.

As so often happens in science, nothing like the expected occurred. Bohn realized that the beautiful blue crystals represented a far more important product. Not only was the first synthetic vat dye, Indanthrene, prepared but also, by studying the reaction at higher temperature, a useful yellow dye, Flavanthrone, could be produced.

By 1907, Roland Scholl had showed unambiguously that the structure proposed by Bohn for Indanthrene was correct, and a major new area of theoretical and practical importance was opened for the organic chemist.

Impact of Event

Bohn's discovery led to the development of many new and useful dyes. The list of patents issued in his name fills several pages in *Chemical Abstracts* indexes.

The true importance of this work is to be found in a consideration of all synthetic chemistry, which may perhaps be represented by this particular event. There are more than two hundred dyes related to Indanthrene in commercial use. The colors represented by these substances are a rainbow making nature's finest hues available to all. The dozen or so natural dyes have been synthesized into more than seven thousand superior products through the creativity of the chemist.

Despite these desirable outcomes, there is doubt if there is any real benefit to society from the development of new dyes. This must be considered as a result of limited natural resources. With so many urgent problems to be solved, scientists are not sure whether to search for greater luxury. A more germane question may be if scientists can afford *not* to continue the search. If the field of dye synthesis reveals a single theme, it must be to expect the unexpected. Time after time, the search for one goal has led to something quite different. No one can say with any certainty where the important clue to the solution of a problem will come. Certainly no one would predict that cancer will be cured through dye research. On the other hand, improved dyes for staining tissue are always needed.

It is perhaps noteworthy in this context that fifty years after Bohn's discovery, an English team headed by Professor William Bradley began a long series of studies aimed at determining how Indanthrene is formed. Their efforts were directed not at the synthesis of new dyes but at obtaining a fundamental understanding applicable in other fields of research.

Several textbooks on dye chemistry have pointed to the importance of knowledge of that field to other sciences—one more assertion that basic research often reveals the unexpected.

Bibliography

Fieser, Louis F., and Mary Fieser. *Topics in Organic Chemistry.* New York: Reinhold, 1963. Excellent historical perspective and a clear explanation of the chemistry and background about the people involved.

Gordon, P. F., and P. Gregory. *Organic Chemistry in Colour.* New York: Springer-Verlag, 1983. Contains an excellent historical introduction to the general field, along with a specific description of Bohn's work on several dyes.

Ihde, Aaron J. *The Development of Modern Chemistry.* New York: Harper & Row, 1964. Provides access to this highly technical world for the nonchemist. Superbly written, and gives authoritative information on the scientific and commercial aspects of the dye industry.

Rys, P., and H. Zollinger. *Fundamentals of the Chemistry and Application of Dyes.* New York: Wiley-Interscience, 1972. Some historical material and an adequate presentation of Bohn's synthetic efforts. Brief but effective; pays unusual attention to the central idea and general properties.

Thorpe, Jocelyn Field, and Christopher Kelk Ingold. *Synthetic Colouring Matters: Vat Colours.* London: Longmans, Green, 1923. An extremely detailed and valuable history of the entire field only a few years after it became well established. Especially important historically in showing the close relationship between organic and dye chemistry as seen through the eyes of Thorpe's student Ingold.

Zollinger, Heinrich. *Color Chemistry: Syntheses, Properties, and Applications of Organic Dyes and Pigments.* New York: VCH, 1987. A brief historical introduction coupled with an extensive and technical discussion of a wide range of modern chemical ideas as illustrated by dye chemistry. Written for the specialist, but containing many insights of interest to one attempting to see dye chemistry in the historical perspective of developing modern chemistry.

K. Thomas Finley
Patricia J. Siegel

Cross-References

Baekeland Invents Bakelite (1905), p. 280; Corning Glass Works Trademarks

Pyrex and Offers Pyrex Cookware for Commercial Sale (1915), p. 605; Carlson Invents Xerography (1934), p. 1013; Carothers Patents Nylon (1935), p. 1055; The Artificial Sweetener Cyclamate Is Introduced (1950), p. 1368; The Artificial Sweetener Aspartame Is Approved for Use in Carbonated Beverages (1983), p. 2226.

GRIJNS PROPOSES THAT BERIBERI IS CAUSED BY A NUTRITIONAL DEFICIENCY

Category of event: Medicine
Time: 1901
Locale: Javanese Medical School, Batavia, Java

Grijns' proposal that beriberi was caused by a nutritional deficiency in a diet of polished rice led to the concept of vitamins

Principal personages:

CHRISTIAAN EIJKMAN (1858-1930), the Dutch physician and Nobel laureate who showed that a diet of polished rice caused beriberi

GERRIT GRIJNS (1865-1944), the Dutch physician and assistant to Eijkman who proposed that the deficiency of a protective substance in polished rice caused beriberi

ROBERT KOCH (1843-1910), the German physician, bacteriologist, and Nobel laureate who taught Eijkman about bacteriology

Summary of Event

Beriberi is a disease caused by a deficiency of vitamin B_1, thiamine, in the diet. The name of the disease is Sinhalese for "I cannot." It resulted from the fact that people afflicted with severe beriberi are too sick to do even the simplest things. The disease was endemic to the Far East in the nineteenth and early twentieth centuries. Its symptoms include stiffness of the lower limbs, paralysis and severe pain, gradual breakdown of the muscles, anemia, mental confusion, enlargement of the heart, and death resulting from heart failure.

In the 1990's, the incidence of severe beriberi is much lower, found mostly in undernourished people in the rice-eating nations of Asia, Indonesia, and Africa. In industrialized nations, beriberi is seen most often in chronic alcoholics. They get the disease because their limited diets—which consist mostly of alcohol—are very deficient in vitamins. This deficiency includes insufficient amounts of thiamine. Very severe thiamine deficiency in alcoholics, or Wernicke-Korsakoff syndrome, is an irreversible disease. It causes psychosis and memory loss and requires hospitalization.

Despite the fact that the need for thiamine has been known for many years, a significant percentage of the American public is believed to eat barely enough thiamine to meet dietary needs. In recent years, many foods inherently low in thiamine (such as white bread, breakfast cereal, pasta, and white flour) have been enriched with the vitamin as a health precaution. The best natural sources of thiamine are the whole grain cereals, nuts, and fish. Raw fish should be avoided, however, because it contains an enzyme that will destroy thiamine. Excessive consumption of tea also can lead to thiamine-deficiency symptoms because tea contains a chemical that antagonizes the actions of this vitamin.

Beriberi has been known for thousands of years. For example, reference to a disease with its symptoms is found in the Chinese medical literature of 2700 B.C. The Orient is the main beriberi zone because of the dietary combination of high tea consumption, use of rice as the main cereal grain, and the common practice of eating raw fish.

In 1887, an important early observation on the dietary origin of beriberi was reported by Takagi Kanehiro, the director-general of the Japanese navy. Kanehiro became interested in beriberi because one-third of all Japanese sailors had the disease. After careful study, Kanehiro became convinced that the cause of naval beriberi was the standard navy diet—polished rice and fish—which was low in protein. He ordered the addition of red meat, vegetables, and whole-grain wheat to their diet. After Kanehiro's change was made, beriberi became rare in the navy.

At this time, the work of Christiaan Eijkman and his assistant, Gerrit Grijns, began. The Dutch East Indies (now Indonesia) was becoming a dangerous beriberi zone. The disease was spreading in epidemic proportions in the armed forces, prisons, and general population. For example, conditions were so bad in Javanese prisons that a jail sentence was considered to be almost a death sentence. In some afflicted people, cardiac insufficiency and massive edema (swelling) of the legs, a "wet beriberi," predominated. In many other patients, the progressive paralysis of the legs, "dry beriberi," was the main problem. Because of the apparent epidemic nature of the disease, a bacterial origin was suspected. Consequently, the Dutch government appointed a commission composed of two physicians, Cornelius Pekelharing and Clemens Winkler, to study beriberi firsthand. A third commissioner, Eijkman, was appointed after Winkler and Pekelharing asked Robert Koch to join their expedition. Koch demurred, but suggested the substitution of Eijkman, a physician who had just completed training in bacteriology with him.

The commissioners arrived in the East Indies in 1886 and immediately began their studies. Within two years, they had shown that beriberi was a result of inflammation of the nerves (polyneuritis). Because they isolated a bacterium from the blood of beriberi victims, the commissioners were satisfied that the disease was of bacterial origin. Winkler and Pekelharing returned to The Netherlands. Eijkman stayed behind to continue the research and to direct the Javanese Medical School.

Fortuitously, Eijkman observed a disease similar to beriberi in chickens. The diseased animals developed spontaneously all the beriberi symptoms. Eijkman named the disease polyneuritis gallinarium. In the course of studying the birds, Eijkman had them moved to new quarters. Strangely, the disease disappeared inexplicably. Upon examination, reflection, and study, Eijkman discovered that the food given to the birds had changed in one respect that was to become very important. Originally, the birds had been fed on leftover, boiled, white rice from the officers' ward in the military hospital. After their relocation, the chickens were given unpolished rice because the cook at the new site "refused to give any military rice to civilian fowl."

On the strength of this observation, Eijkman set out to determine whether the polished rice was the cause of the disease. He found this to be the case, because the feedings of unpolished rice cured the disease. Eijkman postulated that a chemical—perhaps a toxin from intestinal bacteria—was the actual causative agent. Although this concept was not correct, Eijkman's efforts began the scientific research that later showed that thiamine taken from the outer layer (the pericarp) of unpolished rice protected against beriberi.

In 1896, Eijkman left Java and returned to The Netherlands because of ill health. At that time, Grijns took over the study of beriberi. In 1901, Grijns proposed that the disease was a result of the lack of some natural nutrient substance that was found in unpolished rice and in other foods (a nutritional deficiency). The work that Eijkman began and Grijns continued is viewed by many as the basis for the modern theory of vitamins.

In 1917, Grijns likewise returned to The Netherlands. By 1921, he had become a professor of animal physiology at the State University in Wagenigen. In 1940, Grijns was awarded the Swammerdam Medal for his concept of nutritional deficiency.

Eijkman was appointed, in 1898, professor of public health at the University of Utrecht. He retained this position for thirty years and completed many other important research efforts during his tenure at the university. In 1929, Eijkman and Frederick Gowland Hopkins shared the Nobel Prize in Physiology or Medicine for their roles in the discovery of vitamins. Eijkman was a member of the Royal Netherlands Academy of Sciences and a foreign associate of the American National Academy of Sciences. His other honors included orders of knighthood and the establishment of an Eijkman medal by the Dutch government.

Impact of Event

Eijkman's observations that beriberi was caused by excessive dietary use of polished rice and that it could be cured by feeding unpolished rice were extremely important to the understanding of nutrition. A sequence of events was set into motion that led to the development of many aspects of modern nutrition theory, to the evolution of biochemical explanations for several very serious nutritional diseases, and to the virtual eradication of the disease.

Eijkman believed incorrectly, however, that a curative material in unpolished rice prevented the action of toxins present in the polished grain. The correct interpretation of the action of this unknown substance was first made in 1901 by Grijns. Grijns proposed that beriberi developed because diets that used polished rice only lacked an essential substance that was required for the appropriate function of the nervous system. Grijns and other researchers soon showed that many foods contained the antiberiberi factor and proved that these foods could be used to treat beriberi.

Another group of studies were carried out in an effort to identify the chemical nature of the substance involved. In 1912, Casimir Funk proposed that beriberi and other nutritional diseases, such as scurvy and rickets, were deficiency diseases caused by substances that were each a "vitamine." Funk admittedly coined the term because it "would sound well and serve as a catch word." Others, including Elmer McCollum and Hopkins, had similar beliefs. Soon the term was both accepted and shortened to its current spelling, vitamin. Subsequently, it was found that there were several types of isolated vitamins, such as A, B, C, and D, named in order of their discovery. A key event was the preparation of a pure antiberiberi factor, thiamine (or vitamin B_1), and determination of its structure by Robert R. Williams, starting in 1935. This effort led to the commercial synthesis of the vitamin by pharmaceutical companies and to its current, wide dissemination.

The availability of the pure vitamin allowed examination of its metabolic rate and actions. Soon, it became clear that thiamine became an essential component (coenzyme) that was required for the biological action of a great many important enzymes (biological catalysts). The lack of function of these catalysts was shown eventually to cause beriberi. Very similar results with other vitamins led to the concept that vitamins were coenzymes or parts of coenzymes and that the deficiency diseases their absence produced were diseases resulting from enzyme inactivation. Therefore, from the tiny acorn of the endeavors of Eijkman and Grijns grew the mighty oak tree of understanding of the roles of vitamins and establishment of the many basic precepts of nutrition.

Bibliography

Bicknell, Franklin, and Frederick Prescott. *The Vitamins in Medicine*. 3d ed. New York: Grune & Stratton, 1953. Chapter 3 describes concisely the history, the chemistry, and many medical aspects of thiamine up to 1952. It is valuable to readers who desire historical perspective on thiamine and its medicinal attributes. Pictures of beriberi victims and 953 references on the vitamin, its isolation, and its use are included.

Lehninger, Albert L. *Biochemistry*. 2d ed. New York: Worth, 1975. Chapter 13 of this college textbook contains a summary of the chemistry and biochemistry of thiamine. Aspects of thiamine enzymology, production of thiamine deficiency, and thiamine isolation are described briefly. The reader who wishes more information will find several useful references at the end of the chapter.

Lindeboom, Gerrit A. "Christiaan Eijkman." In *Dictionary of Scientific Biography*, edited by Charles Coulston Gillispie. New York: Charles Scribner's Sons, 1971. This brief biographical sketch is one of the few sources of information on Eijkman written in the English languge. It focuses on his career and provides Dutch sources of biographical information. Insight into Eijkman's life and his impact on both medicine and nutrition are given.

____________. *Newer Knowledge of Nutrition*. 2d ed. New York: Macmillan, 1922. Chapters 9 and 10 contain useful information on beriberi and on thiamine. The history, symptoms, and treatment of beriberi are covered. The work of Kanehiro, Grijns, and Eijkman; isolation of thiamine by Funk, and aspects of the work of Williams (who later synthesized thiamine) are included. Many important references are cited.

McCollum, Elmer V. *A History of Nutrition*. Boston: Houghton Mifflin, 1957. Chapter 16 of this interesting book updates the information on thiamine in the other

McCollum text. It is especially valuable to readers who wish to trace the evolution of understanding of the chemical nature of the vitamin and the biochemical basis for its actions. Twenty-eight references are cited.

Smith, Emil L., et al. *Principles of Biochemistry.* 7th ed. New York: McGraw-Hill, 1983. Chapter 21 of this excellent biochemistry text contains a brief description of thiamine and beriberi. Aspects of thiamine biochemistry, metabolism, deficiency, and distribution are given. Several references are included for readers who wish more technical coverage.

Sure, Barnett. *Vitamins in Health and Disease.* Baltimore: Williams & Wilkins, 1922. Chapter 21 of this basic nutrition text begins by describing the symptoms of different types of beriberi and their treatment. Aspects of the nutrition, distribution, and medicinal properties of vitamin B_1 are described. The material is most useful for readers who desire a simple approach.

Williams, Robert R. "The Chemistry of Thiamine (Vitamin B_1)." In *The Vitamins, a Symposium*, edited by Morris Fishbein. Chicago: American Medical Association, 1939. The article describes the synthesis and chemical characterization of thiamine. It credits isolation of the vitamin from natural sources, describes Williams' chemical synthesis and characterization of thiamine, and inventories several aspects of the first evidence for the coenzymatic nature of the vitamin. Forty-six references are cited.

Sanford S. Singer

Cross-References

Hopkins Suggests That Food Contains Vitamins Essential to Life (1906), p. 330; McCollum Names Vitamin D and Pioneers Its Use Against Rickets (1922), p. 725; Steenbock Discovers That Sunlight Increases Vitamin D in Food (1924), p. 771; Szent-Györgyi Discovers Vitamin C (1928), p. 857; Krebs Describes the Citric Acid Cycle (1937), p. 1107.

HEWITT INVENTS THE MERCURY VAPOR LAMP

Category of event: Applied science
Time: 1901
Locale: Ringwood Manor, New Jersey

Hewitt developed a light based on gaseous discharge, leading to significant developments such as the fluorescent light

Principal personages:

PETER COOPER HEWITT (1861-1921), an American electrical engineer who invented the mercury rectifier and other devices

HEINRICH GEISSLER (1814-1879), a German instrument maker who developed a pump to produce a high vacuum in a tube

JULIUS PLÜCKER (1801-1868), a German physicist who succeeded in passing current through an evacuated tube

Summary of Event

For all of recorded history, humankind has possessed the desire to bring light into enclosed spaces and to banish darkness with an artificial light. Artificial light was provided in the form of some type of controlled flame, beginning with simple torches and wood fires. Devices such as oil lamps and candles were more effective and indeed are still used today. In the nineteenth century, dramatic advances were made in the field of artificial lighting. The first of these was the development of gas lighting. Flammable gases for use as a fuel could be derived from a multitude of sources. By the 1850's, the streets of several urban areas were illuminated by gas light.

In modern times, light generated in some fashion by electricity has all but supplanted other sources for everyday use. People have long been familiar with electricity in the form of lightning and static electricity and such notables as Benjamin Franklin performed experiments to discover the nature of electricity. A critical observation was that made by Luigi Galvani, an Italian physicist, in 1780 when an electrical spark caused a frog leg to twitch. Alessandro Volta studied this phenomenon, which resulted in a man-made battery, or Voltaic cell, for the production of an electric current in 1794. The battery provided a simple means to construct a reliable source of electric current and prompted an outpouring of research involving electricity.

The first commercially successful electric light was the electric arc lamp, in which an electrical discharge between the slightly separated tips of two electrodes produces a very strong light. While arc lights were not practical for indoor use, they were very effective for outdoor applications, particularly street lighting. The light produced was of good quality, although a fairly elaborate and labor intensive support system was required to maintain the network. In particular, arc lights gave a very favorable appearance to the exteriors of buildings, which led the governing bodies of some cities to dictate their continued use even after suitable alternative systems were available. Of course, construction of these lighting systems was possible only after the development of machinery to provide reliable electric current.

A complete system of electrical supply—municipal and residential lighting—and electrical power for other applications was envisioned by Thomas Alva Edison. While recognized, along with Joseph Wilson Swan, as the developer of the incandescent light bulb, it is important to realize that Edison saw this invention as part of a larger picture. It was his determination that fostered public awareness of and desire for electric lighting, leading to the mammoth electrical generating facilities in existence today. Edison's electric light relied on a filament heating up to incandescence when a current is passed through it. For most domestic lighting, the incandescent bulb is the lighting of choice, even though it is relatively inefficient. Much of the energy provided to a light bulb is lost as heat.

While Edison struggled with the electric light and development of power transmission, other researchers were exploring areas of physics and chemistry that would lead to the development of another type of lighting device. As early as 1675, the French astronomer Jean Picard had observed a discharge of light from a mercury barometer. Subsequently, it was understood that this discharge was stimulated by static electricity, although the cause of the flash of greenish-blue light was unknown. The mercury barometer, invented by Evangelista Torricelli, was simply a long glass tube sealed at one end, filled with mercury, which was then placed open end down into a vat of mercury. When prepared properly, the mercury in the tube drops down to a height which is supported by air pressure. Above the mercury in the sealed end of the tube, a good vacuum is created, which contains only a slight amount of mercury vapor. This type of vacuum is isolated and thus not easily studied.

In 1855, the German inventor Heinrich Geissler developed a simple pump that used a moving column of mercury to generate a good vacuum in a tube. This development came at an opportune time for Geissler, as many physicists at the time were investigating the possibility of forcing electricity through a vacuum. A key experiment in this field was performed by Julius Plücker, a German physicist and contemporary of Geissler. Plücker succeeded in passing current through one of these tubes, and in 1858 reported what came to be known as cathode rays. These rays were studied by other physicists, such as Sir William Crookes, who proposed that the charge flowed through the tube because of movement of some sort of charged particle. Joseph John Thomson conclusively proved this postulate in the late 1890's, and thus is credited with proving the existence of the electron.

It was therefore known by the beginning of the twentieth century that in a near vacuum, current in the form of electrons could flow between two electrodes and that electricity could cause a discharge of light in a near vacuum, as had been observed by Picard. Armed with this knowledge, Peter Cooper Hewitt tried to devise a way to establish a discharge in a glass tube containing mercury vapor at a very low pressure. It had been shown by Plücker that when an element is stimulated, it emits only certain characteristic wavelengths of light, or what is commonly known as a line spectrum. This is in contrast to the light of an incandescent bulb, which is a continuous spectrum. Hewitt hoped to use the electrical discharge through the mercury vapor to generate light.

In principle, the electrons traveling through the tube would collide with mercury atoms. This collision would contribute energy to the mercury atom, causing an electron in the atom to become energized and move to a high-energy state. When the electron returned to its original energy level, light would be emitted with an energy characteristic of the transition. A benefit of this type of light is that only visible light would be emitted so that the light source would generate very little heat. The tube marketed commercially by Hewitt was 2.5 centimeters in diameter and about 127 centimeters in length and was very similar in appearance to modern fluorescent lights. In Hewitt's original lamp, the glass tube was hung at a slight angle and contained a small amount of liquid mercury. To activate the light, the tube was tilted so that a stream of mercury stretched from one electrode to the other. The current flow vaporized the mercury, and the mercury vapor was then stimulated by the current to produce light. Hewitt's first lamp was a direct current lamp. One of the major obstacles to a discharge lamp was that a large potential was needed to start the discharge, but a much lower voltage sufficed to maintain the discharge once established. This was why Hewitt developed the starting system. Further developments by Hewitt incorporated into subsequent versions of his lamp included the addition of an electromagnet to tilt the lamp and an induction coil, which provided a high voltage to start the lamp.

One of the major drawbacks of the light generated was that it produced distortions in the appearance of objects because of the lack of red light in the mercury spectrum. The quality of light was, however, quite useful in photography. Further developments would make the light compatible with alternating current and in subsequent years other inventors would develop commercially successful lights based on the principle of electric discharge through a vapor.

Impact of Event

Hewitt's invention of the mercury vapor lamp was a novel means of obtaining light. Initially, practical applications of the lamp were limited, primarily because of the quality of the light produced. Hewitt developed the mercury rectifier to convert alternating current to direct current and refined the lamp to improve its color characteristics. This improved model was used as a source of illumination in many factories. One property of the mercury lamp is that in addition to certain wavelengths of visible light, large quantities of ultraviolet light are also emitted. Ultraviolet light does not pass through ordinary glass, but quartz is transparent to it. By producing a light with a quartz tube, Hewitt marketed a source of ultraviolet light which is of great utility in biological application. A related modern use of similar quartz tube lamps is in tanning beds.

By placing the electrodes closer together and increasing the pressure of vapor, more intense light is generated. This type of lamp commonly utilizes either mercury or sodium. The sodium vapor lamp, developed in the early 1930's, produces a distinct yellow light. This type of lamp was widely used as an external source of illumination, especially in street lights, all but supplanting mercury lamps until refined models became available. Modern city lighting often includes mercury lamps in the vicinity of downtown areas, where the light produced seems to enhance the appearance of buildings and sodium lights in rural areas and intersections, where the main concern is seeing the road and other automobiles.

A related and equally familiar development in the field of gas discharge lighting was introduced in 1910 by Georges Claude, a French chemist who placed various types of gases in tubes. He found that certain elements produced attractive colors. In particular, neon produced a very attractive discharge. By altering the mixture of gases, a wide variety of colors could be produced. The application of these lights, commonly called neon lights, introduced a minor revolution in advertising that is especially in evidence in such cities as Las Vegas. Another major development of the low-pressure mercury vapor lamp was made in the 1930's. This development took advantage of the fact that mercury emitted ultraviolet light when stimulated. There exist materials known as phosphors, which absorb ultraviolet light and undergo fluorescence. In the process of fluorescence, the ultraviolet light energy absorbed by the phosphor is subsequently emitted in the form of visible light. This discovery prompted development of the fluorescent light in which the interior of a gas discharge tube, very similar to that developed by Hewitt, is coated with a phosphor. The result when the light is turned on is a uniform emission of light from the phosphor. Manipulation of the chemical nature of the phosphor allows for variation in the light produced. Since their initial development, fluorescent lights have become widely used for illumination in office and factory environments, while their domestic use has been somewhat restricted to utility areas in the home. As further refinements to mercury discharge lamps are introduced, they may become the predominant source of external and internal artificial lighting.

Bibliography

Asimov, Isaac. *Asimov's Biographical Encyclopedia of Science and Technology.* 2d rev. ed. Garden City, N.Y.: Doubleday, 1982. Written for the interested reader, this work includes short biographical sketches of 1,510 scientists. Included in that group are several key players in the development of vapor lights. The book is extensively cross referenced so that the personalities of the scientists involved and the overlapping of their interests can be appreciated.

Burke, James. *Connections*. London: Macmillan, 1978. This excellent volume describes how different innovations are often related to one another through subsequent development. Of interest in the context of lighting is an excellent discussion of gas lighting and the production of fuel for gas lighting.

Cayless, M. A., and A. M. Marsden, eds. *Lamps and Lighting*. 3d ed. London: Edward Arnold, 1983. This comprehensive textbook includes chapters on low-pressure mercury lamps, fluorescent lamps, sodium lamps, and the like. A wealth of theoretical material makes this a challenging work for the lay person, but most chapters are well written. Includes diagrams of the design of various types of lamps and specification and construction techniques.

Hall, Stephen S. "The Age of Electricity." In *Inventors and Discoverers*, edited by Elizabeth L. Newhouse. Washington, D.C.: National Geographic Society, 1988. An excellent chapter in this text; a history of the development of electricity is presented. Well illustrated; with both photographs and artwork, it conveys a concise picture of the history. Includes a good biography of Edison and describes the monumental alternating/direct current battle.

Schroeder, Henry. *History of Electric Light*. Washington, D.C.: Smithsonian Institution, 1923. Written only a few years after Hewitt's discovery, this source includes an excellent drawing of Hewitt's lamp. Also contains excellent history of development of other types of lights in use at the beginning of the twentieth century. While obviously a dated text, it contains a wealth of information that cannot be found in current texts.

Craig B. Lagrone

Cross-References

Fleming Files a Patent for the First Vacuum Tube (1904), p. 255; Steenbock Discovers That Sunlight Increases Vitamin D in Food (1924), p. 771; Fluorescent Lighting Is Introduced (1936), p. 1080.

IVANOV DEVELOPS ARTIFICIAL INSEMINATION

Categories of event: Medicine and biology
Time: 1901
Locale: Soviet Union

Ivanov developed practical techniques for the artificial insemination of farm animals, revolutionizing livestock breeding practices throughout the world

Principal personages:

LAZZARO SPALLANZANI (1729-1799), an Italian physiologist who performed the first successful artificial insemination of a mammal

ILYA IVANOVICH IVANOV (1870-1932), a Soviet biologist who developed practical techniques for artificial insemination of animals and taught them to farmers

R. W. KUNITSKY, a Soviet veterinarian who assisted Ivanov with his research on horses

Summary of Event

The tale is told of a fourteenth century Arabian chieftain who sought to improve his mediocre breed of horses. Sneaking into the territory of a neighboring hostile tribe, he stimulated a prize stallion to ejaculate into a piece of cotton. Quickly returning home, he inserted this cotton into the vagina of his own mare, who subsequently gave birth to a high-quality horse. This may have been the first case of artificial insemination, the technique by which semen is introduced into the female reproductive tract without sexual contact.

The first scientific record of artificial insemination comes from Italy in the 1770's. Lazzaro Spallanzani was one of the foremost physiologists of his time, well known for having disproved the theory of spontaneous generation. There was some disagreement at that time about the basic requirements for reproduction in animals. It was unclear if the sex act was necessary for an embryo to develop, or if it was sufficient that the sperm and eggs come into contact. Spallanzani was familiar with the work of Jacobi, who had suggested in 1742 that the sex act was not necessary for reproduction. Jacobi had been able to breed young fish by combining the sex cells of male and female fish in a glass dish. Spallanzani also began by studying animals in which union of the sperm and egg normally takes place outside the body of the female. He stimulated males and females to release their sperm and eggs, then mixed these sex cells in a glass dish. In this way, he produced young frogs, toads, salamanders, and silkworms.

Next, Spallanzani asked whether the sex act was also unnecessary for reproduction in those species in which fertilization normally takes place inside the body of the female. He collected semen that had been ejaculated by a male spaniel and, using a syringe, injected the semen into the vagina of a female spaniel in heat. Two months later, she delivered a litter of three pups, which bore some resemblance to both the mother and the male that had provided the sperm. Spallanzani was justifiably proud of this accomplishment.

Despite Spallanzani's obvious pleasure, he apparently did no other work in this field. John Hunter's report that he had used Spallanzani's technique to impregnate a woman whose husband could not ejaculate directly into the vagina was not uniformly believed. In an era when obstetrical care was provided by midwives, a male physician in the bedchamber sounded suspicious. James Marion Sims, who tried this technique on an infertile patient in 1866, was likewise discouraged by his American colleagues.

It was in animal breeding that Spallanzani's techniques were to have their most dramatic application. In the 1880's, an English dog breeder, Sir Everett Millais, conducted several experiments on artificial insemination. He was interested mainly in obtaining progeny from dogs that would not normally mate with each other because of a difference in size. He followed Spallanzani's method to produce a cross between the short, low Basset hound and the much larger bloodhound. At about the same time, several French and American horsebreeders were using artificial insemination sporadically to overcome certain types of infertility in their mares.

Ilya Ivanovich Ivanov was a Soviet biologist who had conducted postgraduate research on the accessory sex glands of animals. He may have heard of the success of artificial insemination in horses when he worked at the Pasteur Institute in Paris from 1897 to 1898. Upon returning to the Soviet Union, he published a historical essay entitled "Artificial Impregnation in Mammals," which showed his familiarity with the work of Spallanzani as well as contemporary animal breeders. Although these techniques were widely used for fish breeding in the Soviet Union at the time, they were not used at all in the breeding of farm animals. Ivanov saw the potential to increase the efficiency of horse breeding, and he was commissioned by the government to investigate the use of artificial insemination of horses.

Unlike previous workers who had used artificial insemination to circumvent certain anatomical barriers to fertilization, Ivanov began the use of artificial insemination to propagate thoroughbred horses more effectively. His assistant in this work was the veterinarian R. W. Kunitsky. In 1901, Ivanov founded the Experimental Station for the Artificial Insemination of Horses in Dolgoe village, Orlovskaya *guberniya*. As its director, he embarked on a series of experiments to devise the most efficient techniques for breeding these animals. Not content with the demonstration that the technique was scientifically feasible, he wished to ensure further that it could be practiced by Soviet farmers. If sperm from a male were to be used to impregnate females in another location, potency would have to be maintained for a long time. Ivanov first showed that the secretions from the accessory sex glands were not required for successful insemination; only the sperm itself was necessary. He demonstrated further that if a testis were removed from a bull and kept cold, the sperm would remain alive. More useful than preservation of testes would be preservation of the ejaculated sperm. By adding certain salts to the sperm-containing fluids, and by keeping these at cold temperatures, Ivanov was able to preserve sperm for long periods. With sperm preserved in this way, Ivanov was able to impregnate a female with sperm from a male living at another location.

Because of the novelty of this technique, Ivanov was concerned that the offspring of this artificial breeding might be abnormal. His subsequent observations eliminated this worry; the offspring of artificial breeding grew and developed at the same rate as normally bred animals.

Ivanov also developed instruments to inject the sperm, to hold the vagina open during insemination, and to hold the horse in place during the procedure. In 1910, Ivanov wrote a practical textbook with technical instructions for the artificial insemination of horses. He also trained some three hundred veterinary technicians in the use of artificial insemination, and the knowledge he developed was quickly disseminated throughout the Soviet Union. Artificial insemination became the major means of breeding horses.

An experiment begun in 1901 was to have even further-reaching effects. In that year, Ivanov petitioned the Department of Agriculture for permission to attempt artificial insemination of cows. Initially turned down, Ivanov eventually obtained permission to purchase ten cows, which he housed at the Moscow College of Agriculture. The following year, he produced two calves by artificial insemination of these cows. At the Moscow Zoo, he obtained two sheep, which were also artificially inseminated the same year. In 1910, Ivanov had the opportunity to embark on a large-scale study of sheep breeding by artificial insemination in Askania-Nova. Over the next twenty years, more than 1 million ewes had been artificially inseminated using the techniques he developed.

Until his death in 1932, Ivanov was active in researching many aspects of the reproductive biology of animals. He developed methods to treat reproductive diseases of farm animals and refined methods of obtaining, evaluating, diluting, preserving, and disinfecting sperm. He also began to produce hybrids between wild and domestic animals in the hope of producing new breeds that would be able to withstand extreme climatic conditions better and that would be more resistant to disease. His crosses included hybrids of ordinary cows with aurochs, bison, and yaks, as well as some more exotic crosses of zebras with Przhevalsky's horses. He hoped to use artificial insemination to help preserve species that were in danger of becoming extinct. In 1926, he led an expedition to West Africa to experiment with the hybridization of different species of anthropoid apes.

Impact of Event

Although Ivanov worked on the artificial insemination of a variety of species, his work on cattle and sheep had the greatest impact. By 1936, more than 6 million cattle and sheep were artificially inseminated in the Soviet Union. Despite its success, several decades elapsed before artificial insemination was used on a large scale in other countries. In the United States, Enos J. Perry began the first large-scale facility for artificial insemination in New Jersey in 1938, having learned the techniques while on a tour of Denmark.

Research subsequent to Ivanov's work has improved further the methods used to collect sperm and to preserve it. The greatest beneficiaries of this technique have been dairy farmers. Some bulls are able to sire genetically superior cows that produce exceptionally large volumes of milk. Under natural conditions, such a bull could father at most a few hundred offspring in its lifetime. Using artificial insemination, a prize bull can inseminate ten to fifteen thousand cows each year. This use of superior sperm can continue even after a male's death if the semen is kept frozen. Artificial insemination also prevents the possibility that a female infected with a venereal disease will spread the infection to a prize male. For dairy farmers, artificial insemination also means that they no longer need to keep dangerous bulls on the farm; frozen sperm may be purchased through the mail. Artificial insemination has become the main method of reproduction of dairy cows, with about 150 million cows produced this way throughout the world.

In the 1980's artificial insemination gained added importance as a method of breeding rare animals. Animals kept in zoo cages, which lack the behavioral repertoire necessary for normal mating, may still produce normal sperm that can be used to inseminate a female artificially. Some species require specific conditions of housing or diet for breeding to occur, conditions not available in all zoos. Such animals can still reproduce using artificial insemination. Elephants also are frequently artificially inseminated, allowing zoos to house only females and their offspring without the more difficult-to-handle males.

Artificial insemination has proved effective in treating various types of infertility in humans. Although it had been used occasionally before Ivanov's time, the widespread use of artificial insemination in the human species was possible only after it was clear that it was successful in animals and that the offspring in animals were not abnormal in any way.

Bibliography

Asdell, S. A. "Historical Introduction." In *Reproduction in Domestic Animals*, edited by H. H. Cole and P. T. Cupps. 3d ed. New York: Academic Press, 1977. A brief overview of the development of the knowledge of reproductive biology as it developed over the centuries. Includes mention of Ivanov and others involved in artificial insemination. Also included are chapters on the biology of sperm, artificial insemination, and reproduction in horses, cattle, pigs, sheep, goats, dogs, cats, and poultry. Somewhat technical, but clearly written. Extensive references to the scientific literature.

Bearden, H. Joe, and John W. Fuquay. *Applied Animal Reproduction*. 2d ed. Reston, Va.: Reston, 1984. An in-depth explanation of the modern techniques used to manipulate reproduction in farm animals. Particularly useful for its liberal illustrations, including photographs of the instruments and techniques used in the artificial insemination of cattle. References.

Perry, Enos J. *The Artificial Insemination of Farm Animals*. 4th ed. New Brunswick,

N.J.: Rutgers University Press, 1968. This volume was published by Perry thirty years after he began the first organization in the United States to perform artificial insemination on a large scale. In addition to chapters on the techniques current at that time, Perry devotes a chapter to the historical background of artificial insemination. References are given to original papers, most of which are not in English.

Polge, C. "Increasing Reproductive Potential in Farm Animals." In *Reproduction in Mammals, Book 5: Artificial Control of Reproduction*, edited by C. R. Austin and R. V. Short. London: Cambridge University Press, 1972. Contains clear descriptions of the technique of artificial insemination, as well as drawings of the procedure being carried out on farm animals. This chapter emphasizes the advances in artificial insemination since the time of Ivanov, including methods for long-term preservation of sperm, which were pioneered by Polge.

Poynter, F. N. L. "Hunter, Spallanzani, and the History of Artificial Insemination." In *Medicine, Science, and Culture*, edited by L. G. Stevenson and R. P. Multhauf. Baltimore: The Johns Hopkins University Press, 1968. An engaging sketch of early workers in the field of artificial insemination. Includes biographical material on Spallanzani and Hunter, as well as a discussion of the controversy engendered in America by the work of James Marion Sims.

Judith R. Gibber

Cross-References

Brown Gives Birth to the First "Test-Tube" Baby (1978), p. 2099; The First Successful Human Embryo Transfer Is Performed (1983), p. 2235.

RÖNTGEN WINS THE NOBEL PRIZE FOR THE DISCOVERY OF X RAYS

Categories of event: Physics and medicine
Time: 1901
Locale: University of Würzburg, Germany

Röntgen discovered that a new kind of radiation was emitted by a cathode ray tube

Principal personages:

WILHELM CONRAD RÖNTGEN (1845-1923), a German physicist who discovered X rays and received the first Nobel Prize in 1901 in Physics

PHILIPP LENARD (1862-1947), a German physicist who did extensive work on cathode rays for which he received the Nobel Prize in Physics in 1905

MAX VON LAUE (1879-1960), a German physicist who used crystals in 1913 to refract X rays

SIR WILLIAM HENRY BRAGG (1862-1942), an English physicist who, along with his son, Sir Lawrence Bragg, developed the precise methods of crystallography, for which they received the 1915 Nobel Prize in Physics

HENRY MOSELEY (1887-1915), an English physicist who used X rays to establish the atomic number of the elements

HEINRICH GEISSLER (1814-1879), a German physicist who invented the cathode ray tube used by many physicists at the end of the nineteenth century and that led to the discovery of X rays

Summary of Event

Although atomism was proposed by ancient Greek philosophers, it was placed on firm scientific foundations in the nineteenth century, beginning with the work of the English chemist, John Dalton. During the nineteenth century, various properties of matter were investigated by scientists; in physics, this period was marked by extensive developments in thermodynamics, electricity, magnetism, and optics. Nevertheless, until the last decade of the century, the atom still remained the ultimate, indivisible particle that its name implies. In the 1890's, there were three discoveries that led eventually to the awareness that atoms had structure. These were the discovery of X rays by Wilhelm Conrad Röntgen in 1895, the discovery of radioactivity by Antoine-Henri Becquerel in 1896, and the discovery of electrons by Sir Joseph John Thomson in 1897.

The steps that led to the discovery of X rays began with the invention of the cathode ray tube by Heinrich Geissler of Bonn, Germany, in 1857. The cathode ray tube is a partially evacuated glass tube inside of which are sealed two metal electrodes. When a high voltage is applied across these electrodes, positive ions are attracted to the negatively charged electrode and the negative ions to the positive electrode. Because of the kinetic energy acquired by the ions, the molecules of the gas in the tube are excited and radiate electromagnetic energy of various wavelengths that are characteristic of the gas in the tube. Julius Plücker, also a professor at Bonn University and a physicist, became interested in Geissler's tubes and suggested a modified form of the tube whereby the luminous discharge could be confined to a capillary part in the middle. Plücker was the first to observe cathodic rays (without identifying them) and their deflection in the presence of a magnetic field. Eugen Goldstein of the Berlin Observatory was the first scientist to use the term "cathode rays" in 1876 to describe the various kinds of radiation detected in the tube. Other scientists who used the Geissler tube were Johann W. Hittorf, Sir William Crookes, Heinrich Hertz, Philipp Lenard, and Thomson. Because Crookes had made significant improvements in the design of the tube, it became known as the Crookes tube.

Röntgen began his auspicious academic career at the University of Utrecht in 1865, but soon enrolled at the Swiss Polytechnic Institute in Zurich, Switzerland. At Zurich, Röntgen worked under two famous physicists, Rudolph Clausius and August Kundt, the latter a brilliant experimentalist. After receiving the degree of doctor of philosophy in 1869 from the University of Zurich, Röntgen worked successively at the Universities of Würzburg, Strasbourg, and Giessen. Although his doctoral dissertation had been on the study of gases, he expanded his experimental work to research on pyro- and piezo-electrical properties of crystals, surface phenomena of liquids, and dielectrics. In 1888, Röntgen returned to the University of Würzburg, having achieved a solid reputation as a physicist. By 1895, Röntgen had published forty-eight papers and had corresponded with eighty of the most prominent physicists of the time. Röntgen, like his contemporaries, decided in the 1890's to investigate phenomena associated with the cathode ray tube.

On November 8, 1895, he was working alone in the physics institute at the University of Würzburg, attempting to determine the effect of covering a cathode ray tube with a black cardboard box. Röntgen noticed that a paper treated with barium platinocyanide fluoresced, even though it was several meters away from the covered tube; as he was working at night and the laboratory was dark, the fluorescence was noticed easily. Lenard had discovered previously that cathode rays could penetrate a thin aluminum window at the end of a tube and cause fluorescence, but only up to a few centimeters from the tube. Aware of this observation of Lenard, Röntgen realized that he was observing a new, previously undetected kind of ray. He was impressed with the intensity of his rays and their penetrating power. Consequently, his first experiments were to place various obstructions in the path of the rays, such as a book, and wooden and metal objects. Röntgen noted that the degree of fluorescence was little reduced by wood or by a book of about one thousand pages but that lead and platinum blocked the rays completely. Röntgen then placed his hand in the path of the rays and saw that there was a clear outline of the bones on the fluorescent screen. Replacing the fluorescent screen with a photographic plate, he produced the first X-ray photograph. Unable to identify clearly these new rays, he called them simply X rays to indicate that their exact nature was as yet unknown; in Germany, they became known as "Röntgen rays."

The first public announcement of Röntgen's discovery was made in a brief note to the Physico-medical Society of Würzburg. Within the medical community, there was immediate recognition of the significance of this discovery for medical diagnosis. In January, 1896, a photograph was made in Paris of a hand with a bullet embedded in it. So impressive was the photograph of a living hand that popular as well as technical journals immediately reproduced it.

After the initial discovery, Röntgen measured the properties of X rays: their penetrating power, their ability to cause fluorescence, the refractive index in various media, and their reflectivity. As a result of further experiments, he distinguished between "hard" X rays, those that would penetrate very dense materials, like bones, and "soft" rays, those that would penetrate materials like human flesh. In addition, he discovered that, while any solid body could be made to generate X rays when they were bombarded by the cathode rays, the most penetrating rays were produced when platinum was the target. When Thomson identified cathode rays as electrons in 1897, the basic mode of production of X rays was clearly understood—that is, by striking targets with high-velocity electrons.

As is the case with so many other scientific discoveries, the discovery of X rays depended on a considerable amount of work done by Röntgen's predecessors, but it is still possible to credit Röntgen with the clear identification of these rays. Nevertheless, because so many of his contemporaries also worked with cathode ray tubes, it is not altogether surprising that some of them had, in fact, detected X rays without identifying them. The three physicists who did this were Crookes, A. W. Goodspeed, and Lenard. In 1879, Crookes noted that photographic plates lying near his cathode tube frequently became fogged, but his only reaction was to return the plates to the manufacturers as being faulty. In 1890, Goodspeed of the University of Pennsylvania had virtually the same experience, but he also failed to investigate the phenomenon further. Lenard, in his extensive work with cathode tubes, undoubtedly dealt with X rays, but he failed to distinguish them from cathode rays, which were later identified as electrons. In view of the widespread reception of the discovery of X rays and its recognized significance, Röntgen was awarded the first Nobel Prize in Physics in 1901.

Impact of Event

Röntgen's discovery made an immediate and worldwide impact. The earliest record to appear in a scientific journal was a report in the *Electrical Engineer* of New York on January 8, 1896, only two months after the discovery. In that year, more than a thousand books, pamphlets, and articles appeared about X rays. Their mode of generation and their properties proved to be such fruitful areas for research that between 1896 and 1910, more than ten thousand publications were devoted to them.

The significance of X rays as a diagnostic tool for medicine was seen by Röntgen and his contemporaries. In addition to examining broken bones, X rays began to be used for other diagnostic purposes, such as photographing unborn children, examination of the spinal column, photographing the skull with X rays to determine the cause of fainting and blind spells, detection of metal objects in the body, and the detection of cavities in teeth. Chronic illnesses that X rays detected include arthritis, tuberculosis, and bone demineralization. Newly developed techniques in fluoroscopy allowed physicians to study the motion of body organs and tissues on television tubes. By 1970, Americans were making an estimated 179 million visits to X-ray facilities each year at a rate of ninety visits per one hundred people. Eventually, it was discovered that X rays could have serious negative effects on the human body. They can damage the reproductive cells, thus causing subsequent genetic diseases in future generations and can indirectly cause leukemia.

Although Röntgen failed to determine the exact nature of X rays, another German physicist, Max von Laue accomplished this in 1913. Von Laue, along with Walter Friedrich and Paul Knipping successfully demonstrated the diffraction of X rays by crystals, and showed that X rays were electromagnetic radiation, with wavelengths between 10^{-7} and 10^{-11} meters. The work of these physicists was developed further by the Cambridge physicist Sir William Henry Bragg and his son, Sir Lawrence Bragg, who used the newly discovered radiation to determine precisely the atomic arrangement of crystals.

When X-ray spectra are produced, they are found to consist of a continuous range of wavelengths, as well as intense specific lines. The continuous portion is caused by the decelerating electrons and is called "braking radiation." The line spectra are given off by the atoms of the target metal as a result of the direct interaction of the bombarding electrons with the electrons of the target. The line spectra also are known as "characteristic spectra," because their wavelengths are characteristic of the different target materials. Because of this fact, X-ray spectra were seen by the English physicist Henry Moseley to provide a clue as to the position of elements in the atomic table.

Bibliography

Bleich, Alan Ralph. *The Story of X-Rays*. New York: Dover, 1960. This work, written for the general reader, gives a brief historical account of the discovery of X rays and their application in medicine, art, crystallography, and industry. It closes with a brief discussion on the hazards of overexposure to this radiation. It has numerous illustrations, with a glossary of technical terms and an index.

Bragg, W. H., and W. L. Bragg. *X-Rays and Crystal Structure*. London: G. Bell & Sons, 1915. This book was written by the father-and-son team who pioneered the use of X rays to study the structure of crystals. Offers much information of historical interest. The authors discuss the basic physics of X rays and crystals, with a minimum amount of mathematics and a considerable number of excellent photographs and illustrations.

Dibner, Bern. *Wilhelm Conrad Röntgen and the Discovery of X-Rays*. New York:

Franklin Watts, 1968. This is an excellent account of the discovery of X rays. Begins with the contributions made by Röntgen's predecessors and the influence other physicists had on him. Written for the general reader, it has no mathematics but has many fine illustrations and photographs.

Klickstein, Herbert S. *Wilhelm Conrad Röntgen on a New Kind of Rays*. Philadelphia: Mallinckrodt Classics of Radiology, 1966. Intended for readers with a scholarly interest in the subject of Röntgen and X rays, this book is a bibliographical study.

Laws, Priscilla W. *X-Rays: More Harm Than Good?* Emmaus, Pa.: Rodale Press, 1977. As the title implies, this work is a rather thorough discussion of the pros and cons of the use of X rays in medicine. It has numerous practical, helpful appendices on such topics as units of radiation exposure, information on breast self-examination, and a glossary and an index.

Thumm, Walter. "Röntgen's Discovery of X Rays." *Physics Teacher* 13 (April, 1975): 207-214. The discovery of X rays by Röntgen has often been called an accident by historians. Thumm shows why this is not really a tenable description when one takes into account the care with which he examined the phenomenon and the failure of his contemporaries, although working with the same equipment, to make the discovery.

Wilfred Theisen

Cross-References

Barkla Discovers the Characteristic X Rays of the Elements (1906), p. 309; X-Ray Crystallography Is Developed by the Braggs (1912), p. 517; Salomon Develops Mammography (1913), p. 562; Aston Builds the First Mass Spectrograph and Discovers Isotopes (1919), p. 660; Compton Discovers the Wavelength Change of Scattered X Rays (1923), p. 746; X Rays from a Synchrotron Are First Used in Medical Diagnosis and Treatment (1949), p. 1336; Giacconi and Associates Discover the First Known X-Ray Source Outside the Solar System (1962), p. 1708.

KIPPING DISCOVERS SILICONES

Category of event: Chemistry
Time: 1901-1904
Locale: Nottingham, England

Kipping discovered silicones, which nearly fifty years later found widespread markets because of their unique properties

Principal personages:

FREDERIC STANLEY KIPPING (1863-1949), a Scottish chemist and professor who was a pioneer in silicones

EUGENE G. ROCHOW (1909-), an American research chemist who carried out the first synthesis of methylsilicon chlorides from elemental silicon and methyl chloride

JAMES F. HYDE (1903-), an American organic chemist who discovered the use of potassium hydroxide as a rearrangement catalyst for making silicones

Summary of Event

Frederic Stanley Kipping, in the first four decades of the twentieth century, made an extensive study of the organic chemistry of silicon. He had a distinguished academic career, and summarized his silicon work in a lecture in 1937 ("Organic Derivatives of Silicon"). Since Kipping did not have any naturally occurring compounds available with chemical bonds between carbon and silicon atoms (organosilicon compounds), it was necessary for him to find methods of establishing such bonds. In this quest, Kipping drew on the discoveries of his predecessors. Previous syntheses of compounds with silicon-carbon links involved the intermediacy of carbon derivatives of other elements. Thus, Charles Friedel and James Crafts, who prepared the first organosilicon compound in 1863, used diethyl zinc as a reagent for transferring ethyl groups to silicon, replacing the chlorine atoms in silicon tetrachloride. Alfred Stock used dimethyl zinc to form silicon-to-carbon bonds and prepared a small sample of methylsilicone oil in 1919 but did not investigate the polymer further. Albert Ladenberg, who was a student of Friedel, used silicate esters as starting materials and employed sodium metal in some of his procedures. A. Polis also used sodium in a procedure to prepare tetraphenyl silicon. Kipping, in his earliest work, also used sodium metal to activate organic halogen compounds, forming organosodium derivatives which could be used to replace the chlorine in silicon tetrachloride by organic groups. In 1901, the French organic chemist Victor Grignard reported the synthetic uses of organomagnesium compounds (the famous "Grignard reagents"), and Kipping soon realized that these reagents were ideal for organosilicon synthesis (indeed, they are probably unsurpassed for this application even today).

While Kipping was probably the first to prepare a silicone and was certainly the first to use the term "silicone," he regarded silicones as a side issue and did not pursue their commercial possibilities. His careful experimental work was a valuable starting point for all subsequent workers in organosilicon chemistry, including those who later developed the silicone industry.

A few years after Kipping's Bakerian lecture, important research was initiated at the General Electric Company's (G.E.) corporate research laboratory in Schenectady, New York. On May 10, 1940, G.E. chemist Eugene G. Rochow discovered that methyl chloride gas, passed over a heated mixture of elemental silicon and copper, reacted to form organochlorosilanes compounds with silicon-carbon bonds. Kipping had shown that these silicon compounds react with water to form silicones. The importance of Rochow's discovery was that it opened the way to a continuous process that did not consume expensive metals like magnesium, nor flammable ether solvents required in the Grignard method. The copper acts as a catalyst, and the desired silicon compounds are formed with only minor quantities of by-products. This "direct synthesis," as it has come to be called, is now realized commercially on a large scale and is the most important source of silicone precursors.

Silicones are examples of what chemists call "polymers"—at the molecular level, they consist of long repeating chains of atoms. In this molecular characteristic, silicones resemble polyethylene or polyvinyl chloride (PVC), which are familiar plastics in everyday life. Although polyethylene or PVC have chains of carbon atoms, silicone molecules have a chain composed of alternate silicon and oxygen atoms. Each silicon atom bears two organic groups as substituents, while the oxygens serve to link the silicon atoms into a chain. The silicon-oxygen backbone of the silicones is responsible for their unique and useful properties such as the ability of a silicone oil to remain liquid over an extremely long temperature range and to resist oxidative and thermal breakdown at high temperatures.

A fundamental scientific consideration with a silicone, as with any polymer, is to obtain the desired physical and chemical properties in a product by closely controlling its chemical structure and molecular weight. Oily silicones with thousands of alternating silicon and oxygen atoms have been prepared. The average length of the molecular chain determines the flow characteristics (viscosity) of the oil. In samples with very long chains, rubberlike elasticity can be achieved by crosslinking the silicone chains in a controlled manner and adding a filler such as silica. High degrees of crosslinking could produce a hard, intractable material instead of rubber.

The action of water on the organochlorosilanes from Rochow's direct synthesis is a rapid method of obtaining silicones, but does not provide much control of the molecular weight. Further development work at G.E. and at Dow-Corning showed that the best procedure for controlled formation of silicone polymers involved treating the crude silicones with acid to produce a mixture from which high yields of an intermediate called D4 could be obtained by distillation. The intermediate D4 could be polymerized in a controlled manner by use of acidic or basic catalysts. Wilton I. Patnode of G.E. and James F. Hyde of Dow-Corning made important advances in this area. Hyde's discovery of the use of traces of potassium hydroxide as a polymerization catalyst for D4 made possible the manufacture of silicone rubber, which is the most commercially valuable of all the silicones today.

Impact of Event

Although Kipping's discovery and naming of the silicones occurred from 1901 to 1904, the practical use and impact of silicones started in 1940, with Rochow's discovery of the direct synthesis. General Electric constructed a silicone plant at Waterford, New York, and Dow-Corning began production in Midland, Michigan. They were joined by Union Carbide in 1957 and Stauffer Chemical in 1965. The German companies Wacker-Chemie and Farbenfabriken Bayer, A. G. offered silicones beginning in the 1950's; by 1968, silicones were being produced also in Belgium, Czechoslovakia, France, England, Japan, and the Soviet Union. Production of pure silicones rose from about 1,000 tons per year in 1950 to 14,000 tons in 1965 in the United States. World production in the late 1980's was estimated at more than 500,000 tons. The major forms of silicone are oils, resins, and elastomers. Approximate percentages of the total production consumed in various industries are: paper and textiles (30 percent), electrical and electronics (25 percent), construction (15 percent), automobiles (10 percent), office equipment (10 percent), and medical and food processing (10 percent).

Production of silicones in the United States came rapidly enough to permit them to have some influence on military supplies for World War II. In aircraft communication equipment, extensive waterproofing of parts by silicones resulted in greater reliability of the radios under tropical conditions of humidity where condensing water could be destructive. Silicone rubber, because of its stability to heat, was used in gaskets for high-temperature use, in search-lights, and in the superchargers on B-29 bomber engines. Silicone grease applied to aircraft engines also helped to protect spark plugs from moisture and promote easier starting. After World War II, the uses for silicones multiplied. Silicone rubber appeared in many products from caulking compounds to wire insulation to breast implants for cosmetic surgery. Silicone rubber boot soles also walked on the moon, where ordinary rubber would have failed. Ordinary materials failed in the space shuttle *Challenger* disaster in 1986; it is possible to speculate that if silicone rubber O-rings had been used, the accident might have been averted. Silicone oils continue to find a multitude of specialized uses. Recording tape is lubricated with silicone oil, and silicone oil aids in preventing tires from sticking in the molds in which they are made. Other uses of silicone oils include hydraulic fluids and transformer oils, waterproofing of paper and cloth, surface films for automobile finishes and razor blades, and foam reduction in paper mills. Silicone resins were originally developed to serve as binders for glass fibers to produce a composite material for electrical insulation. Additional uses now include heat-resistant coatings for mufflers and catalytic converters on automobiles, and electrically insulating varnish. Foamed silicone resins have been used for heat insulation and as components of sandwich-type composite materials.

The discovery of silicone cannot be isolated in time to a single day or even a single year. Moreover, silicones as we now know them owe much to years of patient developmental work in industrial laboratories. Basic research, such as that conducted by Kipping, Stock, and others, served to point the way and catalyzed the process of commercialization.

Bibliography

Challenger, F. "Frederic Stanley Kipping." In *Great Chemists*, edited by Eduard Farber. New York: Interscience, 1961. A biography by one of Kipping's collaborators.

Chedd, Graham. *Half-Way Elements: The Technology of Metalloids*. Garden City, N.Y.: Doubleday, 1969. This popularized account discusses silicone rubber. Includes color photographs of surgical implantation of silicone breast enhancements and discusses other medical uses of silicones.

Hyde, James F. "Chemical Background of the Silicones." *Science* 147 (1965): 829-836. This short article summarizes some of the chemistry involved in making silicones and focuses on the achievements of Dow-Corning researchers.

Kipping, F. S. "Organic Derivatives of Silicon." *Proceedings of the Royal Society of London* A159 (1937): 139-148. Kipping summarizes his work on silicon and places it in context with that of other workers. Many historical accounts draw on this lecture. Kipping ends on a note of pessimism, stating: ". . . the prospect of any immediate and important advance in this section of organic chemistry does not seem to be very hopeful."

Leyson, Burr W. *Marvels of Industrial Science*. New York: E. P. Dutton, 1955. Leyson, an army officer, was impressed by the uses of silicones in the military. Waterproofing of communication equipment and rubber for low-temperature applications is discussed.

Liebhafsky, H. A. *Silicones Under the Monogram*. New York: John Wiley & Sons, 1978. This rambling and entertaining book is enriched with many photographs of individual scientists, laboratories, and samples of silicones. There is extensive discussion of the commercial development of silicones and the patent litigation relating to it. The annotated bibliography is extensive and varied and could serve as an entry into the study of twentieth century industrial chemistry research.

McGregor, Rob Roy. *Silicones and Their Uses*. New York: McGraw-Hill, 1954. McGregor prepared this monograph after a suggestion from Dow-Corning. Physiological response to silicones, applications of silicones in various industries, and the chemistry of silicone preparation are discussed in simple language. Many of the individual products mentioned are produced by Dow-Corning.

Mueller, Richard. "One Hundred Years of Organosilicon Chemistry." *Journal of Chemical Education* 42 (1965): 41-47. Translated from the German by E. G. Rochow. This article originally appeared in a German periodical and takes the form of a lecture held March 27, 1963, in Dresden, Germany. Includes photographs of organosilicon pioneers and contains fifty-six references to their work.

Rochow, Eugene G. *An Introduction to the Chemistry of the Silicones*. 2d ed. New

York: John Wiley & Sons, 1951. This book appeared before any other on the subject, and was translated into five languages. This edition contains a complete listing of Kipping's publications on silicon chemistry as well as a wealth of information on the properties and uses of organosilicon compounds.

___________. *Silicon and Silicones*. New York: Springer-Verlag, 1987. Rochow tells the story of silicon for the general reader. The history of the silicones is told, along with several personal vignettes, including how Rochow treated his own house with silicone coatings to preserve the wooden siding. Silicones are discussed within the wider context of silicon chemistry in general.

John R. Phillips

Cross-References

Carothers Patents Nylon (1935), p. 1055; The Microprocessor "Computer on a Chip" Is Introduced (1971), p. 1938.

MARCONI RECEIVES THE FIRST TRANSATLANTIC TELEGRAPHIC RADIO TRANSMISSION

Category of event: Applied science
Time: December 12, 1901
Locale: Poldhu, England, and St. John's, Newfoundland, Canada

Marconi received the first transatlantic telegraph signal sent without a cable, demonstrating that long-distance electronic communication through open space was a reality

Principal personages:

GUGLIELMO MARCONI (1874-1937), an Italian scientist and inventor of a system of wireless telegraphy who was a cowinner of the 1909 Nobel Prize in Physics

JAMES CLERK MAXWELL (1831-1879), a Scottish scientist who pointed out that electricity and magnetism were really an instance of electromagnetic radiation

SAMUEL F. B. MORSE (1791-1872), an American portrait painter and scientist who invented the Morse code

JOSEPH HENRY (1797-1878), an American physicist who invented an electric relay, a telegraph, but did not patent it

HEINRICH HERTZ (1857-1894), a German physicist who was interested in the wavelength properties of electromagnetic radiation

Summary of Event

On December 12, 1901, Guglielmo Marconi was in St. John's, Newfoundland, Canada, to receive a Morse code signal to be transmitted to him from Poldhu, Cornwall, England, a distance of 3,440 kilometers across the Atlantic Ocean. Humankind was about to enter the era of worldwide electronic communication.

The principles of electric telegraphy had been discovered in the nineteenth century, and for years individuals worked to send the electric signal further and further through various wire conductors. According to David A. Hounshell of the Smithsonian Institution's National Museum of History and Technology: "The word telegraph originally identified a visual, manually operated signaling system, or semaphore, used to communicate information rapidly over a large distance." In the nineteenth century, however, as research into electricity progressed, the term was associated with signals sent and received over wires and then without wires.

Samuel F. B. Morse was an American artist and inventor who had sailed to Europe in 1832 to study art. On the return voyage, a fellow passenger and American, the chemist Charles T. Jackson, introduced Morse to the principles of electromagnetism, the basis for the telegraph. In 1835, Morse created a transmission device and a code, now known as the Morse code (a series of dots and dashes representing agreed-upon alphanumeric symbols), to be transmitted as an electric signal through a wire and received at a distant location. On May 25, 1844, Morse demonstrated the principles of telegraphy by transmitting and receiving over an electric telegraph line (64 kilometers long) from Washington, D.C. to Baltimore, Maryland. From Washington, D.C., Morse transmitted the message in code, "What hath God wrought?"

Morse was not alone in his work in telegraphy; he borrowed ideas from another American, the physicist and inventor Joseph Henry who, in 1835, had invented an electrical relay that was the forerunner of Morse's telegraph. Unfortunately, Henry did not patent his electrical relay, and Morse's name is the common one heard today.

The application of telegraphic principles proceeded on both sides of the Atlantic. In 1837, the Great Western Railway in England used an early telegraphic signal (with wires strung adjacent to railroad lines) to indicate train speeds. In 1852, an agreement was made between Germany and France allowing telegraph wires to cross their borders for messages.

Morse and Henry, as well as the Scottish mathematician and physicist James Clerk Maxwell and the German physicist Heinrich Hertz, all made their contributions to the achievement of Marconi. No researcher works in a vacuum and everyone builds on and borrows from earlier and contemporary works. Maxwell pointed out that electricity and magnetism were really an instance of a single form of electromagnetic radiation, and he demonstrated that what is called "light" results from electromagnetic vibrations of a certain wavelength. What followed from this mid-nineteenth century discovery was that electromagnetic waves other than light waves could be propagated or transmitted through space. (In 1968, the International Astronomical Union adopted the figure that the speed of light was 299,792.5 kilometers per second.)

As electric energy is transmitted through a wire, it does not travel at the speed of light but travels at speeds determined by the properties of the conducting medium and associated equipment. Materials such as copper and silver make excellent conductors, but there is still resistance to the transmission of the electric energy, dictated by the laws of physics. If signals could be transmitted through space, without wires, the signal would be sent at the speed of light; this is what Marconi accomplished across the Atlantic Ocean on December 12, 1901.

After the initial successes of telegraphy by wire, individuals began to conceive of wires all around the globe. In 1850, a well-insulated copper wire was laid between France and England in the English Channel to carry telegraphic signals and in 1854 individuals began to consider laying a wire cable across the Atlantic Ocean. As James R. Chiles stated: "The first Atlantic cable, which took three attempts in 1857 and 1858 to lay, consumed 367,000 miles [590,503 kilometers] of iron wire and 300,000 miles [482,700 kilometers] of tarred hemp." On August 13, 1858, President James Buchanan of the United States and Queen Victoria of England exchanged telegraphic messages via this first Atlantic cable. It took sixteen and one-half hours for Queen Victoria's ninety word message to cross the Atlantic. The cable of 1858, however, only worked for one month and in the 1860's, two additional cables were laid across the Atlantic Ocean. In 1866, there were two functioning cables connecting North America and England and by the end of the nineteenth century, more than ninety million telegrams a year were being transmitted across the Atlantic Ocean. In 1894, Marconi became familiar with Hertz's work on generating electromagnetic waves through space by using a transmitter. Marconi conceived of the idea of using these waves through space as a transmission signal or a form of wireless telegraphy: The signals would go through space at the speed of light and not be impeded by the resistance of any wire conductor. While in Italy, Marconi worked on his transmission and receiving equipment. After he received little support to continue his work on wireless telegraphy in Italy, Marconi was persuaded to go to England to pursue his work. Marconi thought that the transmission of signals through space to be received by ships at sea would be of importance to a maritime nation. Since England was the greatest maritime power, he went to England and on June 2, 1896, he applied for, and received, the first patent for wireless telegraphy in the world.

From 1896 to 1901, Marconi continued to experiment in transmitting and receiving wireless telegraph signals for greater distances, utilizing the code developed by Morse. In 1894, Marconi had succeeded in sending and receiving a wireless telegraph transmission 2.4 kilometers. He gradually perfected his techniques and was soon sending and receiving wireless signals 6.4 kilometers, 14.5 kilometers, 19 kilometers, 50 kilometers, 121 kilometers, and finally in January, 1901, he sent and received a signal on the south coast of England 299 kilometers.

Individuals expressed great interest in the ability to convey messages even greater distances, and Marconi attempted his transatlantic transmission at the end of 1901. Success was not guaranteed, and many thought it was impossible. At the transmission site in England, Marconi erected a transmission antenna 48 meters tall, consisting of fifty copper wires suspended between two towers 60 meters apart. In Canada, during a winter gale, Marconi sent aloft a kite which had a trailing antenna that was 152 meters in length. He received the letter "S" on December 12, 1901, when the signal was sent from Poldhu, Cornwall, England to St. John's, Newfoundland, Canada, a distance of 3,440 kilometers.

Impact of Event

When Marconi received the letter "S," the world was forever changed, since the transmission of information was no longer limited to a distinct physical medium and could now be transmitted through space at the speed of light. Even though Marconi achieved success on December 12, 1901, he was challenged immediately by individuals and corporations who claimed that his transmission from Poldhu to Newfoundland was an impossible achievement.

In February, 1902, Marconi replicated his December test by a series of transmission-reception tests between the origination station in Poldhu, and the S.S. *Philadelphia*, which was 3,232 kilometers away in the Atlantic Ocean, but there were still those who did not think highly of his achievements. Marconi's younger daughter wrote in 1989 that her father's "scientific work has not been without criticism" and that the point which his critics seemed to have in common was that "Marconi, rather than an inventor of new devices, achieved his major successes by incorporating components already invented by others."

Critics and disbelievers notwithstanding, Marconi continued with his experimentations and transmissions and by the end of 1902, the first official messages, and not test transmissions, were being sent across the Atlantic Ocean. By the early part of 1903, newspaper stories from New York City were being sent for publication in *The Times* of London by means of Marconi's telegraphy, which was undeniably built on the work of Henry, Morse, Maxwell, Hertz, and other individuals.

A Nobel Prize, however, is not given to a committee and in 1909, Marconi and Karl Ferdinand Braun shared the Nobel Prize in Physics. Braun had introduced the first cathode ray tube in 1897, which earned for him the shared 1909 Nobel Prize in Physics. As Marconi stated in his 1909 Nobel Prize Physics Lecture: "The results obtained from these tests, which at the time constituted a record distance, seemed to indicate that electric waves produced in the manner I had adopted would most probably be able to make their way around the curvature of the Earth, and that therefore even at great distances, such as those dividing America from Europe, the factor of the Earth's curvature would not constitute an insurmountable barrier to the extension of telegraphy through space."

Marconi's telegraphic achievements coincided with other achievements occurring in electromagnetics. Sir John Ambrose Fleming became a scientific adviser to the company that Marconi founded, and in 1904 Fleming developed a "valve" that could control the flow of electrons in a tube. In 1906, Reginald Aubrey Fessenden invented a system to modulate electromagnetic radio waves that could be transmitted as a form of wireless telegraphy and radio was thus invented. In 1908, A. Swinton published a brief letter in *Nature* entitled "Distant Electric Vision," and television as a form of "wireless telegraphy" came about. The first public demonstrations of television occurred in England in 1926 and in the United States in 1927.

Marconi's contribution evolved into a global telecommunications system, allowing virtually instant access to anyone in the world, providing one has the appropriate technology. Fiber optics now substitute for copper wires and signals are transmitted through space at the speed of light to geosynchronous communications satellites orbiting the equator at 36,000 kilometers. Although not diminished in size since Marconi, the world has definitely "shrunk" in the manner in which information can be exchanged and shared since that first transatlantic telegraphic transmission on December 12, 1901.

Bibliography

Baker, W. J. *A History of the Marconi Company.* New York: St. Martin's Press, 1971. An excellent book about Marconi and the company he created; it also places telecommunications activities into the context of the times, from the beginning of the twentieth century until the 1960's.

Braga, Gioia Marconi. "Marconi and Instant Global Satellite Communications." In *Space Thirty: A Thirty Year Overview of Space Applications and Explorations*, edited by Joseph N. Pelton. Alexandria, Va.: Society of Satellite Professionals, 1989. A short summary of Marconi by his younger daughter, in a volume that looks at global communications today.

Canby, Thomas Y. "Satellites That Serve Us." *National Geographic* 164 (September, 1983): 281-299, 308-334. For an easily accessible item, an excellent overview into satellites, providing information from Arthur C. Clarke to INTELSAT.

Chiles, John R. "The Cable Under the Sea." *American Heritage of Invention and Technology* 15 (Fall, 1987): 34-41. This is an excellent journal on invention and technology. Chiles' article on the copper cables of the nineteenth century concludes with information on contemporary fiber-optic cables across the Atlantic Ocean.

Dunlap, Orrin E., Jr. *Communications in Space: From Marconi to Man on the Moon*. New York: Harper & Row, 1970. This easy-to-read publication, by an individual who built his first wireless station in 1912, is an excellent overview of the time from Marconi to space exploration.

Franco, Gaston Lionel, ed. *World Communications: New Horizons/New Power/New Hope*. Navara, Italy: Franco, 1983. This "coffee-table-style" trilingual publication (English, French, and Spanish) provides an outstanding visual presentation of worldwide communications, with information on basic scientific discoveries which contributed to telecommunications activities. Information is also provided on contemporary organizations that regulate worldwide telecommunications policies.

Hounshell, David A. *Telegraph, Telephone, Radio, and Television*. Washington, D.C.: Smithsonian Institution Press, 1977. This booklet is merely indicative of the tremendous amount of information available in the institution that has been affectionately called "the nation's attic." In May of 1990, The National Museum of American History (part of the Smithsonian Institution) opened a new permanent exhibit entitled the "Information Age" covering the times from the work of Morse to twentieth century computers.

Marconi, Degna. *My Father, Marconi*. New York: McGraw-Hill, 1962. For a warm and personal view, by Marconi's oldest daughter, this book cannot be surpassed. She points out that upon the occasion of Marconi's funeral in 1937, international wireless operators throughout the world halted their transmissions for two minutes in honor of Marconi.

Shiers, George, ed. *The Development of Wireless to 1920*. New York: Arno Press, 1977. Twenty articles are reprinted in this excellent volume, including the 1909 Nobel Lectures in Physics by Marconi and Braun. Contains papers from Fleming, de Forest, and Fessenden. The introductory essay on the "prehistory" period of 1876 to 1920 provide the reader with a very good overview of the technical aspects of broadcasting history.

Charles F. Urbanowicz

Cross-References

Fleming Files a Patent for the First Vacuum Tube (1904), p. 255; Fessenden Perfects Radio by Transmitting Music and Voice (1906), p. 361; Transatlantic Radiotelephony Is First Demonstrated (1915), p. 615; The Principles of Shortwave Radio Communication Are Discovered (1919), p. 669; Zworykin Develops an Early Type of Television (1923), p. 751; Armstrong Perfects FM Radio (1930), p. 939; The First Transatlantic Telephone Cable Is Put Into Operation (1956), p. 1502; The First Commercial Test of Fiber Optic Telecommunications Is Conducted (1977), p. 2078.

CARREL DEVELOPS A TECHNIQUE FOR REJOINING SEVERED BLOOD VESSELS

Category of event: Medicine
Time: 1902
Locale: Lyons, France; Chicago, Illinois

Carrel's precise procedures for aseptically rejoining blood vessels increased the range of surgical procedures and contributed to the realization of organ transplantation

Principal personages:

ALEXIS CARREL (1873-1944), a French surgeon and biologist whose development of microsurgical techniques made possible modern transplantation surgery

CHARLES CLAUDE GUTHRIE (1880-1963), an American surgeon who worked with Carrel and later achieved independent fame as a noted surgeon

Summary of Event

As Alexis Carrel noted in 1908, "the idea of replacing diseased organs by sound ones, of putting back an amputated limb or even of grafting a new limb on a patient having undergone an amputation, is doubtless very old." Despite the antiquity of the desire, however, the dream remained unrealized until surgeons could restore circulation through the grafted organs or limbs. Carrel went on to describe a procedure he had developed in 1902 for rejoining severed arteries; without this procedure, modern transplantation surgery would be impossible.

On one level, it seems like an easy operation—simply take the two ends of severed vessels and sew them together. As Faust noted, however, blood is a special material. Although it normally clots quickly when exposed to air, it usually remains unclotted inside a vessel. The ability to stay unclotted depends upon an intact inner membrane of a vein or artery: the tunica intima. If this surface is unhealthy or has foreign bodies on its surface, blood clots may form in the vessel. This condition is called thrombosis and is fatal if the blockage occludes (closes) the entire vessel.

Other surgeons before Carrel had attempted to rejoin severed vessels, but their attempts had usually ended in thrombosis and death. There were two general methods for these operations. Surgeons either sewed together the middle and outer layers of the vessel (the tunica media and tunica externa or adventitia) or inserted a fine magnesium tube into the lumen and joined the vessel together. The success of these operations depended upon strict asepsis and an accurate union of the ends. There were several experimental successes with these methods, but their translation into surgical practice was difficult to accomplish.

According to most accounts, Carrel became interested in this procedure when

Sadi Carnot, the President of France, was assassinated in 1894. The bullet had severed an artery, and the president, despite the ministrations of surgeons, had bled to death. Carrel, who had demonstrated much scientific acumen as a youth, turned his attention to the problem of rejoining vessels. He mastered fine stitchery and asepsis and introduced paraffin-covered threads and needles. The key to his procedure was that he rolled back the ends of the vessels and sewed these ends together; the blood then continued to flow uninterrupted in the artery's smooth inner surface. This step allowed for the resumption of circulation. He finally succeeded in 1902. In that year, he later announced that he had transplanted kidneys in dogs on more than one occasion. Although all the animals died because of septic complications, the kidneys began to produce urine immediately, demonstrating that the joined vessels had, in fact, restored circulation through the organ.

In 1905, along with Charles Claude Guthrie, Carrel transferred segments of a jugular vein into a carotid artery. They discovered that the vein soon underwent structural changes—the caliber of the lumen enlarged and the walls became thicker—to accommodate increased blood pressure. Vein-to-artery grafts often ended in thrombosis, however, so they devised methods to store arterial sections for later use. They discovered that segments chilled almost to freezing and gently warmed were still viable after days or weeks of such treatment.

In the course of these experiments, Carrel soon discovered the limits of grafting or transplantation. He noted that grafting pieces of blood vessels or organs taken from the same animal (he called them "autoplastic donors") succeeded, whereas tissues from other animals ("heteroplastic donors") were soon rejected by the host. The further away the species was genetically, the more rapid the rejection. He also noted that blood from unrelated donors usually was fatal to the recipient.

Carrel used his surgical techniques to attempt even more daring grafts and transplants. He and Guthrie continued to work on kidney transplants after 1905 and achieved much greater success than in his earlier 1902 attempts. They eventually transplanted other organs and limbs. Although the animals died, he and Guthrie demonstrated the technical feasibility of these operations on humans.

Carrel, for all his surgical finesse, never engaged in clinical practice. All his work, done either at the University of Chicago in 1905 to 1906, or later at the Rockefeller Institute, was entirely research. It depended on other surgeons to apply Carrel's work to human subjects, or more likely, to rediscover his work. In 1912, he was awarded the Nobel Prize for Physiology or Medicine for his work on the surgery of blood vessels.

Impact of Event

Alexis Carrel's development of the method of rejoining severed blood vessels was the necessary, but not sufficient, condition for the realization of the ancient dream of organ transplantation. In order for that to occur, the physiology of transplantation had to be understood as well as its technique. Carrel recognized these problems. He had noted, for example, that blood from nonrelated donors usually was fatal for the recipient. Also, rejection phenomena had demonstrated that the underlying issues of transplants were biological, not mechanical. Once these physiological issues were settled, Carrel's techniques entered clinical surgery.

Carrel's work had to be complemented by many other discoveries before its impact on surgery could be fully and easily realized. Karl Landsteiner's discovery of blood types solved the puzzle of why some transfused blood, but not all, was toxic to the recipient. Several scientists during the interwar years worked to devise reliable methods for the storage of blood. Peter Medawar's insights into immunity explained heteroplastic rejection.

Many of the techniques that Carrel developed in the early years of the twentieth century were not applied clinically until after World War II. The first documented cases of artery storage and transplantation occurred in the United States in 1948 and in France in 1952. Although the development of fabric grafts has rendered otiose many questions of vessel storage, the procedures would be impossible without Carrel's technique. His discovery that veins could be grafted successfully to arteries is the foundation of much of bypass surgery of the 1990's.

Carrel and Guthrie's work on kidney and limb transplantations is a direct ancestor of today's procedures. The dazzling virtuosity of these operations depends upon the careful suturing of vessels—procedures first accomplished by Alexis Carrel.

Bibliography

Carrel, Alexis. *Man, the Unknown*. New York: Harper & Bros., 1935. Carrel's view on humanity's future. Some saw it as prophetic, others as unfounded speculation.

__________. *The Voyage to Lourdes*. Translated by Virgilia Peterson. New York: Harper & Bros., 1950. A third-person narrative written by Carrel about his visit to Lourdes in 1903. Documents his earliest interests in the relationship between religion and science.

Haeger, Knut. *The Illustrated History of Surgery.* New York: Bell, 1988. Clearly describes Carrel's achievement in the context of other surgical advances of the early twentieth century, such as Rodolph Matas' strengthening of arterial walls threatened by aneurysm or Friedrich Trendelenburg's "high ligature" operation for varicose veins. Good overall coverage, but some information is inaccurate; therefore, readers are cautioned.

Hardy, James D. "Transplantation of Blood Vessels, Organs, and Limbs." *JAMA* 250 (August 19, 1983): 954-957. *JAMA* reprinted Carrel's 1908 article, "Results of the Transplantation of Blood Vessels, Organs, and Limbs," and Hardy commented on it. He reiterated what earlier critics of Carrel's work had noted, that "the new investigator beginning arterial transplantation research would do well to check first to see if Dr Carrel did not publish the proposed research at the turn of the century."

Majno, Guido. *The Healing Hand: Man and Wound in the Ancient World.* Cambridge, Mass.: Harvard University Press, 1975. A landmark book in the history of surgery that narrates in clinical and surgical detail wound management in antiquity. Describes how ancient surgeons manipulated vessels during surgery or closed wounds after it to avoid what is now known as shock, inflammation, and infection.

Malinin, Theodore I. *Surgery and Life: The Extraordinary Career of Alexis Carrel*. New York: Harcourt Brace Jovanovich, 1979. A popular account of both aspects of Carrel's life: as a pioneer in surgical techniques and as a visionary for a political utopia.

Nuland, Sherwin B. *Doctors: The Biography of Medicine*. New York: Alfred A. Knopf, 1988. A popular history of medicine told from the perspective of its famous practitioners. The chapter on organ transplantation discusses Carrel's early work as a necessary stage for the successes achieved in kidney transplantations in the 1950's.

Wilson, Leonard G. *Medical Revolution in Minnesota: A History of the University of Minnesota Medical School*. St. Paul, Minn.: Midewiwin Press, 1989. Outside of Boston, most of the early clinical research and attempts at transplantation were done at the University of Minnesota. "The Development of Organ Transplantation in Minnesota" documents this work based on Carrel's surgical procedure.

Thomas P. Gariepy

Cross-References

Landsteiner Discovers Human Blood Groups (1900), p. 56; Crile Performs the First Direct Blood Transfusion (1905), p. 275; Harrison Observes the Development of Nerve Fibers in the Laboratory (1907), p. 380; Abel Develops the First Artificial Kidney (1912), p. 512; McLean Discovers the Natural Anticoagulant Heparin (1915), p. 610; Blalock Performs the First "Blue Baby" Operation (1944), p. 1250; Favaloro Develops the Coronary Artery Bypass Operation (1967), p. 1835; Barnard Performs the First Human Heart Transplant (1967), p. 1866; Gruentzig Uses Percutaneous Transluminal Angioplasty, via a Balloon Catheter, to Unclog Diseased Arteries (1977), p. 2088; DeVries Implants the First Jarvik-7 Artificial Heart (1982), p. 2195.

JOHNSON PERFECTS THE PROCESS TO MASS-PRODUCE DISC RECORDINGS

Category of event: Applied science
Time: 1902
Locale: Camden, New Jersey

Johnson developed a method of duplicating recordings made on discs, enabling the mass production of recordings of popular music

Principal personages:

ELDRIDGE R. JOHNSON (1867-1945), an American inventor and industrialist who created the Victor Talking Machine Company

THOMAS ALVA EDISON (1847-1931), the famous inventor whose record number of patents were the basis of modern industrial research, the electric utility, and motion picture and phonograph industries

EMILE BERLINER (1851-1929), a German inventor who played an important part in the development of telephony and recorded sound by inventing a microphone and the disc-playing talking machine

Summary of Event

Although Thomas Alva Edison invented the phonograph in 1877, the end of the century marked little progress for the industry based on recorded sound. Cheap talking machines powered by spring motors were available in the 1890's, but recordings, in both cylinder and disc format, were expensive and few in number. The problem lay in duplicating the recordings made in the studio. The only method available in the nineteenth century was the pantograph system, a mechanical linkage between two recording machines that copied the spirals of sound from one recording to another. This process was time-consuming and produced duplicates of inferior quality. Such was the difficulty in record duplication that many companies paid musicians to make scores of recordings of the same song.

Edison was the first to believe that the phonograph could be employed to bring popular music to the masses. He developed a simple talking machine that could play prerecorded cylinders, yet he realized that there would be no market for recorded sound until there was a means to make inexpensive copies of a recording. Subsequently, he began experimenting with record duplication in 1887. By this time, the fragile tin foil of his first phonograph had been replaced by a more durable wax cylinder.

The hardness of the wax was an important factor in storing the signal of the sound wave in the phonograph cylinder. When sound waves vibrated the thin diaphragm of the phonograph, the recording stylus attached to the diaphragm cut a microscopic groove in the wax cylinder. The wave was cut horizontally, up and down in the groove, and was therefore called a hill and dale cut. The vibrations of the reproducer stylus as it traveled along the groove were transmitted to the diaphragm to reproduce the original sound.

As the groove cut into the cylinder was the positive impression of the sound wave, a cast of it would make a negative that could be used to make numerous positive copies. The problem was to find a method to make good negative casts from the soft wax of the recorded cylinder. After years of experiments, Edison found that the best cast of a recorded cylinder had to be made of metal, for only metal was strong enough to mold wax duplicates. The first step was to deposit a fine layer of gold onto the recorded cylinder. This was achieved by placing the cylinder in a vacuum and spraying it with gold dust. The gold particles settled into the hill and dales of the groove and assumed the shape of the sound waves engraved into the wax. Then, layers of copper or nickel were slowly electroplated on top of the gold, forming a metal negative—a matrix—of the original. After the original cylinder was removed from the matrix, the cylindrical metal matrix was cut into two pieces to make a mold. Blank cylinders were placed inside this mold and a slight increase in temperature was enough to soften the wax to receive the impression of the groove. As the matrix cooled, it was then possible to free the duplicated cylinder from the matrix.

By 1899, experimenters in Edison's laboratory had succeeded in making duplicates from metal matrices, but it took several more years to perfect the process for commercial use. The metal matrices left a slightly raised line where the two parts of the mold met. A further problem was the poor sound quality of the duplicates, a result of the slight distortions in the grooves of the recorded cylinder caused by heating and cooling. As Edison was determined not to sacrifice any sound quality in record duplication, research work continued.

In the meantime, other inventors succeeded in making acceptable duplicates from disc records. The disc-playing machine—the gramophone—had been invented by Emile Berliner in 1887. Berliner used a lateral cut to inscribe the signal of the sound wave into the groove, the recording stylus moving from side to side rather than up and down as in the Edison system. Berliner made his masters by recording the sound on a zinc disc covered with fatty film and then using acid to etch the sound waves marked in the film. Unfortunately, the duplicates made by this process had a fuzzy sound because the acid ate away some of the sound wave as it etched the signal into the master. The indistinct reproduction of the gramophone made its commercial future doubtful in the late 1890's.

Eldridge R. Johnson, proprietor of a small machine shop in Camden, New Jersey, advanced the technology of sound reproduction and made Berliner's machine the leading form of recorded sound in the twentieth century. Like many other mechanics of the nineteenth century, Johnson considered himself an inventor, obtaining several valuable patents. After winning a contract to manufacture spring motors for the Berliner gramophone, Johnson made some significant mechanical improvements to the machine. Concerned with the poor sound reproduction of the Berliner gramophone, he started to experiment with the process of duplicating recordings. In 1897, Johnson obtained some test masters from Berliner and carefully examined the microscopic grooves made by the sound waves. His experiments convinced him that the way to make a better recording was to engrave the sound wave into the master rather than etch it as Berliner had done. Instead of the steel needle used by Berliner to record the sound, Johnson designed a miniature cutting tool much like those used on the lathes in his shop. A deeper cut meant more distinct sound reproduction and a more resilient impression to endure the wear and tear of the duplicating process.

Johnson replaced the zinc disc covered with fatty film with a solid wax disc. Much care was taken to ensure that the surface of the master was perfectly flat. Where others had used ordinary gramophones to record, merely shouting into the amplifying horn to vibrate the recording stylus, Johnson designed a special recording machine with a rubber hose connecting the recording assembly with the horn used to pick up the sound. A carefully governed electric motor maintained an exact speed of the revolving disc, ensuring that the recording was made at an even pitch; even the slightest change in the speed of the disc affected the pitch of the reproduced sound.

Once an acceptable master recording had been made, Johnson faced the problem of making a matrix from the wax master—the same problem that Edison had confronted earlier. After more than two years of experiments Johnson came up with much the same solution. His masters were first covered with a fine layer of metal dust and then more metal was electroplated onto the primary layer to make the negative matrix. The matrix was then used to stamp the duplicate onto soft wax blanks.

The resulting duplicates had none of the surface noise that spoiled the Berliner discs and a much louder reproduction than an Edison cylinder. Johnson's achievement was to transfer the technique that Edison had perfected for the cylinder to the disc. He borrowed from Edison's prior experiments and used Edison equipment and wax in his duplicating process.

By 1900, Johnson had edged ahead of Edison in the race to mass-produce sound recordings. Johnson quickly formed the Victor Talking Machine Company in 1901 on his patents and those of Berliner. The company began the manufacture of a line of gramophones and the mass duplication of recorded discs. The advantage of using the disc format was that it was much easier and cheaper to stamp out duplicates than to form them in cylindrical molds. Johnson found it easy to set up the equipment to duplicate thousands of discs a day. Within a short time, Johnson designed machines to stamp out the duplicates automatically. By 1902, the Victor Company had successfully established a large plant to duplicate recordings. It produced more than one million duplicates of recorded discs in that year.

Impact of Event

The experiments carried out by Edison and Johnson produced a significant advance in the understanding of sound waves and the various elements in the spectrum of sound. Yet, Johnson's innovation marked no advance in the theory of sound or its reproduction; like Edison, his approach was strictly empirical. Although both men used the scientific method in their research and took advantage of the latest equipment, especially microscopes and measuring instruments, their experiments were directed at commercial goals. They did not publish the results of their research and Johnson did not patent his duplicating process, preferring to keep it a trade secret. The effects of his innovation are to be found in Western society and culture: He opened the way for the masses to enjoy recorded sound.

The most immediate effect of Johnson's innovation was that the trickle of recorded sound in the nineteenth century became a mighty torrent in the twentieth. Once Edison had perfected his duplicating process in 1903, popular music was available in both disc and cylinder formats and at a price that most could afford. Recordings that had once sold for over a dollar each could now be purchased for 50 cents, and the price continued to drop in the years leading up to World War I, when records could be purchased for 25 cents. The drop in price made this technology more accessible. A talking machine, with its library of recordings, became as commonplace as the electric light and telephone.

Both Edison and Johnson were concerned with the quality of recorded sound, for they deemed this to be the critical part of the public's acceptance of the new technology. The excellent sound reproduction of Johnson's duplicates enlarged the scope of recorded sound. Now, classical music made by great artists could be captured on wax. Recorded sound was no longer the preserve of vaudeville artists and minstrel acts. In 1903, the great tenor Enrico Caruso made a series of historic recordings for Victor. Music lovers who had previously spurned the gramophone now rushed to buy one. Caruso was the first performer to sell a million recordings, making a fortune for himself and the Victor Company in the process.

The disc duplicating process was quickly diffused to Europe through Victor's subsidiary companies. It was copied by his competitors but not significantly improved. Johnson's process was used throughout the industry until electric recording and new record compounds in the late 1920's brought changes in the way sound was recorded and records duplicated. Yet, the basic method of coating the master with a thin layer of metal and then electroplating it to form a matrix has remained unchanged. The matrix is still used to stamp out thousands of duplicates. Since 1902, the technology of recorded sound has advanced continuously, providing microgrooved, long-playing records made of vinyl. Yet, the method of duplicating is essentially the same one that Johnson devised in 1902.

Bibliography

Fagan, Ted, and William R. Moran, comps. *The Encyclopedic Discography of Victor Recordings*. Westport, Conn.: Greenwood, 1983. This reference work contains a reproduction of B. L. Aldridge's history of the Victor Company.

Johnson, E. R. Fenimore. *His Master's Voice Was Eldridge Reeves Johnson*. Milford, Del.: State Media, 1974. Written by Johnson's son, this account draws heavily on primary sources and personal recollection. Although less than critical of Johnson, the accounts of his experimental and business activities are detailed and accurate. This volume is the only full-length account of Johnson's career in print.

Josephson, Matthew. *Edison: A Biography.* New York: McGraw-Hill, 1959. A classic account of the life and work of Thomas Edison. It is the best of the many biographies of Edison, avoiding the excessive praise and harsh criticism that has marred other books. Gives a clear account of the duplication experiments.

Millard, Andre. *Edison and the Business of Innovation*. Baltimore: The Johns Hopkins University Press, 1990. This study of Edison's business activities is centered on the phonograph industry and Edison's struggle with Johnson's Victor Company. Compares the achievements of both men.

Read, Oliver, and Walter L. Welch. *From Tin Foil to Stereo*. 2d ed. Indianapolis: H. W. Sams, 1976. Profusely illustrated and written in an entertaining style, this book covers the history of the talking machine from Edison's invention to the coming of stereo in the 1960's. Although criticized by purists and collectors for errors of fact, this volume remains the best history of the talking machine available.

Vanderbilt, Byron. *Thomas Edison, Chemist*. Washington, D.C.: American Chemical Society, 1971. One of the best books written about work in Edison's laboratory, this volume is much more than an assessment of Edison's chemical experiments. The account of the duplication experiments is insightful and detailed.

Yorke, Dane. "The Rise and Fall of the Phonograph." *The American Mercury* 27 (September, 1932): 1-12. An amusing and well-written account of the early years of the phonograph business as seen by a contemporary. This article provides the general story of this technology in condensed form.

Andre Millard

Cross-References

The Cassette for Recording and Playing Back Sound Is Introduced (1963), p. 1746; Compact Disc Players Are Introduced (1982), p. 2200.

LEVI RECOGNIZES THE AXIOM OF CHOICE IN SET THEORY

Category of event: Mathematics
Time: 1902
Locale: University of Turin, Italy

Levi acknowledged and criticized the axiom of choice in set theory and attempted to justify and carry out infinite choices

Principal personages:
BEPPO LEVI (1875-1961), an Italian mathematician
GIUSEPPE PEANO (1858-1932), an Italian mathematician and logician
FELIX BERNSTEIN (1878-1956), a German mathematician
GEORG CANTOR (1845-1918), a German mathematician

Summary of Event

The notion of infinity has been a perennial problem in both mathematics and philosophy. A major question in (meta) mathematics has been whether and in what form infinities should be admitted. The debate of Gottfried Wilhelm Leibniz and Sir Isaac Newton on the use of the "infinitesimal" in seventeenth century differential calculus is only one of many possible examples of this problem. In most pre-twentieth century mathematics, infinities occur only in what might be called a "potential" (inactual, implicit) form. Potential infinity can be illustrated readily by considering a simple example from limit theory. There is no question of an actually infinitely great or infinitesimally small quantity to be computed or otherwise arrived at. As noted set theoretician Abraham A. Fraenkel (*Set Theory and Logic*, 1966) has remarked, many nineteenth century mathematicians such as Carl Friedrich Gauss and Augustin-Louis Cauchy considered infinity in mathematics to be largely a conventional expression showing the limits of ordinary language when pressed into trying to express pure mathematical concepts.

In addition to the less problematic notion of potential infinity, an equally old problem of "actual" infinity had long been faced in speculative philosophy and theology, including Saint Augustine, Saint Thomas Aquinas, René Descartes, and Immanuel Kant, among many others. A major thematic front throughout nineteenth century mathematics and logic was precise clarification of the boundary and relations between actual and potential infinities, in foundational as well as applied mathematics. One of the earliest such efforts was that of Bernhard Bolzano, whose *Paradoxien des Unendlichen* (1851; *Paradoxes of the Infinite*, 1950) cataloged a large number of extant and novel conundrums, as well as unclear and unusual properties of actual infinities in mathematics. A particular emphasis, to recur repeatedly in later set theory and logic, was the apparent paradox of the equivalence of an infinite set to a proper part or subset of itself, implying different levels or kinds of infinity where previously only one infinity was supposed. The term "set" was first employed in Bolzano's text as an important mathematical concept.

The first major developments in forming a consistent and comprehensive theory of actual infinities in mathematics were primarily the work of Georg Cantor. Between 1873 and 1899, Cantor sought to lay systematically the foundations for a new branch of mathematics—called the theory of aggregates or sets—which would not only formalize the mathematically acceptable concepts of infinity but also serve as the foundation for every other mathematical theory and discipline employing infinities. Cantor's set theory did not begin with philosophical speculations about infinity, but with the problem of actually distinguishing finitely and infinitely many "specified" points, for example, the points of discontinuity in the theory of functions. Yet, despite many strenuous efforts, Cantor, as well as Richard Dedekind, Giuseppe Peano, and many other mathematicians, blurred or passed over unawares the distinction of choices implicitly or explicitly made by some kind of a rule or algorithm, from which a given denumerable or indenumerable (uncountable) point set was defined. In particular, although the first n members of a subset were seen clearly as selected specifically by a selection- or generative-rule, neither Cantor nor anyone else explained how such a rule could actually be extended to define likewise the entire (actually infinite) subset. This problem was futher complicated by the fact that many mathematicians of the era frequently made an infinity of arbitrary selections, not only independently but also where each given choice depended on the choices previously made. These questions not only undercut the proof method of using "one to one" mappings or correspondences between different sets but also underscored the then-formulating foundational questions of whether "in the last analysis" mathematical entities such as sets are discovered or defined/created.

In 1882, particularly, Cantor argued to Dedekind and others that his means of extending the (infinite) sequence of positive numbers by introducing symbols for infinity, such as ∞, $\infty+1$, $\infty+2$, and the like, was not merely conventional but a legitimate number choice. Perhaps the simplest example of Cantor's transfinite numbers is the model suggested by Zeno's paradoxes of motion. Here, a runner uniformly traverses a road divided into intervals. Although the number of intervals is ∞, the time taken to traverse them is finite (hence, the paradox). If the first interval is designated the $\omega-$th interval, the subsequent intervals will be the $\omega+1-$th, $\omega+2-$th, and so on. These numbers ω, $\omega+1$, $\omega+2$, and so on, are the transfinite ordinal numbers first designated by Cantor. Also called into question was Cantor's related continuum hypothesis, which asserts that every infinite subset of the real numbers either is denumerable (countable via a finite procedure) or has the degree of infinity of the continuum. In Cantor's sense, the continuum is defined by the assumption that, for every transfinite number t, 2^t is the next highest number. Equivalently, the continuum hypothesis proposes that there is no (transfinite) cardinal number between the cardinal of the set of positive integers and the cardinal of the set of real numbers.

These developments, and other related developments, left early twentieth century set theory with a network of fundamental, interlinked, and unsolved problems, all of which somehow involved the notions of selection or choice, linking well-known finite mathematics with the newer mathematics of actual infinities. Nevertheless, very few, if any, mathematicians explicitly considered all these questions from this viewpoint. Peano, an Italian mathematician responsible for the first symbolization of the natural number system of arithmetic, was also coming up against similar inconsistencies in his investigations of the conditions for existence and continuity of implicit functions. In 1892, one of Peano's colleagues, Rodolfo Bettazzi, investigated conditions under which a limit-point was also the limit point of a sequence. Bettazzi underscored the same underlying issue as Peano. The third mathematician to consider this problem of how to justify or actually carry out infinitely many choices was Peano and Bettazzi's colleague, Beppo Levi. Levi was completing his dissertation research, inspired by the set theory work of French mathematician René Baire, who had developed the novel notion of "semi-continuity" for point sets. In 1900, Levi published a paper extending Baire's investigations of fundamental properties satisfied by every real function on any subset of the real numbers. Without proof, Levi proposed that every subset 1 is equal to the union of subsets 2 and 3 minus subset 4, where 2 is any closed set, and 3 and 4 are "nowhere-dense" sets. Levi also asserted that every uncountable subset of the real numbers has the power of the continuum, essentially Cantor's continuum hypothesis.

In another dissertation of 1901, a student of Cantor and David Hilbert, Felix Bernstein, sought to establish that the set of all closed subsets of the real number system has the power of the continuum. In 1897, Bernstein had given the first proof of what is known as the equivalence theorem for sets: If each of two sets is equivalent to a subset of the other, then both sets are mutually equivalent. In his 1901 work, Bernstein remarked that Levi's 1900 results were mistaken. As a response, in 1902, Levi published a careful analysis of Bernstein's dissertation, in which his use of choices in defining sets came into sharp and explicit critical focus.

In Levi's broad analysis of then-extant set theory, he questioned Cantor's assertion that any set can be well-ordered. Well-orderedness is the property whereby a set can be put systematically into a one-to-one correspondence with elements of another set. Levi pointed out that even though Bernstein had openly abandoned the well-ordering principle, Bernstein had, nevertheless, employed an assumption that it appeared to be derived essentially from the same postulate of well-orderedness. This questionable assumption was Bernstein's so-called partition principle; that is, if a set R is divided or partitioned into a family of disjoint (nonintersecting) nonempty sets S, then S is less than or equal to R. Following Levi's proof that Bernstein's partition principle was valid only whenever R was finite, Levi critically remarked that the example was applicable without change to any other case where all the elements s of S are well-ordered, or where a unique element in each s can be distinguished. This statement is, essentially, a summary of what would be explicitly termed the axiom of choice.

Impact of Event

Although proving a catalyst for subsequent work by Bernstein, Hilbert, Ernst Zermelo, and many other mathematicians, the direct response to and recognition of Levi's 1902 paper was very limited. Basically, although Levi explicitly recognized the axiom of choice embodied in the work of Cantor and Bernstein, he rejected its use in its present form. This led ultimately to Zermelo's 1904 publication, proving the well-ordering principle by use of the axiom of choice.

In his 1910 seminal paper on field theory foundations in algebra, Ernst Steinitz summarized the widespread attitude toward the axiom of choice in algebra, topology, and the like. Thus, although explicit examination of the axiom of choice was largely ignored for some time, from 1916 on, the Polish mathematician Vaclaw Sierpinski issued many studies of implicit as well as open applications of the axiom of choice. Although Levi in 1918 offered what he called a "quasi-constructivist" improved alternative to the axiom, his effort was considered too limited and unwieldy by most mathematicians. In 1927, American logician Alonzo Church sought unsuccessfully to derive a logical contradiction from the axiom. No alternative was developed until 1962, when two Polish mathematicians, J. Mycielski and H. Steinhaus, proposed their axiom of determinateness. Since then, it has been shown that a number of other propositions are equivalent to varyingly weaker or stronger forms of the axiom of choice, as originally recognized by Levi and positively employed as such by Zermelo.

Bibliography

Dauben, J. W. *Georg Cantor: His Mathematics and Philosophy of the Infinite*. Cambridge, Mass.: Harvard University Press, 1979. A good rendering of Cantor's original set theory papers. Undergraduate-level discussion.

Fraenkel, Abraham A. *Set Theory and Logic*. Reading, Mass.: Addison-Wesley, 1966. Intermediate level. Includes math-background material to the debate arising about the axiom of choice, as well as examples of its contemporary use.

Halmos, Paul R. *Naive Set Theory*. Princeton, N.J.: D. Van Nostrand, 1960. A general reader introduction to Cantorian and contemporary intuitive, nonaxiomatized set theory.

Rubin, Herman, and Jean E. Rubin. *Equivalents of the Axiom of Choice*. Amsterdam: North-Holland, 1985. Although technical, it includes accessible discussions of most of the subsequent theorems considered equivalent to the Levi-Zermelo axiom of choice.

Suppes, Patrick. *Axiomatic Set Theory*. Princeton, N.J.: D. Van Nostrand, 1960. Presents the basic principles of more rigorous set theory, in terms of the Zermelo-Fraenkel and Newmann-Bernays-Gödel axioms.

Van Heijenoort, Jean, ed. *From Frege to Gödel: A Source Book in Mathematical Logic, 1879-1931*. Cambridge, Mass.: Harvard University Press, 1967. Includes many English translations of works of Peano, Russell, Zermelo, Hilbert, and others relevant to set theory. A good introduction.

Gerardo G. Tango

Cross-References

Hilbert Develops a Model for Euclidean Geometry in Arithmetic (1898), p. 31; Lebesgue Develops a New Integration Theory (1899), p. 36; Brouwer Develops Intuitionist Foundations of Mathematics (1904), p. 228; Zermelo Undertakes the First Comprehensive Axiomatization of Set Theory (1904), p. 233; The Study of Mathematical Fields by Steinitz Inaugurates Modern Abstract Algebra (1909), p. 438; Russell and Whitehead's *Principia Mathematica* Develops the Logistic Movement in Mathematics (1910), p. 465; Cohen Shows That Cantor's Continuum Hypothesis Is Independent of the Axioms of Set Theory (1963), p. 1751; Mandelbrot Develops Non-Euclidean Fractal Measures (1967), p. 1845.

McCLUNG PLAYS A ROLE IN THE DISCOVERY OF THE SEX CHROMOSOME

Category of event: Biology
Time: 1902
Locale: Chicago, Illinois; Lawrence, Kansas

McClung suggested that the "accessory" or "X" chromosome possibly determined sex

Principal personages:

CLARENCE ERWIN MCCLUNG (1870-1946), an American zoologist who worked in paleontology, technical microscopy and microscopic techniques, and studied chromosomes

WILLIAM MORTON WHEELER (1865-1937), an American entomologist who suggested that McClung study the sperm formation of a long-horned grasshopper

HERMANN HENKING (1858-1942), a German zoologist who contributed to embryology and cytology; he described the movement of a strange body ("X") in the nucleus of the fire wasp *Pyrrhocoris* during sperm formation

NETTIE MARIA STEVENS (1861-1915), an American cytologist who developed a hypothesis of sex determination by chromosomes

EDMUND BEECHER WILSON (1856-1939), an American cytologist who developed a hypothesis of sex determination by chromosomes

WALTER S. SUTTON (1877-1916), an American cytologist whose work in 1902 and 1903 postulated a causal relationship between the chromosomes and heredity

THEODOR BOVERI (1862-1915), a German cytologist who demonstrated that the chromosomes persist when the cell is not actively dividing, providing a basis for the chromosome theory of heredity

GREGOR JOHANN MENDEL (1822-1884), an Austrian monk who performed experiments on garden peas that resulted in four "laws" of heredity

Summary of Event

From ancient times, observers have recognized that although offspring generally resemble their parents, they differ from them in particular characteristics. Most early attempts to explain these similarities and differences were not very successful. In 1865, Gregor Johann Mendel published a classic paper based on the experimental breeding of garden peas, proposing explanations (now known as Mendel's laws) for the inheritance of traits. He referred to "factors" or units of heredity (genes) but did not realize that his "factors" were located on chromosomes in the cell nucleus. Since he published his results in a relatively obscure journal, they went unnoticed until 1900. Meanwhile, cytologists (scientists who study cells) were taking advantage of improved instruments and staining techniques to observe the behavior of the chromosomes (darkly staining bodies in the cell nucleus) during cell division. These early cytologists, however, did not recognize that the behavior of the chromosomes was related to heredity. Experimental breeding and cytology appeared to be separate and unrelated. During the first decade of the twentieth century, the two apparently unconnected strands began to converge. By 1900, when three investigators rediscovered Mendel's paper, it was possible to make such a synthesis, for cytologists had collected a considerable amount of information regarding the behavior of the chromosomes during body cell division (mitosis) and during the maturation of the egg and sperm cells (meiosis). They knew that chromosomes replicate before each cell division and that each daughter cell receives only one representative of each replicated chromosome found in the parent cell. This division results in two identical daughter cells.

Investigators also observed that a second type of cell division occurs when gametes (eggs or sperm) are formed during meiosis. Instead of producing daughter cells identical with the parent, the number of chromosomes is reduced by half (haploid). When the egg and sperm fuse during fertilization, the set of chromosomes contributed by the egg match with the set contributed by the sperm, restoring the double (diploid) number. The chromosomes of the offspring, then, are derived equally from the egg and sperm, assuring that each diploid pair of chromosomes in the offspring is composed of a maternal and a paternal representative. Mendel's "factors" seemed to be distributed in the same way as the chromosomes, suggesting that the factors either were the chromosomes themselves or were located on the chromosomes.

Chromosomes as the physical agents of heredity could explain why offspring show characteristics of both parents (because half of the chromosomes are contributed by the male and the other half by the female). As reasonable as such a hypothesis appeared, if it were to be confirmed, a specific trait would have to be traced from the parent's chromosomes to those of the offspring. Although no trait had been traced from parent to offspring by the beginning of the twentieth century, investigators found a likely candidate in sex inheritance. Clarence Erwin McClung played an important role in this investigation. Although McClung had a wide range of biological interests—including microtechnique, vertebrate paleontology, and histology—he became interested in the role of the chromosomes when, as a graduate student at the University of Kansas, he spent the summer of 1898 working with William Morton Wheeler at the University of Chicago. Wheeler suggested that McClung study chromosome behavior in sperm production in a species of grasshopper. This introduction, as well as the semester he spent at Columbia University studying with cytologist Edmund Beecher Wilson, directed his attention to the behavior of the chromosomes.

McClung's studies resulted in a paper entitled "A Peculiar Nuclear Element in the Male Reproductive Cells of Insects" (1899). This "nuclear element" had been described in 1891 by Hermann Henking, who followed the behavior of an unusual structure (he did not recognize it as a chromosome) in sperm formation of the fire wasp *Pyrrhocoris*. This structure, described by Henking as a "peculiar chromatin-element" ("X"), or an atypical "nucleolus," did not appear to pair up with a partner during meiosis as did the other chromosomes. He noted that the male bug had an uneven number of chromosomes and the female an even number. During sperm formation, one-half of the sperm received the "X" body and the other half did not receive it. The half receiving it ended up with twelve chromosomes while the half without it had only eleven chromosomes. Henking, however, did not recognize the significance of the unpaired chromosome. McClung accomplished this step when, working on the formation of sperm in the grasshopper *Xiphidion*, he observed the "X" element that Henking had described. In his 1899 paper, he coined the term "accessory chromosome" for Henking's "X" structure. In papers of 1901 and 1902, McClung postulated that the accessory chromosome was involved in the inheritance of sex.

During the beginning of the twentieth century, biologists were divided on the issue of whether the inheritance of sex was determined by heredity or environment. One popular theory suggested that the fertilized egg existed in a sort of balanced state. Environmental factors such as the temperature or the amount of food determined whether it would develop into a male or female. Other investigators, however, insisted that external conditions did not affect sex ratios, and that the determination of sex occurred at the time of fertilization or shortly thereafter. McClung's observations were based on the second view.

The "accessory" chromosome was present in all cells of the organism and was present in one-half the gametes of males. Therefore, at fertilization one-half the gametes contained the accessory chromosome and the other half lacked it. Since sex was the only characteristic that McClung could think of that would divide a species into two equal groups, he concluded that the accessory chromosome might be a sex determiner. By miscounting the number of chromosomes in the female *Xiphidion*, he erroneously concluded that spermatozoa possessing an accessory (X) chromosome were male determiners. In fact, the opposite situation is correct. A sperm bearing an accessory (X) chromosome is female determining, and a sperm bearing either no accessory or a Y chromosome is male determining.

Although McClung's misunderstanding of the way in which the accessory chromosome acts to determine sex was soon modified, his results did not resolve the controversy over the role of chromosomes in determining sex. Even McClung considered his ideas about the role of the accessory chromosome as a mere working hypothesis, not proof that sex was chromosomally determined. He recognized that his evidence was largely circumstantial, dependent upon the presence or absence of the accessory chromosome in males or females. Given the tremendous variability within the animal kingdom and the scarcity of data on the subject, McClung was unwilling to generalize from his specific evidence to all animals. Even though he considered his evidence in grasshoppers compelling, he was not convinced of its truth for all species. He agreed with many of his contemporaries that the environment might play a role in the determination of sex in some species. He also believed in selective fertilization: that the egg could choose whether it was fertilized by an egg carrying an accessory chromosome or one lacking it. McClung did not have all the answers. His creative speculations, however, based on accurate observations of a specific group of animals, provided his successors with an important tool with which they could establish a basis for the chromosomal theory of sex determination.

Impact of Event

McClung's suggestion that the accessory chromosome in insects functioned as a sex determiner represented a significant stage in the evolution of a chromosomal theory of inheritance. His interpretation helped give purpose to the apparently purposeless movements of chromosomes in general and specifically of the accessory chromosome in sperm formation. Even though his proposal was tentative, he was the first to suggest the relation of a particular chromosome to a particular characteristic (sex). His work not only explained the 1:1 male-female ratio observed in most animals but also, more generally, provided the groundwork for a chromosomal theory of heredity.

His work was acknowledged soon after it was completed by those who incorporated his work into their own research. Since the function of the behavior of the chromosomes was one of the major topics of interest within the early twentieth century biological "establishment," cytologists were intrigued by new ideas. Thus, McClung's interpretation of the function of the accessory chromosome as a sex determiner was of vital interest to investigators such as Walter S. Sutton and Theodor Boveri who were convinced that the chromosomes represented the physical basis for heredity, but lacked the evidence. Sutton, who had been a student of both McClung at Kansas and Wilson at Columbia University, published his paper "The Chromosomes in Heredity," in 1903, immediately after McClung's major interpretative paper in 1902. Sutton demonstrated that the behavior of the paired chromosomes during egg and sperm formation parallel exactly the segregation of the paired Mendelian factors in heredity.

McClung's contribution on the accessory chromosome's importance in the inheritance of sex was important to the understanding of the way in which sex is determined. His interpretations were considered by Nettie Maria Stevens and Wilson, who collected additional data and provided extensive hypotheses of sex determination by chromosomes.

From McClung's cautious interpretation of his data that sex in certain insects was determined by a chromosome, others recognized the need to gather additional data. They soon realized that the situation could become quite complicated. In many groups of animals, males carry only one unpartnered accessory chromosome (designated XO), but in others they have an accessory chromosome, plus an additional element, the "Y" chromosome (designated XY). Females generally have two accessories (designated XX). In some groups of animals, such as birds and some insects, however, the situation is reversed, with males having two accessories (XX) and females one (XO or XY). McClung's creative interpretation of his data provided other investigators with a starting point to unravel a most complex situation.

Bibliography

Allen, Garland E. *Thomas Hunt Morgan: The Man and His Science*. Princeton, N.J.: Princeton University Press, 1978. Although this book is basically concerned with the life and work of geneticist Thomas Hunt Morgan, it provides a valuable context for understanding McClung's contributions. It contains an excellent bibliographic essay.

Dunn, L. C. *A Short History of Genetics: The Development of Some of the Main Lines of Thought, 1864-1939.* New York: McGraw-Hill, 1965. A useful source to provide a context for McClung's ideas. It covers concisely the central ideas in classical genetics from Mendel's paper up to 1939. Includes a glossary and a bibliography of both primary and secondary sources.

Hughes, Arthur. *A History of Cytology.* London: Abelard-Schuman, 1959. One section of this book considers theories of inheritance. It includes a discussion of the history of ideas on sex determination.

McClung, Clarence E. "The Cell Theory: What of the Future?" *American Naturalist* 74 (1939): 47-53. McClung puts his own work and that of his contemporaries into context and speculates on the direction that future developments in cytogenetics might take.

Wenrich, D. H. "Clarence Erwin McClung." *Journal of Morphology* 66 (1940): 635-688. This work, written for McClung's seventieth birthday, represents the most complete biographical source available. It also includes a complete bibliography of McClung's works up to 1940.

Wilson, Edmund B. *The Cell in Development and Heredity.* 3d ed. New York: Macmillan, 1937. The first edition of this book, published in 1896, was the outgrowth of a series of lectures he gave at Columbia University in 1892 to 1893. A classic source, Wilson revised it as new data, and interpretations emerged. Includes a massive literature list.

Marilyn Bailey Ogilvie

Cross-References

De Vries and Associates Discover Mendel's Ignored Studies of Inheritance (1900), p. 61; Sutton States That Chromosomes Are Paired and Could Be Carriers of Hereditary Traits (1902), p. 153; Punnett's *Mendelism* Contains a Diagram for Showing Heredity (1905), p. 270; Bateson and Punnett Observe Gene Linkage (1906), p. 314; Hardy and Weinberg Present a Model of Population Genetics (1908), p. 390; Morgan Develops the Gene-Chromosome Theory (1908), p. 407; Johannsen Coins the Terms "Gene," "Genotype," and "Phenotype" (1909), p. 433; Sturtevant Produces the First Chromosome Map (1911), p. 486; Avery, MacLeod, and McCarty Determine That DNA Carries Hereditary Information (1943), p. 1203.

SUTTON STATES THAT CHROMOSOMES ARE PAIRED AND COULD BE CARRIERS OF HEREDITARY TRAITS

Category of event: Biology
Time: 1902
Locale: Columbia University Medical School, New York City

Sutton determined that Mendel's inherited traits of organisms are physically located on chromosomes

Principal personages:

WALTER S. SUTTON (1877-1916), an American geneticist and surgeon who discovered evidence leading to the chromosome theory of inheritance

THEODOR BOVERI (1862-1915), a German biologist who independently discovered Sutton's chromosome theory of inheritance

GREGOR JOHANN MENDEL (1822-1884), an Austrian monk and scientist whose studies of garden peas led to his discovery of inherited traits (genes) in living organisms

CARL ERICH CORRENS (1864-1933), a German geneticist who, in 1900, proposed that chromosomes carry genes

THOMAS HUNT MORGAN (1866-1945), an American geneticist and 1933 Nobel laureate in Physiology or Medicine who studied mutations in the fruit fly *Drosophila melanogaster*

Summary of Event

Beginning in 1856, an Austrian monk named Gregor Johann Mendel initiated a series of experiments with garden peas that would revolutionize twentieth century biology. Mendel cross-pollinated different lines of garden peas purebred for certain characteristics (purple or white flower color, wrinkled or smooth seeds, and the like). From his extensive experiments, he concluded that each garden pea plant carries two copies of each characteristic trait (flower color, seed texture, and so on). He called these traits genes.

Diploid organisms have two copies of each gene that may or may not be identical. Different forms of the same gene are called alleles. For example, the gene for flower color may have two alleles, one conferring purple flower color and the other conferring white flower color. In most cases, one allele is dominant over other alleles for the same gene. For example, a plant having two purple flower color alleles has purple flowers; a plant having two white flower color alleles has white flowers. A plant having one purple and one white flower color allele, however, has purple flowers; the dominant purple allele masks the white allele. It was later demonstrated that the encoding of proteins by genes is responsible for dominance relationships among alleles.

Since most plants and animals reproduce sexually by the fusion of pollen (or sperm) with egg, Mendel discovered the pattern of transmission of these genetic traits (that is, inheritance) from parents to offspring. While each individual has two copies of every gene, each individual can transmit only one copy of each gene (that is, only one of two alleles) to each of its offspring. One copy of each gene is transmitted from the male parent and one copy from the female parent to each of their offspring. If a parent has two different alleles for a given gene, only one of the two alleles can be passed on to each of its children. There is a fifty-fifty probability of either allele being transmitted for each of thousands of different genes conferring different characteristic traits. The inheritance of either allele operates by chance, like flipping a coin. Yet, it makes each individual of a species unique. Mendel summed up the random chance inheritance of different alleles of a gene in two principles. The first principle—allelic segregation—maintains that the different alleles for a given gene separate from each other during the formation of germ line cells (that is, sperm and egg). The second principle—independent assortment—maintains that different alleles of different genes arrange themselves randomly during germ cell production. For example, a person will inherit two alleles of the gene for eye color and two alleles of the gene for hair color, one of the two alleles for each gene coming from either parent. Each parent can contribute any combination of the alleles for each gene, such as a brown eye color allele and blonde hair color allele, blue eye color allele and blonde hair color allele, and the like, depending upon which alleles each parent possesses. Independent assortment requires that the genes in question lie on different chromosomes. It was by luck that Mendel studied traits located on different chromosomes because he did not know that chromosomes are the physical carriers of genetic traits. When Mendel published his results in the 1866 *Proceedings of the Brünn Society for Natural History*, it was scarcely noticed. Twenty years would pass before the importance of his research was understood.

From 1885 to 1893, the German biologist Theodor Boveri researched the chromosomes of the roundworm *Ascaris*. Chromosomes are molecules composed mostly of deoxyribonucleic acid (DNA) and protein. They are located within the nuclei of the cells of all living organisms. Boveri studied under Pierre-Joseph van Beneden, who discovered that sperm and egg each carry one-half of the chromosomes needed to begin a new organism.

In the late 1890's, Walter S. Sutton studied chromosomes of the grasshopper *Brachyostola magna* under Clarence Erwin McClung, discoverer of the sex-determining X-chromosome. Sutton constructed very detailed diagrams of *Brachyostola* chromosomes during various phases of organismal development, including chromosomal behavior during mitosis (chromosome doubling and separating prior to cell division) and meiosis (chromosome dividing and splitting in sperm and egg production). Both Sutton and Boveri were independently attempting to understand chromosomal structure and function. The breakthrough came in 1900, when the German biologist Carl Erich Correns rediscovered Mendel's work with garden peas. Correns boldly proposed that the chromosomes of living organisms carried the organisms' inherited traits. Unfortunately, he did not provide an exact mechanism, nor did he provide detailed experimental data to support his hypothesis. Correns' hypothesis eventually caught the attention of both Sutton and Boveri. With Correns' hypothesis, Mendel's data, and his own research on *Brachyostola* chromosome behavior, Sutton began to derive a mechanism for the chromosomal transmission of inherited traits.

Brachyostola magna is a diploid organism; it contains two copies of every chromosome within each cell of its body. During the process of mitosis for a single cell, these chromosomes duplicate so that there are four copies of each chromosome. By the end of mitosis, the four copies of each chromosome have been separated into two groups of two chromosomes each. The cell then divides into two cells, with each of the two newly formed cells having two copies of each chromosome similar to their single parent cell. If inheritable traits are located on chromosomes, then mitosis would appear to be a very effective mechanism for properly duplicating and dividing the genes up into daughter cells during cell division. Likewise, during the process of meiosis for a sperm or egg cell, the chromosome number is halved from two copies to one copy of each chromosome per cell. This would explain van Beneden's observations that sperm and egg cells each contain one copy of each chromosome and Mendel's assertion that each parent can contribute only one copy of each gene to its offspring. The behavior of chromosomes during meiosis explains Mendel's principles of allelic segregation and independent assortment.

Sutton's conclusions were reported in an article entitled "On the Morphology of the Chromosome Group in *Brachyostola magna*," which appeared in the 1902 *Biological Bulletin*. Sutton proposed that chromosomes physically contain genes and that the process of meiosis for sperm and egg production is the basis of Mendel's two principles. He supported his proposals with detailed diagrams of *Brachyostola* chromosomes during various stages of meiosis.

Boveri followed in 1903 with the same conclusion from his earlier observations of *Ascaris* chromosomes. Together, the results of the two scientists culminated in the Chromosomal Theory of Inheritance, one of the basic tenets of genetics and modern biology. The Chromosomal Theory of Inheritance makes four assertions: First, the fusion of sperm and egg is responsible for reproduction—the formation of a new individual. Second, each sperm or egg cell carries one-half of the genes for the new individual, or one copy of each chromosome. Third, chromosomes carry genetic information and are separated during meiosis. Finally, meiosis is the mechanism that best explains Mendel's principles of allelic segregation and independent assortment.

Impact of Event

The Sutton and Boveri Chromosomal Theory of Inheritance represented a major landmark in the history of biological thought. It reestablished Mendelism and provided a definite physical mechanism for inheritance. It demonstrated the molecular basis of life and thereby launched two successive waves of biological research in the twentieth century: the pre-World War II genetic and biochemical revolution, and the postwar molecular biology revolution which continues today. It has also been very useful for the study of human genetic disorders.

Many geneticists began concentrating on chromosomes and how chromosomal genes actually confer individual traits. The English biochemist Sir Archibald E. Garrod had coined the phrase "inborn errors of metabolism" to describe certain inherited genetic disorders such as phenylketonuria and alkaptonuria. He attributed these genetic disorders in certain individuals to the lack of production of specific enzymes (proteins). He correctly hypothesized that a defective gene in affected individuals caused a defective enzyme resulting in the particular disorder. In the 1930's and 1940's, George Beadle, Edward Tatum, and Boris Ephrussi verified that one gene encodes one enzyme by studying eye color mutations in the fruit fly *Drosophila melanogaster* and biochemical/nutritional mutations in the pink bread mold *Neurospora crassa*.

Thomas Hunt Morgan and his associates generated hundreds of mutations in *Drosophila melanogaster* and mapped these mutations to specific chromosome locations, thereby verifying Sutton and Boveri's theory. Mutations were generated using either chemicals or high-frequency ionizing radiation (for example, ultraviolet light and X rays). This work demonstrated that exposing living organisms to radiation and certain chemicals can cause chromosome and gene damage, often resulting in severe abnormalities and/or death in the exposed individuals and their descendants.

Chromosome studies also proved useful as a tool for understanding evolution. Evolution is mutational changes that occur in organisms over time, thereby giving rise to new types of organisms and new species which are better adapted to the environment. Theodosius Dobzhansky, Alfred H. Sturtevant, and other eminent geneticists obtained evidence supporting Darwinian evolution through chromosome studies of many different species, including *Drosophila* and *Zea mays* (corn).

In 1928, Frederick Griffith discovered that some substance, which he called the "transforming factor," could be passed from dead smooth *Diplococcus pneumoniae* to live rough *Diplococcus pneumoniae*, converting the rough bacterium into a smooth bacterium. Obviously, the transforming factor was a chromosome. Yet, the chemical composition of the genetic material in the chromosome was unknown. Some biologists believed that the chromosome genetic material was DNA; others thought that it was protein. In 1944, Oswald Avery isolated and purified the transforming factor from *Diplococci*, discovering it to be DNA. In 1953, Alfred Hershey and Martha Chase provided compelling evidence for DNA as the chromosome's genetic material in experiments using radioactively marked viruses. The same year, James D. Watson and Francis Crick proposed a physical model for DNA and a chromosomal mechanism by which DNA duplicates itself.

The study of chromosomal abnormalities is invaluable in identifying human genetic disorders. An individual's chromosomes can be studied from blood cell samples and fetal chromosomes can be obtained from amniotic cells extracted from the mother's womb (amniocentesis). Many disorders, such as Down's syndrome, can be identified before birth. The Sutton and Boveri chromosomal theory of inheritance has pervaded all areas of biology and medicine.

Bibliography

Gardner, Eldon J., and D. Peter Snustad. 7th ed. *Principles of Genetics*. New York: John Wiley & Sons, 1984. This introductory genetics textbook for undergraduates is a detailed, comprehensive survey of the field. Discussions of key concepts are very clear and are supported by excellent diagrams and illustrations. Chapter 1, "Historical Perspectives," chapter 2, "Mendelian Genetics," and chapter 3, "Cell Mechanics," discuss the scientific details of Sutton and Boveri's work.

Goodenough, Ursula. *Genetics*. 2d ed. New York: Holt, Rinehart and Winston, 1978. This introductory genetics textbook for undergraduates is clearly written, with careful explanations of difficult topics. It has outstanding diagrams, extensive references, and brief historical sketches of prominent geneticists. Chapter 3, "The Meiotic Transmission of Chromosomes and the Principles of Mendelian Genetics," describes the genetic basis of the chromosomal theory of inheritance.

Moore, John Alexander. *Heredity and Development*. New York: Oxford University Press, 1963. This introductory textbook is an uncomplicated presentation of genetics. It is very clearly written and illustrated. The work of Sutton and Boveri is described, along with Mendel's two principles of inheritance and chromosome behavior during meiosis.

Pai, Anna C. *Foundations for Genetics: A Science for Society.* 2d ed. New York: McGraw-Hill, 1984. Pai's genetics textbook is a very down-to-earth discussion of the subject for the layperson. The book includes very clear explanations of chromosome behavior, mitosis, and meiosis. Mendelian inheritance and the Sutton and Boveri Chromosomal Theory of Inheritance are described.

Raven, Peter H., and George B. Johnson. *Biology.* 2d ed. St. Louis: Times-Mirror/Mosby, 1989. This lengthy textbook is an outstanding introduction to biology for undergraduates. The book contains many beautiful photographs and illustrations. It is very clearly written. Chapter 12, "Patterns of Inheritance," discusses the work of Sutton and Boveri.

Sang, James H. *Genetics and Development*. London: Longman, 1984. This textbook for advanced undergraduate and graduate students is a concise but thorough summary of how genes control the development of living organisms. Numerous important experiments are described, and an extensive reference list is provided. Chapter 3, "Active Chromosomes," describes the organization of genes on chromosomes.

Starr, Cecie, and Ralph Taggart. *Biology: The Unity and Diversity of Life*. 5th ed. Belmont, Calif.: Wadsworth, 1989. This introductory biology textbook for nonmajors is a very clear, concise summary of all biological disciplines. It has excellent photographs and illustrations. Chapter 13, "Chromosomal Theory of Inheritance," is an outstanding discussion of Sutton and Boveri's theory and its relationship to Mendelian inheritance.

Swanson, Carl P., Timothy Merz, and William J. Young. *Cytogenetics: The Chromosome in Division, Inheritance, and Evolution*. 2d ed. Englewood Cliffs, N.J.: Prentice-Hall, 1980. This textbook for advanced undergraduate and graduate students is an excellent survey of chromosomal research from Sutton and Boveri to the present. Chapter 1 discusses the history of cytogenetic research, including a list of important events and dates. Chapter 4, "Cell Division: The Basis of Genetic Continuity and Transmission," describes chromosomes during mitosis and meiosis.

David Wason Hollar, Jr.

Cross-References

De Vries and Associates Discover Mendel's Ignored Studies of Inheritance (1900), p. 61; McClung Plays a Role in the Discovery of the Sex Chromosome (1902), p. 148; Punnett's *Mendelism* Contains a Diagram for Showing Heredity (1905), p. 270; Bateson and Punnett Observe Gene Linkage (1906), p. 314; Hardy and Weinberg Present a Model of Population Genetics (1908), p. 390; Morgan Develops the Gene-Chromosome Theory (1908), p. 407; Johannsen Coins the Terms "Gene," "Genotype," and "Phenotype" (1909), p. 433; Sturtevant Produces the First Chromosome Map (1911), p. 486; Avery, MacLeod, and McCarty Determine That DNA Carries Hereditary Information (1943), p. 1203.

ZSIGMONDY INVENTS THE ULTRAMICROSCOPE

Categories of event: Chemistry and applied science
Time: 1902
Locale: Göttingen, Germany

Richard Zsigmondy developed a device called the ultramicroscope, which allowed scientists to measure and identify individual particles in colloid solutions

Principal personages:

RICHARD ZSIGMONDY (1865-1929), an Austrian-born German organic chemist who was a recipient of the 1925 Nobel Prize in Chemistry

H. F. W. SIEDENTOPF (1872-1940), a German physicist/optician who assisted Zsigmondy in his laboratory during the development of the ultramicroscope

MAX VON SMOULUCHOWSKI (1879-1961), a German organic chemist who assisted Zsigmondy in many of his later experiments using the ultramicroscope

Summary of Event

Richard Zsigmondy's invention of the ultramicroscope grew out of his interest in colloidal substances. Colloids are tiny particles of a substance finely dispersed throughout a solution of another material or substance (for example, salt in water). Zsigmondy, whose father, Adolf, was a noted inventor of dental surgical equipment, first became interested in colloids while working as an assistant to the physicist Adolf Kundt at the University of Berlin in 1892. Although originally trained as an organic chemist, in which discipline he took his Ph.D. at the University of Munich in 1890, Zsigmondy became particularly interested in colloidal substances containing fine particles of gold that produce lustrous colors when painted on porcelain. He subsequently abandoned organic chemistry and devoted his career to the study of colloids.

Zsigmondy began intensive research into his new field of interest in 1893, when he returned to Austria to accept a post as lecturer at the Technische Hochschule at Graz. (For readers unfamiliar with the German and Austrian educational systems, the level of instruction is more comparable to that of an American university than to that of an American high school.) At Graz, Zsigmondy became especially interested in gold-ruby glass, the accidental invention of the seventeenth century alchemist Johann Kunckle. Kunckle, while pursuing the alchemist's chimera of transmuting base substances into gold, discovered instead a method of producing glass (called gold-ruby glass) with a beautiful, deep red luster through the suspension of very fine particles of gold throughout the liquid silica before it was cooled to become glass. Zsigmondy also began studying a colloidal pigment called purple of Cassius, the discovery of another seventeenth century alchemist, Andreas Cassius.

Zsigmondy soon discovered that purple of Cassius was a colloidal solution and not, as many chemists believed at the time, a chemical compound. This fact allowed him to develop techniques for glass and porcelain coloring with potentially great commercial value, which led directly to his 1897 appointment to a research post with the Schott Glass Manufacturing Company in Jena, Germany. With the Schott Company, Zsigmondy concentrated on the commercial production of colored glass objects. His most notable achievement during this period was the invention of Jena milk glass, still prized by collectors throughout the world.

While Zsigmondy was studying these substances, many European chemists insisted that purple of Cassius was a chemical compound. Zsigmondy devised experiments that convinced him that Cassius' purple was not a chemical compound but instead a colloidal substance. When he published the results of his research in professional journals, however, they were not widely accepted by the scientific community. Other scientists were not able to replicate Zsigmondy's experiments and consequently denounced them as flawed. The criticism of his work in technical literature stimulated Zsigmondy to make his greatest discovery, the ultramicroscope, developed to prove his theories regarding the colloidal nature of purple of Cassius.

Between 1900 and 1907, Zsigmondy pursued private research. During this period, he completed most of the work for which he won the 1925 Nobel Prize in Chemistry. In pursuing his investigations into colloids, he encountered a number of problems unresolvable with scientific equipment available at the time. Until the time of Zsigmondy's discovery, scientists had attempted to study colloidal particles in organic solutions by observing a cone of scattered light, called the Faraday-Tyndall cone, shone into the solution. This method, however, did not permit the observer to see the individual particles suspended in the solution. Using the facilities and assisted by the staff (especially H. F. W. Siedentopf, an expert in optical lens-grinding) of the Zeiss Glass Manufacturing Company of Jena, Zsigmondy developed an ingenious device that permitted direct observation of individual colloidal particles. This device, which its developers named the ultramicroscope, made colloidal gold particles that were visible in sunlight.

The principle utilized by Zsigmondy and his associates in the development of the ultramicroscope was not new. Sometimes called "dark-field illumination," the method utilizes a light (usually sunlight focused by mirrors) shone through the solution under the microscope at right angles from the observer, rather than directly from the observer into the solution. The resulting effect is similar to a beam of sunlight admitted to a closed room through a small aperture. If an observer stands back from and at right angles to such a beam, even with the unaided eye, many dust motes suspended in the air will be observed, which are not otherwise visible. Zsigmondy and others who assisted him made a number of refinements and improvements in dark-field illumination procedures, and the result was the ultramicroscope.

Zsigmondy's device shines a very bright light through the substance or solution being studied. The microscope then focuses on the light shaft from the side. This process enables the observer using the ultramicroscope to view colloidal particles ordinarily invisible even to the strongest conventional microscope. To a scientist viewing purple of Cassius, for example, colloidal gold particles as small as one ten-millionth of a millimeter in size become visible. After Zsigmondy publicized his findings, scientists in a variety of disciplines adopted the ultramicroscope, finding many uses for it in fields as disparate as medicine and agriculture. It remains an important tool in many fields of scientific research.

Impact of Event

After Zsigmondy's invention of the ultramicroscope and his subsequent observations concerning colloidal solutions, in 1907, the University of Göttingen appointed him professor of inorganic chemistry and director of its Institute for Inorganic Chemistry. Using the ultramicroscope, Zsigmondy and his associates quickly refuted all the critics that had denounced his research on purple of Cassius, proving that it is indeed a colloidal substance. That finding, however, was the least of the spectacular discoveries that resulted from Zsigmondy's invention. In the next decade, Zsigmondy and his associates found that color changes in colloidal gold solutions result from coagulation—that is, changes in the size and number of gold particles in the solution caused by particles bonding together. Zsigmondy found that coagulation occurs when the negative electrical charge of the individual particles is removed by the addition of salts. Coagulation can be prevented or slowed by the addition of protective colloids.

These observations also made possible the determination of the speed at which coagulation takes place, as well as the number of particles in the colloidal substance being studied. With the assistance of the theoretical physicist Max von Smoluchowski, Zsigmondy worked out a complete mathematical formula of colloidal coagulation that is valid not only for gold colloidal solutions but also for all other colloids. Colloidal substances include milk and blood, which both coagulate, thus giving relevance to Zsigmondy's work in the fields of medicine and agriculture. These observations and discoveries concerning colloids—in addition to invention of the ultramicroscope—earned for Zsigmondy the 1925 Nobel Prize in Chemistry.

After his appointment at Göttingen University, Zsigmondy assembled a research team and began intensive research into colloids using the ultramicroscope. He and his team developed another tool for their investigations, which Zsigmondy called "ultrafiltration." Using the ultramicroscope and ultrafiltration, the scientists undertook a detailed study of gels, particularly silica and soap gels (which are essentially coagulated colloidal solutions). This research led to many important discoveries with numerous commercial applications, including the development of products such as shampoo, shaving cream, cold cream, suntan lotion, and deodorant soap.

In later years, other scientists using the ultramicroscope and Zsigmondy's ultrafiltration techniques succeeded in ascertaining the electrical conductivity of colloidal particles. This process led to the development of a method for separating colloidal solutions from the admixed electrolytes, not only in gold sols but also in other colloidal solutions. These new methods have had numerous commercial applications, which have added to the quality of life for humankind.

Bibliography

Farber, Eduard. *Nobel Prize Winners in Chemistry, 1901–1961.* 2d ed. New York: Abelard-Schuman, 1963. Farber's book, suitable for a wide audience, includes a brief biography of Zsigmondy. The account is drawn from Zsigmondy's Nobel lecture.

Hatschek, Emil, ed. *The Foundations of Colloid Chemistry.* London: Ernest Benn, 1925. Hatschek places Zsigmondy in the context of his predecessors. A background in chemistry is needed.

Hauser, Ernst. *Colloidal Phenomena.* New York: McGraw-Hill, 1939. Explores the status of colloid research up to the late 1930's. A working knowledge of chemical terms and concepts is required.

Kruyt, H. R. *Colloids.* London: Ernest Benn, 1930. A standard reference in colloid chemistry. Kruyt discusses Zsigmondy's contributions in the field.

McBain, James W. *Colloid Science.* Boston: D. C. Heath, 1950. McBain includes an extensive and admiring overview of Zsigmondy's work and his contributions to the field of chemistry. Replete with technical terms and scientific terminology aimed at a knowledgeable reader.

Zsigmondy, Richard Adolf. *Colloids and the Ultramicroscope.* Translated by Jerome Alexander. New York: Wiley, 1909. Zsigmondy's account of the development and uses of the ultramicroscope. Technical at times, but a valuable source.

Paul Madden

Cross-References

McLean Discovers the Natural Anticoagulant Heparin (1915), p. 610; Ruska Creates the First Electron Microscope (1931), p. 958; Müller Invents the Field Emission Microscope (1936), p. 1070; Müller Develops the Field Ion Microscope (1952), p. 1434.

PAVLOV DEVELOPS THE CONCEPT OF REINFORCEMENT

Category of event: Biology
Time: 1902-1903
Locale: St. Petersburg, Soviet Union

Pavlov advanced the concept of reinforcement, which provided a physiochemical explanation for learning

Principal personages:

IVAN PETROVICH PAVLOV (1849-1936), a Russian physiologist who developed the theory of classical conditioning

IVAN MIKHAYLOVICH SECHENOV (1829-1905), a Russian physiologist whose reductionist methodology and speculations on the reflex arc influenced Pavlov

EDWARD L. THORNDIKE (1874-1949), an American psychologist whose investigation of instrumental conditioning influenced Pavlov

CHARLES DARWIN (1809-1882), an English naturalist whose evolutionary views implied that animal behavior could reveal the workings of the human brain

Summary of Event

In April, 1903, Ivan Petrovich Pavlov delivered a surprising address to an International Medical Conference in Madrid. It was thought that the noted Russian physiologist would discuss his research on digestion; instead, he described new investigations into the links between psychic and physiological processes. The focal point of these inquiries was the experience that begins, perpetuates, or eliminates a given kind of behavior. While he did not formally name that concept until the following year, he was referring to reinforcement, the key physiochemical event in the learning process.

Physiologists had a general understanding of the structure and functions of the nervous system before Pavlov advanced the notion of reinforcement. They had not, however, answered the crucial question of how learning occurs—that is, how behavior changes on the basis of experience. One group, variously described as mentalists, vitalists, or subjectivists, maintained that thoughts and emotions were not subject to physical laws and that experimental attempts to penetrate the boundary between mind and body were useless. Their materialist opponents, generally called monists, insisted that concepts such as spirit, mind, or soul clouded the issue of higher brain functions and maintained that explanations of all nervous activity should be sought in the laws of physics and chemistry.

Before Pavlov's work on reinforcement, however, monists lacked a cohesive theory supported by a convincing body of experimental data. They observed animals and drew valuable inferences, and they removed parts of an animal's nervous system, noted the consequences, and reached conclusions. But real insight into learning required a clear approach combined with ways to measure the physical reactions of healthy subjects. Pavlov provided the solution to both problems.

The quantifying technique Pavlov used to address learned behavior was an offshoot of his 1889-1897 investigations into the effect of the nervous system on digestion. In that project, which won for him the 1904 Nobel Prize in Physiology or Medicine, he diverted the duct of a salivary gland (the parotid) to the outside of a dog's muzzle so that saliva could be collected from a funnel. This procedure, which produced measurable secretions from a healthy animal, underpinned his research into reinforcement.

Pavlov's basic approach to the exploration of brain functions also grew from his experiments on digestion. While monitoring the results of sham feeding (the removal of food from a dog's slit esophagus after it was swallowed but before it entered the stomach), he observed unexpected gastric secretions. Since the vagus nerve (which originates in the medulla oblongata and ennervates the abdominal cavity) had been severed during surgery, he concluded that part of the stomach's neural control lay in the cerebral cortex. This led him to believe that secretions could provide information about higher neural activity. Thus, by 1897, Pavlov possessed a powerful leading idea and a technical procedure that could provide measurable data on the activity of the brain.

He also drew upon a number of working assumptions. Pavlov was influenced by Ivan Mikhaylovich Sechenov, a mechanical reductionist who argued that all behavior could be described in terms of reflexes (motor responses to stimulation), reflexes could be explained entirely in physiochemical terms, and complicated human reflex behavior could best be approached through the study of simpler animal reflex behavior. Pavlov also knew that Charles Darwin had advanced a complementary viewpoint. Darwin's evolutionary theory implied that mind had evolved through natural selection—and if mind had evolved, the brain and behavior had also evolved. Thus, Darwin's conclusions supported the reductionist idea that animal behavior could provide insights into the workings of the human brain.

Finally, information drawn from recent observation sharpened Pavlov's interest in the study of learned behavior. He closely studied Edward L. Thorndike's experiments on instrumental conditioning, which required a cat to operate an instrument such as a lever or bobbin in order to escape from a puzzle box. Also, while doing research on the digestive system, he noticed an intriguing occurrence: When the presentation of food to a dog was regularly and closely preceded by an alerting signal such as approaching footsteps, the signal alone caused salivation. Pavlov directed preliminary investigations into that type of behavior (the conditional reflex) as early as 1897.

In 1902, Pavlov applied himself completely to the study of the conditional reflex. The outlines of his theory of classical conditioning immediately took shape, for the basic ideas, much of the terminology, and the investigative methods were almost ready to present. Working in the well-equipped physiological laboratory of the St. Petersburg Institute of Experimental Medicine, he issued preliminary findings at Madrid in 1903 and presented formal results of his research in 1905.

Although the evidence for Pavlov's conclusions was drawn from experiments on dogs, he extended them to the human nervous system. He maintained that all animals had an inborn neural capacity (the unconditional reflex) to react to events necessary for survival (the unconditional stimulus). For example, the sight of food normally produced a certain type and amount of salivation in a dog. As long as the dog was allowed to eat food presented to it, the unconditional reflex endured. Eating the proffered food reinforced the unconditional reflex, for the act of eating maintained (excited) a neural association between the unconditional stimulus and the unconditional reflex. Yet, if a dog was repeatedly shown food it was not allowed to eat, salivation weakened and eventually disappeared. Pavlov interpreted this to mean that the association between the sight of food and salivation was actually a temporary neural connection that could be broken through lack of reinforcement. On the other hand, an unreinforced (inhibited) connection could be reactivated; resumption of eating overrode inhibition and restored the extinguished link between salivation and the sight of food.

That link could also be redirected. If an unrelated stimulus (for example, the sound made by a bell) immediately preceded the delivery of food over a long enough period, it could provoke salivation. In Pavlovian terminology, the coupling of a conditional stimulus, a bell tone, with an unconditional stimulus, the sight of food, produced a conditional reflex. In other words, the bell tone made the dog think of food and then salivate. Moreover, a dog was able to discriminate between similar types of reinforced and unreinforced food. For example, it would stop salivating at the sight of white bread it was forbidden to eat but continue salivating at the sight of brown bread that was regularly fed to it.

Thus, the linchpin of Pavlov's theory was the notion of reinforcement, which held that animals could associate any two occurrences if the first one promptly and reliably predicted the onset of the second. Reinforcement had a clearly adaptive function, for it provided an animal with a wide range of anticipatory responses to its changing environment. Pavlov's experimental data, composed of precise, replicable measurements of the volume and rate of saliva in response to a variety of timed stimuli, confirmed the importance of that validating mechanism. The presence or absence of reinforcement, which created conditional reflexes and maintained, extinguished, or resurrected both conditional and unconditional reflexes, permitted an animal to modify its behavior on the basis of experience. Learning took place amid the forging and breaking of neural connections in reaction to changing circumstances.

Impact of Event

The importance of reinforcement was first recognized in the Soviet Union, where it affected the debate between monistic biologists and state-supported subjectivistic psychologists. The split between government and society during the late czarist period worked to Pavlov's advantage: While state ideology opposed all brands of materialism, including mechanistic reductionalism, private funding allowed Pavlov to expand investigations into reinforcement. For example, in 1911, the Ledentsov Society financed a series of soundproofed chambers, known as the tower of silence, in which he could study the effect of time intervals on the coupling of conditional and unconditional stimuli. His reputation spread rapidly, and by 1914 he headed an influential school of experimental physiology.

The Revolutionary period disrupted Pavlov's work, but after 1921 the Soviet leadership vigorously supported his research. They did so for several reasons: Pavlov's mechanical reductionism was compatible with the materialistic orientation of the new Marxist regime, his doctrine of reinforcement strengthened the arguments of nurture over nature at a time when the government wished to reeducate its citizenry, and the scientific triumphs of the Pavlovian school had propaganda value for socialism. With state sponsorship, Pavlov and his successors directed large research projects on complex aspects of association through reinforcement.

Advances in cybernetics (the comparison of the communication and control systems of human-made devices with those in the nervous systems of animals) have caused some Soviet scientists to question the validity of Pavlov's conclusions. Nevertheless, his school dominates experimental biology, psychology, and cognitive theory in the Soviet Union, and his environmentally based concept of reinforcement undergirds Soviet educational practice.

Largely because of the language barrier, non-Russians were unable to absorb Pavlov's concept of reinforcement until the 1920's. European reaction was unenthusiastic; continental investigators tended toward Gestalt psychology, a holistic approach to perception that emphasized complexity and was at odds with reductionism. The British identified largely with Sir Charles Scott Sherrington's efforts to confine reflex action to the spinal cord.

Pavlovian reinforcement had a different reception in the United States, where it was compatible with behavioralism, an indigenous tradition opposed to introspection. Pavlov's definition of reinforcement had to compete with Thorndike's, however, who had put forward a different one in 1913 while developing his theory of instrumental conditioning. To Thorndike, reinforcement was any action that caused the appearance of an unconditional stimulus (such as pushing a lever to release a food pellet); according to Pavlov, it always followed the unconditional stimulus. The sharply differing terminology led to widespread confusion, with Pavlovian reinforcement sometimes described accurately, sometimes as the unconditional stimulus, and sometimes as rewards and punishments.

Generally, Americans and Western Europeans admire Pavlov's meticulous investigations into the conditions effecting reinforcement and associate his original concept with a specific kind of simple learning (classical, or passive, as opposed to Thorndike's instrumental, or purposeful). Pavlovian reinforcement has constantly combined and recombined with various forms of behavioralism, and some of its aspects have been incorporated into theories of language formation and information processing. Yet, it is only one of many currents affecting the course of Western psychological and cognitive theory.

Bibliography

Babkin, Boris P. *Pavlov: A Biography.* Chicago: University of Chicago Press, 1949. An indispensable primary source rich in anecdotal material, this is a fond memoir by a close colleague. Includes otherwise inaccessible personal and professional information on Pavlov. Although poorly organized and somewhat polemical, Babkin's work is comprehensive in scope. It is sparsely illustrated and contains notes, a bibliography, and an index.

Frolov, Y. P. *Pavlov and His School: The Theory of Conditioned Reflexes*. Translated by C. P. Dutt. New York: Oxford University Press, 1937. A key work by a Soviet physician who worked with Pavlov, which treats Pavlov's career in a systematic fashion. Frolov's biography, however, contains ideologically motivated inaccuracies typical of the Stalinist period. It should be read in conjunction with Babkin's biography. Contains charts, diagrams, photographs, and an index.

Gantt, W. Horsley. "Russian Physiology and Pathology." In *Soviet Science*, edited by Ruth C. Christman. Washington, D.C.: American Association for the Advancement of Science, 1952. A somewhat dated but still valuable work by one of Pavlov's American students. Contains interesting information on Pavlov's attitudes toward contemporary scientists and scientific problems. Gantt presents a somewhat argumentative but solid treatment of Pavlov's school and its influence on various aspects of Soviet scientific and educational thought. Heavily footnoted.

Graham, Loren R. *Science, Philosophy, and Human Behavior in the Soviet Union*. New York: Columbia University Press, 1987. This expanded version of Graham's *Science and Philosophy in the Soviet Union* (1972) offers an excellent treatment of Pavlov's contributions to Russian physiology and psychology. Includes chapters on the nature-nurture debate, biology and human beings, and cybernetics and computers. Lengthy, with notes, an extensive bibliography, and an index.

Gray, Jeffrey A. *Ivan Pavlov.* New York: Viking Press, 1979. An excellent biography that draws extensively from Babkin and Frolov (previously cited). Particularly strong on Pavlov's intellectual background and research methodology. Pavlov's work is evaluated in the light of physiological and psychological developments of the 1960's and 1970's. Contains diagrams, graphs, and an index.

Martin, Irene, and A. B. Levey. "Learning What Will Happen Next: Conditioning, Evaluation, and Cognitive Processes." In *Cognitive Processes and Pavlovian Conditioning in Humans*, edited by Graham Davey. New York: John Wiley & Sons, 1987. An important and provocative article which discusses the influence of Pavlovian reinforcement on contemporary cognitive theory. Particularly useful for readers interested in information processing. Traces Pavlov's impact on behavioralism and outlines new directions in research. A bibliography is provided.

Pavlov, Ivan P. *Experimental Psychology, and Other Essays*. New York: Philosophical Library, 1957. A useful primary source that includes Pavlov's autobiography, the 1903 Madrid speech that contained the first public reference to reinforcement, and the 1904 Nobel Prize speech which first outlined the major elements of classical conditioning. The work also includes material on Pavlov's later attempts to apply aspects of reinforcement to human personality disorders. Contains 653 pages, with notes but no index.

Vucinich, Alexander S. *Science in Russian Culture, 1861-1917.* Stanford, Calif.: Stanford University Press, 1970. Especially useful to readers interested in history. Places Russian science in the cultural and social context of the postemancipation era. Vucinich presents Pavlov's school as part of a wide, uniquely Russian current in experimental biology. Should be read in conjunction with Graham's book. Includes notes, an extensive bibliography, and an index.

Michael J. Fontenot

Cross-References

Ramón y Cajal Establishes the Neuron as the Functional Unit of the Nervous System (1888), p. 1; Bayliss and Starling Discover Secretin and Establish the Role of Hormones (1902), p. 179; Sherrington Delivers *The Integrative Action of the Nervous System* (1904), p. 243; Moniz Develops Prefrontal Lobotomy (1935), p. 1060; Cerletti and Bini Develop Electroconvulsive Therapy for Treating Schizophrenia (1937), p. 1086; Sperry Discovers That Each Side of the Brain Can Function Independently (1960's), p. 1635; Hounsfield Introduces a CAT Scanner That Can See Clearly into the Body (1972), p. 1961; Janowsky Publishes a Cholinergic-Adrenergic Hypothesis of Mania and Depression (1972), p. 1976.

THE FRENCH EXPEDITION AT SUSA DISCOVERS THE HAMMURABI CODE

Category of event: Archaeology
Time: January, 1902
Locale: Susa (Persepolis) in Iran (then called Persia)

Discovery of the code of laws attributed to the Babylonian King Hammurabi placed the origins of Judeo-Christian concepts of justice farther back in the chronological record of the ancient world

Principal personages:

M. J. DE MORGAN, a French archaeologist who was responsible for supervising the French government-sponsored excavations at Susa

VINCENT SCHEIL (1858-1940), a professor at the French School of Advanced Studies (*École des Hautes Études*) who undertook the task of translating the Code of Hammurabi

GEORG FRIEDRICH GROTEFEND (1775-1853), a teacher in Göttingen, Germany, who first deciphered the ancient cuneiform writing system

GEORGE SMITH (1840-1876), a self-trained assistant in the Egyptian-Assyrian section of the British Museum who deciphered the so-called *Gilgamesh* epic (c. 2000 B.C.), which contained Babylonian cross-references

Summary of Event

In January, 1902, French archaeologists, working under the supervision of Monsieur M. J. de Morgan in the ruins of the ancient Elamite capital of Susa, unearthed several detached pieces of a black diorite block of stone. This block, once reassembled, reached a height of nearly 4 meters and was covered on several sides by carved inscriptions. One side bore a representation of the Babylonian King Hammurabi, who reigned over the lands stretching from the Mesopotamian Tigris and Euphrates river valleys to the coast of the Mediterranean between approximately 1792 B.C. and 1750 B.C. King Hammurabi was depicted receiving a corpus of laws from the seated Babylonian Sun God Shamash. The other sides of the massive stone bore the inscriptions, in cuneiform script typical of the literate cultures of ancient Mesopotamia, of the law code that would be known as the Hammurabi Code.

The fact that the stone block was found in the ruins of Susa, and not in Babylon or a dependent city of the great Babylonian kings, seemed to have a logical explanation. Babylon and Susa were the seats of rival dynasties who were frequently at war with each other, both before and after Hammurabi's reign. It was assumed that, as a result of one or another of the military campaigns pitting Elamites against Babylonians some time after the middle of the eighteenth century B.C., the former took the possessions and monuments of the latter and carried the enormous stone monument to their home capital of Susa.

Up until this date, archaeologists had uncovered very little direct historical information concerning King Hammurabi's reign. There were a few surviving written records, particularly from the archives at Mari, or Tell Hariri in Mesopotamia, which was the political seat of Zimri-lim, a contemporary and rival of King Hammurabi. Before Hammurabi conquered Mari in the thirty-third year of his reign and destroyed its main palace (probably along with additional archival materials), Mari's scribes had described certain aspects of the Babylonian king's reign. This source provided incidental evidence of political and commercial relations and some information on Babylonian laws. With de Morgan's expedition's discovery, the importance of Babylonian laws, and particularly Hammurabi's Code, would capture the attention of archaeologists.

On the obverse sides of Hammurabi's monument, opposite the side bearing the inscription of the king and Babylon's Sun God, was a series of ordered columns of writing, apparently representing a list, or code, of laws. One side contained sixteen columns of writing with a total of 1,114 lines. Five columns had been obliterated from this face of the stone monument, presumably by the Elamite "victors" over Babylon who intended to inscribe (but never did) a dedication to their own glory. An additional twenty-five hundred lines in twenty-eight columns were carved into the remaining sections of the surviving monument.

Even before the work of concise transliteration and translation could be undertaken, it was apparent that an important section of the text (containing about seven hundred lines) was devoted to a description of Hammurabi himself, his titles, his qualities as a ruler, his dedication to the Gods of Babylon, and, most important for archaeologists' records, enumeration of the cities and districts under his rule. The text of Hammurabi's Code was most important, not only in terms of numbers of columns and lines of writing but also for its impact on archaeologists' knowledge of the main lines of Babylonian civilization. The content of the code was diverse, but its main concerns were for contracts of sale or business (deposits, liabilities, and the like), farming and animal husbandry practices, and dowry and marriage contracts.

Impact of Event

When the French expedition to Susa discovered King Hammurabi's code of laws in January of 1902, a translation, by Father Vincent Scheil of the *École des Hautes Études,* was published in Paris in October of 1902. Archaeologists' knowledge of ancient Babylonian or later Mesopotamian societies was not necessarily revolutionized; it did, however, take on a much more comprehensive form in a short time.

Evidence relating to Babylonia's place in the history of ancient Mesopotamian civilization had been accumulating steadily, for more than a hundred years before French archaeologists discovered Hammurabi's monument. This evidence should have advanced seriously beyond the realm of massive architectural ruins and pictorial monuments alone when specialists first unlocked the secret of ancient cuneiform writing. Although the first legible transcribed copies of fragments of cuneiform inscriptions had been brought back from the Near East as early as the late seventeenth century (by Cornelius de Bruin), it was not until 1802—exactly one hundred years before de Morgan's find at Susa—that the cuneiform writing system was "decodified." This was done by Georg Friedrich Grotefend, a modest schoolteacher from Göttingen, Germany, working from one of de Bruin's transcriptions of ancient texts from Persepolis. Grotefend did not have any of the advantages of word-for-word comparative language texts that Jean-François Champollion had used when he deciphered Egyptian hieroglyphics from the Rosetta stone in the same generation.

Grotefend's findings were announced formally before the Academy of Sciences in Göttingen late in 1802 and published by his own hand as an appendix to a book on the ancient world authored by a recognized German scholar. The results of Grotefend's work, employed to expand archaeological knowledge of neighboring and older Babylonian and Elamite civilizations, were not noted for nearly a generation.

When attention turned eventually to the study of Babylon, it was the result of English exploration missions interested in major architectural monuments, rather than philologically oriented scholars, that serious advances were made. Before his death in 1821, the amateur archaeologist Claudius James Rich had carried out the first serious excavations that attempted to reconstruct the history of the site of the Mesopotamian capital of Babylon. The nature of archaeological finds during Rich's lifetime and for some years following shows the beginning of a significant, albeit gradual and uneven, trend that would culminate only when the French expedition to Susa uncovered the legal texts inscribed in the name of King Hammurabi. Rich left behind in the collection of the British Museum a series of cuneiform tablets gathered in the ruins of Babylon, plus some drawings that would be used to estimate the topographical layout of the ancient city and to reconstruct the presumed appearance of the fabled capital of King Nebuchadnezzar II, Hammurabi's distant successor (605-562 B.C.). Meanwhile, the magnificence of pictorial stone reliefs on the sides of public buildings and tombs attracted the attention of archaeologists. The meaning of cuneiform inscriptions that often accompanied such reliefs seemed to be less immediately pressing as items of scholarly interest to report to European archaeological enthusiasts.

Despite general concentration on nontextual vestiges and later Assyrian, not Babylonian, ruins, a significant debate over philological, or historical linguistic, approaches to the ancient past arose by the late 1850's that would pave the way for recognition of the importance of the discovery of Hammurabi's Code. A turning point came in 1857, when conflicting scholarly interpretations of an extensive cuneiform inscription attributed to Tiglath-pileser III, King of Assyria, were debated in the Royal Asiatic Society in London. This rather rarefied event generated so much interest in the importance of understanding the textual details that portions of a newly discovered treasure of several thousand clay tablets representing the content of Assyrian King Ashurbanipal's royal library at Nineveh were reexamined carefully. This was done by George Smith, an assistant in the Egyptian-Assyrian section of the British Museum, who discovered the so-called *Gilgamesh* epic, a mixture of Babylonian religious mythology and history that altered the world's view of a number of key ancient historical reference points. Primary among these was the *Gilgamesh* epic's reference, inscribed long before comparable citations in the Old Testament, to the Great Deluge and the role of Ut-napishtim, whom Smith identified as the Babylonian equivalent of the biblical Noah. After Smith undertook a successful search for missing sections of the *Gilgamesh* tablets amid the rubble at Nineveh, his thesis stood solid throughout the rest of the nineteenth century: Elements of the Bible had not been "tailored" without precedent to fit the original needs of Old Testament founders of the Jewish religion; some could be shown clearly to have been fashioned from preexisting Semitic traditions.

Here, then, albeit in literary and religious, not yet legal, terms, lay one of the most important preparatory contributions that would increase the impact of the French 1902 expedition's discovery of the Hammurabi Code. It was shown, through translations, retranslations, comments, and recommendations, that the Mosaic Code, as well as a number of other elements of legal practice associated with the Old Testament, probably had very distant precedents in ancient Mesopotamian history that had been reexamined and then reproduced in later stages in the history of the same region or other regions. Therefore, the sources for twentieth century scholars' study of the origins of Judeo-Christian legal principles and practices were, after the discovery of Hammurabi's Code, necessarily placed farther back chronologically and farther east, both geographically and culturally. It remains somewhat ironic that, despite a near-continuous eighteen-year on-site excavation project carried out in the ruins of Hammurabi's and other later monarchs of Babylon (between 1899 and 1917) by the German architect Robert Koldewey, no landmark textual discovery (as distinct from the uncovering of buildings and monuments) comparable in importance to that of the Hammurabi Code was made in Babylon in the first part of the twentieth century.

Bibliography

Godbey, A. H. "Chirography of Hammurabi's Code." *American Journal of Semitic Languages* 20 (January, 1904): 137-148. Like other editorially authored scholarly announcements concerning the written accuracy of the inscriptions found on Hammurabi's legal monument (see, for example, "Scribal Errors in Hammurabi's Code," in the same volume of *American Journal of Semitic Languages*, pp. 116-136), this study examines the peculiarities of writing techniques in Mesopotamia during the second millennium B.C.

Hammurabi. *The Babylonian Laws.* Translated and annotated by G. R. Driver and John C. Miles. Oxford, England: Clarendon Press, 1955. This annotated translation of Hammurabi's Code is considerably more extensive than the original English translation by the citation listed below. The wealth of detailed information comes not only in attention to the text itself but also in an extensive section offering philological commentary on the words and phrases it contains.

____________. *The Oldest Code of Laws in the World.* Translated by C. H. W.

Johns. Edinburgh, Scotland: T. & T. Clark, 1903. This is the first English translation of Hammurabi's Code, undertaken shortly after the original French translation in 1902. It contains a valuable introductory section describing the actual circumstances surrounding the discovery of the stone bearing the Hammurabi inscriptions.

"Hammurabi and Women." *Harper's Bazaar* 37 (April, 1903): 403. Perhaps the earliest popularly oriented article providing factual information from the original French translation published a few months earlier. Includes an interesting subsection of the nearly four-thousand-year-old Hammurabi Code.

Wooley, Sir Leonard. *Digging Up the Past.* 2d ed. London: Ernest Benn, 1954. Wooley's wide-ranging knowledge of Mesopotamian archaeology, based on his pioneer excavations at Sumerian Ur, earned for him "senior status" among archaeologists who not only carried out scientific research but also provided very readable accounts of significant stages in the progress of archaeology. This book contains a historical introduction to the field.

Byron Cannon

Cross-References

Evans Discovers the Minoan Civilization on Crete (1900), p. 67; Carter Discovers the Tomb of Tutankhamen (1922), p. 730; Archaeologists Unearth Ancient Scrolls (1947), p. 1298.

KENNELLY AND HEAVISIDE PROPOSE THE EXISTENCE OF THE IONOSPHERE

Categories of event: Earth science and physics
Time: March and June, 1902
Locale: Cambridge, Massachusetts, and Newton Abbot, England

Kennelly and Heaviside independently proposed the existence of an electrified layer in the upper atmosphere that would reflect radio waves around the curved surface of the earth

Principal personages:

ARTHUR EDWIN KENNELLY (1861-1939), a British-American electrical engineer and Harvard professor who contributed to the theory of electrical circuits and radio propagation

OLIVER HEAVISIDE (1850-1925), an English physicist and electrical engineer who made significant contributions in applied mathematics and electromagnetic theory

BALFOUR STEWARD (1828-1887), a Scottish physicist whose study of terrestrial magnetism led him to the suggestion of the existence of a charged layer in the upper atmosphere

GUGLIELMO MARCONI (1874-1937), an Italian electrical engineer and cowinner of the 1909 Nobel Prize in Physics for demonstrating radio transmission across the Atlantic Ocean

SIR EDWARD VICTOR APPLETON (1892-1965), an English physicist who won the 1947 Nobel Prize in Physics for his discovery of the ionosphere in 1924

HEINRICH HERTZ (1857-1894), a German physicist who is best known for his discovery of radio waves in 1887

Summary of Event

On December 12, 1901, only thirteen years after Heinrich Hertz discovered radio waves, Guglielmo Marconi succeeded in transmitting radio signals from Cornwall, England, to Newfoundland, Canada. This historic event was difficult to explain since it was known that radio signals consisted of electromagnetic waves like light, which travel in nearly straight lines. Thus, there was considerable discussion among scientists as to how Marconi's signals could propagate around the curved surface of the Atlantic Ocean. Several scientists tried to show that electromagnetic waves of sufficiently long wavelength could bend around the earth's curvature by diffraction (the small tendency of waves to spread around obstacles). Calculations showed, however, that diffraction effects were inadequate to explain Marconi's results.

The correct explanation of radio propagation around the curved surface of the earth was suggested almost simultaneously in 1902 by Arthur Edwin Kennelly in the

United States and Oliver Heaviside in England. They independently postulated the existence of an electrically conducting layer in the upper atmosphere that would reflect radio waves back to Earth. Successive reflections between this conducting layer and the surface of the earth could guide the waves around the earth's curvature. Heaviside also suggested that the conductivity of this region might result from the presence of positive and negative ions in the upper atmosphere caused by the ionizing action of solar radiation.

It is interesting to note that this hypothetical reflector was usually called the Heaviside layer, although it was Kennelly who first published the idea under the title "On the Elevation of the Electrically-Conducting Strata of the Earth's Atmosphere" in the *Electrical World and Engineer* in March, 1902. Heaviside was a brilliant but self-educated scientist who often had difficulty getting his work published. His ideas were more fully developed than Kennelly's, but when he submitted an article to *The Electrician*, it was rejected and then appeared in the *Encyclopædia Britannica* in June, 1902. In 1912, W. H. Eccles published the first theory of how charged particles affect the propagation of radio waves. Since he was aware of Heaviside's rejected article, he apparently attempted to set the record straight by calling the postulated reflecting region of the atmosphere the Heaviside layer.

In fact, the idea of such a conducting shell in the upper atmosphere had already been suggested by Balfour Stewart twenty years earlier in 1882. In his study of terrestrial magnetism, Stewart had proposed that electrical currents flowing high in the atmosphere could explain the small daily changes in the earth's magnetic field. Such variations would be caused by the tidal movements in the surrounding "sea of air," which were caused by solar and gravitational influences. This explanation of fluctuations in the earth's magnetism, combined with many peculiar features of short-wave radio propagation, seemed to confirm the existence of the Kennelly-Heaviside layer, but these phenomena provided only indirect evidence.

After Marconi's demonstration, commercial radio links across the Atlantic were established. It was then noticed that the strength of the signals varied in a regular way during the day and night, and over seasonal and solar cycles. Furthermore, the regular daily variations were disturbed during magnetic storms. These observations suggested that the scientific study of radio propagation could lead to new knowledge about the upper atmosphere. It soon became clear that daily and seasonal variations could be explained by the changing aspects of the sun and its effect on the charge concentration in the postulated Kennelly-Heaviside layer in what came to be called the ionosphere.

The first direct evidence for the existence of the ionosphere was obtained by Sir Edward Victor Appleton, with the assistance of Miles A. F. Barnett in 1924. At the University of Cambridge, Appleton had studied radio signals from the new British Broadcasting Company station in London and noticed the typical variations in their strength. When he took up a new position at the University of London in 1924, Appleton arranged to use the new British Broadcasting Company transmitter at Bournemouth after midnight, with receiving apparatus located at the University of Oxford.

By varying the transmitter frequency, he hoped to detect a changing intensity at the receiver caused by changing interference (canceling of wave crests by the troughs of other waves) between the direct waves (along the ground) and the waves that they assumed would be reflected from the ionosphere.

On December 11, 1924, Appleton and Barnett observed the regular fading in and out of the signal as the frequency of the transmitter was slowly increased. From the lowest and highest transmitted wavelengths and the corresponding number of intensity oscillations at the receiver, they calculated that the reflection was from a height of about 100 kilometers. This confirmation of the Kennelly-Heaviside prediction was published in 1925 in *Nature* under the title "Local Reflection of Wireless Waves from the Upper Atmosphere." The result was confirmed by measuring the angle of incidence of the reflected wave as it reached the ground from a comparison of simultaneous signal variations on two receivers with different antenna orientations.

In later experiments, Appleton used an improved technique of rapid frequency changes so that variations in signal strength could be distinguished more clearly from natural fading because of changes in the ionosphere. In 1926, he found that the ionization of the Heaviside layer (E layer) was sufficiently reduced before dawn by recombination of electrons with positive ions to allow penetration by radio waves. Reflection, however, was still observed from a higher layer where the air was too thin for efficient recombination. The height of what is now called the Appleton layer (F layer) was measured at about 230 kilometers above the earth. This result was published in *Nature* in 1927 under the title "The Existence of More than One Ionized Layer in the Upper Atmosphere." Observations during a solar eclipse in 1927 showed that the height of the Heaviside layer changed, revealing that ionization was caused by solar radiation, as suggested by Heaviside.

Impact of Event

The prediction and discovery of the ionosphere were important in stimulating new scientific understanding and advances in technology. Appleton's magnetoionic theory of the ionosphere showed that electron density and the magnetic field at any layer in the ionosphere can be calculated from the critical frequency for penetrating that layer. In 1931, systematic experiments were begun to determine the variation of electron densities in the ionosphere, revealing an increase in ionization as the sun was rising and low ionization at night, except for sporadic increases possibly caused by meteoric activity. Noon ionization was found to increase as the sunspot maximum of 1937 was approached, suggesting a correlation between sunspots and increases in the ultraviolet radiation that ionizes the upper layers of the atmosphere. This made it possible to measure the ultraviolet radiation from the sun even though little of it reaches the ground. During the sunspot maximum of 1957-1958, the International Geophysical Year was established to study geophysical phenomena on a worldwide scale, including their relation to ionospheric variations. Thus, study of the ionosphere has contributed to developments in other sciences, such as astronomy, meteorology, and geophysics.

Appleton's methods proved especially valuable in the development of radio communications, radar systems, and their applications in meteorology. Long-distance radio communications, in which the waves are guided around the earth by the ionosphere, became especially important during World War II. A worldwide network of more than fifty stations was established to monitor the ionosphere and to determine the most suitable frequencies for radio transmissions as ionospheric conditions changed. Discovery of the influence of the sunspot cycle made it possible to forecast ionospheric weather and thus to improve the reliability of radio communications. On the standard broadcast band (500-1,500 kilohertz), ground waves travel about 500 kilometers, while sky waves are absorbed during the day but travel by reflection several thousand kilometers at night. At frequencies between 5 and 25 megahertz and distances greater than about 100 kilometers, radio transmission depends almost entirely on ionospheric reflections. On the 20-meter amateur radio band (14 megahertz), it is possible to reach some part of the world at almost any time of the day or night.

The early development of radar was closely associated with studies of the ionosphere. The most powerful radar systems use the over-the-horizon technique of reflecting radar signals from the ionosphere to cover distances up to about 3,000 kilometers, about ten times farther than conventional radar. For reliability, over-the-horizon radar systems depend on computers to chart the constantly changing intensity and thickness of the ionosphere and determine where conditions are best and which frequencies are needed for maximum performance. Thus, the ionosphere has become an indispensable tool for both communications and national security.

Bibliography

Aitken, Hugh G. J. *Syntony and Spark: The Origins of Radio*. New York: John Wiley & Sons, 1976. A good history of the early development of radio from the discovery of radio waves by Hertz to the work of Marconi. Only brief mention is made of the ionosphere proposal of Kennelly and Heaviside, but more than one hundred references are given on the work of Marconi and many useful diagrams of early apparatus.

Craig, Richard. *The Edge of Space: Exploring the Upper Atmosphere*. Garden City, N.Y.: Doubleday, 1968. This small book is in the Science Study Series written for secondary students and the lay public. It has a good chapter on the discovery of the ionosphere, along with helpful diagrams.

Davies, Kenneth. *Ionospheric Radio Propagation*. Washington, D.C.: U.S. Government Printing Office, 1965. This highly authoritative volume contains most of what is known about the ionosphere from both theory and worldwide measurements, including many technical details, three hundred references, and thirty pages of indexes.

Mitra, S. K. *The Upper Atmosphere*. Calcutta: Asiatic Society, 1952. A thorough treatment of the physics of the upper atmosphere with many graphs and diagrams. Chapter 6 provides an extensive discussion of the ionosphere, including a brief

BAYLISS AND STARLING DISCOVER SECRETIN AND ESTABLISH THE ROLE OF HORMONES

Categories of event: Biology and medicine
Time: April-June, 1902
Locale: Cambridge, England

Bayliss and Starling proved unequivocally that chemical integration can occur without assistance from the nervous system

Principal personages:

ARNOLD BERTHOLD (1803-1861), a German physiologist in Göttingen who demonstrated that the testes release a substance into the blood that influences male behavior

IVAN PETROVICH PAVLOV (1849-1936), a Soviet physiologist who repeated the experiments of Bayliss and Starling in 1910, reaffirming the mode of action of secretin

SIR WILLIAM MADDOCK BAYLISS (1860-1924), an English physiologist who demonstrated the existence and manner of action of the hormone secretin

ERNEST HENRY STARLING (1866-1927), an English physiologist who collaborated with Bayliss on the discovery of secretin and who coined the term "hormone"

SIR FREDERICK GRANT BANTING (1891-1941), a Canadian physiologist who succeeded in preparing highly potent extracts of insulin, a hormone required for carbohydrate metabolism, and was a corecipient with J. J. R. Macleod for the 1923 Nobel Prize in Physiology or Medicine for the discovery of insulin

Summary of Event

Much information is available in the 1990's concerning the control of secretin release and its targets, but this awareness has not always been the case. At the beginning of the twentieth century, two English physiologists (Sir William Maddock Bayliss and Ernest Henry Starling) were interested in ascertaining what triggered the pancreas to pour out digestive juices as soon as food arrived in the first part of the intestine. A nerve signal from the intestine could order the pancreas to turn on the juice. To prove that this was indeed the case, Bayliss and Starling set up an experiment. The investigators dissected the nerves of an animal's upper intestine and injected stimulating material like food from the stomach. To their astonishment, pancreatic juices poured promptly into the intestine. It should not have happened since all nerves had been cut; yet, some mysterious signal had reached the pancreas and roused it to action. Bayliss and Starling discovered that the signal was chemical in nature, not nervous. Arrival of acid-laden food had caused the intestinal wall to secrete a substance called secretin, which oozed into the bloodstream.

Physiologists know that secretin causes the pancreas to secrete a solution high in bicarbonate, which has the effect of buffering hydrochloric acid. Secretin also tends to inhibit motility of the stomach but to stimulate the secretion of pepsin (an enzyme that breaks specific internal peptide bonds in a protein). Secretin is the single most important inhibitor of gastrin release and therefore of acid secretion.

Starling first used the word "hormone" (Greek *hormon*, exciting, setting in motion) in 1905 with reference to secretin. Today, physiologists know that hormones may inhibit as well as excite; that they do not initiate metabolic transformations but merely alter the rate at which these changes occur. Bayliss and Starling also developed the concept of powerful chemical messengers in the common pool of blood and lymph.

Secretin has been obtained in crystalline form and is a basic polypeptide. The hormone disappears rapidly from circulation owing to the destructive action of an enzyme called secretinase. Small amounts of the hormone are excreted in the urine. Many materials other than hydrochloric acid stimulate the release of secretin by the duodenal mucosa; water, alcohol, fatty acids, partially hydrolyzed protein, and certain amino acids are all effective.

The crude extract prepared by Bayliss and Starling caused the pancreas to secrete enzymes as well as water and bicarbonates. Because purified secretin seemed to stimulate only the output of water and bicarbonate, a search was made for the material in the extract that stimulated secretions of the enzymes. In 1943, A. A. Harper and H. S. Raper succeeded in separating a fraction, which did not increase the volume of pancreatic juice but greatly increased the enzyme concentrations. This substance was called pancreozymin. In 1966, E. Jopes and V. Mutt, in the course of their purification of cholecystokinin (CCK), demonstrated that this hormone and pancreozymin were the same polypeptide. Because CCK was discovered first, the hormone is referred to as cholecystokinin, although some researchers still refer to it as cholocystokinin-pancreozymin. It is now known that actions of secretin and CCK are not as completely separate as previously believed. Secretin augments the action of CCK on pancreatic enzyme secretions, and CCK augments the action of secretin on bicarbonate ion secretions. This augmentation is important in the stimulation of pancreatic fluid when the pH of the duodenum, which is well buffered, does not fall significantly with the entry of a meal into the upper gastrointestinal (GI) tract. Since secretin and pancreozymin in injectable form have become available, tests of pancreatic function using these hormones as stimuli have been developed.

Secretin was available first in a form pure enough for use in humans, and more experience has accumulated with it. The secretin test was introduced by G. Agren et al. in 1936 and has been used extensively in the United States and abroad. To perform the test, a double-lumen tube is passed through the patient's nose or mouth, into the stomach for draining gastric juice and into the duodenum for collecting the pancreatic and duodenal secretions. The gastric and duodenal secretions are collected separately. Gastric juice is usually aspirated continuously by a low-pressure vacuum pump, whereas the duodenal contents are aspirated periodically. Volume, pH, bicarbonate, and enzyme determinations are made on the samples collected from the duodenum. With normal function, there is a rapid increase in flow rate following secretin injection, the maximum flow rate usually being reached in twenty minutes. A flow rate of 1 to 2 milliliters per kilogram of weight per thirty minutes is usually obtained in normal individuals. The bicarbonate concentration increases with the flow rate and may go as high as 140 to 150 milligrams per liter. As the flow rate increases, there is an associated decrease in enzyme concentration. When pancreozymin is used as a stimulus for pancreatic secretion to test pancreatic function, the procedure of the test is the same as that described for the secretin test.

Secretin increases the secretion of bile by the liver, although it does not cause the gallbladder to contract. The site of action of secretin is thought to be the small bile ducts, or ductules. The effect results in increasing the volume of bile by addition of water and markedly increasing the bicarbonate concentration. The onset of effect is similar to that of the pancreas. Some evidence is suggested that an electrolyte fraction may be secreted that is dependent on the active transport of sodium and is inhibited by inhibitors of ATPase. Secretin causes only a moderate increase in bile flow, and none at all if the liver is secreting vigorously already.

The pancreas is under the control of both nervous and humoral mechanisms. The nervous control of pancreatic secretion is provided by both the sympathetic and parasympathetic divisions of the autonomic nervous system. Humoral control of pancreatic secretion is provided by the hormones secretin, CCK, and gastrin. Atropine is said to block the effects of sympathetic stimulations, causing an increase in flow rate, indicating that cholinergic secretory fibers are present in the sympathetic innervation of the pancreas. Stimulations of the vagus, or parasympathetic innervation of the pancreas, result in the secretion of enzymes but have little or no effect on the secretions of bicarbonate. Acetylcholine is the mediator of the vagal effect. Vagotomy and atropine markedly depress the secretions of enzymes by the pancreas. Cholinergic drugs, such as pilocarpine, stimulate pancreatic secretion, whereas atropine and anticholinergic agents produce inhibitions.

Impact of Event

The discovery of Bayliss and Starling relative to the existence and manner of action (at the level of the whole organism) of the hormone secretin has been far-reaching. Ivan Petrovich Pavlov, a Soviet scientist and great pioneer in the study of conditioned reflexes, repeated the work of Bayliss and Starling in 1910 and obtained similar results. Subsequently, S. Kopec demonstrated in 1917 that a hormone from the brain controlled pupation in certain invertebrates (insects), which illustrated for the first time that central nervous structures could perform endocrine roles.

The islands of Langerhans are patches of tissue located in the pancreas that produce the hormone insulin. In 1921, Sir Frederick Grant Banting, Macleod, Charles H. Best, and James B. Collip, working in Toronto, Canada, isolated this hormone from the embryos of animals. The insulin was next injected experimentally into dogs and then into humans; it was found to be effective in relieving the symptoms of diabetes. Individuals suffering from this disease are unable to oxidize sugar in their tissues. The sugar tends to accumulate in the tissues, often with fatal results. It is known that insulin regulates the storage of sugars in the liver and the oxidation of sugar by the body. Thus, diabetes is caused by the inability of the islands of Langerhans to produce adequate quantities of insulin.

In 1949, Thomas R. Forbes (1911-   ) reviewed the work of Arnold Berthold. His paper was read before the Beaumont Club at Yale University on May 13, 1949. Berthold related how he caponized six young cockerels, then returned single testes to the body cavities of some of the birds. The grafts, placed among the intestines, became vascularized. Berthold observed that the host birds continued to exhibit the sexual behavior and accessories of normal young roosters. At autopsy, he found that the nerve supply of the grafted testes had not been reestablished. Hence, Berthold concluded that, since maintenance of sexual behavior and appearance could not have been accomplished by the nerves (which were severed), the results must have been caused by a contribution of the testes to the blood and then to the action of the added substance throughout the entire body.

In November, 1962, Donald G. Cooley (1943-   ), an American physiologist, published a manuscript entitled "Hormones: Your Body's Chemical Rousers" in which he reviewed the experiments of Bayliss and Starling and presented an updated, salient summary concerning the mechanism of hormone action. The article appeared in the November, 1962, edition of *Today's Health*. Strong evidence that a virus can cause juvenile onset diabetes was reported in May, 1979, by scientists at the National Institute of Dental Research in Bethesda, Maryland. The most serious variety of the disease, juvenile onset diabetes, is characterized by a lack of insulin. Ji-won Yoon, Marchall Austen, and Takashi Orodern isolated the virus, called Coxsackie B4, from the pancreas of a ten-year-old boy who died of a sudden and severe case of diabetes. The researchers grew the virus in cultures and injected it into mice. Some strains of mice then developed diabetes. This evidence indicated that the Coxsackie virus can be a causal factor in some cases of diabetes. Nevertheless, the fact that not all of the mice developed the disease indicated that a genetic factor may be needed to trigger its development.

Bibliography

Grossman, Morton I. "Vitamins and Hormones." In *Advances in Research and Application*, edited by Robert S. Harris. New York: Academic Press, 1958. Treats vitamins and their sources, in addition to the discovery, distribution, and modes of action of the hormone secretin. Further discusses the fate of cholecystokinin (CCK), produced by the duodenum.

Jenkin, Penelope. "Discovery of Hormones." In *Monographs on Pure and Applied Biology*, edited by J. E. Harris. New York: Pergamon Press, 1962. A panoramic view of the salient experiments of Berthold, Starling, Kopec, and others. Discusses the mechanism of hormone action as well as the role of the central nervous system.

Morgan, Howard E. "Endocrine Control Systems." In *Best and Taylor's Physiological Basis of Medical Practice*, edited by John R. Brobeck. 10th ed. Baltimore: Williams & Wilkins, 1980. An excellent reference text for students in the health professions. Discusses all aspects of hormonal control systems, including hormonal control of growth and protein metabolism. Hormonal control of permeability is well documented for ions, glucose, and amino acids.

Schauf, Charles L., et al. *Human Physiology: Foundations and Frontiers*, edited by Deborah Allen. St. Louis, Mo.: Times Mirror/Mosby, 1990. A conceptual approach to a complex subject. Presents concepts systematically to demonstrate how organ systems interact with one another. Includes diagrams where required for clarification of the reciprocal fields of histology (the study of tissues) and physiology (the study of functions).

Stille, Darlene R. "Science and Technology: Internal Medicine." In *Science Year: The World Book Science Annual*, edited by Harrison Brown et al. Chicago: Scott and Fetzer, 1980. Includes Science File, a section that provides the "News of the Year" in about forty-five short articles alphabetically arranged by subject matter from agriculture to zoology. The genetics section describes the research work of a team of scientists at City of Hope Medical Center who created a bacterial strain able to produce human insulin, a protein hormone used in treating diabetes.

Nathaniel Boggs

Cross-References

Banting and Macleod Win the Nobel Prize for the Discovery of Insulin (1921), p. 720; The Artificial Sweetener Cyclamate Is Introduced (1950), p. 1368; Du Vigneaud Synthesizes Oxytocin, the First Peptide Hormone (1953), p. 1459; Sanger Wins the Nobel Prize for the Discovery of the Structure of Insulin (1958), p. 1567; The Artificial Sweetener Aspartame Is Approved for Use in Carbonated Beverages (1983), p. 2226.

RUSSELL DISCOVERS THE "GREAT PARADOX" CONCERNING THE SET OF ALL SETS

Category of event: Mathematics
Time: June 16, 1902
Locale: Friday's Hill, Haslemere, England

Russell discovered a paradox concerning the set of all sets: Is the set of all sets which are not members of themselves a member of itself?

Principal personages:

BERTRAND RUSSELL (1872-1970), an English philosopher, mathematician, and social reformer who was the winner of the 1950 Nobel Prize in Literature

GEORG CANTOR (1845-1918), a German mathematician and logician who created set theory, which now constitutes one of the principal branches of mathematics

GOTTLOB FREGE (1848-1925), a German mathematician and philosopher and the founder of modern mathematical logic

CESARE BURALI-FORTI (1861-1931), an Italian mathematician who formulated a famous paradox concerning the greatest ordinal

HENRI POINCARÉ (1854-1912), a French mathematician and philosopher of science who made important contributions to the solution of paradoxes

Summary of Event

The study of the foundations of mathematics can be traced as far back as Greek antiquity. Since then, there have been three major crises that have brought into question a previously established area of mathematics. Each of these crises challenged mathematicians either to revise their position or leave mathematics resting on unstable ground. The first major crisis occurred in the fifth century B.C. and was brought about by the discovery that not all geometrical magnitudes of the same kind are commensurable with one another. It was resolved about 370 B.C. by Eudoxus of Cnidus, whose revised theory of proportion and magnitude may be found in the fifth book of Euclid's *Stoicheia* (*Elements*).

More than two millennia passed before the second crisis in the foundations of mathematics arose. After the discovery of the calculus by Sir Isaac Newton and Gottfried Wilhelm Leibniz in the late seventeenth century, weaknesses in the foundations of analysis soon became evident. Only after the arithmetization of analysis in the nineteenth century was the second crisis thought to be resolved. After the resolution of the second crisis in the foundation of mathematics, the whole structure of mathematics seemed to be redeemed and placed on unshakable ground. By 1900, mathematicians had seemingly imparted to their subject the ideal structure that had been delineated by Euclid in his *Elements.* Several branches were founded on rigorous axiomatic bases, the terminology had been subjected to close scrutiny, and deductive proofs replaced intuitively or empirically based conclusions. When the third crisis in the foundations of mathematics struck, its suddenness was surprising.

The third crisis in the foundations of mathematics was brought about by the discovery of paradoxes or antinomies in the fringe of Georg Cantor's general theory of sets. Since so much of mathematics is permeated by set concepts and, for that matter, can actually be made to rest upon set theory as a foundation, the discovery of paradoxes in set theory naturally cast into doubt the validity of the whole foundational structure of mathematics.

In 1897, the Italian mathematician, Cesare Burali-Forti, brought to light the first publicized paradox of set theory. The essence of the paradox can be conveyed by a description of a very similar paradox found by Cantor two years later. In attempting to assign a number to the set whose members are all possible sets and to the set of all ordinals, Cantor discovered a fundamental paradox about sets. The basic idea is that for any given set, there is always a larger one; the set of all subsets in a given set is larger than the original one. He showed that there were larger and larger transfinite sets and corresponding transfinite numbers. Cantor's theory of sets proved that, for any given transfinite number, there is always a greater transfinite number, so that just as there is no greatest natural number, there is no greatest transfinite number. How can there be a transfinite number greater than the transfinite number of the set whose members are all possible sets, given that no set can have more members than this set? By 1899, Cantor thought that one could not consider the set of all sets or its number. This caused mathematicians to recognize that they had been using similar concepts not only in their newer creations but also in the supposedly well-established older mathematics. They preferred to call these contradictions "paradoxes" because a paradox can be resolved and the mathematicians wanted to believe these could be resolved. The technical word commonly used now is "antinomy."

When Bertrand Russell first discovered Cantor's conclusion about the set of all sets, he did not believe it. He wrote in an essay of 1901: "There is a greatest of all infinite numbers, which is the number of things altogether, of every sort and kind. It is obvious that there cannot be a greater number than this, because, if everything has been taken, there is nothing left to add. Cantor has a proof that there is no greatest number, and if this proof were valid, the contradictions of infinity would reappear in sublimated form. But in this one point, the master must have been guilty of a very subtle fallacy, which I hope to explain in some future work." Russell meditated on this matter and added to the problems of the times his own "paradox." Sixteen years later, when Russell reprinted his essay in *Mysticism and Logic* (1918), he added a footnote apologizing for his mistake, stating that Cantor's proof that there is no greatest number is valid, and that the solution of the puzzle is complicated and depends on a theory of types, which is explained in Russell and Alfred North Whitehead's *Principia Mathematica* (1910-1913). Russell had studied the paradox of Cantor, and instead of finding a fallacy in it, had generated his own version.

Whereas the Burali-Forti and Cantor paradoxes involve results of set theory, the "great paradox," which Russell discovered in 1902, depends on nothing more than the concept of set itself. In order to understand Russell's paradox, one must realize that sets either are or are not members of themselves. For example, the set of all sets is itself a set, but the set of all books is not itself a book. Likewise, the set of all comprehensible things is itself a comprehensible thing, whereas the set of all persons is not itself a person. For example, take the set of all sets that are members of themselves and call it *S*, and take the set of all sets that are not members of themselves and call it *N*. Is *N* a member of itself? If *N* is not a member of itself, then *N* is a member of *N* and not of *S*, and *N* is a member of itself. On the other hand, if *N* is a member of itself, then *N* is a member of *S* and not of *N*, and *N* is not a member of itself. Russell's great paradox lies in the fact that in either case, there is a contradiction. When Russell first discovered this contradiction, he thought the difficulty lay somewhere in the logic, rather than in the mathematics. At first, Russell did not realize that this paradox strikes at the very notion of a set, a notion used throughout mathematics. It was the first of a number of troublesome mathematical paradoxes.

Although Russell first published this paradox in his *The Principles of Mathematics* (1903), he had communicated it a bit earlier to the mathematician Gottlob Frege. Russell had sent a letter to Frege on June 16, 1902, in which he told him about the paradox. Frege received the letter while the second volume of his *Grundgesetze der Arithmetik* (1893, 1903; the basic laws of arithmetic) was at the printer. In this treatise on the foundations of arithmetic, Frege had used the theory of sets, which involved the very paradox that Russell had noted in his letter. In effect, Russell's letter broke the foundation of Frege's work just as his treatise was finished—work on which Frege had labored for more than twelve years.

Russell's great paradox has been popularized in many forms. One of the best known of these forms was introduced by Russell himself some years after his discovery of the paradox; it is known as the "barber paradox." This paradox concerns the plight of a village barber who has advertised that he does not shave those persons who shave themselves, but he does shave all those persons and only those persons of the village who do not shave themselves. The paradoxical nature of this situation is revealed when the barber asks himself whether he should shave himself. If he does shave himself, then he should not because he has advertised that he shaves all those persons and only those persons of the village who do not shave themselves. If he does not shave himself, then he should, according to his claim to shave all those who do not shave themselves. The barber is in a logical predicament.

Impact of Event

Although many mathematicians of the early 1900's tended to disregard the paradoxes because they involved set theory, which was new and peripheral at the time, others were disturbed, recognizing that the paradoxes affected not only classical mathematics but also general reasoning. Russell's great paradox had a catastrophic effect on the world of mathematics. Since its discovery, additional paradoxes in set theory have been produced in abundance. The existence of paradoxes in set theory clearly indicates that something is wrong. Much literature on the subject has been published, and numerous attempts at a solution have been offered.

As far as mathematics is concerned, there seems to be an easy way out. One has merely to reconstruct set theory on an axiomatic basis sufficiently restrictive to exclude the known antinomies. The first such attempt was made by Ernst Zermelo in 1908, and subsequent refinements have been made. This procedure has been criticized as merely avoiding the paradoxes; certainly it does not explain them. Moreover, this procedure does not guarantee that other kinds of paradoxes will not surface in the future.

In 1905, Russell thought that the paradoxes of set theory arose from a fallacy he called the vicious circle principle. Simply stated, Russell's principle says that whatever involves all of a collection must not be one of the collection. In other words, if to define a set of objects, one must use the set of objects itself in the definition, then the definition is meaningless. This was accepted by Henri Poincaré in 1906, who called these definitions impredicative. Poincaré argued that impredicative definitions define an object in terms of a set of objects that contains the object being defined. These definitions are illegitimate; by restricting them, at least the known paradoxes of set theory will be avoided.

The restrictions of Russell and Poincaré amount to restrictions on the notion of set, and, although heeding these restrictions provides a solution to the known paradoxes of set theory, there is at least one serious objection to this solution: Although some parts of mathematics contain impredicative definitions that may be circumvented, many others contain impredicative definitions that cannot be circumvented, and mathematicians are reluctant to discard these parts. For example, the least upper bound of a given nonempty set of real numbers is the smallest member of the set of all upper bounds of the given set. This is an impredicative definition that mathematicians are reluctant to give up.

There have been various attempts to solve the paradoxes of set theory. Some, for example, search for the cause of the paradoxes in logic, and this has brought about a rigorous investigation into the foundations of logic. In effect, Russell's discovery of the great paradox has led to a search for the causes and solutions to such paradoxes in an effort to restore the foundations of mathematics.

Bibliography

Jager, Ronald. *The Development of Bertrand Russell's Philosophy.* London: George Allen & Unwin, 1972. A fine survey of the development of Russell's philosophy. Part of the book is labeled "The Theory of Logic," and section D of this part, entitled "Set Theory and Paradoxes," provides a good general account.

Kneale, William, and Martha Kneale. *The Development of Logic*. Oxford, England: Clarendon Press, 1962. The best historical treatment of its subject. Philosophically oriented and accessible to the general reader. Chapter 11, entitled "The Philosophy of Mathematics After Frege," is an in-depth treatment of the paradoxes of the theory of sets and Russell's theory of types, as well as other solutions to the paradoxes of set theory. Highly recommended as a scholarly treatment of Russell's paradox.

Russell, Bertrand. *My Philosophical Development*. New York: Simon & Schuster, 1959. An excellent basic account of the events surrounding the discovery of the paradox and of the paradox in general. Russell clearly articulates his influences and his attempts to resolve the paradox. Essential for the general reader.

____________. *The Principles of Mathematics.* 2d ed. Cambridge, England: Cambridge University Press, 1938. The work in which Russell first published his paradox. The fundamental thesis is that mathematics and logic are identical; that is, mathematics is merely later deductions from logical premises. The book does not contain much mathematical and logical symbolism, and as such, is accessible to the diligent general reader.

Schilpp, Paul Arthur, ed. *The Philosophy of Bertrand Russell*. Evanston, Ill.: Northwestern University Press, 1944. A valuable collection containing essays by distinguished scholars on many aspects of Russell's work. One may find useful information about the discovery of the paradox in a number of places throughout the volume, including statements by Russell on his discovery of the paradox and his attempts to resolve it.

Van Heijenoort, Jean. *From Frege to Gödel: A Source Book in Mathematical Logic, 1879-1931*. Cambridge, Mass.: Harvard University Press, 1967. An important collection of classic selections on mathematical logic with introductory notes to each selection. Although some of the selections are difficult, many are of interest to the general reader. The reader will find Russell's 1902 letter to Frege and Frege's response to Russell. Other selections, such as Cantor (1899), Burali-Forti (1897), and Russell (1908), are particularly relevant to those interested in Russell's paradox.

Jeffrey R. DiLeo

Cross-References

Brouwer Develops Intuitionist Foundations of Mathematics (1904), p. 228; Zermelo Undertakes the First Comprehensive Axiomatization of Set Theory (1904), p. 233; Russell and Whitehead's *Principia Mathematica* Develops the Logistic Movement in Mathematics (1910), p. 465; Gödel Proves Incompleteness-Inconsistency for Formal Systems, Including Arithmetic (1929), p. 900; Cohen Shows That Cantor's Continuum Hypothesis Is Independent of the Axioms of Set Theory (1963), p. 1751.

TSIOLKOVSKY PROPOSES THAT LIQUID OXYGEN BE USED FOR SPACE TRAVEL

Category of event: Space and aviation
Time: 1903
Locale: Kaluga, Soviet Union

Tsiolkovsky determined by mathematical calculation the necessity for the use of liquid propellants to launch rockets into space

Principal personage:
KONSTANTIN TSIOLKOVSKY (1857-1935), a Russian scientist who was one of the founders of modern astronautics

Summary of Event

Konstantin Tsiolkovsky was known by the nickname "Bird" for his characteristic flitting movements and lightheartedness. When he was ten, however, he contracted scarlet fever, which left him deaf. From that deafness, he developed a severely limited view of life, and buried himself in books. He was one of the three men most credited with the formulation of rocket theory; Robert H. Goddard, an American, and Hermann Oberth, a German, share that distinction with him. Tsiolkovsky and Goddard both suffered a severe childhood disease, both became teachers, and both maintained a total dedication to spaceflight.

Tsiolkovsky credited his deafness for his dedication, and, therefore, for his success with rocket theory formulation. His isolation from other people meant that he never participated in most of the social aspects of life. Tsiolkovsky discovered that because of literary limitations, it was easier to prove his theories physically than to perform research with the available resources. The young scholar was sent to Moscow to study mathematics and physics; he then added astronomy. Tsiolkovsky believed that, through spaceflight, humans could also claim the heavens. It was partially the lack of funds for materials for experimentation that caused the young scientist to prove his theories mathematically, with meticulous journals of his thinking. The experiments Tsiolkovsky performed were financed by cutting back on his food allowance, which, after a time, caused him to weaken and to become ill. Before he was sent home for recuperation, however, he had caught the vision of the theory of spaceflight. He was sitting in a city park, watching a crowd of teenagers hopping off a hay wagon. As each youngster jumped off, the wagon would lurch forward slightly. It was at that moment that Isaac Newton's law of action and reaction came to life for Tsiolkovsky, and he realized what would be needed to propel a rocket into space.

In 1879, Tsiolkovsky took and passed his teacher's certification test, without ever attending school. As his deafness was a problem, most of his teaching consisted of lectures. When he asked a rare question of one of his students, the pupil would have to stand at his side and yell into his left ear. His dedication extended to expending his meager salary on experiments material for his classes. Of course, his theorizing never ceased. He would rise early to study, teach classes, perform familial duties, then experiment in the evenings.

The Society for Physics and Chemistry in Petersburg heard of him through one of his papers. Unfortunately, they became enraged because he had seemingly presumed to write a paper on a subject already proven, calling it original. They later realized, however, that his work was indeed original; he apparently did not know that the theory was already proven. He was saved by the fact that he had proved the theory using an entirely different approach. In other words, he had been reinventing a concept because of his nonsystematic self-teaching, which was not always well-balanced, and left gaps in his knowledge. Dmitry Ivanovich Mendeleyev was impressed with Tsiolkovsky and led the drive to grant membership unanimously in the elite scientific society.

In 1885, Tsiolkovsky dedicated himself to aviation and for two years spent every moment theorizing and creating an all-metal, piloted balloon. He presented his first public lecture, entitled "The Theory of the Aerostat," at the Polytechnical Museum in Moscow in 1887. Following his lectures and heavy two-year workload, he contracted a serious illness and lost his voice for a year. He also lost his library and models when his house burned. The only work saved was the lecture he had presented in Moscow. In 1892, Tsiolkovsky moved to Kaluga, where he took up teaching posts at a high school and at a school for the daughters of clergy. This was an exciting move for him, as Kaluga was a busy, industrial city, with communications to the outside world. For the first time since his illness, he regained his earlier vigor, and his scientific research escalated. In 1894 he presented "Aeroplan ili ptitsepodobnaya (aviatsionnaya) letatelnaya mashina" (the airplane or a birdlike flying machine), in which he had designed a monoplane, a project that did not appear for another twenty years. He also constructed the first Soviet wind tunnel in order to study aerodynamic principles of flight. In 1898, the same year in which the Wright brothers performed their first powered flight, he presented "Issledovanie mirovykh prostranstv reaktivnymi priborami" (exploration of space with reactive devices). Unfortunately, it was not published until 1903, partially because of its highly technical nature. There was little reaction, although Goddard's paper, "A Method of Reaching Extreme Altitudes," published in the same period of time, created a furor.

Between 1903 and 1933, ideas on the fundamentals of space travel flowed from Tsiolkovsky. His was a unique genius, in that he recognized the truth of an idea, not requiring an experimental testing. While Goddard translated his ideas into a tangible product, Tsiolkovsky created endlessly, usually with only his notebooks for proof. Throughout his years of study and research, he repeatedly returned to the idea of an aerostat, a metal-skinned balloon.

Tsiolkovsky's written works are numerous; however, "Issledovanie mirovykh prostrantsv reaktivnymi priborami" proposed several unique, advanced ideas. One of the most important theories was that only reaction devices would function both within the atmosphere and in space. He also proposed that the black powder rockets in use at the time were not sufficient to carry a rocket into space. He went on to suggest the best liquid propellants.

Tsiolkovsky decided that hydrocarbons were the best explosives, citing problems with hydrogen and the limitations of oxygen. Hydrogen would be difficult to store, since it evaporates quickly. It also absorbs energy when going to the gaseous state, when, at that point, energy is needed. Oxygen is useful, in that it would be used as a coolant, in addition to providing air for pilots. It would be necessary in upper atmospheric flight; it is also extremely combustible, giving off heat, not absorbing, like hydrogen. He discarded the notion of using high-pressure gases, because of the need for heavy, sealed containers that would increase the launch weight. He concluded that, while hydrocarbons would produce 20 percent less thrust, their volatility and lower container weight requirements made them the best fuel. He went on to suggest the use of superheated water for initial tests, cutting costs of launches, and predicted a maximum range of 60 kilometers.

Tsiolkovsky described the parameters of liquid fuel rocket components, including the "explosion tube," which, he predicted, should be a cone. In a cone with at least a 1 degree angle, the combustion pressure would be exerted on the entire inner surface. The cone should be as long as possible, and, when needed, the length could be assisted with bends in the structure. He also stated the need to overcome the pressure of explosion by pumping explosives into the combustion chamber. The fuel pump would actually require limited power, and could be an airplane-type engine. The control system was also detailed, including three rudders that would function on take off, in space, and for a gliding reentry: horizontal rudder, direction rudder, and lateral stability rudder. Their sizes would correspond to the size of airplane rudders. In short, Tsiolkovsky described, in 1903, the fundamental systems which were used many years later for early rocket space flight.

Impact of Event

Tsiolkovsky's theories of spaceflight were the basis for actual early flight. Unfortunately, the scientific community did not recognize his work, even after the initial publication of his 1903 paper on liquid-fuel rockets and their design. It was not until 1924, when Oberth republished his work, in both German and Russian, that Tsiolkovsky was termed "the father of space travel." It was his persistence and dedication that made the world examine the possibilities of spaceflight. This same persistence has characterized all efforts, no matter the age of those who hoped to conquer space.

Because of the remarkable accuracy of Tsiolkovsky's theories on liquid-fuel rocketry and the success of the technology developed on his theories, his more far-reaching, somewhat fantastical ideas have not been entirely dismissed. For example, his papers "Will the Earth Ever Be Able to Inform the Living Beings on Other Planets of the Existence on It of Intelligent Beings?" and "The Unknown Intelligent Forces" have influenced modern scientists to pursue the likelihood that there may be other living beings in the universe. At least two major projects have dealt with communication with other life forms.

Tsiolkovsky also predicted the possibilities of a space station. It would serve as an interplanetary way station, servicing rockets en route to other planets. This space station could be constructed in sections carried aloft by large rocket-powered spacecraft, and assembled on location. The visionary scientist described his impression of weightlessness and was remarkably accurate. He surmised that plants grown on a space station, in support of planetary cities and colonies, would grow quickly because of lack of gravity. Solar hothouses on the space station would be closer to the sun and could grow food and regenerate air for colonization.

Liquid-fuel rocket theory made possible the spaceflight which continues still. The reduction of mass in relation to velocity made escape from the earth's gravity a reality. Using liquid as coolant then as fuel lowered the lift-off weight as well. Both Tsiolkovsky's "rocket train" and "rocket squadron" multistage configurations, using liquid fuel, made possible the immense thrust necessary to launch a manned spaceflight. From Tsiolkovsky's observation of Soviet teenagers at play, and his subsequent vision of reaction motion, locomotion of spacecraft within the vacuum of space is now a reality for the world. Tsiolkovsky's description of a gliding reentry has been used routinely for the safe return of the American space shuttle.

The memorial built in Tsiolkovsky's honor following his death quotes Tsiolkovsky's belief that "Mankind will not remain bound to the earth." Just three years before his death, Tsiolkovsky finally won reknown as a Soviet national hero. His determination that humans should not remain earthbound and his persistence in proving the possibilities of space travel are still affecting modern space technology as the very cornerstone. In fact, the effects of his remarkable insight and genius have not yet come to fruition, as humankind continues to explore his visions.

Bibliography

Ley, Willy. *Rockets, Missiles, and Space Travel*. New York: Viking Press, 1951. The author is known as probably the most prolific writer of space history. His style is very readable on a college level, yet it is dense with information. This volume details the history of rocketry from the early conceptions. Contains technical tables.

Ordway, Frederick J., III, and Mitchell R. Sharpe. *The Rocket Team*. New York: Thomas Y. Crowell, 1979. This is a detailed history of the development of the V-2, with developmental references to Tsiolkovsky. College-level reading, with fascinating black-and-white photographs. An excellent resource for early rocket development history.

Thomas, Shirley. *Men of Space*. Vol. 1. Philadelphia: Chilton, 1960. This is a detailed biography of selected men, written to bring to light their extraordinary work. Those featured in this collection were chosen by their colleagues for their contributions to space. While some are well known, others are more acknowledged by the scientific community than the general public.

Tsiolkovsky, K. E. *Works on Rocket Technology*. Washington, D.C.: National Aeronautics and Space Administration, NASA TT F-243, 1965. This is a NASA technical translation of the compilation of Tsiolkovsky's work. The information is easy to follow; written for the college level. Charts and some technical information require a physics background.

Winter, Frank H. *The First Golden Age of Rocketry.* Washington, D.C.: Smithsonian Institution Press, 1990. This is a history of the Congreve rocket and practical developments in the field of rocketry. This is another side of the story of rocketry, necessary for a complete understanding of the developmental history. Tsiolkovsky is referred to, in passing, as a dreamer.

Ellen F. Mitchum

Cross-References

Goddard Launches the First Liquid Fuel Propelled Rocket (1926), p. 810; The Germans Use the V-1 Flying Bomb and the V-2 Goes into Production (1944), p. 1235; The First Rocket with More than One Stage Is Created (1949), p. 1342.

HALE ESTABLISHES MOUNT WILSON OBSERVATORY

Category of event: Astronomy
Time: 1903-1904
Locale: Mount Wilson, California

Hale built several telescopes that served as tools for observation, and developed new techniques for building more powerful telescopes

Principal personage:
GEORGE ELLERY HALE (1868-1938), an American scientist of astronomy; inventor and principal force in establishing observatories at Mount Wilson and Palomar, California

Summary of Event

In the late nineteenth century, observational astronomy had not advanced far beyond Galileo's first telescope in either power or design. The majority of the powerful telescopes were still small, refracting designs. These telescopes were very similar to those used by ship captains; they contained a fixed eyepiece lens and an adjustable lens that could be positioned to focus the light of an image onto the eyepiece. Larger, more powerful reflecting telescopes were being designed and built; these telescopes used a concave mirror to collect light from an object and focused it onto a series of lenses or an adjustable eyepiece. The telescopes of this period were still individual instruments, which were primarily powerful eyes without an ability to analyze the visual information they received. Other instruments of physics and chemistry had not yet been coupled with these telescopes to exploit their more powerful view of the universe.

George Ellery Hale, after graduating from the Massachusetts Institute of Technology, moved to Chicago. At MIT, Hale had access to the 38-centimeter refracting scope at Harvard and the experience of its astronomers. In Chicago, Hale became involved in the building of a 102-centimeter reflecting telescope. The telescope was built at Lake Geneva, Wisconsin, in conjunction with President William Rainey Harper of the University of Chicago; major funding was provided by Charles Yerkes. In 1903, with the success of the Yerkes telescope, Hale left Chicago to establish an observatory at Mount Wilson, California.

The building of an observatory at Mount Wilson was initially funded by the Carnegie Institution of Washington. The decision to build the observatory atop Mount Wilson was based upon its elevation, terrain, and proximity to Pasadena. The elevation was considered ideal because it provided the observatory with a means of moving above the clouds which at lower elevations would interfere with viewing. This location, more than a mile above the valley, afforded the observatory additional clear days for observing. Another key feature of the mountain was the large amount of trees and foliage growing on the mountain sides. The trees and foliage help to absorb the radiant heat of the mountain; most of the heat reflected from the earth was absorbed by the growth. The mountain's ability to absorb heat allowed Mount Wilson to be designed for both stellar and solar observation. The solar observation is easily hindered by the waves of heat that reflect back from the earth and distort the rays of light coming into the solar telescope. The third advantage of the Mount Wilson site was its proximity to Pasadena. The new California Institute of Technology was based in Pasadena; it was also a major city with access to rail transportation, many large manufacturing companies, and industrial shops. Pasadena was also the location of the workshops and business offices of Mount Wilson.

The first project Hale undertook was the moving of the Snow Solar Telescope to Mount Wilson. The telescope was disassembled, shipped to Pasadena, and then brought up the mountain in pieces. While the railroads brought the pieces to Pasadena, the trip up the mountain was a more primitive journey involving mules and wagons struggling up winding dirt roads, which was delayed by accidents along the route. The Snow Telescope was rebuilt and began observations near the end of 1903. Hale immediately encountered his first major design problem. The Snow Telescope was a horizontally designed scope that rested upon the ground in the midst of the reflected heat waves of the earth. The heat became so intense that the mirrors of the telescope would expand and distort during the hot, middle part of the day; therefore, observing was limited to early morning and evening. Hale was disappointed by the limited observation and the quality of the images that the telescope was producing. He had a hypothesis that elevating the telescope would bring it out of the intense waves of heat and provide a cleaner image. To test his hypothesis, he climbed a tall evergreen with a small telescope in hand and compared the image he saw from the tree top with the one he had observed at the foot of the tree. Convinced that he was correct in his theory, Hale began construction of a trial elevated solar telescope.

This new telescope was built as an 18-meter tower. The tower was the telescope, and its immovable nature required Hale to rethink traditional telescope designs. The final version of the telescope consisted of a rotating dome atop the tower which housed a mirror for gathering light. This light was reflected down a shaft to the base of the tower where it was focused by a series of lenses and mirrors. Hale's innovative design was a great success, allowing solar observers to track the sun across its path for the duration of the day. In addition, the reduction in the waves of heat that reached the gathering mirror 18 meters above the earth provided the observers with more detailed images. The success of this tower led to the construction of a more powerful 46-meter tower.

Hale built the solar towers, which increased the duration of observation and the detail of the images, and he added a spectrograph to analyze the data collected by the telescopes. The spectrograph is capable of breaking light into its component parts which scientists can then analyze to determine the nature of the chemical reactions and components of the light-producing bodies. The spectrograph allowed the scientists to observe the sun and to see what it was made of. In order to maintain a constant temperature for the spectrographs critical to their ability to function, Hale housed the spectrographs in wells dug beneath each tower. Not only did the wells provide a constant temperature but they also provided sufficient room for the large spectrographs needed to do the work.

Solar observation, however, was not the only role of Mount Wilson observatory. The success of the 102-centimeter telescope at Lake Geneva inspired Hale to build a 152-centimeter telescope at Mount Wilson. This telescope was built for its value in observation and to test designs for the building of the 254-centimeter telescope, the largest of its day. In 1906, Los Angeles businessman John D. Hooker provided the funds to build a 213-centimeter telescope. A short time later, Hooker increased his funding so the 254-centimeter telescope could be built. During a visit to Mount Wilson in 1910, Andrew Carnegie announced that he would provide half a million dollars to mount and house the telescope. With adequate funding and the 152-centimeter telescope complete, Hale began the construction of the 254-centimeter telescope. Using a modified design of the 152-centimeter telescope housing, Hale was able to construct quickly the building which would house the telescope. In fact, the building of the telescope was rather uneventful except for the most critical aspect of the telescope: its glass mirror.

The glass that would be used for the 254-centimeter mirror was poured and imported from France. The glass was shipped to the United States without being inspected by anyone until its arrival at Mount Wilson. Hale and his colleagues were shocked and disappointed not because the glass was made from green wine bottles but because it was full of bubbles. Hale was afraid that the bubbles would cause the glass to warp or crack when it expanded or shrunk as the temperature changed. The group did not have many options, however, so they decided to take a chance with the glass and proceeded with the grinding and coating. The bubbles turned out to be of no harm, and the construction of the telescope was completed without incident.

Impact of Event

Hale's innovative designs for solar telescopes, coupled with the introduction of the spectrograph attached to the telescopes, provided the foundation for intense solar research. There was now available a means to observe solar flares, solar storms, and other solar disturbances and a method to understand the internal chemical changes that occurred with and precipitated these disturbances.

The 254-centimeter telescope provided scientists with a view 200,000 times greater than that of the human eye. This great power of sight was exploited by astronomer Edwin Powell Hubble. At the time the 254-centimeter telescope was completed, there was controversy among astronomers about the nature of the Milky Way galaxy: What were the clouds that could be seen in the Milky Way? Using the 254-centimeter telescope, Hubble discovered that these clouds were actually distant galaxies and there were hundreds of galaxies that the naked eye had not previously observed. Furthermore, Hubble analyzed the light spectrum of these galaxies and discovered that these galaxies were traveling away from the Milky Way. After carefully analyzing the data, it was determined that the galaxies farther away were traveling at a greater speed than those closer to the Milky Way. This was the first hard evidence for what would become known as the big bang theory. With the data collected using the larger telescope, the discovery of stellar distances, an expanding universe, Doppler effect changes in the light spectrum, and new galaxies was possible.

The success of this telescope and the success of Mount Wilson as a premier sight from which to observe inspired Hale to build a larger telescope. He soon discovered that the 254-centimeter telescope provided information that provoked more questions which could only be solved by a deeper look into space; it proved that large telescopes could successfully be built and the increased cost of construction was rewarded with an equivalent increase in scientific data. Because of this success, Hale began work on building a 508-centimeter telescope in Palomar, California.

The location of Mount Wilson had as much to do with the success of Mount Wilson as the genius of Hale and the brute force of the instruments. The location attracted other astronomers throughout the years after the initial construction of the telescopes. The first infrared interferometer research was conducted at Mount Wilson because of its ideal location. Using two telescopes to observe the same star, the interferometer gives scientists a look at the physical changes a star progresses through as it burns itself out.

Bibliography

Hale, George Ellery. *Signals from the Stars*. New York: Charles Scribner's Sons, 1931. This is Hale's account of the events that led to the founding of Mount Wilson Observatory and his later involvement in the construction of the Palomar Observatory. This work synthesizes the major accomplishments, both solar and stellar, in design and construction of the telescopes. Hale gives an overview of the knowledge gained from the work by the many scientists at Mount Wilson.

____________. *Ten Years' Work of a Mountain Observatory*. Washington, D.C.: Carnegie Institution of Washington, 1915. This book was written as a defense for the construction of the 254-centimeter telescope. Hale goes into much detail about the advances in scientific knowledge, in construction techniques, and methods of observing in the hope of convincing the scientific community of the practicality of this telescope.

Macpherson, Hector. *Makers of Astronomy*. Oxford, England: Clarendon Press, 1933. Written while Hale was still alive and active on the construction of Palomar, this book places Hale in historical context with the great astronomers. Provides a brief insight into how Hale's contemporaries viewed him.

Trefil, James S. *The Moment of Creation*. New York: Charles Scribner's Sons, 1983. An overview of the origins of the universe focused on the big bang theory. Details Hubble's work at Mount Wilson and explains the Doppler effect, the expansion of the galaxies, and how the stellar distances were measured. Illustrations, with a bibliography.

Woodbury, David O. *The Glass Giant of Palomar*. New York: Dodd, Mead, 1941.

An excellent book detailing the creation of Hale's two giant telescopes: the 254-centimeter and the 508-centimeter. A brief biography of Hale, history of the problems encountered at Mount Wilson, and the history of Palomar through its completion. This book examines the foundation work done at Mount Wilson. Illustrated, with a bibliography.

Charles Murphy

Cross-References

Hubble Demonstrates That Other Galaxies Are Independent Systems (1924), p. 790; Hubble Confirms the Expanding Universe (1929), p. 878; Oort and Associates Construct a Map of the Milky Way (1951), p. 1414; Parker Predicts the Existence of the Solar Wind (1958), p. 1577.

BECQUEREL WINS THE NOBEL PRIZE FOR THE DISCOVERY OF NATURAL RADIOACTIVITY

Category of event: Physics
Time: September 10, 1903
Locale: Stockholm, Sweden

Becquerel discovered the phenomenon of natural radioactivity and was awarded the Nobel Prize in Physics

Principal personages:

ANTOINE-HENRI BECQUEREL (1852-1908), a leading French physicist who discovered naturally existing radioactivity

MARIE CURIE (1867-1934), a Polish-French physicist and chemist who pioneered research in radioactivity

SIR JOSEPH JOHN THOMSON (1856-1940), an English physicist who discovered the electron

HEINRICH HERTZ (1857-1894), a German physicist who devised an instrument to measure the shape of both electrical and magnetic waves

ERNEST RUTHERFORD (1871-1937), a leading physicist who contributed toward radioactivity and the structure of the atom

FREDERICK SODDY (1877-1956), an English chemist who coined the term "isotope"

Summary of Event

The discovery of radioactivity is only part of the larger story of scientific investigations of the electromagnetic spectrum and is intimately connected with the discovery of X rays. During the final two decades of the nineteenth century, building on the theoretical foundations of James Clerk Maxwell, scientists explored the nature of a variety of electromagnetic radiation. If Maxwell's theories were correct, it would be possible to produce electromagnetic radiation through a moving electrical charge. In 1887, Heinrich Hertz attempted such an experiment and was the first to identify radio waves. This largely completed the electromagnetic spectrum to include visible light, infrared radiation (discovered earlier by William Herschel), and ultraviolet radiation (discovered by Johann Wilhelm Ritter). By 1890, the spectrum of electromagnetic radiation extended from the long-wavelength radio waves to the extremely short-wavelength ultraviolet rays. The missing member of this family was X rays, which completed the electromagnetic spectrum. In 1895, Wilhelm Conrad Röntgen, while experimenting with a cathode ray apparatus, discovered a powerful ray that penetrated thick paper as well as thin metals. Called "X ray" because of its unknown composition, this radiation had the ability to penetrate materials, and its power and mystery captured the imagination of both the public and other scientists. Antoine-Henri Becquerel was one of those scientists whose imagination was triggered by this event.

For several generations, the Becquerel family had been involved closely with French science. Both Becquerel's father and grandfather were distinguished scientists, and he followed in their footsteps. Becquerel's early research focused on optics; later, he shifted to infrared radiation and finally to the absorption of light in crystals. He completed his doctoral dissertation in 1888, and in 1899, was elected to the Academy of Sciences. During the following years, Becquerel undertook few research projects but moved actively through the academic ranks. He held an unprecedented three chairs of physics at different institutions at the same time and became chief engineer for the Department of Bridges and Highways. By 1896, Becquerel was one of the most successful and powerful scientists in France, and yet none of his accomplishments could be considered major contributions to the history of science. His interest in research was revived by Röntgen's discovery of X rays. He concluded that given that X rays were generated through the use of the cathode ray tube along with visible light, and that both X rays and light were found at the place where the beam produced a fluorescent pattern, therefore, X rays and light may be produced by the same mechanism. (It turned out that fluorescence was purely incidental to the production of X rays.) Becquerel began to search for materials other than glass that could be made to fluoresce through the use of ultraviolet light. His earlier work of crystals brought to mind a uranium-potassium-sulfate compound that possessed such qualities. He exposed to sunlight a photographic plate wrapped in light-proof papers with the uranium crystals placed on top. He had hoped that the ultraviolet rays of sunlight would trigger the crystals to produce X rays and expose the photographic plate. On developing the plate, Becquerel found it blackened by some penetrating rays.

One week later, Becquerel decided to set a control for his experiment by repeating the procedure in total darkness. Much to his surprise, the photographic plates were exposed again. This result violated his working hypothesis: that ultraviolet rays triggered the production of X rays in the uranium crystal. In the weeks that followed, Becquerel considered a number of alternative explanations for this contradictory result. He thought there was a possibility of residual effects from the initial experiment and so isolated his crystals in the dark for some time. The effect, however, remained the same. Becquerel tested other luminescent crystals but could not reproduce the results. This led him to test nonluminescent uranium compounds, which produced penetrating rays, and finally he tried a disk of pure uranium and found the penetrating radiation several times as intense. On May 18, 1896, Becquerel announced the discovery of radioactivity.

In 1897, Becquerel completed a series of studies to separate other possible sources for this radiation and to establish the properties of this form of radiation. He was able to show that the presence of the penetrating rays was a particular property of uranium. He also found that the emission of the rays from the uranium was continuous and independent of any external source. The discovery of radioactivity produced a response equal to that of the discovery of X rays. Laboratories immediately began an extensive examination of this new phenomenon. After his discovery of radioactivity, Becquerel continued his pioneering work in nuclear physics with the identification of electrons in the radiation of radium and produced the first evidence of the radioactive transformation. In 1903, Becquerel and his student Marie Curie were awarded the Nobel Prize in Physics.

Impact of Event

The discovery of radioactivity opened a new world of nuclear physics. At the Cavendish Laboratory in England, Sir Joseph John Thomson and his colleague Ernest Rutherford began to study the ionizing powers of the new rays. Rutherford discovered that there were two kinds of radioactive rays from uranium: alpha-rays, which had low penetrating powers, and beta-rays, which possessed greater abilities to penetrate metal foils. In 1898, Marie and Pierre Curie found that thorium had radioactive properties, and in quick succession they added polonium and radium to the list of radioactive elements. Marie Curie also gave the name to the process whereby uranium gave off rays: radioactivity. As the list of radioactive elements grew in confusing profusion, Rutherford and Frederick Soddy produced the theory of radioactive transformation through "decay" of one element to another. The concept of isotopes also simplified the number of elements on the list. Rutherford utilized the alpha particles to investigate the structure of the atom, which revolutionized the model of the atom. In 1900, Paul Villard observed a third ray more powerful than X rays, called gamma rays, which were emitted by radioactive substances.

By the first decade of the twentieth century, the initial exploration of radioactivity had opened the door to the study of nuclear physics, which changed the very concept of the nature of the material world. In a series of papers published in 1914, Rutherford described the hypothesis of a new atomic model that included the proton as the nucleus of the atom. The Rutherford model of the atom gained prominence through the work of Niels Bohr. Bohr saved this new model of the atom by proposing a quantum theory, whereby the energy of the spinning electron emitted energy only in specific quanta. This meant that during stable orbits of the electron, there was no emission of radiation, but when the electron jumped from one orbit to another through increasing or decreasing energy of the atom, radiation would be emitted. Furthermore, the quantum numbers for the electrons were whole numbers. Although this model of the atom was later referred to as the "Bohr atom," credit must be given to Rutherford for providing the basic structure of the atom. The stage was set to investigate further the structure of the atom, a process that involved a multinational effort through the twentieth century. In the history of science, there exists no single revolution that is comparable to the discovery of radioactivity.

Bibliography

Broglie, Louis de. *The Revolution in Physics: A Non-Mathematical Survey of Quanta*. Translated by Ralph W. Niemeyer. New York: Noonday Press, 1953. A highly recommended text for those seeking nontechnical information on quantum mechanics. Attempts to provide a popular explanation of the rapidly changing world of physics and makes the difficult subject of quantum mechanics accessible to the general reader.

Crowther, J. G. *The Cavendish Laboratory, 1874-1974*. New York: Science History Publications, 1974. Though this work covers the history of the laboratory, some of the text focuses on the person who provided the foundations and research directions for the future. Various chapters cover Thomson's early years at the laboratory, his initial work as director, his assistants and students, his work on the electron, and the later period through the end of World War I. In addition, two chapters cover his predecessor Lord Rayleigh, and eight chapters discuss his successor, James Rutherford.

Einstein, Albert, and Leopold Infeld. *The Evolution of Physics: The Growth of Ideas from Early Concepts to Relativity and Quanta*. New York: Simon & Schuster, 1938. This is probably one of the most accessible single volumes of history on the development of modern physics available to the general reader. There are few technical terms, and no knowledge of mathematics is required. The sections on the decline of the mechanical view and on quanta are highly recommended.

Jammer, Max. *The Conceptual Development of Quantum Mechanics*. New York: McGraw-Hill, 1966. This work traces both the physics and the conceptual framework of quantum theory. The sections covering the formative development quantum are moderately accessible for the general reader.

Romer, Alfred. *The Restless Atom: The Awakening of Nuclear Physics*. Garden City, N.Y.: Anchor Books, 1982. This standard 1960 text on the early days of nuclear physics had been out of print for some time but was republished by Dover. Romer makes the details of this part of history easily accessible, and it is highly recommended.

Segrè, Emilio. *From X Rays to Quarks: Modern Physicists and Their Discoveries*. San Francisco: W. H. Freeman, 1980. Segrè was one of the handful of physicists who participated directly in nuclear physics, received a Nobel Prize for his work, and wrote a number of popular accounts on the history of physics. The earlier sections of this volume cover the discoveries and theories of those who produced a coherent picture of the atom. Consequently, it is possible to appreciate the full significance of Becquerel's contribution to the field of nuclear physics.

Victor W. Chen

Cross-References

Röntgen Wins the Nobel Prize for the Discovery of X Rays (1901), p. 118; Thomson Wins the Nobel Prize for the Discovery of the Electron (1906), p. 356; Bohr Writes a Trilogy on Atomic and Molecular Structure (1912), p. 507; Rutherford Presents His Theory of the Atom (1912), p. 527; Rutherford Discovers the Proton (1914), p. 590; Lawrence Develops the Cyclotron (1931), p. 953; Chadwick Discovers the Neutron (1932), p. 973.

THE WRIGHT BROTHERS LAUNCH THE FIRST SUCCESSFUL AIRPLANE

Category of event: Space and aviation
Time: December 17, 1903
Locale: Kitty Hawk, North Carolina

The Wright brothers invented, built, and flew the world's first power-driven, heavier-than-air device, opening the aviation age

Principal personages:

WILBUR WRIGHT (1867-1912), an American inventor and engineer who, along with his brother, Orville, designed, built, flew, and further developed the world's first airplane

ORVILLE WRIGHT (1871-1948), an American inventor and engineer who, along with his brother, Wilbur, designed, built, flew, and further developed the world's first airplane

OCTAVE CHANUTE (1832-1910), a French-born American civil engineer and aviation pioneer who provided valuable aid and counsel to the Wright brothers in their aviation researches

Summary of Event

On the cold, windy morning of December 17, 1903, along the sandy shores of Kitty Hawk, North Carolina, Orville Wright piloted a Wright Flyer a distance of 37 meters for twelve seconds. In doing so, he became the first person to operate and control a machine which took off on its own power for a sustained flight. By noon that day, Orville and his brother, Wilbur Wright, conducted a total of four trial flights; the last flight that day, piloted by Wilbur, covered 260 ground meters and lasted for 59 seconds. These successful trials demonstrated to the Wright brothers, and eventually to the world, that powered, manned, sustained heavier-than-air flight was possible. These mechanical geniuses from Dayton, Ohio, had realized a centuries-old dream of human flight and began an era of aviation.

Although people had dreamed of human flight from the time of the ancient Greeks, experimenters achieved little success until the late eighteenth and nineteenth centuries. By 1875, however, work with hot-air balloons, kites, and gliders provided a foundation of scientific and technical studies for meaningful flight research. By the time Wilbur Wright and Orville Wright began their aeronautical investigations in the 1890's, they drew on the pioneering work of men such as George Cayley, Octave Chanute, Samuel Langley, and Otto Lilienthal. Their own experiences in the printing business, in building and repairing bicycles, their talents as research engineers and innovators, and their technical insight and keen intuition combined to give them the prowess necessary to address successfully the problems of powered, controlled human flight.

Wilbur Wright and Orville Wright were skilled investigators, not amateur tinkerers who discovered the successful path to aviation by chance. Drawing on their own considerable talents, they approached the problems of manned flight in a very rational, orderly way. As skilled research engineers, they applied the scientific method to their thorough, systematic investigations. In the tradition of engineering excellence, they began their analysis with a meticulous search of contemporary developments in aeronautics. Wilbur wrote the Smithsonian Institution in 1899 requesting information on flight and received pamphlets and a reading list. This material set in motion a literature search and a considerable amount of reading and study. The Wright brothers communicated with several other aeronautics pioneers, especially Chanute, and then built on that work with their own experiments and analyses. This organized research included various tests and investigations. The brothers studied birds in flight, experimented with kites, gliders, and various wind foils, built their own wind tunnel, and finally, tested their own airplane designs. The process allowed them to apply their own logic and imagination in their investigations. Never fully trusting others' results, the Wright brothers produced their own quantitative data and evaluations, many of which they shared and discussed with Chanute, who offered much advice and counsel. Seeing the world in tactile and visual images, the Wright brothers extended existing aeronautical knowledge to the challenges of manned flight. These talents played a key role in their success as research engineers.

Of equal importance in the quest for a flying machine was their experience with bicycle repair and manufacture. The bicycle business honed their mechanical skills in building and testing airplanes; the tools they used and the skills they acquired in that enterprise gave them the wood- and metal-working acumen needed to build a lightweight, yet sturdy, machine. The Wright brothers noted the similarity between a bicycle and an airplane: Both needed to be lightweight, fast machines, and both were inherently unstable devices requiring constant control to maintain balance. Bicycles taught them the importance of human control in more than one axis. They also realized that piloting an airplane paralleled riding a bicycle: One learned to do so only after much practice and adjustment.

Orville Wright and Wilbur Wright approached airplane design in a unique manner. Instead of building a basically stable device, they created a flying machine in which the pilot provided complete and constant control over the airplane; it had to be managed by the pilot while aloft. This meant that the pilot was an essential part of the machine's design. This design included a forward elevator to control pitch (up and down movement of the nose), a wing-warping device to achieve a helical twist along the wing to control roll (rotation of the body of the airplane), and a rudder to control yaw (left and right movement of the nose).

Their many hours of careful research and experimentation, their extensive flights with gliders from 1900 to 1902, their astute use of wind tunnel tests, and their own engineering expertise allowed them to design a powered aircraft with engine and propellers of their own crafting. The result was their 1903 Wright Flyer, a biplane (two parallel wings) with skids rather than wheels, a 12-meter, 10-centimeter wingspan, a 12-horsepower motor driving two propellers, a biplane elevator in front, and a double rudder in the rear. Confident that their design, research, and previous experimental work were correct, they set out for the outer shores of North Carolina in December, 1903, to put their confident notions to the test of experience.

Success did not occur immediately. On December 14, 1903, Wilbur Wright attempted the first trial of the Wright Flyer. Shortly after take-off, Wilbur erred in his control of the elevators, and he and his plane dropped quickly into the sand. Although he escaped serious injury, the airplane sustained enough damage that it required a few days of repair work. By the morning of December 17, 1903, however, both airplane and its builders were ready for another test flight.

This time, Orville Wright lay in the pilot's perch. The eight years of determined research and testing produced success. At 10:35 A.M. on that December morning, Orville Wright flew the Flyer for 12 seconds more than 37 meters of ground and more than 152 meters of distance in the air. The brothers then alternated piloting their successful airplane that morning, each one flying twice. On their fourth trial at noon that day, Wilbur flew for 59 seconds over 260 meters of ground and an air distance of more than 0.8 kilometer. After Wilbur had landed the plane, with some minor damage, the brothers were discussing the day's trials and a sudden gust of wind struck the aircraft. Despite their efforts to protect the flying machine from wind damage, the brothers watched helplessly as their historic craft was damaged beyond repair for further flight.

On December 17, the Wright brothers achieved a first: powered, pilot-controlled sustained flight of a heavier-than-air craft. No one bettered their record for three years. Further, their pilot-controlled design allowed them to perform intricate maneuvers nearly impossible with other, inherently stable flying machines. Although they continued their development work for several years beyond these Kitty Hawk flights of 1903, Wilbur Wright and Orville Wright were the pioneers in human flight and opened an era of aviation technology.

Impact of Event

Initially, the Wrights' flights of 1903 generated little publicity. Confidants such as Octave Chanute, however, realized the great significance of their achievement. Through a careful system of design, using their own research results and experience in flight tests, the Wright brothers had built and flown the world's first successful airplane. From 1903 to 1905, they continued their developmental work with flight tests at Huffman Prairie near Dayton. By 1905, they had created the world's first practical airplane, the Wright Flyer III. Although Chanute urged them to enter public exhibitions and flying contests, the brothers instead chose to limit their publicity and sought patent protection and contract sales of their successful flying machine.

Gradually, news of the Wrights' achievement spread throughout the United States and Europe. As it did, other aviation pioneers took news of the Wrights' success as encouragement for their own continual efforts at flight. Just as gradually, they began public demonstrations of their Flyer, which, in turn, stimulated more advances in aviation technology. By 1910, the Wrights engaged in exhibition flying; their efforts were equaled by another American aviation pioneer, Glenn Hammond Curtiss, who began matching their feats with his exhibition and contest flying. Their endeavors had begun an era of commercial aviation and provided the stimulus for further technological innovation of the airplane.

The Wright brothers also saw the potential for military aircraft and stimulated activities in that technology of warfare. They successfully negotiated a contract with the U.S. Army Signal Corps to purchase and use an airplane for military purposes. As part of this arrangement, they agreed to train military pilots and, thus, began an era of military aviation in the United States in the 1910's.

Beyond these specific aviation developments, the Wright brothers affected aviation technology with their commitment to systematic inquiry, to thorough investigations of problems, and to hands-on experimentation in their design and development of a flying machine. Such excellent methods of research and development became hallmarks of American aviation science and technology in the twentieth century. That legacy of meaningful, purposeful, and highly organized original research is as great as the pioneering airplane, the 1903 Wright Flyer. Indeed, these brothers from Dayton, Ohio, taught the world how to fly and how to design aircraft. In doing so, they opened the era of aviation for the world.

Bibliography

Bilstein, Roger E. *Flight in America, 1900-1983: From the Wrights to the Astronauts.* Baltimore: The Johns Hopkins University Press, 1984. A thorough history of aviation in the United States by a leading historian in the field; a chapter places the Wrights in the context of the era of early flight.

Combs, Harry B., and Martin Caidin. *Kill Devil Hill: Discovering the Secret of the Wright Brothers.* Boston: Houghton Mifflin, 1979. A good source of information about aviation technology and the Wrights' contributions to the field with several helpful drawings and illustrations.

Corn, Joseph J. *The Winged Gospel: America's Romance with Aviation, 1900-1950.* New York: Oxford University Press, 1983. An interesting and lively account of many activities in aviation and the personalities, including the Wright brothers, involved in America during the first half of the twentieth century.

Crouch, Tom D. *The Bishop's Boys: A Life of Wilbur and Orville Wright.* New York: W. W. Norton, 1989. An invaluable and enjoyable biography of the Wright brothers written by a foremost historian of aviation technology; this is a thorough account depicting the lives and work of the Wrights and discusses other developments in aviation of the time; a first-rate book.

Hallion, Richard P., ed. *The Wright Brothers: Heirs of Prometheus.* Washington, D.C.: Smithsonian Institution Press, 1978. This informative anthology celebrates the seventy-fifth anniversary of the Wright brothers' Kitty Hawk flights; includes a first-hand account by Orville Wright. A chronology of the Wright brothers and photographic essays add to the great value of this volume.

Harry J. Eisenman

Cross-References

Zeppelin Constructs the First Dirigible That Flies (1900), p. 78; Blériot Makes the First Airplane Flight Across the English Channel (1909), p. 448; Lindbergh Makes the First Nonstop Solo Flight Across the Atlantic Ocean (1927), p. 841; The First Jet Plane Using Whittle's Engine Is Flown (1941), p. 1187; The First Jumbo Jet Service Is Introduced (1969), p. 1897.

ELSTER AND GEITEL DEVISE THE FIRST PRACTICAL PHOTOELECTRIC CELL

Category of event: Applied science
Time: 1904
Locale: Wolfenbüttel, Germany

Elster and Geitel's pioneering work on photoelectric effect and photoelectric cells was of decisive importance in the electron theory of metals

Principal personages:

JULIUS ELSTER (1854-1920), a German experimental physicist and teacher of mathematics and physics

HANS FRIEDRICH GEITEL (1855-1923), a German physicist and collaborator of Elster who originated the concept of atomic energy

WILHELM HALLWACHS (1859-1922), a German physicist who discovered the external photoeffect in 1888

Summary of Event

The photoelectric effect was known to science in the early nineteenth century when Alexandre-Edmond Becquerel in France wrote of it in connection with his work on glass-enclosed primary batteries. He discovered that the voltage of his batteries increased with intensified illumination and that green light produced the highest voltage. Since Becquerel researched batteries exclusively, however, the liquid type photocell was not discovered until about ninety years later when, in 1929, the Wein and Arcturus cells were introduced commercially. These cells were miniature voltaic cells arranged so that light falling on one side of the front plate generated a considerable amount of electrical energy. The cells were of short life, unfortunately; when subjected to cold, the electrolyte froze, and when subjected to heat, the gas generated would expand and explode the cell.

What came to be known as the photoelectric cell, a device connecting light and electricity, had its beginnings in the 1880's. At that time, scientists noticed that a metal plate charged negatively lost its charge much faster when it was subjected to light (especially to ultraviolet light) as opposed to darkness. Several years later, researchers demonstrated that this phenomenon was not an "ionization" effect because of the air's increased conductivity, since the phenomenon took place in a vacuum but did not take place if the plate were positively charged. Instead, the phenomenon had to be attributed to the light that excited the electrons of the metal and caused them to fly off: A neutral plate even acquired a slight positive charge under the influence of strong light. Study of this effect not only contributed evidence to an electronic theory of matter—and as a result of some brilliant mathematical work by Albert Einstein, later increased knowledge of the nature of radiant energy—but also further linked the studies of light and electricity. It even explained certain chemical phenomena, such as the process of photography. It is important to note that all the experimental work on photoelectricity accomplished prior to the work of Julius Elster and Hans Friedrich Geitel was carried out before the existence of the electron was known.

After Sir Joseph John Thomson's discovery of the electron in 1897, investigators soon realized that the photoelectric effect was caused by the emission of electrons under the influence of radiation. The fundamental theory of photoelectric emission was put forward by Einstein in 1905 on the basis of Max Planck's quantum theory (1900). Thus, it was not surprising that light was found to have an electronic effect. When the longer radio waves were known to shake electrons into resonant oscillations, and the shorter X rays could detach electrons from the atoms of gases, the intermediate waves of visual light would have been expected to have some effect upon electrons—such as detaching them from metal plates and so setting up a difference of potential. The photoelectric cell, developed by Elster and Geitel in 1904, was a practical device to make use of this effect.

In 1888, Wilhelm Hallwachs observed that an electrically charged zinc electrode loses its charge when exposed to ultraviolet radiation if the charge is negative, but is able to retain a positive charge under the same conditions. The following year, Elster and Geitel discovered a photoelectric effect caused by visible light; unlike Hallwachs, however, they used the alkali metals potassium and sodium for their experiments instead of zinc.

The Elster-Geitel photocell (a vacuum emission cell, as opposed to a gas-filled cell) consisted of an evacuated glass bulb containing two electrodes. The cathode consisted of a thin film of a rare, chemically active metal (such as potassium) that lost its electrons fairly readily; the anode was simply a wire sealed in to complete the circuit. This anode was maintained at a positive potential in order to collect the negative charges released by light from the cathode. The Elster-Geitel photocell resembled two other types of vacuum tubes in existence at the time: the cathode-ray tube, in which the cathode emitted electrons under the influence of a high potential, and the thermionic valve (a valve that permits the passage of current in one direction only) in which it emitted electrons under the influence of heat. Like both of these vacuum tubes, the photoelectric cell could be classified as an "electronic" device.

The new cell, then, emitted electrons when stimulated by light, and at a rate proportional to the intensity of the light. Hence, a current could be obtained from the cell. Yet, Elster and Geitel found that their photoelectric currents fell off gradually; they, therefore, spoke of "fatigue" (instability). It was discovered later that most of this change was not a direct effect of a photoelectric current's passage; it was not even an indirect effect, but was caused by oxidation of the cathode by the air. Since all modern cathodes are enclosed in sealed vessels, that source of change has been completely abolished. Nevertheless, the changes that persist in modern cathodes often are indirect effects of light that can be produced independently of any photoelectric current.

The chief sources of instability in the Elster-Geitel cell arose from changes in the cathode caused by change of temperature or bombardment by positive ions and to changes in the field as a result of charges on the walls of the cell. These changes are connected with the incidence of light and the passage of a photoelectric current because light is usually accompanied by heat and because a photoelectric current may generate positive ions. As long as the constitution of a cathode is unchanged, its emission is independent of temperature within wide limits, except in the neighborhood of the threshold; however, change of temperature may alter its constitution by causing the evaporation or deposition of surface films. Positive ion bombardment may produce the same effects as rise of temperature, as well as other effects.

Those cathodes that are least stable are ones whose emission depends most closely on the presence of volatile surface layers. Plain metals are relatively stable. Metals sensitized by the Elster-Geitel process are very unstable. First, these metals are subject to chemical change. The alkali metal absorbs hydrogen on its surface during sensitization; this hydrogen tends to diffuse into the unchanged potassium below. If a cell is badly prepared, the hydrogen may diffuse away entirely, leaving the surface bright once more; although this diffusion does not occur in a well-prepared cell, progressive change in sensitivity—usually a loss—might occur over long periods of time. Second, heating of the cathode causes the alkali metal to distill to the cooler parts of the cell and produces an irreversible change in its sensitivity. Cooling of the cathode (less usual) may cause the reverse change, but not a restoration of the original sensitivity. Such changes of temperature of the cathode relative to the rest of the cell are most likely to occur when the cathode is supported in the center and is therefore highly insulated.

Impact of Event

The Elster-Geitel photocell was for some twenty years used in all emission cells adapted for the visible spectrum, and throughout the twentieth century, the photoelectric cell has had a wide variety of applications in numerous fields. For example, if products leaving a factory on a conveyor belt were passed between a light and a cell, they could be counted as they interrupted the beam. Personnel entering a building could be counted also, and if invisible ultraviolet rays were used, they could be detected without their knowledge. Simple relay circuits could be arranged that would automatically switch on street lamps when it grew dark. The sensitivity of the cell with an amplifying circuit enabled it to "see" objects too faint for the human eye, such as minor stars or certain lines in the spectra of elements excited by a flame or a discharge. The fact that the current depended on the intensity of the light made possible photoelectric meters that could judge the strength of illumination without human errors—for example, in order to get the right exposure for a photograph. Throughout the history of science, men and women have searched for devices more reliable than their own faculties for estimating size, weight, temperature, loudness, and so on; now light joined the list.

A further use for the cell was that it made talking films possible. The earlier systems for these had depended on gramophone records, but it was very difficult to keep the records in time with the film. Now, the waves of speech and music could be recorded in a "sound track" by turning the sound into current through a microphone and then into light with a neon tube or magnetic shutter and photographing the variations in the intensity of this light on the side of the film. By the reverse process of running the film between a light and a photoelectric cell, the visual signals could be converted back to sound.

John Logie Baird also used the photoelectric cell in developing the television process in the early 1920's. In his "scanning" system, light from each part of the transmitted picture had to be taken in turn, sent through a single set of instruments, and built up again in the same order. That this gave a steady-looking final image, despite the fact that at any one time only a single small area of it was lit, was caused by the persistence of vision of the human eye. If each picture succeeded the one before about sixteen times a second, the eye would have no more sense of discontinuity than when watching a film. Finally, the use of the photoelectric cell in transforming the energy of the sun into other forms of energy has been a major development in the search for alternative energy sources.

Bibliography

Campbell, Norman Robert, and Dorothy Ritchie. *Photoelectric Cells: Their Properties, Use, and Applications*. 3d ed. London: Pitman, 1934. Although dated, this informative book discusses the theory of photoelectricity, contemporary technical advances in the field (including Elster and Geitel's photocell), and applications that introduce or illustrate important principles. Includes illustrations, graphs, and references following each chapter.

Lange, Bruno. *Photoelements and Their Application*. Translated by Ancel St. John. New York: Reinhold, 1938. This book, although dated, is a digest of knowledge in the field of photoelectricity. Written for the layperson as well as the engineer or scientist, the author gives a useful historical introduction to the development of photoelectricity. Part 1 discusses the various theories and physical properties of the photoeffect; part 2 focuses upon the construction and performance of photocells, as well as their applications in various fields. Includes graphs, charts, illustrations, and a bibliography after parts 1 and 2.

Morgan, Bryan. *Men and Discoveries in Electricity*. London: John Murray, 1952. This brief work written for the layperson gives an overview on the history of electricity and magnetism. It focuses in depth upon the work of those scientists whose work has shaped that history—primarily Michael Faraday, Thompson, and Baird. The work of these scientists is discussed in the larger context of the history of ideas that led up to and followed their breakthrough discoveries. Elster and Geitel's contributions are also discussed. Work includes copious illustrations and an appendix of the names and dates of the most notable scientists in the field of electricity up to the 1950's.

Sommer, A. *Photoelectric Cells*. Brooklyn, N.Y.: Chemical Publishing, 1947. This very brief work is devoted entirely to photoelectric cells of the emission type, as opposed to cells of the barrier-layer and photoconducting types. A brief survey of the principles of photoelectric emission is followed by a more detailed description of the manufacture of photocathodes. The book's final chapter discusses the purely photoelectric applications of these cells. Book contains numerous graphs and illustrations; also contains a bibliography.

Summer, W. *Photosensitors*. London: Chapman & Hall, 1957. This work contains useful information on the wide range of photoelectric devices that were currently in use at the time of publication. Also of interest is the discussion of the many applications of these devices to industry. Work contains many illustrations and a bibliography.

Genevieve Slomski

Cross-References

Warner Bros. Introduces Talking Motion Pictures (1926), p. 820; Bell Telephone Scientists Develop the Photovoltaic Cell (1954), p. 1487.

HARTMANN DISCOVERS THE FIRST EVIDENCE OF INTERSTELLAR MATTER

Category of event: Astronomy
Time: 1904
Locale: Potsdam, Germany

Hartmann discovered the first indications that there is matter in between the stars

Principal personages:

JOHANNES FRANZ HARTMANN (1865-1936), a German astronomer who first discovered evidence of interstellar matter in the spectrum of the star system Delta Orionis

HENRI-ALEXANDRE DESLANDRES (1853-1948), a French astronomer who first discovered the changes in spectral lines of double stars, which led Hartmann to his discovery

VESTO MELVIN SLIPHER (1875-1969), an American astronomer who confirmed the interpretation of Hartmann's discovery as being the result of interstellar matter

EDWIN BRANT FROST (1866-1935), an American astronomer who found evidence similar to Hartmann's in the spectra of other stars

EDWARD EMERSON BARNARD (1857-1923), an American astonomer who identified dark patches in the Milky Way as clouds of dark interstellar matter

SIR ARTHUR STANLEY EDDINGTON (1882-1944), an English astronomer who arrived at the accepted interpretation of Hartmann's results

Summary of Event

Astronomy changed in the late nineteenth and early twentieth centuries. Rather than studying only the motions and positions of heavenly bodies, astronomers utilized new tools that enabled them to learn about the physical composition of the heavens, from stars to the galaxy itself. Johannes Franz Hartmann found the first evidence that the galaxy contains diffuse interstellar matter as well as the stars and planets, which form discrete objects. The discovery that there is matter in between the stars had implications for many areas in astronomy, from the debate over the nature of the Milky Way galaxy to the question of the beginnings of the solar system. The work of many astronomers contributed to the knowledge of interstellar matter.

Spectroscopy was among the new tools that were important to astronomy in this time period. The spectroscope is an instrument that breaks starlight down into its component parts (its spectrum) by passing it through a prism or reflecting it from a finely ruled grating. Sir Isaac Newton was the first to realize that white light is composed of light of many colors, or wavelengths, which blend together and appear white. Astronomers now can use the colors and dark bands of an object's spectrum to determine its chemical composition and many other properties. In particular, the spectrum of a star reveals its velocity along a line of sight to the observer (its radial velocity) and the velocity with which it turns about its axis (its rotational velocity). The dark lines appearing in a star's spectrum reveal the presence of particular elements, each of which produces its own distinctive patterns of lines at particular wavelengths in the spectrum. The atoms of each element can absorb light at several specific wavelengths only, and as light from a radiating object passes through a gas, the atoms of various elements in the gas will each absorb light at their own peculiar wavelengths. This absorption of light causes the dark bands, and once the wavelength at which a band occurs is measured, the element responsible for the band can be determined. Thus, it is possible to identify the elements in a star's atmosphere, or in any other cloud of gas intervening between the observer and the rest of the star.

In 1900, Henri-Alexandre Deslandres discovered that lines in the spectrum of the star Theta Orionis were undergoing rapid changes in their positions in relation to the rest of the spectrum. Hartmann followed up this observation at the Potsdam Astrophysical Observatory, where state-of-the-art spectrographs were well suited for studying this new phenomenon. Hartmann found changes in position of spectral lines for the star Delta Orionis. He found slow shifts, however, rather than the rapid shifts described by Deslandres. He also discovered that some of the lines did not shift in the same way as the rest of the lines, but instead demonstrated a different type of motion relative to the motion of the other spectral lines.

The reason for the motion of all the spectral lines turned out to be the Doppler effect. As an object that is emitting waves (for example, light waves) approaches an observer, the waves appear bunched together, and the wavelength appears shorter than it really is (light appears to have shorter wavelengths, and looks bluer, while sounds seem to be higher pitched). Conversely, as a wave-emitting object recedes from an observer, the waves appear stretched out and the wavelength appears longer than it really is (light appears to have longer wavelengths, and looks redder, while sounds seem to be lower pitched). The amount by which the light or sound is shifted in wavelength is related to the velocity of the moving object. This effect can appear to shift the lines in a star's spectrum either redward or blueward, and is the tool used in measuring the velocity of an individual star along the line of sight to the star (its radial velocity). In some double-star systems, such as Delta Orionis, the system appears edge-on or partially edge-on in the sky, and thus the stars in the system, as they orbit one another, are moving alternately toward and away from Earth. This motion results in Doppler shifts alternately toward the redward and the blueward, such as those observed by Hartmann. This information about Doppler shifts in the spectra of double stars was used later by Antonia Caetano Maury at Harvard College Observatory as a tool for identifying as double stars those that display the Doppler shifts but visually do not appear double.

The motion of the double stars explained the type of shift (redward and then blueward) observed for most of the spectral lines in Delta Orionis. It did not, however, explain the lines that had a different Doppler motion, one that did not move the lines from redward to blueward and back but moved the lines in the same direction and by the same amount in a constant manner. Hartmann called these lines "stationary" because they must be caused by matter, which is stationary relative to the rest of the double-star system. The obvious explanation was that the lines were caused by a massive component of the double-star system that was so heavy relative to the other components that, for practical purposes, it could be considered the center of mass about which the other stars revolved; at the same time, it remained essentially motionless with respect to the rest of the system. The unchanging Doppler shift would represent, then, the motion of the entire system with respect to Earth. Hartmann carefully ruled out this explanation, however, as well as the possibility that Earth's atmosphere was responsible for producing the lines. The possibility remained that there was an immobile cloud of matter that was producing the lines. Because the lines were of the appropriate wavelength for the element calcium, Hartmann concluded that there was a cloud of calcium gas causing the lines. Yet, Hartmann was not sure where the calcium was located. He believed it could be situated somewhere in between Earth and the double-star system and unconnected with the double stars. Or, it could be part of the double-star system. These questions were addressed and resolved years after Hartmann's original discovery.

Hartmann also had observed stationary spectral lines in the spectrum of Nova Persei, a nova that occurred in 1901. He suggested that perhaps these lines were the product of clouds of matter that bore some relation to the dark clouds that Edward Emerson Barnard had observed in photographs of the Milky Way. It took Barnard many years of doubting and confusion before he could come to the conclusion finally that the dark spots he saw in his photographs indicated the presence of cold dark matter that blocked out the light of stars behind them, rather than merely gaps in the sky where no stars were to be found. Barnard's work was a long chapter in the discovery of interstellar matter; Hartmann's discovery of stationary lines provided the first direct evidence for clouds of gaseous matter independent of any stars or star systems. Today, several areas of astronomy are aided by knowledge of gaseous and particulate interstellar matter and the ramifications of this matter. Unfortunately, the importance of Hartmann's work essentially went unrecognized for many years.

Impact of Event

The discovery of the stationary lines was only one step in a long story. Hartmann's work was confirmed by Vesto Melvin Slipher at the Naval Observatory in Flagstaff and by Edwin Brant Frost at Yerkes Observatory. Slipher made more observations of Delta Orionis and was the first to suggest that it was truly interstellar matter that was responsible for the lines. (Hartmann's supposition had been that the matter was associated with the double-star system, although not linked to it.) Frost, in 1909, observed other stars whose spectra also displayed stationary lines.

Some astronomers disagreed with Hartmann, Slipher, and Frost about the nature of the lines and used several observations to argue that the lines were, in fact, connected with the stars in question rather than being interstellar in nature. Reynold

Kenneth Young noted in 1920 that stationary lines appeared only in the spectra of relatively young stars and argued from this relationship between stellar type and presence of the lines that there must be some connection between the star and the calcium gas causing the lines. Also, Oliver Justin Lee presented a Ph.D. dissertation at the University of Chicago on the orbit of a double star. Lee interpreted some of his data as evidence that the calcium gas was connected with the double-star system and could, in fact, explain some peculiarities in the way the velocities of the stars changed over time. This seemed to indicate that the lines were stellar rather than interstellar in origin.

Sir Arthur Stanley Eddington and Otto Struve, in the late 1920's, presented their work, which finally settled the nature of the stationary lines and resolved the question of the existence of interstellar matter. Struve wrote several papers on the calcium lines in stellar spectra, which stimulated much discussion and led to the acceptance of interstellar matter as the most satisfactory explanation for the stationary lines. Eddington explained why the lines appeared only in relatively young stars by showing that older stars have spectra that would not reveal the presence of the stationary lines easily, even when they were present. Eddington ignored Lee's work, as no one had been able to confirm his findings.

More than twenty years after Hartmann's original discovery, astronomers accepted the existence of clouds of interstellar calcium. The presence of interstellar matter had ramifications for various studies in astronomy. For example, when interstellar matter appears in the form of dust, it absorbs some of the light from stars and makes stars appear dimmer (and more distant) than they really are. This knowledge affected the distance scale astronomers devised for the universe. Also, interstellar matter plays an important role in theories of how stars form because it provides the material for star formation. Because these clouds move about the galactic center, their motion, when measured, can give indications of the speed and direction of galactic rotation. Thus, this one seminal discovery had far-reaching effects on the science of astronomy.

Bibliography

Asimov, Isaac. "Stellar Evolution." In *The Universe from Flat Earth to Quasar.* Rev. ed. New York: Walker, 1971. In Asimov's usual enjoyable style, he tells of Hartmann's work in the context of stellar formation from interstellar gas and dust. Explains Hartmann's work clearly and gives a diagram. Contains a list of further readings. A concise summary of interstellar matter's detection and implications for theories of star formation.

Berendzen, Richard, Richard Hart, and Daniel Seeley. *Man Discovers the Galaxies.* New York: Columbia University Press, 1984. Focuses on early twentieth century discoveries about the nature of the Milky Way galaxy and other galaxies. Presents some information on Hartmann's work, set in the context of galactic studies at the time. Discusses the work done in following up on Hartmann's original discovery. Uses original documents to help bring the events described to life. Includes photographs, drawings, and graphs.

Mitton, Simon, ed. *Cambridge Encyclopedia of Astronomy.* New York: Crown, 1977. Several chapters of this standard handbook on astronomy discuss interstellar matter, both gaseous and dusty, its detection, and its role in star formation. Gives an explanation of how absorption lines (such as the stationary lines) are formed. Includes many drawings, charts, and lovely photographs, as well as an appendix containing a brief explanation of some of the physical concepts discussed in the text.

Struve, Otto, and Velta Zebergs. *Astronomy of the Twentieth Century.* New York: Macmillan, 1962. Cowritten by Struve, who participated in some of the events connected with Hartmann's work. Contains a chapter on interstellar matter and how it was discovered, including Hartmann's work. One of the best sources for material on the history of twentieth century astronomy. Contains photographs (some of spectrographs in which "stationary lines" appear), graphs, drawings, and a bibliography.

Verschuur, Gerrit L. *Interstellar Matters: Essays on Curiosity and Astronomical Discovery.* New York: Springer-Verlag, 1989. An excellent history of the circuitous path astronomers took to arrive at their present knowledge of interstellar matter, gaseous and dusty. Engagingly written; gives details about Hartmann's work and how it was interpreted by other astronomers, as well as about many other astronomers' contributions to the story of interstellar matter. Includes photographs (some of spectrographs in which "stationary lines" appear), graphs, and drawings.

Mary Hrovat

Cross-References

Hertzsprung Uses Cepheid Variables to Calculate the Distances to Stars (1913), p. 557; Michelson Measures the Diameter of a Star (1920), p. 700; Eddington Formulates the Mass-Luminosity Law for Stars (1924), p. 785; Eddington Publishes *The Internal Constitution of the Stars* (1926), p. 815; Oort Proves the Spiral Structure of the Milky Way (1927), p. 830; Supernova 1987A Corroborates the Theories of Star Formation (1987), p. 2351.

KAPTEYN DISCOVERS TWO STAR STREAMS IN THE GALAXY

Category of event: Astronomy
Time: 1904
Locale: Groningen, The Netherlands

Kapteyn discovered that the proper motions of stars were not randomly distributed but tended in two opposite directions, implying the rotation of the galaxy

Principal personages:

JACOBUS CORNELIS KAPTEYN (1851-1922), a Dutch astronomer who devoted most of his life to the detailed study of the Milky Way by means of star counts

HARLOW SHAPLEY (1885-1972), an American astronomer whose work with globular clusters within the Milky Way led him to dispute the Kapteyn universe and propose a much larger size for the galaxy

HEBER DOUST CURTIS (1872-1942), an American astronomer who supported Kapteyn in a major debate with Shapley in 1920

Summary of Event

Following several years of study of the structure of the Milky Way, Jacobus Cornelis Kapteyn concluded, on the basis of exhaustive star counts in sampled portions of the sky, that the proper motions (perpendicular to the line of sight) of the stars were not randomly distributed. To his own satisfaction, Kapteyn confirmed previous perceptions of the galaxy as a flattened ellipsoid (lens-shaped). Yet, because gas and dust toward the center of the galaxy distorted his counting in that direction, he incorrectly placed the solar system relatively near the center of the galaxy.

Having constructed what he believed was a correct view of the galaxy, Kapteyn then measured the proper motions of many stars for the purpose of verifying that these were randomly distributed, as theory indicated. Instead, he discovered systematic departures from randomness, with the stars showing movement in opposite directions in different parts of the sky. The two streams of movement of the stars implied that the galaxy was not static with the sun near the center of random individual stellar motions. Kapteyn's discovery had major cosmological implications, most obviously that the Milky Way was a rotating galaxy.

The originator of this great discovery was well prepared both in temperament and in education to contribute to the development of modern astronomy. Kapteyn, one of fifteen children, was the son of a boarding school proprietor who stimulated his interest in physics. He studied at the University of Utrecht and was awarded a doctorate in physics in 1875. Following completion of his studies, he pursued a position at the Leiden Observatory, where he became a successful astronomer. By 1878, had won a professorship in astronomy at the University of Groningen. Upon arrival at

Groningen, however, he was unable to raise funding for equipping an observatory.

Undaunted, he became a world leader in arranging collaboration with other astronomers, notably with Sir David Gill at the Cape of Good Hope in South Africa. From 1885 to 1899, he analyzed photographic plates taken by Gill and assisted in producing a star catalog with nearly a half million entries. He later continued studies as a visiting astronomer at the Mount Wilson Observatory in Pasadena, California.

While still at Leiden, Kapteyn became interested in the structure of the universe, a topic he pursued through his career. His primary method consisted of counting stars at the different levels of brightness, his underlying assumption being that the measured brightness of stars was a statistically reliable means of assessing their distance. Through this method he made the first major advance since the work of Sir William Herschel a century before. Herschel resolved some "nebulae" into star clusters and consequently believed that all nebulae would eventually be resolved, with some located outside our galaxy. Kapteyn also accepted the nebulae or unresolved patches of light in the sky as "island universes" outside our own galaxy. The thinning of star counts away from the plane of the Milky Way implied to Kapteyn an ellipsoid-shaped galaxy. In the midst of these studies, he found the two star streams.

If the galaxy were essentially a static collection of stars, then theoretically their motions across the line of sight (proper motions) ought to be randomly distributed. Only those stars relatively close to the Earth have measurable proper motions, but there were enough of these by the early part of the twentieth century that Kapteyn could analyze them statistically. The result was that he found consistent displacement in two opposite directions for neighboring stars. The motions, toward the constellations Orion and Sagittarius, implied what he called the streaming of stars or a pattern of movement that indicated that stars were moving in opposite directions. In 1904, Kapteyn made the announcement of his discovery at the International Congress of Science at the World's Fair in Saint Louis.

Desiring to follow up these early discoveries, Kapteyn devoted much of his time after 1904 to attempts to organize the worldwide astronomical community in cooperative efforts to photograph and analyze 206 different portions of the night sky. These studies focused upon magnitude, color, proper motion, radial velocity (along the line of sight), and spectral type. Before he was able to arrange for the full implementation of his plan, World War I largely ended international cooperation. He did, however, generate enough information to convince him that his ellipsoidal shape for the galaxy was correct. He estimated that the major axis and the minor axis of the ellipsoid were related by the ratio of five to one, the major axis measuring approximately 52,000 light years. His main concern about the accuracy of his model was the central location of the sun, a point which also caused other astronomers to question the model because of this apparent favored position.

Kapteyn was also deeply concerned about the effects upon his model of absorption of light by gas and dust. He realized that if there were appreciable absorption in the direction of the galactic center, it would have caused his central position for the sun to be wrong and, more important, would have caused him to underestimate the size of the galaxy. As a result, he repeatedly sought some means of measuring the amount of absorption. Having failed to measure any sizable amounts by 1918, he convinced himself that it was negligible and his model accurate.

His inability to measure the effects of obscuration of light in the direction of the galactic plane did cause him to underestimate seriously the size of the galaxy. Harlow Shapley, who was studying many of the same problems as Kapteyn at Mount Wilson in California, proposed a much larger universe on the basis of his study of clusters of stars (globular clusters) and Cepheid variables (stars of fluctuating brightness). The superior equipment that had resolved the clusters into individual stars led Shapley to conclude that all the nebulosities might eventually be resolved and that they could be a part of the galaxy. By 1914, Shapley had established that all the globular clusters were in one direction; thus, he concluded that they surrounded the center of the galaxy, and he necessarily placed the sun at the edge. If the universe were the size Kapteyn indicated, the clusters would have been outside the galaxy, but Shapley was confident in his distance measurements based on the brightest stars and the Cepheid variables. He was also confident in his interpretation of the cause of their asymmetrical location. The implication followed that the streaming of the stars, which was firmly established, would have to be reinterpreted in a new fashion to fit the data.

The divergence of Shapley's views, including his much larger estimates of the scale of the universe, led to conflict with Heber Doust Curtis and was publicly debated at the annual meeting of the National Academy of Sciences on April 26, 1920. Shapley presented his views while Curtis defended the perspectives of Kapteyn's followers. Both sides, on the basis of later discoveries, were partially right and partially wrong. Kapteyn's view of the "nebulae" as island universes was eventually established (even by 1914 Curtis had already noted the high radial velocity of some of the spirals, implying their distance and recession from the Earth), and Shapley's views of the dimensions of the Milky Way were found to be more accurate. This debate served to place Kapteyn at the center of the cosmological questions of the day. While his interpretation of the physical cause of the streaming effect turned out to be only partially correct, it does not minimize the significance of his achievement in recognizing the reality of the streaming.

Impact of Event

The efforts of Kapteyn in the early part of the twentieth century demonstrate clearly that careful observation and experimentation have exceptional value, even if the interpretation is erroneous. On the basis of methodical observation and careful statistical analysis of star counts from sampled regions of the sky, Kapteyn constructed a model of the galaxy in the shape of a lens. The galaxy was rotating as demonstrated by his discovery of the two star streams, with the earth in a position near the center. The Kapteyn universe provided a framework for the explanation of the astronomical observations that served well at the beginning of the century.

Kapteyn's view of the universe dominated the first two decades of the twentieth century. His work significantly affected the cosmology of the day, especially his view of the nebulae as "island universes" beyond the bounds of the Milky Way. When this perspective was combined with the high radial velocities of the nebulae, it eventually contributed to Georges Lemaître's cosmology of the mid-1920's, which was, in turn, an antecedent of the currently accepted big bang cosmology. His discovery of the two star streams contributed to the establishment of the rotation of the galaxy, although he incorrectly placed the sun at the center and believed the streams were the result of passage of stars on either side of the earth.

Kapteyn's model was inaccurate for a variety of reasons, the most important of which was the obscuring effect of gas and dust toward the center of the galaxy. His failure to measure obscuration successfully caused him to be overly confident in his star counts in that direction. Thus, he searched for alternative explanations for Shapley's research and disputed it to his death. Shapley, in turn, erred in his interpretation of the nebulae because he depended upon erroneous rotational measurements for the nebulae. In less than a decade, however, further research resolved the conflict of the early 1920's.

By the late 1920's, as the result of work by Bertil Lindblad, Jan Hendrik Oort, and Edwin Powell Hubble, a more accurate picture emerged of our galaxy as a spiral rotating in the same fashion as the other exterior galaxies. In 1927, Oort, a former student of Kapteyn, demonstrated (once the sun's position near the edge of the galaxy was established) that the apparent motion in one direction was the result of the sun lagging behind stars nearer the center. The motion in the opposite direction was the result of slower outer stars falling behind the sun. Kapteyn's universe was superseded by research it had stimulated.

At the beginning of the twentieth century, Kapteyn's influence greatly enhanced interest in galactic astronomy, which had been somewhat eclipsed by emphasis upon planetary studies at the end of the nineteenth century. He also set a fine example of the advantages of cooperative efforts among astronomers. Though utterly confident in the value of his star counts and overly committed to his method, (soon superseded by better methods and newer, more powerful equipment), he made a contribution essential to later cosmological studies.

Bibliography

Berendzen, Richard, Richard Hart, and Daniel Seeley. *Man Discovers the Galaxies.* New York: Columbia University Press, 1984. A major survey of historical cosmology. The first five chapters contain a fairly complete picture of Kapteyn's work as it relates to the work of other astronomers of the period.

Kapteyn, Jacobus C. "Discovery of the Two Star Streams." In *Source Book in Astronomy, 1900-1950*, edited by Harlow Shapley. Cambridge, Mass.: Harvard University Press, 1960. This is an excerpt from a paper presented to the South African meeting of the British Association in 1905. It recounts briefly his procedures for discovering the two star streams.

____________. "First Attempt at a Theory of the Arrangement and Motion of the

Sidereal System." *Astrophysical Journal* 55 (1922): 302-327. Significant because it is a last and best statement of Kapteyn's perspective on the significance of his discovery of star streaming. Unfortunately, since Shapley turned out to be closer to the truth about the size of the galaxy, it was already out of date when it was published. Somewhat technical, but intelligible to the general reader.

____________. "On the Absorption of Light in Space." *Astrophysical Journal* 29 (1909): 47-54, and 30 (1909): 163-196. Here he summarized his efforts to establish the amount of absorption of light by gas and dust. Presents the evidence that encouraged him to maintain that the sun was near the center of the galaxy, and that his interpretation of the two streams was correct. Moderately technical, but worth the effort to understand the seriousness with which he regarded the problem.

____________. *Plan of Selected Areas.* Groningen: Hoitsema Brothers, 1906. An explanation of what Kapteyn wished to accomplish by means of cooperative surveys of 206 regions of the sky by observatories in various parts of the world. The project was to form a data base for his statistical studies that have caused some to call Kapteyn the father of statistical astronomy.

Kapteyn, Jacobus C., and Pieter J. Van Rhijn. "The Proper Motions of the Cepheid Stars and the Distances of the Globular Clusters." *Bulletin of the Astronomical Society of the Netherlands* 1 (1922): 37. Kapteyn's main effort, an attempt to establish two classes of Cepheids, one dimmer than the other, to defend his position against the newer evidence Shapley was marshaling against Kapteyn's model of the size of the universe and the location of the sun, which affected the interpretation of the two star streams.

Paul, E. Robert. "J. C. Kapteyn and the Early Twentieth-Century Universe." *Journal for the History of Astronomy* 17 (August, 1986): 155-182. A major historical study of the significance of Kapteyn's work. Presents an almost wistful regret that one who contributed so much to the development of statistical astronomy could have been incorrect about the shape and size of the galaxy.

Ivan L. Zabilka

Cross-References

Shapley Proves the Sun Is Distant from the Center of Our Galaxy (1918), p. 655; Oort Proves the Spiral Structure of the Milky Way (1927), p. 830.

GORGAS DEVELOPS EFFECTIVE METHODS FOR CONTROLLING MOSQUITOES

Category of event: Medicine
Time: 1904-1905
Locale: Panama Canal Zone

Employing recent discoveries on the role of mosquitoes in the transmission of malaria and yellow fever, Gorgas applied strict sanitary controls within the Panama region, thereby enabling construction of the canal

Principal personages:

WILLIAM CRAWFORD GORGAS (1854-1920), an American army surgeon and sanitarian who was appointed, in 1904, the Panama Canal Commission's chief sanitary officer and eliminated yellow fever within two years

SIR RONALD ROSS (1857-1932), an English physician who proved conclusively that mosquitoes transmit malaria; awarded the 1902 Nobel Prize in Physiology or Medicine for investigations of the parasite and its various stages in mosquito's salivary glands

WALTER REED (1851-1902), a bacteriologist who was largely responsible for the eradication of yellow fever in Havana by screening patients and by waging a war on the mosquitoes that transmitted the disease

JOHN FRANK STEVENS (1853-1943), the chief engineer of the Isthmanian Canal Commission who played a crucial role in the design of the canal

Summary of Event

With the discovery of gold in California in 1848, considerable interest quickly developed in the United States concerning the construction of a transoceanic canal through Central America that would shorten the time and distance then necessary to travel from the East to West coast. Flushed with his success in building the Suez Canal during the 1860's, Ferdinand de Lesseps of France initiated efforts in the 1870's to construct such a canal by organizing the necessary financial backing and arranging for preliminary surveys. After evaluating a number of potential routes, de Lesseps and the leaders of his Panama Canal Company decided upon a sea-level canal cutting through the narrow Panamanian isthmus between Colón and Panama City. It was initially envisioned that the proposed canal would be 9 meters deep, 30 meters wide, and that it would cost a total of 658 million francs to build.

The challenges facing the highly skilled French engineers arriving in Panama at the start of the project, in 1881, were formidable. They included the taming of the unpredictable Chagres River and excavation difficulties at Culebra, where geological formations resulted in a continual problem with slides. Yet, the most difficult and, indeed, insurmountable obstacle confronting French engineers and workers during the 1880's was the high incidence of various diseases, especially malaria and yellow fever. In 1889, a virtual epidemic on several occasions decimated the work force and ultimately ended de Lesseps' venture. After eight years in Panama, the French had lost an estimated two thousand workers to yellow fever and more than fifty-five hundred to other illnesses. A second brief attempt at continuing excavations ensued during the 1890's; but, in the end, de Lesseps and the French had failed. The episode was a tragic testimony to a misguided, optimistic faith in science and technology that was incommensurate with the realities of the project.

Despite these failures, American interest in the canal intensified after the Spanish-American War. Beginning in 1904, the United States embarked on a canal construction program that succeeded where the French had not; the reason for this achievement can be traced to advances in tropical medicine related to the eradication of malaria and yellow fever.

Throughout the nineteenth century, most physicians thought that diseases were spread by odors; Panama, with its abundance of decomposing vegetation, filth, and decomposing animal carcasses, possessed conditions that supported this idea. This so-called miasma theory of disease was gradually refuted. In the case of malaria, it was disproved by the English physician, Sir Ronald Ross, working in isolated Secunderabad, India. During the late 1890's, Ross showed that the *Anopheles* mosquito spread the disease after it had fed on an infected patient, and he carefully investigated the malaria parasite, *Plasmodium falciparum*, in the mosquito's stomach and salivary glands. It became obvious to Ross that in order to stamp out malaria, one had to isolate the mosquito from those infected; his sanitary ideas were systematically outlined in his *Mosquito Brigades and How to Organise Them*, published in 1901.

Concurrent with the publication of *Mosquito Brigades*, a team of United States Army doctors, which included Walter Reed and William Crawford Gorgas, were eliminating yellow fever in Havana, Cuba, using similar public health techniques. Reed, borrowing from the work of Cuban physician Carlos Finley, had concluded that only one type of mosquito, *Stegomyia fasciata*, was responsible for the dreaded "yellow jack." Once the particular pattern of incubation was determined, it was clear that to eliminate yellow fever, one had to prevent the propagation of the insect by keeping the female *Stegomyia fasciata* from laying its eggs.

Gorgas would employ this theoretical understanding of yellow fever in his practical public health efforts in Havana. Fresh water that had been left standing typically had been present inside the sick rooms of French engineers and workers at the Panama Canal Company hospital at Ancón during the 1880's. Tragically, the water had served as a mosquito-breeding ground. It was immediately disposed of or sealed off with wooden lids or screens. Further, following a suggestion by entomologist Leland O. Howard, water left standing was covered with a thin film of kerosene or oil whenever this method was feasible. In addition, adult mosquitoes were killed by the fumigation of every house in Havana where a case of yellow fever had appeared.

After doors and windows were tightly sealed, sulfur or powdered pyrethrum was burned in pots designed specifically for this purpose.

Reed's discoveries and Gorgas' yellow fever campaign in Cuba were great achievements of the day, and the results of the sanitary program were dramatic. Since the insect had only a ten-day life cycle, the *Stegomyia* population diminished rapidly. With this significant accomplishment, the stage was set for Gorgas' work in Panama, beginning in 1904. As a deeply religious man, Gorgas had long thought of his life experiences as ordered and ordained by God with the purpose of readying him for this one great task.

Somewhat ironically, Gorgas' success in Cuba was derived from Reed's scientific work; yet, Gorgas had been extremely resistant to the idea that a mosquito transmitted yellow fever. Born in Mobile, Alabama, in 1854, he received his undergraduate education at the University of the South, located in Sewanee, Tennessee, and his medical degree at New York City's Bellevue Hospital Medical College. Interested in medicine more as a means to enter the army than as a profession, Gorgas joined the United State Army Medical Corps in 1880, and he initially served on a number of outposts in the West. With the onset of the Spanish-American War, Gorgas was sent to Cuba on the Santiago expedition, and later to Havana, where he was first placed in charge of yellow fever patients before his appointment as chief sanitation officer.

In 1904, Gorgas was sent to Panama as the Isthmanian Canal Commission's chief sanitary officer, charged with the elimination of two major obstacles to the completion of the American project there: malaria and yellow fever. Yet, despite the work of Ross and efforts in Cuba, most of the political leadership in Washington and key members of the commission, including Admiral John G. Walker and Governor George W. Davis, did not believe in the mosquito theory. Gorgas' arguments fell on deaf ears until a deadly epidemic hit the Canal Zone in the spring and early summer of 1905 and the subsequent appointment of John Frank Stevens as chief engineer.

Stevens, a railroad engineer with considerable experience, recognized that if the canal project were to succeed it had to be designed using locks and gates rather than constructed at sea level. Most significant, excavating the Culebra cut hinged upon the removal of dirt using an extensive rail network. Above all, Stevens clearly perceived that before construction could begin effectively, diseases such as yellow fever and malaria had to be eliminated and that Gorgas had to receive unequivocal support. Thus, in 1905, Stevens' engineering department stood behind Gorgas, who now had first priority in terms of men and materials. By the fall of 1905, Gorgas had more than four thousand men engaged in sanitation work, and his budget was increased dramatically. Supplies necessary for the eradication of mosquitoes were ordered and received in unprecedented quantities, including the requisition of 120 tons of pyrethrum powder, 300 tons of sulfur, and 50,000 gallons of kerosene per month. Fumigation pots, screens, buckets, garbage cans, and brushes were soon in abundance; this equipment was used in the subsequent house-by-house campaign. As a result of these efforts, cases of yellow fever in the Canal Zone fell from sixty-two in June, 1905, to twenty-seven in August and one in December, with no further outbreaks in 1906. By using similar techniques to combat the *Anopheles* mosquito, malaria was reduced, although not totally eradicated, since the species had a much broader range of flight and bred in a more widespread area.

Impact of Event

Gorgas' measures to control the spread of mosquito-transmitted diseases in Panama had both short- and long-term significance. Although malaria proved to be much more difficult to contain than yellow fever, construction on the Panama Canal progressed steadily to its completion and the opening of the waterway to commercial traffic on January 1, 1915. For his pioneering efforts, Gorgas received numerous honors, including honorary degrees from the University of the South, the University of Alabama, Harvard University, and The Johns Hopkins University. In 1907, he was appointed by President Theodore Roosevelt as a member of the Isthmanian Canal Commission and the following year was elected president of the American Medical Association. Promoted to the rank of brigadier general and named surgeon general of the United States Army in 1914, Gorgas played an influential role in sanitation work during World War I, retiring in 1918. He then became the director of the yellow fever program for the International Health Board of the Rockefeller Foundation and traveled to Central America and Peru during the last two years of his life. He implemented procedures in other areas that were often subject to periodic outbreaks of yellow fever.

Gorgas' methods were quickly implemented throughout South America and Africa. Considering the deadly nature of yellow fever, or *vomito negro*, an illness that typically claimed the lives of more than 50 percent of those afflicted, it was remarkable that, between 1910 and 1925, no major epidemic occurred in temperate regions or in the traditional endemic regions located in Ecuador, Mexico, and Brazil. Furthermore, Gorgas' practical sanitation techniques were subsequently complemented by scientific advances related to yellow fever. In 1918, researchers identified the microbe, *Leptospira ecteroides*, that caused the disease, a minute spiral organism that was studied extensively during the 1920's.

Perhaps the most enduring legacy to Gorgas' work is the Panama Canal itself, an undertaking that could not have been completed without his sanitation measures. During the first ten years of operation, between 1915 and 1925, commercial transits increased from 1,072 to 4,673 and tolls from $4.3 million to $21.6 million. Because of the canal, shippers saved almost 12,900 kilometers sailing from New York to San Francisco, and almost 11,300 kilometers from New York to Honolulu. While these figures increased steadily before 1939, it was in the post-World War II era that traffic and tolls escalated dramatically as global trade and a dynamic international economy took on enhanced significance during the last quarter of the twentieth century.

Bibliography

Bishop, Joseph Bucklin. *The Panama Gateway.* New York: Charles Scribner's Sons, 1913. Written by the secretary of the United States Panama Canal Commission, this authoritative and comprehensive work covers the history of the region and canal construction efforts from the Spanish colonial era to the completion of construction. Chapters describing Gorgas' sanitation efforts and the technology employed in the canal's design are extremely detailed. The author often relies on an abundance of tabulated factual data, including expenditures.

Gorgas, Marie D., and Burton J. Hendrick. *William Crawford Gorgas: His Life and Work*. Garden City, N.Y.: Doubleday, Page, 1924. Although this hagiographic account must be read critically, it is an important source on Gorgas' life. Sensitively written, it includes many intimate family details. Several revealing letters are contained within the narrative.

Gorgas, William Crawford. *Sanitation in Panama*. New York: D. Appleton, 1913. A definitive work describing the efforts of Gorgas and his sanitation department in Panama, this book is particularly valuable for its description of the campaign in Havana and the published correspondence between Walter Reed and William Gorgas. Details concerning practical methods, hospital facilities, and quarantine procedures are contained in many of the chapters.

McCullough, David. *The Path Between the Seas: The Creation of the Panama Canal, 1870-1914*. New York: Simon & Schuster, 1977. The best single historical work on the construction of the Panama Canal. Well researched and written, the author brilliantly portrays the main figures involved in the canal's construction, including Gorgas, and perceptively explores broader historical themes related to the most massive engineering undertaking of the early twentieth century.

Ross, Ronald. *Memoirs, with a Full Account of the Great Malaria Problem and Its Solution*. London: John Murray, 1923. A fascinating autobiography reflective of the role of public health efforts during the Age of Imperialism. Containing many letters and personal insights, it is a valuable source for one desiring to understand both Ross's scientific achievement and the late Victorian society in which he lived.

____________. *Mosquito Brigades and How to Organise Them*. New York: Longmans, Green, 1901. Written by the physician who won a Nobel Prize for his efforts in determining the role of mosquitoes in the transmission of malaria, this classic in the history of tropical medicine includes methods used to identify mosquitoes and organizational techniques aimed at waging a public health campaign to eradicate them. The work concludes with a history of nineteenth century attempts to combat the disease and case studies of malaria control programs in Hong Kong, Nigeria, New York City, Havana, and Sierra Leone.

John A. Heitmann

Cross-References

Reed Establishes That Yellow Fever Is Transmitted by Mosquitoes (1900), p. 73; Theiler Introduces a Vaccine Against Yellow Fever (1937), p. 1091.

BROUWER DEVELOPS INTUITIONIST FOUNDATIONS OF MATHEMATICS

Category of event: Mathematics
Time: 1904-1907
Locale: Amsterdam, The Netherlands

Brouwer pioneered the intuitionist reformulation of the logical foundations of mathematics

Principal personage:
L. E. J. BROUWER (1881-1966), a Dutch mathematician, logician, and epistemologist

Summary of Event

The first decade of the twentieth century was a crisis period in many areas of mathematics, particularly in geometry, arithmetic, set theory, formal logic, and their interrelations. These crises arose predominantly from a number of paradoxes in Georg Cantor's theory of sets, from the attempted formalization of logic and arithmetic by Richard Dedekind and Giuseppe Peano, and from methodological disputes about the validity of different types of mathematical proof. From the mid-1880's, these contradictions indicated to many mathematicians that there were unforeseen defects arising from various attempts to reformulate classical geometry, arithmetic, and logic in ways noncontradictory with respect to linguistic uses and ontological commitments about mathematical terms and referents. As a response to these crises, three main schools of mathematical philosophy arose: logicism, formalism, and intuitionism.

Philosophical discussions of arithmetic and geometry in Immanuel Kant's *Kritik der reinen Vernunft* (1781; *Critique of Pure Reason*, 1838) has been cited often as the earliest progenitor of mathematical intuitionism. Kant explicated his theory of the innate forms of spatial and temporal perception, respectively, as the intuitional basis of geometry and arithmetic. Kant argues that the way we first learn that $7 + 5 = 12$ is precisely by a process like counting progressively from 7 to 12 via successive additions of 1, operating or constructed with a particular example of objects first given in mental intuition. In an algebraic sense, the set of numbers generated by the above, or related procedures, is united with the natural counting numbers insofar as it has an initial element 0 and a successor relation 1. The algebraist, according to Kant, gets arithmetic results by manipulating representational symbols according to certain rules of construction, which Kant claims cannot be obtained without these prior intuitions. Kant posited this construction with intuitions as the chief source of clarity and evidence in the fundamentals of mathematics.

In the mid-nineteenth century, German mathematician Leopold Kronecker published what many historians and philosophers of mathematics consider to be pre-intuitionist queries about the "existential proofs" of anti-Kantian mathematician-philosopher Bernhard Bolzano in the theory of differential equations. In particular, Kronecker doubted the utility and validity of proofs that assume without demonstrating the existence of solutions to classes of differential equations, insofar as formalists like Bolzano sought to establish the existence of mathematically permissible solutions by indirect deductive arguments, without either explicitly constructing or providing a finite procedure for finding such a solution. Between 1880 and 1906, a more extensive constructivist school of French mathematicians, including Émile Borel, Henri-Léon Lebesgue, and notably Henri Poincaré, likewise stressed the need of basic intuitions to define the sequence of integers and the general content upon which all mathematics depends.

Many of these issues came to a focus with the publication of David Hilbert's *Grundlagen der Geometrie* (1899; *The Foundations of Geometry*, 1902), on the one hand, and Edmund Husserl's *Logische Untersuchungen* (1900-1901; logical investigations) on the other, together with the subsequent acrimonious sessions on "Logic and Philosophy of Science" at the Second International Congress of Philosophy in Paris in 1904. With this background, Gerrit Mannoury's views on the relations of ordinary language, psychology, and Kantian philosophy laid the foundations of mathematics. As a graduate student in 1904, L. E. J. Brouwer chose the foundations of mathematics as his dissertation topic. Brouwer closely reexamined the ongoing (1901 to 1906) debates among Poincaré, Louis Couturat, and Bertrand Russell on the relations of priority and (in)dependence between (symbolic) logic and (axiomatic) mathematics, and on the criticality of logical deduction versus psychological intuition in mathematical creation and proof. In his first semitechnical publication on the origins of mathematical thinking and its "objects," Brouwer sought a way to keep the methodological autonomy, purity, and rigor of Hilbert, together with the links to philosophical and everyday thinking of Husserl, while avoiding the logical and linguistic puzzles of Russell.

With Kant, Brouwer believed the ultimate foundation of mathematics is the subjective awareness of time, divided into past/before and future/after. This, for Brouwer, gives rise to a basic intuition of "bare twoness," not itself a number, but an absolutely basic and simple mental intuition from which it is possible to construct all finite ordinal numbers. As developed in his doctoral dissertation of 1907, Brouwer regarded the starting point for mathematics to be neither abstract logical rules nor functional mathematical axioms, but the individual's originative form of understanding, without language or logical concepts, of the sequence of positive integers vis-à-vis repeated duplication of temporally-sequential unit constructions with |, ||, |||, and so on (which might be indexed books on a shelf, sticks in a row, or any other heuristically convenient placeholder concept). Some scholars consider Brouwer's constructions with mental intuitions similar to "thought experiments" of his contemporary, the English philosopher, Francis Herbert Bradley, whereas others see Brouwer's as an implicit psychological theory of mentation. The paradoxes Brouwer claimed exist only because of logic and language and the fact that these symbolic and discursive concepts can never be made commensurate with temporal intuitions. In Brouwer's intuitionist view, mathematics is apodictically certain and true precisely because it rests on this simple and direct mental awareness of self-evident temporal intuitions and constructs by the individual. Brouwer's intuitionism sought to restrict mathematical knowledge to only that which can be actually and directly constructed in finite proofs: To know something mathematically is to have a specifically finite-implementable proof based on temporal intuitions without existence assumptions. Moreover, Brouwer denied Hilbert and Russell's claim of being able to circumscribe the full extent of any formal systems of mathematics, radically implying that the previously timeless, absolute, and impersonal realm of logic and mathematics in some ways exhibits the character of historical time, factual incompleteness, and personal mental activity.

Together with Friedrich Ludwig Gottlob Frege, Hugo Dingler, and Husserl's philosophies of mathematics, Brouwer's intuitionism takes the component statements of a mathematical theory to be really meaningful. Also, Brouwer did not agree with Poincaré's neointuitionism or Hilbert's formalism, which denied genuine intrinsic content or meaning to given mathematical propositions apart from the complete set of axioms of which it is a part. Brouwer concurs with Russell's logicism, insofar as intuitionism identifies a particular axiom (versus those dependent only on an entire system of axioms). In 1906, Russell's paper "The Paradoxes of Logic" acknowledged that logicism cannot eliminate but can regulate only the use of mathematical intuition. Thus, for Brouwer, although propositions in and about mathematics can be ordered and ranked in an asymmetric hierarchy according to their complexity of derivation, on explicitly philosophical grounds, Brouwer denied that the possibilities of mathematical construction, what Brouwer called an "open system," can be confined a priori, within the bounds of any formal closed system of statements, conceptually anticipating some of the implications of Kurt Gödel's later undecidability-incompleteness theorems. Brouwer likewise demonstrated at least nine pre- and meta-levels with respect to operational mathematics and logic (in contrast to Hilbert's subsequently published three-level system).

In his famous 1908 paper *Over de Onbetrouwbaarheid der logische Principes* (on the untrustworthiness of logical principles), Brouwer rejected radically the traditional belief (held by Aristotle, Kant, Bolzano, Russell, Frege, Hilbert, Husserl, and many others) that classical logic has unquestionable validity. This rejection included denying the universally accepted laws of double-negation (for example, that "the opposite of the opposite of a true statement is a false statement"), which Russell and Hilbert associated with the equation $-(-1) = 1$. Also rejected was the law of the excluded middle (that no options are possible but true or false statements), which latter concept Russell identifies with the law of noncontradiction in his *Principia Mathematica* (1910-1913). To illustrate the not infrequent inapplicability of the law of the excluded middle, Brouwer cited a number of constructionally incompletable sequences, such as algorithms to compute the final digit of pi or the second-to-last valid example of Pierre de Fermat's conjecture, arguing that unless and until a proof is constructed to deduce their truth or falsity, the law of the excluded middle simply cannot be applied.

Impact of Event

Following Brouwer's initial publications between 1904 and 1911, both formalist and logicist camps reexamined their criticisms, consistency, and opportunities in the light of intuitionism. Initially, it was objected that intuitionist rejection of the pure axiomatic approach of Hilbert and the classical and symbolic logics of Aristotle and Peano and Russell left unaccounted for a large body of well-established pure and applied mathematics. In response to attacks against intuitionism's alleged impoverishment, in 1918, Brouwer set out explicitly to reconstruct classical mathematics in accord with his own intuitionist tenets, seeking greater natural clarity in construction without too great an increase in mental labor, proof duration, and complexity. Brouwer's own positive contributions included a new theory of rational numbers, an intuitionist theory of sets and topology, and a theory of finite and transfinite ordinal numbers. Arend Heyting, Brouwer's principal student, later contributed to extending intuitionist reconstruction into the theory of functions, as well as formal and some classes of symbolic logic.

Although working intuitionist analogs are unavailable still for many important areas of mathematics (for example, the implicit function theorem and the fundamental theorem of calculus), certain powerful and perplexing results can be proved in intuitionist theory that cannot be demonstrated in classical mathematics. Perhaps the most notable example of the latter is the intuitionist theorem that any real-valued function, which is defined everywhere on a closed interval of the real number continuum is uniformly continuous on that interval. Since the 1950's, a number of philosophical (metamathematical) objections have been raised to some intuitionist claims; notably, to the status of negative propositions referring to objects (like a square circle) that do not exist and to occasional intersubjective conflicts between intuitionist mathematicians on whether a given intuitive proof is sufficiently self-evident. Some intuitionist-influenced mathematicians—such as Hermann Weyl, George Polya, Ferdinand Gonseth, and Evert Willem Beth, among others—have sought to reformulate Brouwer's introspective notion of intuition, from a purely subjective psychology of mental experience to a more general genetic, or heuristic, model for the historic or individual development of concrete empirical concepts facilitating understanding and using abstract mathematics. Many mathematicians and logicians do not consider Brouwer's or subsequent intuitionist mathematics capable of ever being a self-sufficient and fully successful renovation of classical mathematics and logic. Nevertheless, in addition to its continuance as an identifiable school-of-thought with its own proceedings and journal, most acknowledge intuitionism's many and lasting influences on mathematics; for example, in the work of Gödel's later recursive function theory and Paul Lorenzen's efforts to develop Dingler's notion of constructivistic protologic for arithmetic, geometry, and other pure and applied mathematics.

Bibliography

Beth, Evert W., and Jean Piaget. *Mathematical Epistemology and Psychology.* Translated by W. Mays. Dordrecht, Holland: D. Reidel, 1966. A complementary approach, which preserves and updates Brouwer's psychology of mental intuitions.

Brouwer, L. E. J. *Collected Works*. 2 vols. New York: Elsevier, 1975-1976. A critical complete edition of Brouwer's collected and translated works. Volume 1 includes key sections of Brouwer's 1907 thesis, as well as several book and essay reviews on Kant, Frege, Russell, and others.

Cassirer, Ernst. *The Problem of Knowledge*. Translated by William H. Woglom and Charles W. Hendel. New Haven, Conn.: Yale University Press, 1950. Offers a shorter but similarly minded neo-Kantian interpretation.

Dummett, Michael. *Elements of Intuitionism*. Oxford, England: Clarendon Press, 1977. This intermediate-level monograph is a philosophical reconstruction of basic concepts by Brouwer and Heyting by a longtime student of Russell and Frege's logic. Offers an interpretation of Brouwer's position.

Fischbein, Efraim. *Intuition in Science and Mathematics*. Dordrecht, Holland: D. Reidel, 1987. Offers a thorough synopsis of the diversified possible meanings of intuition in mathematics.

Kleene, Stephen C., and Richard E. Vesley. *The Foundations of Intuitionistic Mathematics*. Amsterdam: North Holland, 1965. Requires more background in formal logic and proof theory. Spells out intuitionist efforts, as well as the relations of Brouwer's formal insights to Gödel's undecidability theorem.

Van Heijenoort, Jean, comp. *From Frege to Goedel*. Cambridge, Mass.: Harvard University Press, 1967. Offers a carefully detailed outline of several stages in Brouwer's early thought.

Gerardo G. Tango

Cross-References

Hilbert Develops a Model for Euclidean Geometry in Arithmetic (1898), p. 31; Lebesgue Develops a New Integration Theory (1899), p. 36; Levi Recognizes the Axiom of Choice in Set Theory (1902), p. 143; Russell Discovers the "Great Paradox" Concerning the Set of All Sets (1902), p. 184; Zermelo Undertakes the First Comprehensive Axiomatization of Set Theory (1904), p. 233; Russell and Whitehead's *Principia Mathematica* Develops the Logistic Movement in Mathematics (1910), p. 465; Gödel Proves Incompleteness-Inconsistency for Formal Systems, Including Arithmetic (1929), p. 900; Turing Invents the Universal Turing Machine (1935), p. 1045; The Bourbaki Group Publishes *Éléments de mathématique* (1939), p. 1140; Cohen Shows That Cantor's Continuum Hypothesis Is Independent of the Axioms of Set Theory (1963), p. 1751.

ZERMELO UNDERTAKES THE FIRST COMPREHENSIVE AXIOMATIZATION OF SET THEORY

Category of event: Mathematics
Time: 1904-1908
Locale: University of Göttingen, Germany

Zermelo undertook the first comprehensive axiomatization of (Cantor's) set theory, establishing key axioms as the basis of subsequent mathematics

Principal personages:
Ernst Zermelo (1871-1953), a German mathematician
Georg Cantor (1845-1918), a German mathematician
David Hilbert (1862-1943), a German mathematician
Bertrand Russell (1872-1970), an English philosopher and logician
Abraham Adolf Fraenkel (1891-1965), a German mathematician
Thoralf Albert Skolem (1887-1963), a Swedish mathematician
Kurt Gödel (1906-1978), an Austrian mathematician and logician

Summary of Event

Between 1895 and 1897, Georg Cantor published his chief papers on ordinal and cardinal numbers, the culmination of three decades of research on aggregates, collections, or sets. Because of its novel treatment of topics such as transfinite numbers, Cantor's set theory also was effectively a new theory of the infinite in mathematics as a legitimate and consistently definable entity. Yet, many critics from logic, mathematics, and several schools of philosophy refused to accept Cantor's set theory, not only because of several omissions on his part (for example, inadequate discussion of his well-ordering principle) but also because of a number of ambiguities and contradictions that apparently could be deduced from set theory.

One of the earliest paradoxes arose about 1895 directly from one of Cantor's theorems. According to his theorem, for every set of ordinal numbers (numbers that designate the order and position of numbers in a series: first, second, and so on), there is one ordinal number larger than all ordinal numbers of the set. An apparent contradiction arises when one applies this theorem to consider the set of all ordinal numbers, as Cantor described to David Hilbert. Greater problems resulted from Bertrand Russell's antinomy of 1903, involving the set of all sets which do not contain themselves as a subset. Many aspects of these and other paradoxes arose because Cantor's works gave no clear and consistent definition of the concept of ordinal versus cardinal numbers (integers one, two, and so on). The paradoxes were seen as extremely threatening by Russell, Giuseppe Peano, and Hilbert, among others. Further problems with Cantor's "naive" set theory concurrently arose, when Cantor lost confidence that the well-ordering principle might be a necessary universal law of thought and concluded, instead, that it was in need of clarification and proof, a conclusion with which Hilbert strongly agreed. The well-ordering principle basically states that it is possible to select simultaneously or choose—in an abstract yet valid sense—sets with infinite elements, without having to specify or carry out this selecting by actual operations or calculations. In his international conference review paper of 1900 on the major unsolved problems then facing mathematicians, Hilbert placed Cantor's set theory at the top of his famous list of twenty-three problems.

By 1902, Russell had convinced fellow logician Gottlob Frege that there were numerous other contradictions inherent in Cantor's set theory, some generalized criticisms of which appear in Russell's 1903 *The Principles of Mathematics*. In 1903, Hungarian mathematician Jules König presented a paper that challenged Cantor's claims for being able to order and proceed between different orders or levels of infinity, as well as attacking the more general claim that any set can be well-ordered. Within a day of König's presentation, a student of Cantor and Hilbert, Ernst Zermelo, prepared a counterdemonstration vindicating Cantor's system, by proving that König had improperly used a previous result of Felix Bernstein, which had been examined independently and attacked by Beppo Levi in his paper on Cantor's axiom of choice. Nevertheless, in the face of questions from mathematicians such as Hilbert, Russell, Émile Borel, Henri-Léon Lebesgue, W. H. Young, and others, Zermelo (at Hilbert and Cantor's urging) sought to establish more strongly the overall validity of Cantor's set theory.

Zermelo had already become well-acquainted with many aspects and problems of Cantor's theory. In 1899, Zermelo discovered Russell's paradox and informed Hilbert. In his spring, 1901, lectures at Göttingen, Zermelo expressed doubts on extant validations by Cantor of some of his set-theoretic axioms on cardinality. Assisted by Erhard Schmidt, Zermelo developed the background to what in 1904 would be his published proof of Cantor's well-ordering principle, using the controversial axiom of choice. In this much-criticized paper, Zermelo proved that in every set, an ordering principle can be introduced, in the form of the relation "a comes before b." Central to controversies around Zermelo's proof was his argument that for every subset M, a corresponding element m can be imagined. Many doubted that such formal imaginatory postulation had any real meaning, unless, for example, a concrete procedure or algorithm for carrying out the one-to-one correspondence between members of two sets could be given. For Zermelo, Hilbert, and other "formalists," however, such principles were irreducible and self-evident.

The other key aspect of Hilbert's formalism influencing Zermelo's efforts to clarify and reorder set theory was that of axiomatics. Hilbert's *Grundlagen der Geometrie* (1899; *The Foundations of Geometry*, 1902) first made clear the benefits and economy derived from using a minimum system of primitive assumptions, combined through explicit axioms to redefine naive/intuitive areas of mathematics such as geometry. Hilbert was interested in seeing a rigorous proof that the real number system is a consistent set, since in showing the completeness and consistency of Euclidean geometry, Hilbert had assumed the consistency of the real numbers. In 1900-1901,

Hilbert attempted a preliminary axiomatization of set theory. Proposing the basic concept of number as a given, Hilbert sought to determine the relation between this and other primitive concepts by introducing axioms such as the operations of arithmetic (+, −, ×, ÷), the commutative and associative laws, several ordering relations, the "Archimedean" axiom of continuity, and a completeness axiom. Hilbert believed (but did not show in detail) that his systematization of set theory was consistent, complete, and noncontradictory with regard to all the then-known theorems about the real numbers.

Zermelo's efforts at the axiomatization of set theory were not motivated solely by the logical paradoxes, but by continuing Hilbert's efforts. In mid-1908, Zermelo published two papers in Hilbert's journal *Mathematische Annalen*, which set out his own more comprehensive set theory axiomatics. Zermelo began like Hilbert, with a specific domain of mathematical objects, including sets, with the defined relations of membership between objects of his domain as a key set theory relation. Zermelo then stated seven central axioms, which he asserted were mutually independent and consistent. Without exhaustive technical details, Zermelo's seven axioms included axioms of extensionality, elementary sets, separation, power sets, union, choice, and infinity. The axiom of choice validated a major means of proving inductively that a given property holds for all set elements if it holds for one element. Zermelo outlined how many extant axioms of arithmetic could be derived from these axioms, including operations of union, intersection, product, equivalence, and functionality. Zermelo devoted much of his discussion to developing a theory of cardinals (transfinite numbers) in terms of specific ordinal numbers, showing that the earlier Frege-Russell definition of cardinal numbers was incompatible with his and Hilbert's system of arithmetic. Zermelo devoted minimal discussion to the paradoxes of Cantor, Russell, and the like, since he believed that his "condition of definiteness of subsets" avoided any contradictory construction. Definiteness, or a definite property, E, of a set S is said to be one for which the seven axioms permit one to determine whether E holds for any element of S.

Impact of Event

Zermelo's axiom system for Cantor's set theory was not acknowledged at first. Most initial attention focused almost exclusively on its weak points: for example, Russell and Henri Poincaré's suspicions that Zermelo's formal system was inconsistent or employed methods or ideas improper for mathematics. Poincaré, in particular, objected to formal axiomatization as contrary to the constructive intuition and creation of mathematical entities. Russell's criticisms were that set theory, with arithmetic, was a part of logic and could be made consistent only when recast into symbolic logic. Russell and Alfred North Whitehead, in their *Principia Mathematica* (1910-1913), presented their theory of "types" to resolve the paradoxes of set theory. This established a hierarchy of types of sets, which includes the real numbers, but also contains the unintuitive axiom of irreducibility. This states that for any property of types higher than order −0, there is a property over the same range of order −0 that can be shown as equivalent to the assertion that impredicative definitions are equivalent to predicative definition. (An impredicative definition is one made only in terms of the totality of which it is a part.)

Despite Zermelo's concise and clear exposition, his axiomatic for set theory was largely ignored in favor of nonaxiomatic set theory, until 1921. In this year, Abraham Adolf Fraenkel, in his efforts to demonstrate the independence of the axiom of choice from other axioms, first pointed out some shortcomings in Zermelo's system, as well as their remedy. Fraenkel objected that Zermelo's axiom of infinity was too weak, that his property of definiteness was too vague, and that his total system was insufficiently categorical to encompass all of ordinal arithmetic and the transfinite numbers. This led Fraenkel to add what is now known as the axiom of replacement, which states that if each element of a set is associated with one and only one set, then the collection of associated sets is a set. In place of Zermelo's notion of a definite property, Fraenkel proposed the notion of a new function. In 1923, the Swedish mathematician Thoralf Albert Skolem independently arrived at a similar result, regarding "definiteness" as a property expressible in first-order logic. Skolem proposed that within any predicate calculus (first order logic), it is impossible to establish an ultimate categorical system of axioms for the natural numbers by means of a finite set of axioms, a result suggestive of Kurt Gödel's later theorems. With these modifications, Zermelo's theory has become known as the Zermelo-Fraenkel system for set theory.

In 1923 and 1925, John von Neumann published papers presenting, respectively, a new definition of ordinal number sets and an alternate axiomatic for set theory. Paul Bernays in 1937 and Gödel in 1940 published alternatives to the Zermelo-Fraenkel theory of sets. A system of ten other axioms was developed, taking "classes" as the basic variable, for which sets are those classes adequate for arithmetic with natural/real numbers, and proper sets are those nonset classes for which contradictions can arise. Although others, such as Willard van Orman Quine in 1958 offered further alternative axiomatizations for set theory, Zermelo-Fraenkel and von Neumann-Bernays-Gödel theories to date remain the only working options within mathematics.

Despite the avoidance of earlier paradoxes, axiomatic set theory is by no means a closed subject. In 1963, Paul Cohen proved that each of Zermelo's axioms is logically independent, a result that opened several avenues of continuing inquiries. Although the relative consistency of both set theories can be demonstrated with respect to other mathematical theories, any would-be absolute validation of axiomatic set theory is as yet unavailable. Because Gödel's inconsistency and indecidability theorems apply to both set theories, such absolute verification may never be forthcoming. Most contemporary emphasis is directed to providing at least a more complete and comprehensive set theoretic description, efforts that were first pioneered by Zermelo.

Bibliography

Cohen, Paul J. *Set Theory and the Continuum Hypothesis*. New York: W. A. Benjamin, 1966. An advanced and technical account of Cohen's independence proof of the axiom of choice in the context of the Zermelo-Fraenkel set theory.

Dauben, J. W. *Georg Cantor: His Mathematics and Philosophy of the Infinite*. Cambridge, Mass.: Harvard University Press, 1979. A rigorous, yet nontechnical, account of naive set theory as Zermelo found it.

Fraenkel, Abraham A., and Yehoshua Bar-Hillel. *Foundations of Set Theory*. Amsterdam: North-Holland, 1958. Fraenkel's main work on axiomatic set theory. Contains key material and a discussion of his 1921 and 1922 papers.

Halmos, Paul R. *Naive Set Theory*. Princeton, N.J.: D. Van Nostrand, 1960. A general undergraduate introductory-level text to what is essentially Cantor's original set theory.

Hamilton, A. G. *Numbers, Sets, and Axioms: The Apparatus of Mathematics*. Cambridge, Mass.: Harvard University Press, 1983. An introductory survey of naive set theory and the basics of axiomatic set theory in the context of prior and subsequent mathematics.

Moore, Gregory H. *Zermelo's Axiom of Choice: Its Origins, Development, and Influence*. New York: Springer-Verlag, 1982. An exhaustive historical-conceptual treatment of some background and developmental aspects of Zermelo's set theory axiomatics.

Stoll, Robert Roth. *Set Theory and Logic*. San Francisco: W. H. Freeman, 1960. A logical representation of Cantor's set theory. Points out sources of the paradoxes and their avoidance by axiomatic set theory.

Suppes, Patrick. *Axiomatic Set Theory*. Princeton, N.J.: D. Van Nostrand, 1960. An intermediate-level account of set theory from a logician's perspective.

Watson, S., and J. Steprans, eds. *Set Theory and Its Applications*. New York: Springer-Verlag, 1989. An intermediate-advanced text on set theoretic applications in computer and engineering science.

Gerardo G. Tango

Cross-References

Hilbert Develops a Model for Euclidean Geometry in Arithmetic (1898), p. 31; Lebesgue Develops a New Integration Theory (1899), p. 36; Levi Recognizes the Axiom of Choice in Set Theory (1902), p. 143; Brouwer Develops Intuitionist Foundations of Mathematics (1904), p. 228; The Study of Mathematical Fields by Steinitz Inaugurates Modern Abstract Algebra (1909), p. 438; Russell and Whitehead's *Principia Mathematica* Develops the Logistic Movement in Mathematics (1910), p. 465; Noether Publishes the Theory of Ideals in Rings (1921), p. 716; Cohen Shows That Cantor's Continuum Hypothesis Is Independent of the Axioms of Set Theory (1963), p. 1751; Mandelbrot Develops Non-Euclidean Fractal Measures (1967), p. 1845.

BRANDENBERGER INVENTS CELLOPHANE

Categories of event: Applied science and chemistry
Time: 1904-1912
Locale: Thaonles, Vosges, France

Brandenberger cast viscose cellulose from solution into thin, continuous, transparent sheets, the start of the modern packaging era

Principal personages:

JACQUES EDWIN BRANDENBERGER (1872-1954), a Swiss chemist who invented cellophane

CHARLES FREDERICK CROSS (1855-1935), an English chemist who, with Bevan, invented the process for modifying cellulose

EDWARD JOHN BEVAN (1856-1921), an English chemist who, with Cross, invented the viscose process

Summary of Event

The invention of cellophane, while representing a tremendous amount of perseverance and insight by Jacques Edwin Brandenberger, was not a major theoretical achievement. Instead, it was merely a modification of an existing process, one of several processes for making cellulose plastics. Nor was the amount of cellophane, on a weight basis, particularly significant relative to other materials made from wood; at its peak, the production of cellophane was only 0.5 percent of the production of paper and less than 20 percent of the production of all cellulosic plastic materials. It did, however, revolutionize the packaging industry as the first clear, plastic film.

The first successful cellulose plastic—indeed, the first synthetic plastic ever made—was made in 1846 by Christian Friedrich Schönbein of Switzerland when he formed nitrocellulose (also called cellulose nitrate depending on its use) by the action of sulfuric and nitric acids on cotton. Nitrocellulose soon became very important and had a variety of uses. It was used as a propellant/explosive by the Austrian army in 1852 (called guncotton or smokeless powder, the first plastic explosive), as the first modern lacquer in the United States in 1882 using amyl acetate as a solvent, in early photographic films (termed celluloid as is cellulose acetate, which later replaced cellulose nitrate for this use), and in textiles as the first artificial silk by Count de Chardonnet in 1884 in France. Because of the presence of nitrate groups (the same functional group that supplies the oxygen in gun powder), cellulose nitrate is highly flammable and is no longer used in textiles. Its unsuitability for use in textiles became apparent when the dress of a woman attending a cocktail party disappeared in a flash of smoke after ashes from her escort's cigar ignited it. Shortly thereafter, in 1869, cellulose acetate was first made in France. Technical difficulties using this material were finally overcome, and by 1908, the Celluloid Company was producing acetate film for use around the world for photography. In 1914, the Lustron Company was set up by the Arthur D. Little Company to start manufacturing cellulose acetate textile fibers, which are called acetate rayon, or rayon for short. Cellulose nitrate was quickly replaced by cellulose acetate and other materials for all uses except explosives and propellants.

In 1892, Charles Frederick Cross and Edward John Bevan developed the viscose rayon process. Except for a modification in the method of casting the solution into usable form, it is the basis for cellophane manufacture. In this process, cellulose from pulped wood fibers (very similar to cotton fibers, except the cellulose chains are about one-tenth the length) is treated with sodium hydroxide solution (also called caustic soda) to swell the fibers in a process called mercerization. Mercerization was invented by the Englishman John Mercer in 1844 as a method to shrink cotton and make it glossy. The mercerized, shredded pulp, which has aged for several days, is treated with carbon disulfide, which reacts with cellulose to form the viscous, orange-colored cellulose xanthate solution known as viscose. The viscose is aged for several days, filtered, and then forced through a thin hole into an acid bath containing salts that convert the cellulose xanthate back into cellulose. The regenerated cellulose is no longer soluble and forms a solid filament known as viscose rayon. Cross and Bevan used this process as a means of forming continuous, carbon filaments for use in the newly discovered light bulb. In 1911, the American Viscose Company produced 165,000 kilograms of yarn in its first year of production, a total production rate that cellophane would take several years to match in the United States. Worldwide production of acetate and viscose rayon is about 2.7 billion kilograms per year, having been fairly constant since 1965. Less than 10 percent of this is used in cellophane.

In 1904, Brandenberger, apparently frustrated by sloppy tables in a local French café, desired to develop a protective material for tablecloths, which in those days were made only of cloth and, therefore, were susceptible to staining. His employer, the Thaon Textile Company, encouraged him over the eight years it took to develop cellophane. At first, Brandenberger used cellulose nitrate solutions, but these gave a film that was much too brittle and rigid to be useful for tablecloths. Later, he worked with the new viscose material that had been perfected by Cross and Bevan. He was able to peel a transparent film from the treated tablecloth. While viscose was not suitable for use in tablecloths, despite numerous attempts, Brandenberger pursued the idea of casting viscose into thin films from solution. In 1908, he filed a patent on regenerated cellulose films from viscose using a water bath containing ammonium sulfate to coagulate the viscose and an acidic water bath to regenerate the cellulose. The final bath usually contains glycerol, which acts as a plasticizer; that is, it makes the cellophane much less brittle. The only significant difference from the viscose rayon process is that in the cellophane process, the viscose solution is cast through long, thin slots rather than small holes. It is much more difficult to handle the thin sheet compared to handling the filament; this is where most of the development took place. By casting it into a ring-shaped slot, a continuous tube forms, which is used in sausage casings. Until 1912, Brandenberger worked to perfect the machine that would produce a continuous film of cellophane. He filed for additional patents to protect his invention and tried to find uses for cellophane. Cellophane was used in World War I as unbreakable eyepieces in gas masks. After the war, it was used as trimming for hats. With the backing of France's largest rayon company, Brandenberger started the company La Cellophane, literally translated as "the clear cellulose." In 1920, du Pont, recognizing an important material, bought the North American rights to cellophane. On April 4, 1924, the first cellophane was manufactured in the United States at the du Pont plant in Buffalo, New York.

The original cellophane, while waterproof, was not water-vapor proof. Bread and other moist materials would dry out over time, while dry items would pick up moisture and become soggy. Much research and development went into solving this problem; in 1928, after trying hundreds of compounds, William Hale Church of du Pont solved this problem by using a very thin wax coating to prevent "breathing" by cellophane. Also, since the original cellophane was not heat-sealable, packaging had to be done by hand, so only expensive items such as perfume could be wrapped. This problem was overcome with a coating that allowed cellophane to stick to itself with the application of heat and pressure, allowing it to be used in automated packaging machines. Ironically, this coating contained cellulose nitrate, the same material replaced by these new cellulose materials.

Impact of Event

At first, cellophane was very expensive, almost six dollars per kilogram. Vendors were said to store their stocks frequently in safes. As the process was dramatically improved within the first year, however, prices dropped to two dollars per kilogram. Cellophane became popular very quickly. It replaced waxed paper and glassine paper as wrapping materials; it was, then, the first wrapping material that was clear and allowed the contents to be plainly visible. This was an extremely important marketing tool that created the modern packaging industry. Soon, people were wrapping everything in cellophane, and cellophane received high amounts of publicity. Sales of food items such as baked goods and dried pasta increased by as much as ten-fold or more as people could now easily see the product in the package. Even through the depression of the 1930's and World War II, production of cellophane continued to increase.

Another product created with cellophane was the clear tape used to mend paper and tape items to each other. Now, other materials are used and what was called "cellophane tape" is called "Scotch tape." Indeed, the courts ruled that the world "cellophane" was in such common usage that no one could license it as a trademark. Cellophane is also used in semipermeable membranes (which allow small molecules to pass but block large molecules) for applications such as kidney dialysis.

As cellophane became popular, production increased quickly in the United States. About 50 million kilograms per year were manufactured by 1940 and 200 million kilograms per year by 1960, its peak year of production. Competition by polyethylene, saran wrap, and other materials derived from petroleum was very intense. Cellophane production declined over the next twenty years until the rate of production of about 50 million kilograms per year was assumed in 1983. Cellophane is about 0.03 millimeter thick, so each kilogram corresponds to about 40 square meters of cellophane.

True, other materials replaced cellophane as they had improved properties and were cheaper to manufacture, but they filled the niche that cellophane created; therefore, replacement films did not come into the market with the same excitement that cellophane did. While the amount of cellophane is still quite significant, it has lost its relative importance as the packaging industry used many other types of materials in large amounts. The impact of cellophane is measured only partly by its production over time. Its long-term significance is its impact as the original transparent film of the modern packaging industry.

Bibliography

Battista, O. A. "Transparent Bonanza." *Science Digest* 27 (April, 1950): 84-87. This brief, interesting article chronicles how Brandenberger invented cellophane, the early days of cellophane, and cellophane's impact on packaging and marketing.

"Cellophane: A Substitute for Paper." *The Literary Digest* 80 (February 2, 1924): 61-62. This article, written early in the year that cellophane was to be produced in the United States, describes its importance and its manufacture. It is described as a replacement for paper, as it replaced waxed paper and glassine as packaging materials; also discussed is the creation of a new industry.

"Competition Slows Cellophane's Growth." *Chemical and Engineering News* 39 (March 6, 1961): 32-33. This article was written at the peak of cellophane's success. The article discusses the importance of polyethylene as the most inexpensive clear film and other materials that would capture more and more of the packaging film industry. It predicts the replacement of cellophane as the dominant clear packaging material with other materials.

Haynes, William. *Cellulose: The Chemical That Grows*. Garden City, N.Y.: Doubleday, 1953. This is a fascinating, well-written account of the cellulose industry from the early 1800's through the early 1950's. It tells of the important discoveries, how they relate, and how they interact. Only a few illustrations are included.

Peck, A. P. "Cellophane Is Born." *Scientific American* 158 (May, 1938): 274-275. A two-page spread with nine pictures showing the process of making cellophane with captions describing the process. The best available pictorial on the process.

Turbak, Albin. "Rayon." In *Encyclopedia of Polymer Science and Engineering*, edited by Herman F. Mark. 2d ed. New York: John Wiley & Sons, 1985-1986. This is a fairly technical, in-depth discussion of viscose rayon and acetate rayon. It includes figures for United States and worldwide production and use. Other articles such as "Cellulose" would be of interest also to some readers.

Christopher J. Biermann

Cross-References

Baekeland Invents Bakelite (1905), p. 280; Carothers Patents Nylon (1935), p. 1055; Land Invents a Camera/Film System That Develops Instant Pictures (1948), p. 1331.

SHERRINGTON DELIVERS *THE INTEGRATIVE ACTION OF THE NERVOUS SYSTEM*

Category of event: Biology
Time: April-May, 1904
Locale: Yale University, New Haven, Connecticut

Sherrington clarified the role of the nervous system in molding adaptive hierarchies of reflexes into purposeful behavior

Principal personages:

Sir Charles Scott Sherrington (1857-1952), an English neurologist who was a cowinner of the 1932 Nobel Prize in Physiology or Medicine

Marshall Hall (1790-1857), an English physician and neurologist who created the standard account of reflexes until Sherrington's work

René Descartes (1596-1650), a French philosopher whose attempt to create a mechanical model of the body introduced some of the basic mechanisms of reflexes

Camillo Golgi (1843-1926), an Italian histologist whose innovative techniques clarified neuroanatomy; he shared the 1906 Nobel Prize in Physiology or Medicine with Santiago Ramón y Cajal

Santiago Ramón y Cajal (1852-1934), a Spanish histologist whose interpretation of his work differed significantly from Golgi's

Sir Michael Foster (1836-1907), an English physiologist at Cambridge and instructor of Sherrington who is considered the founder of English physiology

Ivan Petrovich Pavlov (1849-1936), a Soviet physiologist who pioneered reflexology but who disagreed with Sherrington's philosophy

Edgar Douglas Adrian (1889-1977), a student of Sherrington who investigated the electrical activity of nerves; he shared the 1932 Nobel Prize in Physiology or Medicine with Sherrington

Sir Henry Dale (1875-1968), a student of Sherrington who investigated the chemical activity of nerves

Summary of Event

The Integrative Action of the Nervous System (1906) consummated centuries of speculation and experimentation about reflexes and created a research agenda for another generation of neurophysiologists. A reflex is a stereotypical motor response to sensory stimuli. Sir Charles Scott Sherrington had the task of understanding how purposeful behavior arose out of hierarchies of reflexes.

Early ideas about the reflex were speculative, not experimental, and appeared as ancient philosophers and physicians tried to comprehend puzzling observations about humans in health and disease. They believed that injured organs affected other organs even though no connection was seen between them. Aristotle, for example, had noted that sometimes there were "motions of the heart and of the penis; for often upon an image arising and without express mandate of the intellect these parts are moved." Involuntary or sympathetic motions were distinct from nonvoluntary motions such as falling asleep, waking up, or respiration, in which imagination or desire played no role. Heartbeats, on the other hand, were both sympathetic and nonvoluntary. One did not have to will one's heart to beat, but passions affected its rate. Galen, a Greek physician, kept Aristotle's distinctions but added other examples, such as that the contraction of pupils when exposed to light was independent of the will. The regularity of these responses to stimuli puzzled ancient philosophers.

René Descartes inherited these ideas about sympathy when he developed his mechanical model of the animal body in the seventeenth century. In the treatise *L'Homme* (1664; *Treatise of Man*), he explained sympathy as well as all nervous activity in terms of a network of hollow nerves filled with a refined subtle matter, the "animal spirit." A well-known illustration from the *Treatise of Man* closely approximates the definition of the reflex. A young boy stuck his foot into a fire; particles of fire forced animal spirit up a nerve to the brain; the brain redirected animal spirit down the nerve to the muscles of the foot; and the foot was withdrawn. Descartes concluded from this and several other "explanations" that animals, including humans, were capable of a large repertoire of complicated and purposeful behavior without the intervention of the will or intelligence.

Stephen Hales was one of the earliest experimenters, as opposed to speculators, on reflexes. In the early 1730's, he decapitated frogs and noted that irritating their rear legs still provoked hopping. If the spinal cord was damaged, however, motion ceased. Clearly, the brain was not responsible for all types of involuntary or reflex motions. His work led to the reevaluation of the functions of the spinal cord. Previously, it had been considered only a trunk line for communications to and from the brain and periphery, but now it was demonstrated to have independent functions. Robert Whytt advanced Hales's work considerably. In *An Essay on the Vital and Involuntary Motions of Animals* (1751) and *Philosophical Essays* (1755), he systematized the previous work on reflexes and first defined modern terminology such as "stimulus" and "response."

Most of the major work leading to the modern understanding of the reflex followed from the Bell-Magendie law, named after Sir Charles Bell and François Magendie, which stated that the anterior root of the spinal cord governed motor functions and the posterior root, sensory functions. Marshall Hall, who demonstrated conclusively that the spinal cord had a functional unity of its own, widened the scope of reflexes to include such activities as vomiting and parturition and studied the effect of poisons on reflexes.

As important as these physiological discoveries were, the microanatomy of the reflex remained hidden. Progress was impossible until theoretical advances such as cell theory and technological advances such as staining techniques and oil-immersion lens microscopes were applied to neuroanatomy. Camillo Golgi and Santiago Ramón y Cajal used these new technologies to uncover the details of nerve cells, but they interpreted their results differently. Golgi claimed that nerves formed a patterned network and that nerve impulses spread wherever neural nodes overlapped (reticularist model). Ramón y Cajal, on the other hand, saw individual nerves as approaching each other but never touching (neurone model). Each model had strengths and weaknesses that complemented the other: The neurone offered clear pathways but no communication between nerves; the network offered communication but no clear pathways.

By the 1890's, physiological and anatomical pieces to the reflex puzzle abounded, but no one had crafted a synoptic vision to place them into a coherent whole. Sherrington confronted this confusion when he began his work. He was well prepared for this task. He was educated as a physician, trained in physiology under Sir Michael Foster, and exposed to the best European physiologists. Despite his preparation, the enormity of the task astounded him. He eventually delineated the sensorimotor pathways of the macaque monkey. Foster had told his students that the nervous system functioned as a unity, not as individual pieces, and Sherrington was determined to discover that unity.

Yale University invited Sherrington to deliver the Silliman lectures in the spring of 1904, and this opportunity allowed him to synthesize his ideas on the unity of nervous action. The lectures were published two years later as *The Integrative Action of the Nervous System*. Sherrington's thesis in the Silliman lectures and in *The Integrative Action of the Nervous System* was that "the nervous synthesis of an individual from what without it were a mere aggregation of commensal organs resolves itself into co-ordination by reflex action." In other words, how are distinct activities such as scratching, blinking, or maintaining posture coordinated with other activities like breathing or walking to compose ultimately an individual animal? Sherrington stated this question more clearly in his 1933 Rede lecture at Cambridge University, "The Brain and Its Mechanisms": "How a limited set of agents of the outside world working through the nerve and brain of the animal can produce from its muscles the thousand and one dexterous acts of normal behaviour is itself a problem. . . ." His answer began with an analysis of the three functions of the nerve cell: to support its own life, to communicate with other cells, and to integrate sensorimotor activity into complex behavior. Only nerve cells carried on this last function, so the nervous system regulated the interplay between any animal and its environment. Reflexes, he argued, were the most basic units of sensorimotor activity, and it was their coordination into purposeful behavior that defined the integrative action of the nervous system. Sherrington's simplicity of experimental technique and clarity of argument ordered reflexology's pieces into a coherent pattern.

Sherrington introduced one of the most basic concepts in neurology: the synapse. He accepted Ramón y Cajal's neurone theory and stipulated that nervous transmission occurs only across the small gap—synapse—between nerves. He did not know yet how the impulse "jumped" between nerves; however, he knew from his own work that the neuronal theory and the synapse were true.

Reflexes, Sherrington asserted, played an evolutionary, or adaptive, role. In terms he coined himself, he claimed that proprioceptive reflexes, which gave an animal a sense of place, and nociceptive, which gave a sense of pain, were necessary for the survival of any individual. Despite the progress represented in *The Integrative Action of the Nervous System*, Sherrington acknowledged that he had hardly begun to understand the complexities of human behavior.

Impact of Event

One reviewer of *The Integrative Action of the Nervous System* likened Sherrington to Sir Isaac Newton; a former student compared him to William Harvey. They praised Sherrington not only because he cleared up many neuroanatomical and neurophysiological questions but also because he raised those sciences to new levels and suggested new research.

His endorsement of Ramón y Cajal's neurone theory and the postulation of the synapse stimulated his students to discover unsuspected chemical and electrical activities in the transmission of the nerve impulse across synapses. The newer vacuum tube technology allowed Edgar Douglas Adrian, along with others, to amplify minute electrical charge variations in nerves. This tool enabled them to explore the afferent nerve impulse and to discover the electrical coding of nerve signals. Sir Henry Dale, who later became director of the Wellcome Physiological Research Laboratory, pioneered with others the recognition of acetylcholine as a neurotransmitter. Adrian's work led to Hans Berger's development of the electroencephalogram (EEG) in 1929; Dale's work helped spawn the rapidly expanding field of neurochemistry.

The Integrative Action of the Nervous System enhanced Sherrington's already high reputation and helped create the Sherrington School of Physiology. From all over the world, students flocked to the University of Oxford to work with Sherrington and, after returning to their home institutions, continued the research begun with him. Several Nobel laureates were nurtured under Sherrington's tutelage.

Sherrington's interpretation of reflex action raised old questions again about the relation of philosophy or theology to the scientific understanding of human behavior. Sherrington often said that after twenty-five hundred years of examining the physiological foundation of behavior, researchers were still in basically the same position as Aristotle. Sherrington and several of his contemporaries had been trained in a materialist physiology that viewed advances in reflexology as pushing back the domain of vitalist or idealist doctrines. As he matured, however, he moved from materialistic monism to a body/mind dualism. He accepted the reality of an independent mind, for although he could not explain it physiologically, neither could he explain the complexity of human behavior—as opposed to animal behavior—only in terms of reflexes. In later essays, Sherrington feared the political and social ramifications of a too-hasty solution to these ancient problems. He deeply influenced John Eccles, one of his students, who has continued to reflect upon these issues.

Certainly not all physiologists shared his views. Ivan Petrovich Pavlov, Sherrington's contemporary and also a Nobel laureate, remained staunchly materialistic. For him, all behavior, including human, was reducible to reflexes. He concluded from his experiments on dogs that learning itself was a variety of sensorimotor activity. Surprised by Sherrington's dualism in "The Brain and Its Mechanisms," he wondered aloud to colleagues if Sherrington had become ill or senile.

When in the 1930's Sherrington looked back on his work, he admitted candidly that his reflex studies were limited and had served their purpose. Twentieth century neuroscience has bloomed luxuriantly in fields with applications Sherrington could never have imagined, yet hardly any of its aspects do not owe something of their roots to Sherrington's *The Integrative Action of the Nervous System*.

Bibliography

Clarke, Edwin, and L. S. Jacyna. *Nineteenth-Century Origins of Neuroscientific Concepts*. Berkeley: University of California Press, 1987. Modern scholarship that places Sherrington's achievement in the context of nineteenth century neurology.

Eccles, John C. *The Human Mystery*. New York: Springer-Verlag, 1979. Eccles' Gifford lectures at the University of Edinburgh from 1977 to 1978 represent his reflections on themes Sherrington presented in his Gifford lecture, *Man on His Nature* (1941).

Eccles, John C., and William C. Gibson. *Sherrington: His Life and Thought*. New York: Springer-Verlag, 1979. Eccles, a former student, and Gibson, a historian of science, go beyond Swazey's account (cited below) by examining Sherrington's scientific work after 1906 and his philosophical essays written in later life. Uses memoirs and interviews of his colleagues to bring to life what it was like to work with Sherrington.

Granit, Ragnar. *Charles Scott Sherrington: An Appraisal*. London: Thomas Nelson, 1966. A personal memoir by one of Sherrington's students. The material on *The Integrative Action of the Nervous System* has a more personal approach.

Jeannerod, Marc. *The Brain Machine: The Development of Neurophysiological Thought*. Translated by David Urion. Cambridge, Mass.: Harvard University Press, 1985. A frankly materialistic historical interpretation of neurophysiology that contrasts with Sherrington's view.

Liddell, Edward George Tandy. *The Discovery of Reflexes*. Oxford, England: Clarendon Press, 1960. One of the earliest scholarly historical accounts of the background of Sherrington's achievement, written by a member of his laboratory; covers only as far as *The Integrative Action of the Nervous System*.

Ramón y Cajal, Santiago. *Recollections of My Life*. Translated by E. Horne Craigie with Juan Cano. Cambridge, Mass.: MIT Press, 1989. The original Spanish edition was reprinted often in Madrid between 1901 and 1917; the first English edition appeared in 1937. Considered by many to be the finest example of scientific autobiography.

Sherrington, Sir Charles Scott. *The Integrative Action of the Nervous System*. New

Haven, Conn.: Yale University Press, 1906. A complete account of the Silliman lectures, expanded with charts, experimental protocols, and an immense bibliography. Often reprinted but still hard to find.

____________. *Man on His Nature*. New York: Macmillan, 1941. Originally presented as the Gifford lectures at the University of Edinburgh from 1937 to 1938; depicts a mature Sherrington reflecting upon his life work and its significance.

Swazey, Judith P. *Reflexes and Motor Integration: Sherrington's Concept of Integrative Action*. Cambridge, Mass.: Harvard University Press, 1969. The best scholarly examination of Sherrington's work, especially *The Integrative Action of the Nervous System*. Includes lengthy sections on the background of his work, clear diagrams—some from Sherrington—and abundant documentation.

Thomas P. Gariepy

Cross-References

Ramón y Cajal Establishes the Neuron as the Functional Unit of the Nervous System (1888), p. 1; Pavlov Develops the Concept of Reinforcement (1902), p. 163; Berger Develops the Electroencephalogram (EEG) (1929), p. 890; Sperry Discovers That Each Side of the Brain Can Function Independently (1960's), p. 1635.

CONSTRUCTION BEGINS ON THE PANAMA CANAL

Category of event: Applied science
Time: Summer, 1904
Locale: Central America, Republic of Panama

The Panama Canal was a ten-year engineering project that cut off a 12,874 kilometer trip for ships going around South America

Principal personages:

GEORGE WASHINGTON GOETHALS (1858-1928), a military engineer who supervised the canal project from 1907 until its completion in 1914

WILLIAM CRAWFORD GORGAS (1854-1920), an Army surgeon and sanitary officer who eliminated yellow fever and greatly reduced malaria in the Canal Zone by a systematic program of mosquito eradication

JOHN F. STEVENS (1853-1943), the chief engineer from 1905 to 1907 who resigned after a dispute with President Theodore Roosevelt

WALTER REED (1851-1902), an Army surgeon and bacteriologist who showed in 1900 that yellow fever is caused by a mosquito that carries the virus from person to person

FERDINAND DE LESSEPS (1805-1894), a French politician who successfully built the Suez Canal in 1869 and started excavation for a Panama canal in the 1880's but failed

Summary of Event

The isthmus of Panama is a narrow strip of land in Central America between Costa Rica and Colombia. The separation of the Atlantic and Pacific oceans is less than 80 kilometers. During the California gold rush in the early 1850's, a railroad was built across Panama. Passengers from the eastern United States could go by ship to the port of Colón on the Atlantic side, take a short train ride to Panama City on the Pacific, and continue by ship to San Francisco. In comparison, going cross-country to California by wagon train was only for the hardiest travelers.

A canal across Panama was viewed as very desirable for shipping because it would cut off a long and dangerous trip around South America. France made the first serious attempt to build "the big ditch" under the leadership of Ferdinand de Lesseps, who had successfully built the Suez Canal through the Egyptian desert in 1869. The Suez Canal made it possible to travel from Europe to India and the Far East without going around the continent of Africa. Lesseps, a popular hero in France, made an inspection trip to Panama in 1879 and announced that a canal connecting the Atlantic and Pacific oceans should be easier to build than the Suez. Money from private investors was raised for the project and construction started in 1881. The magnitude of the task had been greatly underestimated, however, and the work came to a halt in 1889.

The French effort failed because the French machinery was too small and lightweight for the job. Suez had sand, whereas Panama had rocks and mud. Second, the death toll from yellow fever and malaria was high, with more than twenty thousand deaths in eight years. The role of mosquitoes in transmitting these tropical diseases was unknown, sanitation was inadequate, and working conditions were poor. Finally, the idea of a sea-level canal was not practical in Panama. In the middle of the isthmus, the mountains are more than 90 meters high and 11 kilometers wide. In order to make a channel down to sea level, a large amount of rock would have to be blasted with dynamite. The French did make considerable progress with the Culebra Cut through the highest terrain, but unfortunately, frequent mud slides during the rainy season refilled some of the excavated area.

In the United States, popular support for building a canal in Central America developed during the Spanish-American War from 1898 to 1900. The battleship *Oregon* was stationed in San Francisco in March of 1898 when the captain received orders to bring his ship to Cuba as quickly as possible to participate in the blockade. With a maximum speed of 16 knots, the *Oregon* traveled south along the coast of Chile, came around the tip of Argentina, and headed north toward the Caribbean. Along the way, there were delays because of illness among the crew, a four-day storm, and Spanish warships. After ten long weeks, the battleship finally arrived off the coast of Cuba. This dramatic voyage was publicized widely and later became a rallying point in Congress for those who advocated federal funding for a canal because of its military necessity.

The United States government under President Theodore Roosevelt negotiated an agreement in 1902 with the French canal company to take over their unfinished project. The land belonged to Colombia, however, and they turned down an American offer of $10 million for the strip of land along the canal route. A so-called phony revolution took place on November 4, 1903, when the Republic of Panama seceded from Colombia. Within three weeks, the United States gave recognition to the new country of Panama and obtained rights to a 16-kilometer wide "Canal Zone" at the original price that had been offered to Colombia. Twenty years later, the United States paid an indemnity of $25 million to Colombia in compensation for this subterfuge.

Chief Engineer John Findley Wallace was appointed to coordinate the building project in 1904, but he lasted only for about a year. Progress in excavation was too slow, and an unfortunate outbreak of yellow fever in the spring of 1905 took many lives. President Roosevelt replaced Wallace with Chief Engineer John F. Stevens and appointed William Crawford Gorgas to head the mosquito eradication project.

Walter Reed had shown only four years earlier that both yellow fever and malaria are transmitted by mosquitoes. Previously, these dreaded illnesses were thought to be the result of tropical climate or poor sanitation. Because Dr. Gorgas had worked as sanitation officer with Reed in Cuba, he knew what had to be done. One technique was to spray larvicide oil on water surfaces. Mosquito larvae must come to the surface periodically to breathe and are killed on contact. Other preventive measures were to install a sewage system where workers lived, to train exterminators for fumigating buildings, and to place sick patients in an isolation cage so mosquitoes could not get to them. The campaign against mosquitoes was so effective that yellow fever was completely eliminated in the Canal Zone by 1906, and malaria gradually was reduced to less than 10 percent among the workers.

Under Chief Engineer Stevens, the design of the canal was changed from excavating a sea-level route to using locks. The idea was to create a large lake well above sea level, to raise the ships up to the lake with three locks on one side, and then to lower the ships back down to the ocean with locks on the other side. The design was appropriately called "a bridge of water" going from ocean to ocean. Gatun Dam took more than six years to construct, eventually forming Gatun Lake with an area of 264 square kilometers. Ships could cross the lake under their own power, and it provided the waterway for more than half of the canal route. The locks to raise and lower the ships were a mammoth project, requiring large quantities of cement for their 305-meter length and huge steel doors that had to swing open and shut to hold the water.

President Roosevelt visited Panama for three days in 1906, the first time in history that an American president made a trip to a foreign country. Although Stevens was popular with the workers, he had some disagreements with Roosevelt and resigned in 1907. Roosevelt immediately replaced him with Colonel George Washington Goethals, whose forceful leadership carried the canal project through to its successful conclusion.

Colonel Goethals had to deal with the difficult problem of the Culebra Cut through the mountains, which was to be the passageway from Lake Gatun to the locks on the Pacific side. Goethals recognized that the speed of excavation was limited by the rate of rock removal. Up to thirty trains per hour carried rocks from Culebra to be dumped as a breakwater in the ocean. Workers had to shift the tracks daily as the steam shovels advanced. The biggest tragedy of the project was a premature dynamite explosion that killed twenty-three workers in 1908. Occasional mud slides, especially during the rainy season, sometimes spoiled a month's work.

The labor force was segregated by Colonel Goethals into a hierarchy with three distinct categories. First came the "gold employees," about fifty-six hundred white Americans who were paid in gold with an average salary of several thousand dollars per year. Next came the "silver employees," about four thousand European laborers, mostly from Spain and Italy, who were paid about five hundred dollars per year in silver. At the bottom of the pay scale were about thirty thousand blacks from Jamaica and the Caribbean area, who received about three hundred dollars per year. Each group of people received a higher income than they could have earned at home. Strict segregation was maintained in housing, quality of food, medical care, and transportation to the job site. To maintain good morale, entertainment was provided in the form of motion pictures, band concerts, and baseball games. A weekly newspaper called *The Canal Record* spurred on competition between work crews.

In the fall of 1913, Gatun Dam was finished and the lake behind it began to fill with water, extending into the Culebra Cut. On August 15, 1914, ten years after the start of construction, the first ship made the transit from ocean to ocean in slightly under ten hours. Goethals received a congratulatory telegram from Washington: "A stupendous undertaking has been finally accomplished, and a perpetual memorial to the genius and enterprise of our people has been created."

Impact of Event

The Panama Canal Act of 1912 provided for the operation, maintenance, and protection of the canal. The toll for commercial ships to pass through was set in proportion to their tonnage. American vessels went for free at first because the construction costs had been paid by tax money. During the first seventy-five years of operation, an average of ten thousand ships per year went through the canal. In 1990, the fee for the 62,000-ton *Star Princess* of the Princess Cruise Line was $120,000 for one transit. All operation and maintenance costs have to be covered by the tolls, including underwater dredging, which is necessary to maintain the proper depth of water.

In the early years of canal operation, three major benefits were attributed to the canal project. First was the major saving in distances traveled by commercial ships that previously had to go around South America. Second was the military advantage of moving United States Navy ships rapidly to protect either coast, thus reducing the cost of a two-ocean Navy. Finally, the improvements in sanitation and the reduction of tropical diseases in Panama served as an example for what could be done to improve living conditions in other tropical countries.

After the end of World War II, various countries that had been part of colonial empires struggled to win their independence. Among these were India, Indonesia, and Algeria. The Suez Canal, which had been administered by France and England for nearly one hundred years, was taken over by Egypt in the Suez crisis of 1956. In Panama, resentment was building up against the United States for its "ownership" of the Canal Zone on Panamanian territory. The spirit of nationalism came to a climax in 1964 when Panamanian rioters entered the Canal Zone and clashed with United States troops. The fighting caused loss of lives and destruction of property and led to a break in diplomatic relations.

Over the next several years, difficult negotiations for a new canal treaty were conducted. The treaty of 1904 had promised "perpetual jurisdiction" of the Canal Zone to the United States, which, however, clashed with the idea of national sovereignty and the goals of the "good neighbor" policy. In 1978, President Jimmy Carter signed a new treaty with Panama, which the United States Senate barely ratified by a two-thirds majority. The new treaty provided for a period of joint administration with a gradually increasing share of the leadership positions given to Panama until the year 2000, when all United States participation would cease. Some opponents of the treaty described it as a pure "give-away." The majority, however, took the long-range view that good relations with our Latin-American neighbors are important for national security.

Bibliography

Barrett, John. *Panama Canal: What It Is, What It Means*. Washington, D.C.: Pan American Union, 1913. Written for the American public shortly before the canal was opened. Resembles an enthusiastic tourist guide book of that era, telling the reader how to get there, where to stay, and which sights are the most spectacular.

Cameron, Ian. *The Impossible Dream: The Building of the Panama Canal*. New York: William Morrow, 1972. Describes the personalities and contributions of the key figures, from de Lesseps to Goethals. Quotes extensively from the newspaper *The Canal Record*; gives a sense of an unfolding sequence of successes and setbacks.

Goethals, George W. "The Panama Canal." *National Geographic* 22 (February, 1911): 148-211. Text of an address presented by Colonel Goethals to the National Geographic Society; a progress report on the canal project that includes many interesting photographs. Emphasizes the military necessity of the project.

Jacobs, David, and Anthony E. Neville. *Bridges, Canals, and Tunnels*. New York: American Heritage, 1968. Discusses the fact that the Panama Canal was engineering on a colossal scale. Well-written; gives a historical perspective to this major engineering accomplishment.

Keller, Ulrich, ed. *The Building of the Panama Canal in Historic Photographs*. New York: Dover, 1983. The National Archives in Washington, D.C., has a collection of more than ten thousand photographs showing the Panama Canal construction. About 160 of these historic pictures are reprinted, each with informative and interesting commentary by Keller. An outstanding documentary record.

McDowell, Bart. "The Panama Canal Today." *National Geographic* 153 (February, 1978): 279-294. Describes a transit through the canal, with historical notes and observations along the way. Even after more than seven decades of operation, only a few United States Navy ships and some supertankers are too large for the locks.

Ryan, Paul B. *The Panama Canal Controversy*. Stanford, Calif.: Hoover Institution Press, 1977. In 1977, President Carter negotiated a new treaty that eventually would give Panama control over the canal. At the time of this book, the Senate was beginning its debate on whether to ratify the treaty. Presents questions of United States national security versus Panama's sovereignty from the conservative political point of view.

Salt, Harriet. *Mighty Engineering Feats*. Philadelphia: Penn, 1937. Describes ten major American engineering accomplishments, including the canal. Filled with interesting factual details.

Hans G. Graetzer

Cross-References

Reed Establishes That Yellow Fever Is Transmitted by Mosquitoes (1900), p. 73; Gorgas Develops Effective Methods for Controlling Mosquitoes (1904), p. 223;

Construction Begins on the Empire State Building (1930), p. 906; The Completion of Boulder Dam Creates Lake Mead, the World's Largest Reservoir (1936), p. 1075; Theiler Introduces a Vaccine Against Yellow Fever (1937), p. 1091; The St. Lawrence Seaway Is Opened (1959), p. 1608; The Verrazano Bridge Opens (1964), p. 1782.

FLEMING FILES A PATENT FOR THE FIRST VACUUM TUBE

Category of event: Applied science
Time: November 16, 1904
Locale: London, England

Fleming found an application for the Edison effect as a detector for radio waves, starting the electronics industry

Principal personages:

SIR JOHN AMBROSE FLEMING (1849-1945), an English physicist, professor of electrical engineering, and electrical adviser to the Marconi Wireless Telegraph Company

THOMAS ALVA EDISON (1847-1931), the famous American inventor of the light bulb and discoverer of the Edison effect

LEE DE FOREST (1873-1961), an American scientist and inventor who transformed the Fleming valve into the audion

Summary of Event

A number of events occurred in the twenty years before the invention of the vacuum tube by Sir John Ambrose Fleming that are relevant to its discovery. Often, numerous events that contribute to a discovery are overlooked or forgotten. This is especially true with the development of the vacuum tube. Since the mid-1800's, many scientists were working with the properties of electricity in glass apparatus with most of the air removed. Because the concept of the electron had not been formulated, and the idea that electricity could consist of particles with one two-thousandth the mass of hydrogen was completely foreign, there was no obvious way of tying all these observations together and making predictions for systems not yet assembled. At this time, electricity was thought of only as a force and wave phenomenon and not as a particle.

Among the earliest notable work leading to the discovery of the vacuum tube was that of Alexandre-Edmond Becquerel, who, as early as 1853, had worked out many of the conductivity properties of gases at various temperatures and pressures but could offer no explanation for these properties. Certainly the most important discovery leading to the invention of the vacuum tube was the Edison effect by Thomas Alva Edison in 1884. While studying why the inner glass surface of light bulbs (which, during this period, used a carbon thread for the filament) blackened, Edison inserted a metal plate near the filament of one of his light bulbs. He discovered that electricity would flow from the positive side of the filament to the plate, but not from the negative side to the plate. Like other workers, he made the observation but offered no explanation.

Edison had, in fact, invented the first vacuum tube, which was later termed the diode; however, at that time there was no use for this device. Therefore, the discovery was not recognized for its true significance; it was a discovery much ahead of its time. A diode converts electricity which alternates in direction (alternating current) to electricity that flows in the same direction (direct current). Since Edison was a proponent of producing direct current in generators, he essentially ignored this aspect of his discovery. Like many other inventions or discoveries which were ahead of their time—such as the laser—for a number of years, the Edison effect was "a solution in search of a problem." The fact that the amount of electricity was proportional to the intensity of the filament, however, made it potentially useful as a control device for electric generators; Edison was granted a patent on October, 1884, for this use of the Edison effect, although Edison did not patent the use of his device as a diode.

The explanation for why this phenomenon occurred would not come until after the discovery of the electron by Sir Joseph John Thomson. In retrospect, the Edison effect can be identified as one of the first observations of thermionic emission; that is, electrons that are "boiled off" of hot surfaces. Electrons were attracted to the positive charges and would collect on the positively charged plate, thus providing current; but they were repelled from the plate when it was made negative, with no current produced. Since current flowed in only one direction, it was compared to a check valve used to allow a liquid to flow in only one direction. This analogy is popular since the behavior of water has often been used as an analogy for electricity, and this is the reason that the term "valves" became popular for vacuum tubes.

Another interesting, related invention was that of Arthur Wehnelt, who was working with thermionic emission. On January 15, 1904, he applied for a German patent on one of his tubes which converted alternating current into direct current. He did not mention its use at high frequency (that is, for radio waves), only for low-frequency, power generation applications such as charging storage batteries. He was the first to apply for a patent for a vacuum tube that had an application for the rectification of alternating current to direct current. Unfortunately, since Wehnelt did not mention the applicability of his device for detection of radio waves, he was unable to sell it for use in radio sets after Fleming applied for his patent, although it was quite suitable for application in radio sets.

Fleming, acting as adviser to the Edison Electric Light Company, had studied the light bulb and the Edison effect starting in the early 1880's before the days of radio. Many years later, he came up with an application for the Edison effect as a radio detector when he was a consultant for the Marconi Wireless Telegraph Company. Detectors (devices that conduct electricity in one direction but not another, just as the diode does, but at higher frequencies) were required to make the high-frequency radio waves audible by converting them from alternating current to direct current. Several types relying on chemical reactions, physical actions, and properties of crystals were available, but they were not overly effective at high frequency.

Fleming was able to detect radio waves quite effectively by using the Edison effect. Ironically, Fleming used essentially the identical device that Edison patented, but for a different purpose and with a different external circuit: as a radio detector. Indeed, he even had the Edison & Swan Electric Light Company make up twelve units to his specifications for testing. Like Edison, Fleming offered no explanation of how or why his device functioned. Fleming's device was also essentially identical, although not as refined, as that of Wehnelt's device, except that Wehnelt did not apply his invention to radio waves. Fleming applied for a patent in England on November 16, 1904. He called his device the thermionic valve; thermionic tubes are still called "valves" in England, whereas the American name is "vacuum tube." The latter name reflects the fact that these devices require an internal vacuum in order to operate.

In 1906, Lee de Forest refined Fleming's invention by adding a zigzag piece of wire between the metal plate and the filament of the vacuum tube. The zigzag piece of wire was later replaced by a screen called a grid. The grid allowed a small voltage to control a larger voltage between the filament and plate. It was the first complete vacuum tube and the first device ever constructed capable of amplifying a signal—that is, taking a small voltage signal and making it much larger. He named it the "audion" and was granted a U.S. patent in 1907. In 1907-1908, the American Fleet carried radios equipped with de Forest's audion in its goodwill tour around the world. While useful as an amplifier of the weak radio signals, it was not useful at this point for the more powerful signals of telephone. Other developments were made quickly as the importance of the emerging fields of radio and telephony were realized.

Impact of Event

While it took the world twenty years to understand the significance of the Edison effect, it took only a short time after that to develop the electronics industry, which was especially pushed by the need for communications during World War I. With many industrial laboratories working on vacuum tubes, improvements came quickly. For example, tantalum and tungsten filaments quickly replaced the early carbon filaments. In 1904, Wehnelt discovered that if metals were coated with certain materials such as metal oxides, they emitted far more electrons at a given temperature. These materials have a lower "work function"; that is, the electrons escape the surface of the metal oxides more easily than those of metals. It would be like pumping water from a shallow well rather than from a very deep well: It is much easier. Later work by Bell Telephone Laboratories showed that oxides of barium or strontium are particularly effective; all vacuum tubes soon used these coatings. Thermionic emission and, therefore, tube efficiencies increased by a factor of one hundred by this method.

Another important improvement in the vacuum tube came with the work of Irving Langmuir of the General Electric Research Laboratory starting in 1909 and Harold D. Arnold of Bell Telephone Laboratories. They used new techniques such as the mercury diffusion pump to achieve higher vacuums. Original tubes had about one ten-thousandth the pressure of atmosphere, whereas new techniques introduced after 1910 allowed vacuums below one-millionth atmospheric pressure. Working independently, Langmuir and Arnold discovered that very high vacuum used with higher voltages increased the power these tubes could handle from small fractions of a watt to hundreds of watts. The de Forest tube was now useful for the higher power audio signals of telephone. This resulted in the introduction of the first transamerican speech transmission in 1914, followed by the first transatlantic in 1915.

In 1916, a long-standing conflict between the Marconi Wireless Telegraph Company, owner of the Fleming patent, and the de Forest Radio Company, came to an end when the United States District Court ruled that the de Forest audion when used in any aspect of radio, violated the Fleming patent, although the audion could be used for telephony purposes. This gave Marconi a virtual monopoly in radio. In 1919, the domestically owned Radio Company of America (RCA) was formed to buy out the foreign-owned Marconi Company because radio was so important to the United States' national security.

Over time, vacuum tubes became more and more complicated with additions of a second grid to produce the oscillation tube to accomplish the complicated purposes of future inventions. Vacuum tubes made the development of television, radar, radio, and numerous other inventions possible. From 1945 to 1950, vacuum tube sales in the United States were more than two hundred million dollars each year.

The invention of the transistor in 1948 by William Shockley, Walter H. Brattain, and John Bardeen ultimately led to the downfall of the tube. With the exception of the cathode ray tube, such as those used in television picture tubes, transistors could accomplish the jobs of nearly all vacuum tubes much more efficiently. Also, the development of the integrated circuit allowed small, efficient, highly complex devices that would be impossible with radio tubes. By 1977, the major producers of vacuum tubes had all ceased production.

Bibliography

Bowen, Harold G. *The Edison Effect*. West Orange, N.J.: Thomas Alva Edison Foundation, 1951. This is a well-written, captivating account on the development of the vacuum tube. It is not overly technical, and most of the information is accessible to the layperson.

Fleming, John A. *A Handbook for the Electrical Laboratory and Testing Room*. 2 vols. London: "The Electrician" Printing and Publishing Co., 1901-1903. This laboratory manual is useful to understand the state of the art at the time the vacuum tube was invented. While technical for its time, mostly in terms of equipment, the underlying theory should be accessible for anyone with an introductory physics course.

____________. *The Thermionic Valve and Its Development in Radio-Telegraphy and Telephony*. 2d ed. London: Wireless Press, 1924. The history, development and use of vacuum tubes in the early days is given by the inventor of the first vacuum tube. Some parts are very technical, others historical. Many figures and photographs are included.

Stokes, John W. *Seventy Years of Radio Tubes and Valves*. Vestal, N.Y.: Vestal Press,

1982. A practical description of the earliest vacuum tubes of the world to those of the 1960's. It identifies tubes by manufacturer, types, and physical descriptions, with hundreds of tubes photographed. This is very useful for collectors.

Tyne, Gerald F. J. *Saga of the Vacuum Tube*. Indianapolis: Howard W. Sams, 1977. This is a comprehensive volume on the development of the vacuum tube including relevant observations of the 1700's, 1800's, and early 1900's about the properties of electricity, the development of vacuum technology, and other related studies. While not as good as seeing the actual original devices and notes, many of the first vacuum tubes, vacuum pumps, pages from laboratory notebooks, and other items are pictured, adding to the thrill of discovering new knowledge, as the reader is vicariously standing side-by-side with these inventors and observing their work.

Christopher J. Biermann

Cross-References

Kipping Discovers Silicones (1901), p. 123; Marconi Receives the First Transatlantic Telegraphic Radio Transmission (1901), p. 128; Thomson Wins the Nobel Prize for the Discovery of the Electron (1906), p. 356; Fessenden Perfects Radio by Transmitting Music and Voice (1906), p. 361; Zworykin Develops an Early Type of Television (1923), p. 751; Shockley, Bardeen, and Brattain Discover the Transistor (1947), p. 1304; The Microprocessor "Computer on a Chip" Is Introduced (1971), p. 1938.

EINSTEIN DEVELOPS HIS THEORY OF THE PHOTOELECTRIC EFFECT

Category of event: Physics
Time: 1905
Locale: Bern, Switzerland

Einstein postulated that the process by which electrons are liberated from a metal surface by incident light can be understood if the light is considered to be composed of particles called light quanta, which later became known as photons

Principal personages:

ALBERT EINSTEIN (1879-1955), an American physicist who developed theories about relativity, Brownian motion as evidence for atoms, and the photoelectric effect for which he was awarded the 1921 Nobel Prize

MAX PLANCK (1858-1947), a German physicist who proposed that light was emitted from a glowing body in discrete bundles which he called light quanta, for which he was awarded the 1918 Nobel Prize

HEINRICH HERTZ (1857-1894), an experimental physicist who was considered to be the discoverer of the photoelectric effect

SIR JOSEPH JOHN THOMSON (1856-1940), an English physicist who discovered the electron, for which he won the 1906 Nobel Prize, first identified the photoelectric effect as the emission of electrons

Summary of Event

The photoelectric effect is the process by which electrons are ejected from a metal surface when light of the appropriate frequency is shined on that surface. Since it requires energy to remove an electron from a metal, it is clear that this energy is coming from the incident light. Yet, in 1905, there were a number of mysteries associated with the process, all of which were solved by Einstein's explanation of the photoelectric effect in terms of light quanta.

By 1905, a variety of physical phenomena associated with the photoelectric effect had been discovered. In 1887, Heinrich Hertz had discovered that light incident on a metal surface can produce visible sparks if that surface is in the presence of an electric field. The sparking action demonstrated that something was being removed from the metal surface, although nobody knew what it was. In 1888, Wilhelm Hallwachs showed that shining light on a surface can cause an uncharged body to become positively charged. In 1899, Sir Joseph John Thomson, who had discovered the electron two years earlier, stated that the photoelectric effect involved the emission of electrons from the metal. This explained Hertz's observation: The emitted electrons were accelerated by the electric field until they gathered enough energy to create a spark. It also explained Hallwachs' results: The emitted electrons were carrying negative charge away from the metal body, thus leaving it with a net positive charge. In 1902, Philipp P. Lenard showed that the energy of the ejected electrons or, equivalently, their speed, did not depend on the intensity of the incident light. It was shown, however, in 1904 that the energy of the ejected electrons depended on the frequency, or color, of the light: The higher the frequency of the incident light, the greater the speed of the escaping electrons.

The photoelectric effect posed serious problems for classical physics. According to the classical theory, light was an electromagnetic wave that carried energy based on its intensity. When this energy was transmitted to the irradiated body, the electrons in the body would gain energy gradually, or "heat up," until eventually they became energetic enough to escape from the body. If the incident light was very intense, the electrons should be escaping with a large supply of energy. The experimental observations however, were inconsistent with this explanation; they showed that the energy of the ejected electrons depended on the frequency of the incident light but not on its intensity. Yet, the question of why a dim source of high-frequency light resulted in the ejection of electrons with a higher energy than a bright source of low-frequency light still remained.

In 1905, Albert Einstein published three revolutionary papers. The most famous was on relativity, one was on Brownian motion as evidence for the existence of atoms, and the third was on the photoelectric effect. Einstein postulated that the photoelectric effect could be understood by discarding certain key concepts from classical physics and replacing them with radical new ideas, which were to become known as modern physics. One of these radical ideas was the concept of the light quanta which had been tentatively proposed in 1900 by Max Planck to explain the distribution of radiation from hot bodies.

As an aid to understanding, Einstein postulated that the incident light of the photoelectric effect should not be viewed as a classical wave but rather as a collection of particles which he called light quanta, later to be renamed "photons." These photons each carried a discrete amount of energy which was proportional to their frequency, $E = hf$, where E is the energy of the photon, f is the frequency, and h is a proportionality constant that had been discovered by Planck. In Einstein's conception, a beam of light is more analogous to a flock of birds than a stream of water.

By viewing the incident light as a collection of photons, Einstein was able to explain the photoelectric effect as follows: When a photon is incident on a metal surface, there is a strong chance that it will penetrate through the surface and encounter the free electrons that are known to lie within the metal. When a photon encounters an electron, it will typically transfer all of its energy to the electron. In the language of modern physics, it is said that it was "absorbed" by the electron. In general, an electron can only absorb one photon, but it will always absorb this photon in its entirety. The electron, which has very little energy before it absorbed the photon, now has an amount of energy hf. If this energy is high enough, the electron will be able to escape out of the metal. An electron will typically expend a certain amount of energy as it escapes. This energy is characteristic of the specific metal and is known as its work function, P. The work function is the "energy cost" of the escape. If the electron is going to escape, then the energy provided by the photon, hf, must be greater than P. Einstein proposed the following formula $E = hf - P$, which states that the energy possessed by a photoejected electron is equal to the energy of the incident photon minus the energy of the work function. By analogy, one could say that the money possessed by an escaping convict is equal to the money that was smuggled in to arrange his escape, less the amount he spent bribing the relevant prison officials who arranged his escape.

Einstein postulated the curious relationship between the energy of the ejected electrons and the incident light: First, the intensity is irrelevant if an electron absorbs only a single photon. Higher intensity means more photons that might eject more electrons but will not increase the energy of any specific electron since, at most, one photon is absorbed. Second, the energy of the ejected electrons increases with the frequency of the incident light, since higher-frequency photons impart more energy to the electrons. The photoelectric effect was no longer a mystery.

Einstein was also able to make several predictions: The energy of a photoejected electron can never exceed hf, the energy of the photon; and, if hf is less than the work function of the metal P, no electrons will be ejected no matter how intense the incident light. Einstein's predictions were experimentally verified.

Einstein's explanation for the photoelectric effect came at a time when classical ideas were still strong and the notion of "light quanta" seemed radical and mysterious. Even Planck had never fully accepted the reality of the quantum that he had discovered in 1900. Einstein, by calling his ideas "heuristic," indicated that he had reservations about the physical reality of the light quanta. In fact, it was almost two decades before these important ideas were universally accepted.

Impact of Event

The first three decades of the twentieth century witnessed the overthrow of classical physics and the birth of quantum mechanics, which has been the foundation for most of the physics that has developed since then. Like many revolutions in science, the quantum revolution was accomplished through a series of small steps that eventually led to an entirely new way of looking at the universe. Einstein's explanation of the photoelectric effect in 1905 was one of the small steps along the road to the radically new world into which physics was about to enter. The light quanta hypothesis became an important part of several larger theories. In 1911, Niels Bohr began to use the idea of the light quanta to account for the emission spectra of atoms. It was known that atoms, when excited, gave off light with certain characteristic frequencies which differed from one atom to the next. The famous "Bohr model of the atom" stated that these frequencies could be understood as the frequency of the light quantum, or photon, given off by an atom when an electron jumped from a large orbit to a smaller one. The energy of the emitted photon would be equal to the energy difference between the two orbits.

In 1923, Arthur Holly Compton performed some very significant experiments in which he studied the collision of photons with electrons. By treating the photon as if it were a particle rather than a wave, he was able to demonstrate the transfer of energy and momentum from a particle of energy to a particle of matter. These experiments helped to confirm the existence of photons, which still were not universally accepted. At the same time that Compton was colliding photons with electrons, Louis de Broglie was pondering the apparent wave-particle duality of light. De Broglie recognized that light, which certainly had been demonstrated to behave like a wave, also behaved like a particle at times. If this "dual character" were true, then should not electrons, which had always been understood as particles, also behave like waves? So, de Broglie proposed his famous wave-particle duality, which stated that light and matter had both a wave and a particle character.

The idea of light quanta was slow to catch on, however. In 1911, Einstein still was calling attention to the tentative nature of his hypothesis. Even as the explanation was verified experimentally and became widely accepted, there was still hesitation about the physical reality of light quanta. As late as 1924, Bohr coauthored a paper that still argued that the photon was not real and would eventually be replaced by an improved understanding of matter and radiation. Nevertheless, it was rapidly becoming clear to the physics community that the quantum revolution had arrived. Eventually, all the opposition ceased, and the photon became one of the most important concepts in physics, universally accepted, and a model for later developments in other areas of physics.

Bibliography

Gamow, George. *Thirty Years That Shook Physics: The Story of Quantum Theory.* Garden City, N.Y.: Doubleday, 1966. This classic book by one of the principal physicists of this century is a charming and insightful survey of the revolution that produced the modern theory of quantum physics.

Halliday, David, and Robert Resnick. *Fundamentals of Physics: Extended Version.* New York: John Wiley & Sons, 1988. This popular general physics text is typical of the many excellent books that discuss the photoelectric effect in their treatment of elementary modern physics.

Pais, Abraham. *"Subtle is the Lord . . .": The Science and the Life of Albert Einstein.* New York: Oxford University Press, 1982. This highly acclaimed book is written by a physicist who knew Einstein very well. Somewhat technical in parts, it nevertheless has an absolutely authoritative discussion of Einstein and his ideas.

Rosenthal-Schneider, Ilse. *Reality and Scientific Truth: Discussions with Einstein, von Laue, and Planck.* Detroit: Wayne State University Press, 1980. Consists of correspondence and discussions with three eminent scientists: Albert Einstein, Max von Laue, and Max Planck. Topics covered include "The Universal Constants of Nature," "Concepts of Substance and Conservation," and "The Smallest Length."

Wolfson, Richard, and Jay M. Pasachoff. *Physics: Extended with Modern Physics.* Glenview, Ill.: Scott, Foresman, 1989. Chapter 36, "Light and Matter: Waves or

HERTZSPRUNG NOTES THE RELATIONSHIP BETWEEN COLOR AND LUMINOSITY OF STARS

Category of event: Astronomy
Time: 1905
Locale: Denmark

Hertzsprung discovered that the color of a star is related to its luminosity, which led to his presentation of the first Hertzsprung-Russell diagram

Principal personages:

EJNAR HERTZSPRUNG (1873-1967), a Danish astronomer and photographer, who established that there is a relationship between a star's color and its luminosity (brightness)

HENRY NORRIS RUSSELL (1877-1957), an American astronomer who discovered the color-luminosity relationship and codeveloped the Hertzsprung-Russell diagram that led to his theory of stellar evolution

Summary of Event

In the late nineteenth and early twentieth centuries, the science of astronomy was changing. Astronomy up until that time had been devoted chiefly to the study of the motions of heavenly bodies and the mechanical laws and forces that governed those motions. With the advent of spectroscopy and photography in the late nineteenth century, astronomers began to study the nature of the stars and other bodies, rather than merely their motions.

Spectroscopy is the study of the spectra of stars. A star's spectrum is the band of colored light and dark lines that results when the star's light is spread out by a prism or grating. The patterns of lines can be used to classify the stars and provide clues to the star's properties, such as what it is made of, how hot it is, whether it is rotating and at what speed, and more. Photography was used to record these spectra so that they could be studied and compared and the relationships between a star's spectrum and its properties could be worked out. As data were gathered for large numbers of stars, scientists could begin to search for order and patterns in the data and for relationships among various properties of stars. Ejnar Hertzsprung's work was a part of this quest to understand the nature of stars as revealed in their light.

Hertzsprung was well-suited to work in the emerging study of stars as revealed through their spectra, as he had studied chemistry and specialized in photochemistry. He began the study of astronomy in 1902, after several years of working as a chemist. He was drawn to astronomy by his interest in observing how black body radiation related to the radiation of stars. (Black body radiation is the radiation emitted by a body that absorbs all wavelengths of light and radiates all wavelengths when heated. It shows a relationship between the temperature of the body and the wavelength at which the most radiation is emitted; the wavelength is related to the color of the body.) He studied stellar spectra photographed in Denmark, star classification work done at Harvard College Observatory, and wrote two papers, in 1905 and 1907, both entitled *Zur Strahlung der Sterne* ("The Radiation of Stars"). He presented several results that would prove important to the growing science of astrophysics. His research contained the discovery that there are two different types of stars: the giants and the dwarfs. Hertzsprung had studied the colors, brightnesses, motions, and distances of stars to arrive at this discovery, which led to his development of a diagram plotting the intrinsic brightness against the temperature for a group of stars. This type of plot, developed independently in the United States by Henry Norris Russell and presented in 1913, is known as a Hertzsprung-Russell, or H-R, diagram and is a basic tool of astrophysics today. Hertzsprung first made such a plot in 1906, and did further work to investigate the relationship between the color of a star and its brightness or luminosity.

Hertzsprung used open-star clusters in his determination of the relationship between a star's brightness and its color. An open-star cluster is a group of stars clustered together in one place in the sky; it seems reasonable to assume that the stars in a cluster are in roughly the same spatial location. When observing a star's brightness, one usually cannot know what the intrinsic brightness of the star is; one only knows how bright the star appears from Earth. To know the intrinsic brightness, one must apply a correction for the distance of the star; stellar distances are not easy to discover. Yet, stars in a cluster can be assumed to be at about the same distance. Thus, one does not need to worry about making the distance correction to arrive at the absolute brightness, since the distance correction will be roughly the same for all members of the cluster.

Hertzsprung studied the colors of stars in several clusters whose spectra he had photographed at the Urania Observatory in Copenhagen. He was able to measure the wavelength of peak light emission from individual stars, and then used this wavelength as an index to the stars' colors. (The wavelength is the distance from crest to crest of each electromagnetic wave making up the light; these wavelengths are very small. Red light has a longer wavelength than green light, which, in turn, has a longer wavelength than blue light: The wavelength is related to the color.) Once he had indexed the stars' colors, he constructed a diagram in which he plotted the color versus the magnitude (brightness) for the stars in two star clusters, the Pleiades and the Hyades.

In 1911, these diagrams were the first of this type of diagram to be published, when Hertzsprung was a senior staff astronomer at the Astrophysical Observatory at Potsdam, East Germany. On these two plots, Hertzsprung made the important observation that stars do not appear in all possible combinations of color and brightness (that is, the points are not scattered all over the plot), but that there is a relationship between how bright the star is and what color it is. Thus, the brightnesses and colors of stars in the two clusters form a narrow diagonal band across the plot. The band stretches from bright blue stars in the upper left corner to dim red stars in the lower right corner. This band was called the main sequence, or sometimes the dwarf sequence, to differentiate it from the giant sequence of large stars, which was revealed on other H-R diagrams. The discovery of this relationship gave astronomers major data about the population of the heavens. The task of determining the relationship between color and brightness was to keep astronomers occupied for years. The H-R diagram is still an essential tool of the astrophysicist today.

Another area in which Hertzsprung contributed was stellar distance determination. This has traditionally been a difficult problem in astronomy. Hertzsprung made a discovery which led to the use of a star's spectrum to determine its absolute brightness and thence its distance.

At Harvard College Observatory in the late nineteenth century, hundreds of photographic plates of spectra were studied and hundreds of thousands of stars were classified. In particular, Antonia Caetano Maury developed a sophisticated system involving twenty-two categories of stars, with three subdivisions for each category designating the width of spectral lines. Hertzsprung used her classifications in his spectral work. He found that stars which had Maury's designation "c," for their narrow spectral lines had high absolute brightnesses. A star's apparent brightness as seen from Earth depends on its distance; its absolute brightness is a measure of how bright it actually is. Once the absolute brightness of a star is known, it can be compared to the apparent brightness and its distance can be calculated. (The process is similar to the way in which the distance of a light bulb would be estimated, based on how bright it appears and the knowledge that it is a 100-watt bulb.) This method was elaborated upon by later studies and became the method of spectroscopic parallax, which enables one to determine a star's distance from its spectrum.

Impact of Event

Hertzsprung continued his work on the color-luminosity relationship by comparing the H-R diagrams for the Pleiades, Hyades, and Praesepe star clusters, and noticed differences in the types of stars in the three clusters. Astronomers today interpret these differences as indications that the Pleiades are younger than the other two clusters and are able to use H-R diagrams as a tool for the determination of ages of clusters. It is now known that the brighter and bluer a star is, the shorter its lifetime; therefore, a cluster with bright blue stars still left in it must be younger than the maximum age such stars would reach. In general, by seeing the brightness and color of the stars highest on the main sequence (closest to the bright blue end) for a given cluster, one can estimate an upper limit to the age of the cluster. In addition, a method was later developed in which the brightness-color diagram for a cluster was compared with that of another cluster at a known distance; this comparison can yield an estimate of the distance to the first cluster. In working on the Pleiades cluster, Hertzsprung did related work on their spectra which involved making many measurements of wavelength and that led to estimates of the total mass of the cluster. H-R diagrams, thus, provide a rich source of information about star clusters.

Hertzsprung's plot of stars on the main sequence, together with his demonstration that a giant sequence exists, provided support and information to Russell, which he used in formulating his theory of stellar evolution based on the H-R diagram. Russell's evolutionary picture turned out to be mistaken and to require drastic overhauls in the light of later knowledge of nuclear processes. Yet, it was an important first step in the use of H-R diagrams to plot the distribution of stars and to determine the significance of this distribution for stellar evolution. Hertzsprung also showed that there is a gap, today called the Hertzsprung gap, separating main sequence stars from giant sequence stars on an H-R diagram.

The discovery of a spectral type-luminosity connection was followed up by Walter Sydney Adams and Arnold Kohlschütter at Mount Wilson Observatory in California. They found differences in the ratios of intensities of certain lines in the spectra of giants and dwarfs and used these differences for stars of known difference to calibrate a plot of ratio versus magnitude (brightness). Therefore, if one measured the ratio for a star, one could consult the plot and determine its absolute brightness and could then use the apparent brightness to arrive at an estimate for its distance. This distance technique proved to be a powerful one.

Hertzsprung's work with stars' spectra and their colors yielded results that were vital for the further studies conducted in astrophysics in the twentieth century. The recognition of a color-luminosity relationship was a first step toward understanding the reasons behind the relationship, reasons that have had profound implications for ideas on stellar formation and evolution.

Bibliography

Abell, George O. *Realm of the Universe*. 3d ed. New York: Saunders College Publishing, 1984. Introductory college textbook. Contains sections on properties of stars and development of the H-R diagram, and on the uses of the diagram in studying stellar evolution, in particular on the application of the diagram to the determination of the ages of star clusters. Glossary and bibliography, numerous diagrams and drawings, and color plates. Each chapter also includes exercises for the student.

Degani, Meir H. *Astronomy Made Simple*. Rev. ed. Garden City, N.Y.: Doubleday, 1976. Chapter 7 contains useful information on H-R diagrams and on the stellar properties, such as brightness, which are used in constructing the diagrams. Contains drawings and graphs, glossary, and exercises for the self-motivated learner.

Moore, Patrick. *Patrick Moore's History of Astronomy*. 6th rev. ed. London: Macdonald, 1983. A chapter "Exploring the Spectrum" gives background to some of Hertzsprung's spectral work and introduces the reader to the complexities of analyzing a star's radiation. The chapter "The Life of a Star" introduces the H-R diagram and its applications to stellar evolution studies. Written for the average reader. Contains a list of landmarks in the history of astronomy.

Pannekoek, A. *A History of Astronomy*. New York: Barnes & Noble Books, 1961. The chapter, "Common Stars," discusses the development and uses of spectroscopy and the construction of H-R diagrams by Hertzsprung and Russell. Also discusses the uses of H-R diagrams in studies of stellar evolution. A classic in the history of astronomy.

Rigutti, Mario. *A Hundred Billion Stars*. Translated by Mirella Giacconi. Cambridge, Mass.: MIT Press, 1984. Part 2 contains useful information on stellar classification and on the H-R diagram. Part 3, on stellar evolution, discusses the application of the H-R diagram to the problem of stellar evolution in general, and to the ages of star clusters in particular. Includes diagrams and some black-and-white photographs.

Struve, Otto, and Velta Zebergs. *Astronomy of the Twentieth Century.* New York: Macmillan, 1962. Cowritten by an astronomer who has direct knowledge of some of the astronomical events described, this book offers a good overview of the development of the H-R diagram and the understanding of the relationships between a star's spectrum and its other properties. Contains drawings and illustrations. End-papers contain a timeline of twentieth century astronomy. Glossary and bibliography.

Vaucouleurs, Gérard de. *Discovery of the Universe*. London: Faber & Faber, 1957. Sections 6 through 9 contain an overview of astronomy from the late nineteenth century ownard. Gives more information on applications of the H-R diagram and follow-up to Hertzsprung's work, including spectroscopic parallax. Contains diagrams, drawings, and an annotated bibliography.

Mary Hrovat

Cross-References

Russell Announces His Theory of Stellar Evolution (1913), p. 585; Eddington Formulates the Mass-Luminosity Law for Stars (1924), p. 785.

PUNNETT'S *MENDELISM* CONTAINS A DIAGRAM FOR SHOWING HEREDITY

Category of event: Biology
Time: 1905
Locale: Cambridge, England

Punnett published his book, Mendelism, *which included a diagram to explain how traits are inherited*

Principal personages:

REGINALD CRUNDALL PUNNETT (1875-1967), an English geneticist who collaborated with William Bateson to expand, interpret, and popularize Mendel's laws

WILLIAM BATESON (1861-1926), a British geneticist who collaborated with Punnett and made important contributions toward the establishment of the Mendelian conception of heredity and variation

HUGO DE VRIES (1848-1935), a Dutch plant physiologist whose experimental work led to the rediscovery of Mendel's laws and a theory of mutation

CARL ERICH CORRENS (1864-1933), a German botanist whose experimental work on the xenia question led to an independent rediscovery of Mendel's laws

ERICH TSCHERMAK VON SEYSENEGG (1871-1962), an Austrian botanist who rediscovered Mendel's laws while working on hybridization experiments

Summary of Event

In 1905, Reginald Crundall Punnett's landmark book, *Mendelism*, was published. It was the first textbook on the subject and was so popular that the *Westminster Gazette* listed it along with a novel by Marie Corelli as the best-seller of the week. The small book stated clearly the principles of heredity espoused by Austrian monk Gregor Johann Mendel in his classic paper of 1865. As understanding of the application of Mendel's ideas increased, the book's size expanded in subsequent editions. In this small book, Punnett included a simple graphic device to demonstrate results expected when a single set of characteristics (monohybrid) or two sets of characteristics (dihybrid) are considered in a cross. The diagrams are explicit, simple, and visually convincing. The representations, called "Punnett squares" after their inventor, soon became a valuable tool for beginning genetics students.

When *Mendelism* first appeared, the subject was unknown to most investigators. Mendel's original paper of 1865, long ignored because it was published in an obscure journal, had been independently rediscovered in 1900 by three botanists: Hugo de Vries, Carl Erich Correns, and Erich Tschermak von Seysenegg. Although Mendel's ideas were intriguing, early twentieth century biologists did not agree on the universality of his "laws" of heredity. They searched for an understanding of the significance of the Mendelian patterns. Punnett's book became an excellent learning tool for the rediscovered Mendelism.

Punnett formed a fruitful partnership with William Bateson, whose rejection of traditional Darwinism caused him to be out of favor with the academic establishment. In 1910, Bateson resigned from the inadequately funded, temporary Chair of Biology at the University of Cambridge. The Chair became adequately endowed in 1912 and was given to Punnett, who became the first Arthur Balfour Professor of Genetics at the University of Cambridge.

Undoubtedly, Punnett's work on the Mendelian explanation of sex determination, sex linkage, complementary factors and factor interaction, autosomal linkage, and mimicry were important scientific accomplishments. His practical achievement during World War I, when he devised a scheme to distinguish sex in very young chickens was ingenious. (By using sex-linked plumage-color factors that would appear only in male chicks, researchers could distinguish and destroy the unwanted males so that food would not be wasted on them.) Important, but in a different way, the *Journal of Genetics*, founded jointly by Bateson and Punnett in 1911, was significant to the advancement of genetics. The simple diagrammatic scheme, now called the Punnett square, helped assure the success of Mendelism during the first decade of the twentieth century and has contributed to the subsequent understanding of classical genetics.

Mendel had used the pea plant, *Pisum*, to investigate the inheritance of individual characteristics such as seed pod color, height, flower color, and pod shape. In one set of experiments, he investigated only the heights of the plants. When he crossed pure tall (1 to 2 meters) and pure dwarf (0.5 to 1 meter) plants, all of the offspring (Fl generation) were tall. Mendel applied the term "dominant" to the tall plants and "recessive" to the dwarf. Next, he collected the seeds from the Fl plants. When he sowed them the following year, both tall and dwarf plants appeared in the offspring (F2 generation), but each individual was either tall or dwarf, with no intermediates. In one set of experiments, out of a total of 1,064 of these F2 plants, 787 were tall and 277 were dwarf, approximately a ratio of 3:1. The next year, Mendel planted the seeds of the F2 generation. He found that although the seed from the dwarf plants produced only dwarfs, the tall plants produced two different kinds of seed, some giving rise to tall plants only and others producing both talls and dwarfs in a 3:1 proportion. By breeding subsequent generations, Mendel demonstrated that pure dominants and recessives breed true, but that impure dominants always produce dominants and recessives in a constant proportion of 3:1. Since the pure dominants are only half as numerous as the impure dominants, when impure dominants are crossed with each other, they will produce pure dominants, impure dominants, and recessives in the ratio of 1:2:1.

The case considered above involves only a single set of characteristics (in this example, height) and is known as a monohybrid cross. When the original parents differ in two pairs of characters, such as seed color and height (a dihybrid cross), Mendel ascertained that the two traits were assorted independently of each other. When a tall yellow-seeded pea is crossed with a dwarf green-seeded plant, the F1 plants all exhibit the dominant character of each pair and are tall yellows. In the F2 generation, talls and dwarfs appear in the ratio of 3:1 and yellows and greens also appear in a 3:1 ratio. Consequently, out of the total sixteen plants, twelve will be tall and the other four, dwarf. Out of every four talls, three will be yellows and the other, green. Out of the twelve talls, therefore, nine will be yellow and three will be green. Out of the four dwarfs, three will be yellow and one will be green. Consequently, the F2 generation arising from the cross will result in nine yellow talls, three green talls, three yellow dwarfs, and one green dwarf. For every sixteen plants, nine will show both dominants, two classes of three each will show the dominant character of one pair and the recessive of the other, and one plant will show both recessive characters. This principle holds true for three sets of characters and can be extended indefinitely for any number of pairs of characteristics.

After describing the crossing procedure and results, Punnett showed his simple graphic way of viewing the probabilities from these crosses. In Punnett's square, he allowed Aa and Bb to represent two sets of unit characteristics (in the above example, height and seed color). The capital letters designate dominant and the lower case recessive characteristics.

A = dominant	B = dominant
a = recessive	b = recessive

If a parent pure (homozygous) for the dominant "A" characteristic is crossed with a parent pure (homozygous) for the recessive "a" character (AA × aa), all of the offspring (F1 generation) resemble the dominant "A" parent but carry the recessive "a" (Aa). If two of these F1 individuals are crossed (Aa × Aa), the ratio of their offspring (F2 generation) will be 1 (AA), 2 (Aa), and 1 (aa). To demonstrate this 1:2:1 ratio, Punnett drew four large squares. One "A" and one "a" is assumed to appear across the horizontal axis and one "A" and one "a" down the vertical axis (Punnett did not explicitly complete this step). When the "A" and the "a" from the horizontal axis are multiplied by the "A" and the "a" from the vertical axis, the four large squares indicate one individual with AA, two with Aa, and one with aa, as Mendel predicted.

If a second set of characteristics Bb × Bb is considered, the result will be the predicted 1:2:1. To demonstrate the disposition of the "B" and the "b" factors, Punnett divided each of the four original squares into four again. In each of the new squares, one individual is BB, two are Bb, and one bb—again, the predicted 1:2:1 ratio for a monohybrid cross.

Punnett's square can illustrate also a dihybrid cross, in which two sets of characters are considered simultaneously. If in the first parental cross (P1), one parent is pure (homozygous) for both dominants (AABB) and the other parent is pure (homozygous) for both recessives (aabb), the Fl generation will be heterozygous, having both A and a and B and b.

Pl cross	AABB × aabb
Fl generation	AaBb
Fl cross	AaBb × AaBb

Of the sixteen squares necessary to illustrate this cross, nine contain both A and B; three, A but not B; three, B but not A; and one, neither A nor B. The ratio illustrated by the diagram is the 9:3:3:1 expected from a dihybrid cross.

Although the method can be used to consider three or more characteristics, the number of squares required makes it cumbersome. To illustrate a trihybrid cross, sixty-four squares would be necessary and to consider four different characteristics, 256 squares would be needed.

Impact of Event

Punnett's book *Mendelism*, in which he created a simple, visual way to illustrate Mendel's ratios, played a vital role in popularizing Mendel's work. Through this book, he launched an aggressive propaganda campaign for Mendel's ideas. Even after 1900, when Mendelism was rediscovered, it was by no means universally known or, if known, universally accepted. Rules that worked for the inheritance of color in pea plants did not seem to apply to the inheritance of comb types in chickens or to apparent cases of blending inheritance in certain traits. Using his clear diagrams, Punnett engaged in a crusade for Mendelism, showing how apparent exceptions could be explained within the framework of Mendel's conclusions.

Punnett's emphasis on the importance of Mendelism inspired a biological revolution, but one which he never joined. During the latter part of the nineteenth century, cytologists (those who study cells) observed the peculiar way that chromosomes were distributed during ordinary body cell division (mitosis) and in the special divisions that produce egg and sperm. They proceeded in their microscopic studies, totally unaware of the significance of Mendel's work to their own observations. In the meantime, botanists such as Correns and de Vries, cast in the Mendelian mold of plant hybridizers, continued with their breeding experiments equally unaware of the studies on the chromosomes that were being undertaken at the same time.

Although some late nineteenth century cytologists suspected that the chromosomes might play a part in heredity, they had no evidence as to how traits were distributed among offspring. Mendel, of course, had determined already some of these relationships, but his ideas were unavailable to cytologists until after 1900.

In the period of time before the rediscovery of Mendel's laws, many theories of heredity were competing. No one theory could be taken as the "paradigm," the term used by Thomas Kuhn in his book *The Structure of Scientific Revolutions* (1962), to denote the accepted "establishment" point of view. Mendel's theory on its own was not sufficient to represent a complete paradigm until it was fused with the cytological ideas that placed Mendel's "factors" on the chromosomes. Between 1900 and

1910, evidence mounted to indicate that the chromosomes represented the physical basis for Mendel's observed ratios.

Punnett worked only on one side of this problem, the Mendelian. Throughout his career, he remained a Mendelian and was not affected by the development of the theory of the gene and of cytogenetics. Nevertheless, his brilliant work on the Mendelian aspects of heredity, and his clear ways of presenting his data (through his diagrammatic approach), made possible the fusion between Mendelian genetics and cytology and the establishment of the new paradigm that occurred around 1910.

Bibliography

Dunn, L. C. *A Short History of Genetics: The Development of Some of the Main Lines of Thought, 1864-1939.* New York: McGraw-Hill, 1965. The book summarizes the general history of theories of heredity, and places Bateson and Punnett's work in context. It includes a glossary and a bibliography of both primary and secondary sources.

Olby, Robert C. *Origins of Mendelism.* New York: Schocken Books, 1966. Olby evaluates Mendel's work and considers the consequences of its rediscovery on early twentieth century biology.

Punnett, R. C. "Early Days of Genetics." *Heredity* 4 (1950): 1-10. In this paper—an address delivered at the hundredth meeting of the Genetical Society, Cambridge, on June 30, 1949—Punnett provides an entertaining look at the early development of Mendelian genetics in England.

"Punnett, Reginald Crundall." *Biographical Memoirs of Fellows of the Royal Society* 13 (1967): 323-326. This memoir contains the most complete biographical treatment of Punnett as well as a complete bibliography of his writings.

Sturtevant, A. H. *A History of Genetics.* New York: Harper & Row, 1965. This book summarizes general theories of heredity and helps put Punnett's work into context.

Marilyn Bailey Ogilvie

Cross-References

De Vries and Associates Discover Mendel's Ignored Studies of Inheritance (1900), p. 61; McClung Plays a Role in the Discovery of the Sex Chromosome (1902), p. 148; Sutton States That Chromosomes Are Paired and Could Be Carriers of Hereditary Traits (1902), p. 153; Hardy and Weinberg Present a Model of Population Genetics (1908), p. 390; Morgan Develops the Gene-Chromosome Theory (1908), p. 407.

CRILE PERFORMS THE FIRST DIRECT BLOOD TRANSFUSION

Category of event: Medicine
Time: 1905-1906
Locale: Cleveland, Ohio

Crile performed the first artery-to-vein blood transfusion and developed a ring device that established the technique as a practical surgical procedure

Principal personages:

GEORGE WASHINGTON CRILE (1864-1943), an American surgeon, author, Brigadier General in the United States Army Medical Officers' Reserve Corps, and a founder of the Cleveland Clinic Foundation

ALEXIS CARREL (1873-1944), a French surgeon who developed the suturing technique to connect blood vessels that enabled Crile to perform his first direct blood transfusion

SAMUEL JASON MIXTER (1855-1923), an American surgeon who gave Crile the idea to use a short-tube device to connect blood vessels

Summary of Event

It is impossible to say when and where the idea of blood transfusion first originated, although descriptions of this procedure are found in ancient Egyptian and Greek writings. The earliest documented case of a blood transfusion is that of Pope Innocent VII. In April, 1492, the Pope, who was gravely ill, was transfused by the blood of three young boys. As a result, all three boys died without bringing any relief to the Pope. In the centuries that followed, there were occasional descriptions of blood transfusions, but it was not until the middle of the seventeenth century that transfusion gained popularity following William Harvey's discovery of the circulation of the blood in 1628. In the medical thought of those times, blood transfusion was considered to have a nourishing effect on the recipient. In many of those experiments, the human recipient received animal blood, usually from a lamb or a calf. Blood transfusion was tried as a cure for many different diseases, mainly those that caused hemorrhages, but also other medical problems and even marital problems.

Blood transfusions were a dangerous procedure, causing many deaths of both donor and recipient as a result of excessive blood loss, infection, passage of blood clots into the circulation of the recipients, passage of air into the blood vessels (air embolism), and transfusion reaction as a result of incompatible blood types. In the mid-nineteenth century, blood transfusions from animals to humans stopped after it was discovered that the serum of one species agglutinates and dissolves the blood cells of other species. A sharp drop in the use of blood transfusion came with the introduction of physiologic salt solution in 1875. Salt solution was simple to use and safer than blood.

In 1898, when George Washington Crile began his work on blood transfusions, the major obstacle he faced was solving the problem of blood clotting during transfusions. He realized that salt solutions were not helpful in severe cases of blood loss, when there is a need to restore the patient to consciousness, steady the heart action, and raise the blood pressure. At that time, he was experimenting with indirect blood transfusions by drawing the blood of the donor into a vessel, then transferring it into the recipient's vein by tube, funnel, and cannula, the same technique used in infusion of saline solution. The solution to the problem of blood clotting came in 1902 when Alexis Carrel developed the technique of surgically joining blood vessels (anastomosis). Crile learned this technique from Carrel and used it to anastomose the peripheral artery in the donor to a peripheral vein of the recipient. Since the transfused blood remained in contact with the inner lining of the vessels, blood clotting did not occur.

Crile thought that blood transfusion would be an effective method of treating hemorrhage and shock, unlike the infusion of saline solution. He started experimenting with blood transfusions on dogs in the Laboratory of Surgical Research of Western Reserve University in 1904. The first human blood transfusion was performed by Crile in December, 1905. The patient, a thirty-five-year-old woman, was transfused by her husband but died a few hours after the procedure. In his autobiography, Crile wrote: "This was the first case in which . . . a transfusion of blood by direct anastomosis of the vascular system of one human being to another was ever attempted." The second, but first successful, transfusion was performed on August 8, 1906. The patient, a twenty-three-year-old male, suffered from severe hemorrhage following surgery to remove kidney stones. After all attempts to stop the bleeding were exhausted with no results, and the patient was dangerously weak, transfusion was considered as a last resort. One of the patient's brothers was the donor. Following the transfusion, the patient showed remarkable recovery and was strong enough to withstand surgery to excise the kidney and stop the bleeding. When his condition deteriorated a few days later, another transfusion was done. This time, too, he showed remarkable improvement which continued until his complete recovery.

For his first transfusions, Crile used the Carrel suture method, which required using very fine needles and thread. It was a very delicate and time-consuming procedure. At the suggestion of Samuel Jason Mixter, Crile developed a new method using a short tubal device with an attached handle to connect the blood vessels. He called this method "the cannula method of blood-vessel anastomosis." By this method, 3 or 4 centimeters of the vessels to be connected are surgically exposed, clamped, and cut, just as he would do under the previous method. Yet, instead of suturing the blood vessels, the recipient's vein is passed through the tube and then cuffed back over the tube and tied to it. Then, the donor's artery is slipped over the cuff. The clamps are open, and blood is allowed to flow from the donor to the recipient. In order to accommodate different-sized blood vessels, tubes of four different sizes were made, ranging in diameter from 1.5 to 3 millimeters.

Still, even after Crile's method was accepted, blood transfusions continued to be a procedure performed only by surgeons in the operating room. It required a full operating room staff with two assistants, three nurses, and an orderly. Both donor and recipient were sedated with morphine, and local anesthesia was applied to the surgery site. Although some of the risks of earlier transfusions were greatly reduced as a result of surgical and sterile technique, blood transfusion was neither a safe nor a simple procedure.

In 1900, Karl Landsteiner first reported the discovery of three different blood groups (A, B, and C) and the agglutinating effect that occurs when mixing incompatible blood groups. It took eleven years before cross-agglutination was first used as a test before blood transfusion. Crile did not apply this test to his transfusion subjects; therefore, transfusion reaction, sometimes fatal, continued to be a serious risk after transfusion. Although Crile made some use of laboratory tests before transfusion, he did not consider them reliable. He wrote, "[i]n the interpretation of results and their clinical application, experience has shown that the occurrence of hemolysis in vitro before transfusion does not necessarily indicate that it will occur in the vascular system of the recipient after transfusion." This conclusion, published in Crile's most authoritative book on transfusion, led many physicians to overlook the importance of compatibility tests before transfusion. There were two additional disadvantages to direct blood transfusion: Crile's method did not allow for an accurate measurement of the amount of blood transfused, thus risking both recipient and donor; and, it necessitated the sacrifice of the donor's artery.

Impact of Event

Crile's method was the preferred method of blood transfusion for a number of years. Following the publication of his book on transfusion, a number of modifications to the original method were published in the medical journals of those times. In 1913, Edward Lindeman developed a method of transfusing blood by simple needle puncture, making it for the first time a nonoperative method. This method allowed one to measure the exact quantity of blood transfused. It also allowed the donor to serve in multiple transfusions. This development opened the field of transfusions to all physicians. The elimination of the need to sacrifice an artery in the process of transfusion rapidly increased the number of willing donors. Lindeman's needle and syringe method also eliminated another major drawback of direct blood transfusion: the need to have both donor and recipient in close proximity.

In April, 1915, the American surgeon Richard Lewisohn developed the citrate transfusion, later known as Lewisohn's method. Lewisohn showed that by mixing low doses of citrate solution with blood, it is possible to prevent blood clotting without making it toxic for the recipient. This development made indirect blood transfusions a practical method. During World War I, the citrate method became the standardized practice in military service. Blood transfusions were applied for the first time by a large but specially trained transfusion team that saved the lives of many wounded soldiers.

In the 1940's, the method of preserving blood with citrate was refined and modified. Acid citrate dextrose was developed, which made possible the preservation of red cells in whole blood for up to twenty-one days. This development made practical the collection, storage, and distribution of blood. It enabled transfusion medicine to fulfill its life-saving potential. Because the importance of blood group compatibility had been scientifically established and quick and reliable tests were developed, transfusion was considered for a long period to be almost free of risk to the patient. As a result, transfusion practice grew rapidly. For the health care system to be able to realize fully the benefits of transfusion, it was necessary to arrange for a supply of donor blood that would be available at all times. This need gave the push to the establishment of blood banks in which large quantities of blood are stored in refrigerators.

Among the major developments that occurred in transfusion medicine following World War II were methods for separating and preserving the formed elements of blood—methods for freezing, thawing, and washing blood cells. These developments enabled the use of transfusions to support patients undergoing more aggressive therapy.

In the 1980's, transfusion medicine may have faced its most serious challenge since the days of Crile with the discovery that the human immunodeficiency virus (HIV) can be transmitted by blood. Major efforts were made to make blood transfusion a safe mode of therapy once again.

Bibliography

Bernheim, Bertram M. *Adventure in Blood Transfusion*. New York: Smith & Durrell, 1942. This is a fascinating book about the personal experience of Bernheim, a professor of surgery at The Johns Hopkins Medical School, with blood transfusion. Bernheim was the first to apply and modify the Crile method at his hospital in 1909. He changed the design of the Crile tube and later developed a two-piece tube that simplified direct blood transfusion even further. Discusses his first attempts at direct blood transfusion, his successes and failures, and the suspicious attitudes of the medical community at the time toward the new mode of therapy. He follows the history of blood transfusion and the developments in its administration, beginning with direct blood transfusion and ending with the establishment of blood banks. Illustrated.

Crile, George W. "Direct Transfusion of Blood in the Treatment of Hemorrhage." *Journal of the American Medical Association* 47 (November, 1906): 1482-1484; and "The Technique of Direct Transfusion of Blood." *Annals of Surgery* 46 (September, 1907): 329-332. These are two short but very influential articles in the modern history of blood transfusion. They are the first published reports of Crile's experiments with animals and his clinical cases. They contain the first detailed descriptions of the Crile method of direct blood transfusion and the patients that he treated.

____________. *George Crile, an Autobiography*. Edited by Grace Crile. Philadelphia: J. B. Lippincott, 1947. This book, edited by Crile's daughter, is recommended for those interested in the life of George Crile, one of the most prominent American surgeons at the beginning of the twentieth century. Among the many other achievements in the life of Crile, it also discusses the first direct blood transfusion. Illustrated and indexed.

____________. *Hemorrhage and Transfusion: An Experimental and Clinical Research*. New York: D. Appleton, 1909. This book is divided into two major parts: The first part is an examination of the problem of hemorrhages, and the second part is dedicated to a detailed study of blood transfusions. It provides a thorough examination of the different aspects of blood transfusion, stressing the pioneering stage at which this technique stood in 1909. It contains many descriptions of experiments with dogs, in addition to exact explanations of the two major techniques of direct blood transfusion. Illustrated.

Maluf, N. S. R. "History of Blood Transfusion." *Journal of the History of Medicine* 9 (January, 1954): 59-107. This is one of the best articles on the history of blood transfusion from ancient history and up to the early decades of the twentieth century. This article provides a detailed description of the different and changing applications of blood transfusions as well as the techniques and tools used. Contains many excellent illustrations and an extensive bibliography.

Gershon B. Grunfeld

Cross-References

Landsteiner Discovers Human Blood Groups (1900), p. 56; Carrel Develops a Technique for Rejoining Severed Blood Vessels (1902), p. 134; McLean Discovers the Natural Anticoagulant Heparin (1915), p. 610.

BAEKELAND INVENTS BAKELITE

Category of event: Applied science
Time: 1905-1907
Locale: Yonkers, New York

Baekeland developed the first totally synthetic thermosetting plastic, which paved the way for modern material science

Principal personages:
LEO HENDRIK BAEKELAND (1863-1944), a Belgian-born chemist, consultant, and inventor
CHRISTIAN FRIEDRICH SCHÖNBEIN (1799-1868), a German chemist who produced guncotton, the first man-made polymer
ADOLF VON BAEYER (1835-1917), a German chemist who synthesized the resinous product of phenol-formaldehyde reactions
GEORGE EASTMAN (1854-1932), an American industrialist and inventor whose purchase of Velox from Baekeland enabled Baekeland to pursue his research

Summary of Event

In the 1860's, the firm of Phelan and Collender offered a prize of ten thousand dollars to anyone producing a substance that could serve as an inexpensive substitute for ivory, which was somewhat difficult to obtain in large quantities at reasonable prices. At the time, this represented a sizable sum of money and many tinkerers and inventors sought to win the prize. Christian Friedrich Schönbein laid the groundwork for a breakthrough in the quest for a new material in 1846 by the serendipitous discovery of nitrocellulose, more commonly known as guncotton, which was produced by the reaction of nitric acid with cotton. The properties of this material were explored by a number of scientists, among them Alexander Parkes. Parkes had hopes of developing rubberlike materials that were more colorful and thus more appealing than rubber. The first public display of his new material, dubbed Parkesine, was at an 1862 exposition in London. While Parkesine was colorful and had potential for a wide variety of uses, it was not a commercial success as various shortcomings of the material could not be overcome by Parkes.

While Parkes struggled with Parkesine, an American inventor named John Wesley Hyatt was tinkering with a similar substance. Hyatt was already in the synthetic materials business, owning a factory that produced dominoes from a shellac and wood pulp material. In his investigations of nitrocellulose, Hyatt found the key that had eluded Parkes. His discovery was that addition of camphor to nitrocellulose under certain conditions led to formation of a white material which could be molded and machined. He dubbed this substance celluloid, and this product is now acknowledged as the first synthetic plastic. Celluloid won the prize for Hyatt, and he promptly set out to exploit his product. Celluloid was used to make baby rattles, collars, dentures, and other manufactured goods. Ultimately, the properties of celluloid led to the realization that it was more suited for other things. As a billiard ball substitute, it was not really adequate for various reasons. First of all, it is a thermoplastic—in other words, a material that softens when heated and can then be easily deformed or molded at high temperatures. It was thus too soft for billiard ball use. Also, it was highly flammable, hardly a desirable characteristic. A widely circulated, perhaps apocryphal, story claimed that celluloid billiard balls detonated when they collided. One of the more interesting uses of celluloid was in the making of motion picture film as it could be produced in a thin layer. It was used almost exclusively, despite its flammability, until displaced by a less flammable substitute.

Celluloids are thermoplastic substances, a related class of materials are thermosetting compounds. As the name implies, thermosetting materials are initially fluid and harden when heated, assuming the shape of their container. These thermosetting reactions are generally nonreversible, meaning the products tend to be stable. The first product of this class was developed in 1897 from the reaction of formaldehyde with milk protein. This was the first example of what are known as casein plastics, and it was widely used in the production of small items. It is significant as the first man-made thermosetting plastic. While celluloid and casein plastics were significant achievements, they both had something in common: Each material could be viewed as a derivative of a natural product and, thus, was not a completely synthetic or man-made substance. Therefore, the stage was set for a breakthrough into a totally new area of manufactured goods: the synthetic plastics.

Leo Hendrik Baekeland is recognized as the first person to produce a completely artificial plastic. Born in Ghent, Belgium, Baekeland emigrated to the United States in 1889 to pursue applied research, a pursuit not encouraged in Europe at the time. Baekeland worked in the photographic industry until his reputation allowed him to start his own consulting firm in 1891. He also pursued his research interests and achieved commercial success by developing Velox, a photographic paper. The rights to Velox were purchased by George Eastman for one million dollars. With this wealth, Baekeland purchased a home in Yonkers, New York, constructed a laboratory there, and continued his research efforts.

One area in which Baekeland hoped to make an inroad was in the development of an artificial shellac. Shellac at the time was a natural product, and there would be a wide market for any reasonably priced substitute. Baekeland's research scheme, begun in 1905, focused on finding a solvent that could dissolve the resinous products from a certain class of organic chemical reaction. The particular resins he used had been reported in the mid-1800's by the German chemist Adolf von Baeyer. These resins were produced by the condensation reaction of formaldehyde with a class of chemicals called phenols. Baeyer found that frequently the major product of such a reaction was a gummy residue that was virtually impossible to remove from glassware. Baekeland focused on finding a material which could dissolve these resinous products. Such a substance would prove to be the shellac substitute he sought. These efforts proved frustrating, as an adequate solvent for these resins could not be found. After repeated attempts to dissolve these residues, Baekeland shifted the orientation of his work. Abandoning the quest to dissolve the resin, he set about trying to develop a resin that would be impervious to any solvent, reasoning that such a material would have useful applications.

Baekeland's experiments involved the manipulation of phenol-formaldehyde reactions through precise control of the chemical proportions, addition of catalysts, and manipulation of the temperature and pressure at which the reactions were performed. Many of these experiments were performed in a 1.5-meter-tall reactor vessel, which he called a Bakelizer. In 1907, these meticulous experiments paid off, when Baekeland opened the reactor to reveal a clear solid that was heat resistant, nonconducting, and machinable. Experimentation proved that the material could be dyed practically any color in the manufacturing process, with no effect on the physical properties of the solid.

Baekeland filed a patent for this new material in 1907. (Ironically, this patent was filed one day before that of James Swinburne, a British electrical engineer who had developed a similar material in his quest to produce an insulating material.) Baekeland dubbed his new creation Bakelite and announced its existence to the scientific community on February 15, 1909, at the annual meeting of the American Chemical Society. The following year, the General Bakelite Corporation was formed, and the Bakelite produced was used in many different industries. Among the first uses was the manufacture of ignition parts for the rapidly growing automobile industry. At the time of Baekeland's death in 1944, annual production of Bakelite was more than 125,000 tons, and the material was being used to produce an incredible variety of manufactured goods.

Impact of Event

Baekeland's synthesis of Bakelite paved the way for the development of a great variety of novel man-made products, some of which are similar to natural materials and some of which have no parallel in nature. Bakelite was introduced at the very time that the automobile industry was in need of parts fabricated from such a material. This provided a firm industrial base for the material that is still retained.

In synthesizing Bakelite, Baekeland drew on both his theoretical knowledge of chemistry and to a large extent his practical knowledge based on his years of experience. Thus, he did not fully understand the structure of his product on the molecular level. Bakelite proved to be the first of a class of compounds called synthetic polymers. Polymers are long chains of molecules chemically linked together. There are many natural polymers, such as cotton. The discovery of synthetic polymers led to vigorous research into the field and attempts to produce other useful artificial materials. These efforts met with a fair amount of success; by 1940, a multitude of new products unlike anything found in nature had been discovered. These included such items as polystyrene and low-density polyethylene. In addition, artificial substitutes for natural polymers, such as rubber, were a goal of polymer chemists. One of the results of this research was the development of neoprene. Industries also were interested in developing synthetic polymers to produce materials that could be used in place of natural fibers such as cotton. The most dramatic success in this area was achieved by Du Pont chemist Wallace Carothers, who had also developed neoprene. Carothers focused his energies on forming a synthetic fiber similar to silk, resulting in the synthesis of nylon. Plastics were widely used in World War II because of their properties, as in the case of Bakelite, and as substitutes for natural products in short supply, as with nylon.

Postwar development in the plastics industry yielded many useful new products, such as high-density polyethylene, for which Karl Ziegler and Giulio Natta shared the 1963 Nobel Prize in Chemistry. Research and development of novel synthetic materials in recent years has produced such innovations as Kevlar, and the area is one of intense competition.

Synthetic polymers constitute one branch of a broad area known as material science. Novel, useful materials produced synthetically from a variety of materials have allowed for tremendous progress in many areas. Examples of these new materials include high-temperature superconductors, composites, ceramics, as well as plastics. These materials find use as structural components of aircraft, artificial limbs and implants, tennis rackets, and garbage bags. Unlike most natural products, plastics are extremely persistent in the environment. Through implementation of recycling technology or the development of agents that facilitate the biodegradation of the materials, this may become less of a problem in the future.

Bibliography

Asimov, Isaac. *Asimov's Biographical Encyclopedia of Science and Technology.* 2d rev. ed. Garden City, N.Y.: Doubleday, 1982. A user-friendly text that includes biographical information on 1,510 scientists and inventors. Very readable text and extensive cross referencing enable the interested reader to learn much about both scientific discoveries and the personalities and accomplishments of the scientists involved.

Canby, Thomas Y. "Reshaping Our Lives: Advanced Materials." *National Geographic* 176 (December, 1989): 746-781. Presents a comprehensive overview of modern material science, including detailed discussions of polymers, ceramics, composites, and other materials. Also contains information on yearly production figures, theoretical background, and manufacturing techniques. Beautiful color photographs by Charles O'Rear are also depicted.

Chiles, James R. "On Land, Sea, and in the Air, Those Polymer Invaders Are Here." *Smithsonian* 16 (November, 1985): 76-86. Illustrated with excellent photographs by Brian Wolff, this informative article gives a brief summary of the history of plastic and an excellent overview of modern plastic materials. Includes several good pictures of plastic manufacturing processes and a detailed color photograph of the Bakelizer.

Friedel, Robert. "A World of New Materials." In *Inventors and Discoverers,* edited by Elizabeth L. Newhouse. Washington, D.C.: National Geographic Society, 1988. This well-illustrated chapter presents a historical survey of materials science, including detailed history of such materials as rubber, celluloid, and plastics. Includes interesting photographs illustrating the broad potential of plastics touted to a receptive public. There is also an excellent biographical profile of Baekeland.

Williams, Trevor I. *A Short History of Twentieth-Century Technology.* New York: Oxford University Press, 1982. An excellent overview of technology from 1900 to 1950, this work compresses an incredible amount of material into a small text. Includes a chapter on chemicals, with a section devoted to plastics that places the twentieth century achievements in synthetic plastic research in perspective.

Craig B. Lagrone

Cross-References

Brandenberger Invents Cellophane (1904), p. 238; Carothers Patents Nylon (1935), p. 1055; Heeger and MacDiarmid Discover That Iodine-Doped Polyacetylene Conducts Electricity (1977), p. 2063.

BOLTWOOD USES RADIOACTIVITY TO OBTAIN THE AGE OF ROCKS

Categories of event: Earth science and physics
Time: 1905-1907
Locale: Yale University, New Haven, Connecticut

Boltwood pioneered the radiometric dating of rocks, giving impetus to the use of nuclear methods in geology and establishing a new chronology of Earth

Principal personages:

BERTRAM BORDEN BOLTWOOD (1870-1927), the first American scientist to study radioactive transformations and a founder of modern geochronology

ERNEST RUTHERFORD (1871-1937), a leading nuclear scientist who won the 1908 Nobel Prize in Chemistry

LORD KELVIN, SIR WILLIAM THOMSON (1824-1907), a prominent physicist who recognized and formulated the second law of thermodynamics and who attempted to use it to determine the age of the earth

Summary of Event

Bertram Borden Boltwood, educated at Yale University and in Germany, began his career as an instructor of analytical chemistry at Yale University. He remained at Yale until 1890, when he left to establish a private laboratory (Pratt and Boltwood, Consulting Mining Engineers and Chemists) in New Haven, Connecticut. He was interested in radium and was fascinated by the theory of radioactive disintegrations proposed in 1903 by two McGill University scientists: Ernest Rutherford and Frederick Soddy. According to that theory, radioactivity is always accompanied by the production of new chemical elements on an atom-by-atom basis. In 1904, Boltwood impressed Rutherford by demonstrating that all uranium minerals contain the same number of radium atoms per gram of uranium. This confirmation of the theory of radioactivity marked the beginning of a close collaboration between the two scientists. Actually, Boltwood's involvement with research in radioactivity can be traced to 1899, when he was supervising a student project on the extraction of radium from uranium ores by Marie Curie's method.

The significance of Boltwood's work can be appreciated better in the light of a chronological controversy raging at that time between geologists and physicists. It had been accepted generally that the earth, at some time in its history, was a liquid ball and that its solid crust was formed when the temperature was reduced by cooling. The geological age of Earth was, thus, defined as the period of time necessary to cool it down from the melting point to the present temperature. Several estimates of that time were made in the second half of the nineteenth century by the famous physicist, Lord Kelvin, Sir William Thomson, who claimed that the age of the earth did not exceed 100 million years. In 1897, he specified that the actual figure was probably close to 20 million and certainly not as large as 40 million. His calculations were mathematically correct but did not take into account radioactivity, which had been discovered the year before. Geologists believed that 40 million years simply was not enough to create continents, to erode mountains, or to supply oceans with minerals and salts. Studies of sequences of layers (stratigraphy) and of fossils (paleontology) led them to believe that the earth was older than 100 million years, but they were not able to prove it. Boltwood and Rutherford proved that geologists were right. This came as a by-product of their research on the nature of radioactivity. The gaseous element helium played an important role in that line of research. Rutherford knew that helium was always present in natural deposits of uranium, and this led him to believe that in radioactive minerals, alpha particles somehow were turned into ordinary atoms of helium. Accordingly, each rock of a radioactive mineral is a generator of helium. The accumulation of the gas proceeds more or less uniformly so that the age of the rock can be determined from the amount of trapped helium. In that sense, radioactive rocks are clocks embodied in Earth's natural environment.

In line with this model and knowing how much helium is produced from each gram of uranium per billion years, Rutherford and his collaborators were able to see that naturally radioactive rocks are often older than 100 million years. The accumulated amounts of helium were simply too large to allow other interpretations. The ages of some samples exceeded 500 million years. Moreover, Rutherford was aware that the assigned ages would have to be revised upward in order to account for helium that was escaping from the rocks. Impressed by these results, and trying to eliminate the uncertainties associated with leakage of helium, Boltwood decided to work on another method of dating. This resulted from his earlier attempts to demonstrate that, as in the case of radium, all uranium minerals contain the same number of atoms of lead per gram of uranium. Chemical data, however, did not confirm this expectation; the measured lead-to-uranium ratios were found to be different in minerals from different locations on Earth. Yet, this still needed to be interpreted. According to Rutherford and Soddy's theory, a spontaneous transformation of uranium into a final product proceeds through a set of steps in which alpha particles and electrons are emitted, one after another. Boltwood realized that lead must be the final product and that its accumulation could be used to date minerals. By focusing on lead, rather than on helium, he hoped to reduce the uncertainties associated with the leakage. Lead, he argued, is less likely to escape from rocks than helium because, once trapped, it becomes part of a solid structure.

Motivated by these ideas, Boltwood proceeded with the development of the uranium-lead method of dating. To accomplish this, he had to determine the rate at which lead is produced from uranium. That rate is about 0.15 gram of lead per 1 gram of uranium, per billion years. Suppose, for example, that a chemical analysis of a rock yields 0.03 gram of lead for each gram of uranium. The age of the rock can be found by setting $0.03 = 0.15 \times t$, where t is the age of the rock, in billions of years. The solution of this simple equation is $t = 0.2$, or 200 million years. A slightly smaller t, 192 million, would be deduced if the exponential nature of radioactive decay were taken under consideration.

Boltwood was aware that even the most carefully performed experiments can lead to highly unrealistic conclusions. In his case, two possible sources of interpretational errors were possible: one having to do with the preexisting (nonradiogenic) lead and another having to do with the fact that chemicals can migrate through rock boundaries. Suppose, Boltwood argued, that 50 percent of the lead found in a sample was already there at the time of solidification, but a researcher was not aware of that fact. In such a case, the derived age would be exaggerated by a factor of two. In other words, the preexisting lead would make rocks appear older than they actually were. On the other hand, the opposite would be true if one-half of the lead generated from uranium were able to escape through the rock boundaries. Recognizing the dangers of interpretational errors, Boltwood was very careful when selecting samples. Only deep portions of samples were analyzed to minimize uncertainties associated with the possibility of lead leakage. Boltwood started his investigations in 1905 and before the end of the year, he had analyzed twenty-six samples. One of them was identified as 570 million years old. In a formal publication, which appeared in 1907, he described forty-six minerals collected in different locations; their reported ages were between 410 and 2,200 million years old.

Similar results were reported earlier by Rutherford, from his laboratory in Montreal, and by Robert John Strutt (also known as Fourth Baron Rayleigh), from the Imperial College in London. It is true that similarities were mostly qualitative, rather than quantitative; however, all data showed that the absolute time scale was undeniably longer than anticipated, even by some geologists. In other words, the large margins of error that characterized early results did not interfere with the basic overall conclusion that many rocks were at least ten times older than what was calculated by Lord Kelvin. It was significant that helium and lead methods, based on different chemical procedures and applied to a large number of very different minerals, yielded similar results in three institutions.

Impact of Event

The main results of Boltwood's pioneering work and his successors' was the realization that geological times must be expressed in hundreds and thousands of millions of years, rather than in tens of millions, as advocated by Lord Kelvin. This was particularly significant for the acceptance of the theory of evolution by biologists and, in general, for better understanding of many long-term processes on Earth. Geochronology has been used, for example, in investigations of reversals of the terrestrial magnetic field. Such reversals occurred many times during the geological history of Earth. They were discovered and studied by dating pieces of lava, naturally magnetized during solidification. The most recent reversal took place approximately 700,000 years ago. It is clear, in retrospect, that the discovery of radioactivity affected geochronology in two ways: by providing tools for radiometric dating and by invalidating the thermodynamic calculations of Lord Kelvin. These calculations were based on the assumption that the geothermal energy lost by Earth is not replenished. The existence of radioactive heating, discovered in 1903 in France, contradicted that assumption and prepared scientists for the acceptance of Boltwood's findings. Lord Kelvin died in the same year in which these findings were published, but he knew about Rutherford's findings as early as spring, 1904. He was very interested in radioactive heating but never came forth with a public retraction of his earlier pronouncements.

In its modern form, the uranium-lead method is very different from what it was nearly a century ago. This results primarily from a better understanding of radioactive transformations and to the existence of instruments that were not available in the early part of the twentieth century. The most important tool of modern geochronologists is a mass spectrometer. Its widespread use after World War II is responsible for the development of new methods of dating. The availability of several methods is important when redundancy can be used to minimize, and sometimes eliminate, interpretational errors recognized by Rutherford, Strutt, and Boltwood.

A large number of samples from different geological strata, all over the world, plus lunar samples and meteorites, have been dated by modern methods. The data show that the oldest known terrestrial rocks solidified 3.7 billion years ago; this is 0.9 billion years less than the ages of lunar samples. The ages of most meteorites, between 4.5 and 4.7 billion years, are essentially the same as the ages of lunar samples. This is consistent with the overall picture, also backed by astronomical data, that Earth's solar system was formed 4.7 billion years ago, and that Earth remained liquid in the first 0.9 billion years of its existence. Boltwood would be proud to be counted among the contributors to that knowledge.

Close collaboration between Rutherford and Boltwood ended after the first decade of the twentieth century. Rutherford moved to England, where he created and directed a modern nuclear research laboratory. Boltwood became a professor of radiochemistry at Yale University, concentrating on teaching and administrative activities. He also developed educational materials and designed new laboratories.

Bibliography

Asimov, Isaac. *Exploring the Earth and the Cosmos.* New York: Crown, 1982. An excellent story describing the ways researchers learn about the world. Includes twenty-three chapters and covers the entire panorama of findings in astronomy, biology, chemistry, geography, and physics. Chapters 13 and 14 are devoted to chronology and to the age of the earth.

Badash, Lawrence. "Bertram Borden Boltwood." In *Dictionary of Scientific Biography,* edited by Charles Coulston Gillispie. New York: Charles Scribner's Sons, 1970. This is a multivolume compilation of biographies of famous scientists, published under the sponsorship of the American Council of Learned Societies. The biography on Boltwood contains a detailed description of his family background, professional activities, and private life. Equally detailed and interesting are entries on Kelvin and on Rutherford.

___________. *Radioactivity in America*. Baltimore: The Johns Hopkins University Press, 1979. A good account of the history of the early (pre-1920) research in radioactivity and of practical applications of that research, primarily in the United States. The life of Boltwood and his professional activities, including work on radiometric dating, are described in chapters 4, 5, and 6. Chapters 9 and 10 are devoted to medical applications of radium and to the marketing of that material.

___________, ed. *Rutherford and Boltwood: Letters on Radioactivity*. New Haven, Conn.: Yale University Press, 1969. Good reading for those who may wish to experience the scientific excitement of two scientists, friends and collaborators. Some correspondence includes details about the instruments and procedures used in research, but this technical information can be skipped easily by readers who are more interested in historical, sociological, and personal issues. Of particular interest is a letter, dated November 18, 1905, in which Boltwood recognizes Rutherford as the initiator of the uranium-lead method of dating minerals.

Burchfield, Joe D. *Lord Kelvin and the Age of the Earth*. New York: Science History Publications, 1975. An excellent book for all who are interested in the evolution of geochronologies, both before and after Lord Kelvin. Dramatic effects of the discovery of radioactivity on geoscience are described in the last chapter. Highly readable; recommended to both the scientist and historian of science. The bibliography is extensive and excellent.

Eicher, Don L. *Geologic Time*. 2d ed. Englewood Cliffs, N.J.: Prentice-Hall, 1976. The author of this short but very informative book, an earth scientist from the University of Colorado, explains the geological time concept in its historical perspective. He focuses on significant discoveries that contributed to the growth of scientific geology. Various methods of dating geological formations are described. The book is rich in photographs, sketches, and maps.

Faul, Henry. *Ages of Rocks, Planets, and Stars*. New York: McGraw-Hill, 1966. Faul is a geophysicist with extensive experience using modern methods of nuclear geochronology. The general scientific background for understanding these methods is provided, together with clear explanations of different techniques and descriptions of interesting findings.

Friedlander, Gerhart, et al. "Nuclear Processes in Geology and Astrophysics." In *Nuclear and Radiochemistry*. 3d ed. New York: John Wiley & Sons, 1981. This is a nuclear science textbook for college students with background in general physics and in chemistry. Chapter 13 should be very useful to those who want to be introduced to the quantitative aspects of various methods of geochronology. Numerous references to other texts, both old and recent, are provided at the end of the chapter.

Ludwik Kowalski

Cross-References

Becquerel Wins the Nobel Prize for the Discovery of Natural Radioactivity (1903), p. 199; Thomson Confirms the Possibility of Isotopes (1910), p. 471; Libby Introduces the Carbon-14 Method of Dating Ancient Objects (1940's), p. 1160; Leakey Finds a 1.75-Million-Year-Old Fossil Hominid (1959), p. 1602; Apollo 12 Retrieves Surveyor 3 Parts from the Lunar Surface (1969), p. 1913; Hominid Fossils Are Gathered in the Same Place for Concentrated Study (1984), p. 2279.

LOWELL PREDICTS THE EXISTENCE OF PLUTO

Category of event: Astronomy
Time: August, 1905
Locale: Flagstaff, Arizona

Lowell instituted an observational search for a trans-Neptunian planet

Principal personages:

PERCIVAL LOWELL (1855-1916), a businessman in textiles, mathematician, astronomer, and founder of the Lowell Observatory in 1894 in Flagstaff, Arizona

GIOVANNI VIRGINIO SCHIAPARELLI (1835-1910), the Italian astronomer who reported seeing "canals" on the surface of Mars and inspired Percival Lowell to pursue a career in astronomy

VESTO MELVIN SLIPHER (1875-1969), the director of Lowell Observatory during the final search for Pluto

EARL CARL SLIPHER (1883-1964), the younger brother of Vesto Melvin Slipher who assisted in observations of Mars and other planets

CARL O. LAMPLAND (1873-1951), an astronomer who assisted with Percival Lowell's search for Pluto from 1914 to 1916

CLYDE W. TOMBAUGH (1906-   ), the staff assistant at Lowell Observatory who made the actual discovery of Pluto on photographic plates

A. LAWRENCE LOWELL (1856-1943), a brother of Percival Lowell and president of Harvard University for twenty-four years who funded the purchase of a 33-centimeter telescope for Lowell Observatory

WILLIAM HENRY PICKERING (1858-1938), a physicist and astronomer at Harvard University who competed with Percival Lowell to be the first person to discover a ninth planet

Summary of Event

Percival Lowell, a member of a prominent New England family, was born in Boston, Massachusetts, on March 13, 1855. Although his family was associated with textile mills, Percival pursued his formal education (at Harvard University) in mathematics with a personal interest in astronomy.

In 1877, the Italian astronomer Giovanni Virginio Schiaparelli began an observational study of the planet Mars, which led to his announcement of complex and detailed surface patterns. He called these features *canali*, a word incorrectly translated into the English word "canals," which proved to be a source of subsequent confusion.

Percival Lowell followed the discoveries of Schiaparelli keenly and imagined many great Martian civilizations. In the early 1890's, Schiaparelli's eyesight had deteriorated and he could no longer make telescopic observations. Determined to continue Schiaparelli's research, Percival founded the Lowell Observatory (opened on June 1, 1894) in Flagstaff, Arizona, and devoted it to a study of Mars and the other planets. Lowell made personal observations of the so-called Martian canals and imagined these canals as irrigation channels built by a desperate, yet intelligent, civilization striving to retain their water supply from the polar caps of Mars. In 1898, Lowell founded a journal that published his thoughts (about Mars) to the scientific sector, where they were met with varied receptions. His theories on Martian life were well received by science-fiction writers and the general public but ridiculed by professional astronomers. Percival Lowell became frustrated and sought ways to improve his credibility among his colleagues.

The mathematical prediction of the planet Neptune and its discovery in 1846 gave rise to the possibility of additional remote and undiscovered planets in the solar system. Lowell thought that if he could predict the location of a ninth planet, beyond Neptune, then it would improve his reputation as an astronomer. In 1905, Lowell inspired his colleagues at the Lowell Observatory to begin the first systematic photographic and visual search for Planet X. He expected this distant planet to be very similar to Neptune in density, size, and magnitude (brightness) and therefore easy to spot.

Lowell began his first search, from 1905 to 1907, with a camera 12.7 centimeters in aperture (a 12.7-centimeter Brashear lens of 89 centimeters focal length) used to photograph star fields. More than two hundred photographic plates were taken by Lowell's assistants. Each sky area was photographed twice to record faint stars. All three-hour exposure plates were centered on the ecliptic (the plane of the solar system) and varied at intervals of 5 degrees longitude. Dr. Lowell personally examined each plate with a hand-held magnifying glass. His technique used the basic assumption that a comparison of two photographs taken of the same star fields would show movements of nonstellar objects such as comets, asteroids, and planets.

Between 1911 and 1912, a sequence of photographic plates was made on the observatory's 107-centimeter reflecting telescope. Lowell used a Zeiss Blink Comparator (a device that superimposes two photographic plates of the same region for direct comparison), which greatly expedited studies of multiple photographic plates. Unfortunately, this intensive search for Planet X also proved unsuccessful.

The search by Lowell and his assistant, Carl O. Lampland, continued from 1914 to 1916 with a 23-centimeter Brashear lens. Almost one thousand photographic plates were examined under the Blink Comparator, but with negative results.

By this time, Lowell realized that it would be difficult to discover observationally the ninth planet and attempted to compute its orbital and physical characteristics through mathematical calculations and statistical methods. He admitted that his theory would be limited and contain some positional uncertainty, as his calculations were based on errors embedded within observations of the distant planets, Uranus and Neptune.

On January 13, 1915, Lowell presented his theory on Planet X to the American Academy of Arts and Sciences. His book, *Memoir on a Trans-Neptunian Planet*, was published in the spring of that year. Using methods of celestial mechanics and mathematics, Lowell studied minute perturbations in the orbits of Uranus and Neptune. He attributed the irregularities in their motions to a trans-Neptunian planet, that is, a planet whose orbit would intersect that of Neptune. Unfortunately, Neptune's orbit was not well known at the time (its revolution had been observed for only fifty years), so aberrations in its motion could not be accurately calculated. Also, any deviations in the motion of Uranus were expected to be further minimized because of its great distance from Planet X; these additional errors were introduced into his calculations. In addition, Lowell studied the perturbations of comets and suggested that a distant planet could gravitationally affect a cometary orbit.

Lowell's rigorous mathematics led him to assume that Planet X would have a mass seven times that of Earth, an inclination (with respect to the ecliptic) of 10 degrees, and an average distance of forty-three times the distance between Earth and the Sun. He also believed that the undiscovered planet would have a low density and a high albedo (the ratio of sunlight received versus sunlight reflected), as in the case of the four largest planets. In addition, Lowell expected the planet to appear faint because of its great distance from Earth and could only generally predict its location to be in or near the constellation of Gemini. Originally, he had predicted the location to be within the constellation of Libra.

Percival Lowell overworked himself in an attempt to locate the ninth planet and died of a stroke on November 12, 1916, at the age of sixty-one. He left an endowment of more than one million dollars for the Lowell Observatory to continue his search program. Unfortunately, his wife challenged the will, and a legal battle ensued for almost a decade. Court fees depleted much of the money and his observatory could not afford to maintain his work. This caused an interruption of thirteen years in the search for Planet X. Lowell died without ever seeing his theoretical and distant world. Yet, his enthusiasm and courage were retained by his staff, and the search continued.

In 1925, Dr. A. Lawrence Lowell, Percival's brother, provided funds for the purchase of a 33-centimeter telescope with a focal length of 168 centimeters. The new telescope began operation in 1929, at which time an assistant, Clyde W. Tombaugh, was employed. On February 18, 1930, Tombaugh used the Zeiss Blink Comparator and discovered Lowell's Planet X on photographic plates taken on January 21, 23, and 29, 1930. Carl Lampland and Vesto and Earl Slipher took additional photographic plates to confirm the sighting. On March 13, 1930, Vesto Slipher sent a telegram to Harvard University announcing the discovery of the trans-Neptunian planet, as originally predicted by Lowell.

Impact of Event

News of the discovery of a ninth planet was made at Lowell Observatory on the seventy-fifth anniversary of Percival Lowell's birth. The press and general public were intrigued by the search and discovery of a ninth planet. Hundreds of telegrams and congratulatory letters were received by the observatory staff.

Before his death, Lowell had more recognition among astronomers in Europe than in America. In the United States, many of his colleagues never acknowledged Lowell as a professional astronomer; they considered him to be an anxious amateur. Unfortunately, the controversy that Lowell had begun earlier with his observations of Martian canals made him and the Lowell Observatory outcasts among much of the astronomical community. With the exception of magnitude, however, the planet's location and basic characteristics were so similar to Lowell's prediction that its discovery was not assumed to be by mere chance. In fact, Vesto Slipher stated that the discovery was simply the conclusion of a search program set forth in 1905 by Lowell and his theoretical work on the dynamical evidence of a ninth planet. Professional astronomers contacted Lowell Observatory and requested specific coordinates of the new planet in order to compute its orbital path around the sun. Early calculations suggested that the planet took several thousand years to orbit the sun. Nevertheless, this value decreased as more observations were recorded. The discovery of a ninth planet prompted astronomers at other institutions to begin their own searches for more distant planets. Tombaugh personally assisted in a continued search program at Lowell Observatory until the start of World War II. Since Pluto's discovery, however, no other distant planets have been detected.

The name, Pluto, was suggested by an eleven-year-old girl in England. It is interesting to note that the planetary symbol for Pluto is the interlocked letters P and L. In this way, people are reminded of Percival Lowell, by the planet he predicted but did not live to see.

Percival Lowell contributed to the advancement of science through his work in astronomy. He established the Lowell Observatory, whose staff made the actual discovery of Pluto. The equipment at this observatory made possible not only a search for Pluto but also observations of Mars, Saturn and its rings, and the atmospheres of Jupiter and Uranus. Similar studies continue among astronomers worldwide. Additional research has been conducted on Pluto since its discovery. In the 1970's, spectral observations suggested that Pluto had frozen methane on its surface. In 1978, an astronomer from the U.S. Naval Observatory examined some photographic plates taken from Flagstaff, Arizona, the location from which Tombaugh originally discovered Pluto. Upon close inspection, some of Pluto's images appeared slightly elongated, which led astronomers to announce the discovery of Pluto's moon Charon. Given a moon and a planet, astronomers are able to calculate the masses of these bodies using Newton's laws. When Pluto was first discovered, its mass was suspected to be about that of Earth; however, new measurements suggest Pluto's mass is only about one four-hundredth that of Earth.

Outside the scientific community, Lowell's early ideas of intelligent life on Mars and his search for a trans-Neptunian planet spawned excitement from the general population. People read many newspaper reports on the new planet. Educators from local observatories gave presentations on observing the night sky. The Adler Planetarium in Chicago opened its doors to the public among the wonder of discovering more planets hidden against the starry background sky. Although the discovery of

Pluto astonished much of the astronomy community, it produced a sense of mystery and fascination that continues to prompt the public to learn more about astronomy and keep looking up toward the stars.

Bibliography

Binzel, Richard. "Pluto." *Scientific American* 262 (June, 1990): 50-58. This article contains the most up-to-date information on the physical characteristics of Pluto. Artwork and line drawings of Pluto and its moon Charon are helpful in explaining why this system is often considered to be a double planet. The actual photograph used to discover Charon in 1978 is reproduced, along with spectra and light curves of Pluto.

Croswell, Ken. "Pluto: Enigma on the Edge of the Solar System." *Astronomy* 14 (July, 1986): 7-22. A concise biography of Percival Lowell accompanies this reference, which features a discussion of Lowell's strategy for the search of Planet X, Clyde Tombaugh's actual discovery of that planet, and a summary of physical data for Pluto. Information on the discovery of Pluto's moon, Charon, is also outlined.

Lago, Don. "The Canals of Lowell." *Griffith Observer* 49 (August, 1985): 2-11. This informative article provides a background on Percival Lowell's interest in the "canals" of Mars, through Schiaparelli's observations. Lowell's family background in Massachusetts is detailed as well as his desire to continue the work of Schiaparelli. The author also speaks of Lowell's interest in the search for a ninth planet.

Lowell, A. Lawrence. *Biography of Percival Lowell*. New York: Macmillan, 1935. This book, written by Percival Lowell's brother, contains minute details of both the professional and personal side of Percival. The text provides an excellent chronology of Percival's career. This is recommended reading if more information is wanted about Percival Lowell, the man and the astronomer.

Lowell, Percival. *Memoir on a Trans-Neptunian Planet*. Lynn, Mass.: Press of T. P. Nichols and Son, 1915. This is the best resource on Percival Lowell's search for Planet X. His descriptive text and mathematical equations are listed in this memoir. A summary lists his predictions on the planet's mass and distance from the sun. Although this text may be considered a technical reference, it is the only resource for Lowell's original assumptions, derivations, and final predictions about the ninth planet.

Shapley, Harlow, ed. *Source Book in Astronomy, 1900-1950*. Cambridge, Mass.: Harvard University Press, 1960. This text is an excellent reference for many great historical moments in astronomy. Chapter 14, "The Discovery of Pluto," is particularly relevant to a review of the search for the ninth planet. It is also an outstanding general reference for key moments in astronomy within the first fifty years of the twentieth century.

Simon, Tony. *The Search for Planet X*. New York: Basic Books, 1962. This book is geared toward primary- and middle-school-aged readers. It tells the story of Pluto's discovery by placing the reader as a detective and providing clues. The illustrations are directed toward a younger audience, but the book still contains many interesting photographs.

Tombaugh, Clyde W. "The Discovery of Pluto: Some Generally Unknown Aspects of the Story." *Mercury* 15 (May/June, 1986): 66-72. Tombaugh shares his personal account of the discovery of Pluto through a series of flashbacks, from his childhood on a Kansas farm to his employment at Lowell Observatory. He recounts his experience studying photographic plates in the search for a ninth planet.

___________. "The Discovery of Pluto: Some Generally Unknown Aspects of the Story, Part II." *Mercury* 15 (July/August, 1986): 98-102. In a continuation of the above reference, Tombaugh conveys the excitement of discovering Lowell's Planet X and his sudden worldwide fame. Tombaugh also discusses his extended search for additional planets, which was interrupted by World War II.

___________. "Reminiscences of the Discovery of Pluto." *Sky and Telescope* 19 (March, 1960): 264-270. This article is an excellent resource, beginning with Percival Lowell's intention to search for Planet X through the investigative years to the final discovery of Pluto. There are photographs of Lowell Observatory, Percival Lowell, the Slipher brothers, Carl Lampland, Clyde Tombaugh, and the Zeiss Blink Comparator.

Noreen A. Grice

Cross-References

Slipher Obtains the Spectrum of a Distant Galaxy (1912), p. 502; Tombaugh Discovers Pluto (1930), p. 944.

EINSTEIN STATES HIS THEORY OF SPECIAL RELATIVITY: $E = mc^2$

Category of event: Physics
Time: Fall, 1905
Locale: Bern, Switzerland

Einstein's theory of special relativity challenged Newtonian physics, replacing the theory of three-dimensional space and one-dimensional time with the widely accepted theory of space-time

Principal personages:

ALBERT EINSTEIN (1879-1955), a German physicist who is best known for his theories of special and general relativity, honored with the 1921 Nobel Prize in Physics for his work with the light quantum and photoelectric effect

SIR ISAAC NEWTON (1642-1727), a renowned mathematician, much revered in scientific circles, whose theory of gravitation Einstein challenged

ALBERT A. MICHELSON (1852-1931), the first American to win a Nobel Prize in Physics (1907) for his work in optics that resulted in the closest approximation of the speed of light, a fundamental requirement for Einstein's theorizing

HENDRIK ANTOON LORENTZ (1853-1928), a Dutch physicist who demonstrated deficiencies in existing theories of Newtonian mechanics that did not differentiate sufficiently between the characteristics of light waves and sound waves, information vital to Einstein's thinking on relativity

JOHANN JAKOB LAUB (1872-1962), a physicist who collaborated with Einstein in his research on black body radiation

Summary of Event

Albert Einstein's special theory of relativity was first articulated in an article in 1905 in the German journal *Annalen der Physik* under the title "Zur Elektrodynamik bewegter Körper" (on the electrodynamics of moving bodies), in which he does not mention the formula now so closely associated with his name: $E = mc^2$ (energy = mass times the speed of light squared). In this article, Einstein points out that time cannot be viewed as an absolute.

He asserts, rather, that time is relative and broaches the question of simultaneity, which must be considered when one contemplates events occurring in time and space. Because it takes an infinitesimal amount of time for the light that illuminates an event that occurs near a person to travel to the perceptor, the event necessarily is perceived after it happens rather than when it happens. It was not until later, in 1905, that Einstein published in the same journal the three-page article that contained his famed formula: "Ist die Trägheit eines Körpers von seinem Energieinhalt abhängig?" (Does the inertia of a body depend on its energy content?)

In the normal course of earthly events, the slight discrepancy between when an event occurs and when it is perceived matters little. The discrepancy usually can be measured in milliseconds. Considering this matter in cosmic terms, however, Einstein clearly demonstrated that the discrepancy may become significant. Related to this observation is a new way of viewing the correlations between time and space.

According to Einstein, the speed of light is a constant; no matter how fast one may pursue it, it always travels at the same speed, now measured to within one hundred-thousandth of 1 percent (0.00001 percent) at 299,792.456 kilometers, or about 186,330 miles, per second. In 1882, Albert A. Michelson's experiments led him to measure the speed of light more accurately than anyone had up to that time, and it has been measured even more accurately since that time. Michelson's findings, concluding that light travels at 299,853 kilometers, or 186,320 miles, per second proved indispensable to Einstein as he moved toward devising his special theory of relativity.

Einstein postulated that nothing can move faster than the speed of light. He further conjectured that as objects accelerate toward the speed of light, they will become shorter in the direction they are traveling, their mass will become greater; and time will pass more slowly for them. This hypothesis has been demonstrated in various ways, notably by physicists who carried finely tuned atomic clocks with them on around-the-world commercial jet flights and, upon their return, checked them against other comparable clocks in their laboratories, only to find that the clocks they had carried with them had, indeed, lost time totally in accordance with their predictions based upon Einstein's theory.

If it were possible physically for an object to travel in excess of the speed of light, that object would predictably move backward in time. Because objects do not move at anything approaching the speed of light, the loss of length and time and the gaining of mass that rapidly moving objects incur is barely perceptible. An object moving at the incredible speed of one-seventh the speed of light—26,614 miles, or slightly more than one circumnavigation of the globe at the equator every second—would change in mass, length, and time measurement by only 1 percent.

Drastic alterations take place, however, as the object approaches the speed of light more closely. When the velocity is six-sevenths the speed of light, the mass is twice what it is when the object is at rest, while the length and time measurements are cut in half. Were the velocity of the object to equal the speed of light, its mass would become infinite. In terms of Einstein's theory of relativity, if a person were in a vehicle capable of coming within 99.995 percent of the speed of light, the mass of the person and the vehicle would be increased one thousand times; and time would pass for them at one one-thousandth of the normal rate so that each year of such travel would be equivalent to one thousand years of time, as scientists normally measure it.

Einstein's findings generally have been verified by various means, ranging from carrying clocks on commercial jet flights, as mentioned above, to experiments with high-speed electrons that, under laboratory conditions, frequently achieve ten thousand times their normal mass as their speed increases.

Central to Einstein's special theory of relativity is his shattering of the Newtonian theory that space is three-dimensional and that time is one-dimensional. Rather than thinking of time and space separately, Einstein viewed space-time as a coordinated, four-dimensional system. Time cannot be conceived of outside the spatial context; space cannot be conceived of outside the temporal context. The two exist simultaneously and interdependently as a coordinated system.

The only major physical phenomenon that Einstein could not explain using his special theory of relativity was gravity; this led him, in 1915, to the articulation of his general theory of relativity. The reason that gravity requires a special explanation is that space-time is curved, so that when scientists consider its vastness, their calculations are necessarily affected by this curvature.

The flat space-time concept of the special theory of relativity, despite the curvature of space-time, is still serviceable when one is making local measurements, such as measuring the distance from one's front door to one's garage. If, however, one wants exact measurements of an area whose boundaries are Montreal, Tierra del Fuego, Auckland, and Seoul, the curvature of space-time must be taken into account for the calculations to be wholly accurate. The inaccuracy of the measurement increases as the size of what is being measured increases if a flat space-time methodology is used. The special theory of relativity is comparable to Euclidean geometry, whereas the general theory of relativity is comparable to the geometry of curved surfaces.

Fundamental to Einstein's concept of relativity is the notion that energy (the E in his equation) and mass (the m) are really the same: All energy has mass; all mass, if conditions are right, can be converted to energy. Indeed, it can be converted into energy sufficient to cause the fission that, with the splitting of the atom, has resulted in the creation and harnessing of nuclear energy.

Einstein's study of Brownian motion led him to the basic conclusion that such moving particles as protons and electrons travel in waves and are intimately connected with photons. This conclusion led to the development of the field of wave mechanics in physics and was a fundamental element in Einstein's work in photoelectric effect, for which he was awarded the Nobel Prize. Although the prize did not come as recognition for his work in relativity, certainly his best-known contribution, it was his work with Brownian motion that contributed substantially to his special theory of relativity and that resulted in his later research on the photoelectric effect. Certainly, the special theory of relativity established Einstein as the most compelling and influential physicist of his day. It led to a total rethinking of the entire field of physics.

Impact of Event

When Einstein entered the Eidenössische Technische Hochschule in Zürich upon the completion of his secondary school education, he followed the advice of his elders and concentrated on mathematics rather than physics because, as they told him, little remained to be done in physics. All that was significant in the field had already been discovered. It was regarded as a scientific specialty in which no new worlds remained to be conquered.

It is, therefore, ironic that Einstein was the person who, more than any other, brought insights to the field that were so compelling as to cause a complete restructuring of the discipline. Because all activity in the various subspecialties of physics—optics, electromagnetism, mechanics—takes place in a space-time continuum, Einstein's postulation of a four-dimensional space-time context affected every aspect of physics and moved the discipline into areas that had important scientific implications but that had even more highly significant philosophical implications as well.

The special theory of relativity left unanswered many of Einstein's salient questions. In moving beyond his early work in relativity in 1905, Einstein formulated his general theory of relativity that, in 1915, responded to many questions regarding gravity that his earlier hypothesis had not answered to his satisfaction.

The two theories corrected some of Sir Isaac Newton's crucial scientific hypotheses that had, to that time, been sacrosanct among many physicists. It freed them to pursue studies that led to the splitting of the atom, to the development of nuclear energy, to space exploration that has involved sending humans into outer space, to the development of a theory of superconductivity, and to countless other accomplishments that have helped humankind explore the universe and understand it in greater depth than had previously been possible.

One can point to no area of modern science that has not been affected by Einstein's work in relativity, the heart of which is found in his initial work on the special theory of relativity. Albert Einstein redefined space and time, and his definition of space-time as a coordinated system has withstood the thousands of tests to which it has been put. His hypothesis has not been diminished by the extensive tests to which researchers have subjected it.

Einstein, as a scientist who continually searched and constantly questioned, certainly never foreclosed the possibility that his hypotheses might one day be disproved. Were he alive today, he would be at the forefront of testing them scrupulously and unfailingly as he did in his work, with Johann Jakob Laub, on black body radiation and in his subsequent work in the direction of evolving a unified field theory of physics. The astrophysical research into black holes, which now teeters on the brink of explaining much about the origins of the solar system and of the entire universe, is one of the direct offshoots of Einstein's early work in relativity. The implications of such astrophysical research are, perhaps, as important as those of any exploration being conducted in the field of physics.

Bibliography

Bernstein, Jeremy. *Einstein*. New York: Viking Press, 1973. This frequently reprinted book is one of the most popular studies of Albert Einstein for laypersons.

Bernstein, a physicist, frequently writes about science for popular audiences. His chapter entitled "$E = mc^2$" is particularly lucid in its presentation of Einstein's theory of special relativity. The bibliography of selected primary sources and of some secondary sources contains valuable annotations, although not all entries are annotated.

Eddington, Arthur. *Space, Time, and Gravitation*. New York: Harpers & Brothers, 1959. Of Eddington's several books about Einstein and relativity, this one deals most valuably with the theory of special relativity and is especially clear in its presentation of how Einstein departed from Newtonian physics and the implications of that departure for the discipline of physics. The book might be daunting to the uninitiated.

Frank, Philipp. *Einstein: His Life and Times*. New York: Alfred A. Knopf, 1947. This biography has not been supplanted by any of the subsequent studies of Einstein that have appeared, although it was published before Einstein's death. The book is well written, and the necessary scientific explanations are as lucid as Frank can make material that is inherently as complicated and complex as Einstein's.

Hoffman, Banesh, and Helen Dukas. *Albert Einstein: Creator and Rebel*. New York: Viking Press, 1972. Helen Dukas, Einstein's secretary for twenty-seven years, has collaborated with Banesh Hoffman to produce an intimate book about Einstein. The documentation is internal, and there is no bibliography; the index is exhaustive, and the illustrations are generous and apt. A reasonable starting point for the beginner.

Holton, Gerald. "Influences on Einstein's Early Work in Relativity Theory." *American Scholar* 37 (Winter, 1968): 59-79. This article is geared to the informed layperson. Holton's style is lively, his information compelling and accurate. Relates Einstein's work in relativity to the work of George Francis FitzGerald, Albert A. Michelson, Hendrik Antoon Lorentz, Edward Williams Morley, and others who worked in the field at the same time Einstein formulated his early hypothesis.

Katz, Robert. *An Introduction to the Special Theory of Relativity*. Princeton, N.J.: D. Van Nostrand, 1964. Katz deals specifically and extensively with Einstein's theory of special relativity in terms that nonscientists may find difficult to understand. An excellent, exhaustive book; definitely not for beginners.

Moszkowski, Alexander. *Conversations with Einstein*. New York: Horizon Press, 1970. Moszkowski, an admiring interviewer of Einstein, lacked the background in physics that would have enabled him to ask penetrating questions. Valuable for its personal insights into the man, who is portrayed as a singularly humane person—a concerned genius with a well-developed social conscience. Easily accessible to beginners.

Pais, Abraham. *"Subtle Is the Lord . . .": The Science and Life of Albert Einstein*. New York: Oxford University Press, 1982. A lively book whose extensive mathematical formulas should not interfere with beginners' efforts to understand it.

Well-chosen illustrations, a dual index, and copious endnotes compensate for the lack of a bibliography. Well researched, carefully documented, and accurate.

R. Baird Shuman

Cross-References

Bohr Writes a Trilogy on Atomic and Molecular Structure (1912), p. 507; Rutherford Presents His Theory of the Atom (1912), p. 527; Einstein Completes His Theory of General Relativity (1915), p. 625; Cockcroft and Walton Split the Atom with a Particle Accelerator (1932), p. 978; Anderson Discovers the Positron (1932), p. 983; Gamow and Associates Develop the Big Bang Theory (1948), p. 1309; The Mössbauer Effect Is Used in the Detection of Gravitational Redshifting (1960), p. 1640; Wheeler Names the Phenomenon "Black Holes" (1968), p. 1881.

ANSCHÜTZ-KAEMPFE INSTALLS A GYROCOMPASS ONTO A GERMAN BATTLESHIP

Category of event: Applied science
Time: 1906
Locale: Kiel, Germany

Anschütz-Kaempfe designed and manufactured the first practical gyroscopic compass, a north-seeking device operating independent of the earth's magnetic field and capable of detecting changes in the orientation of a moving vehicle

Principal personages:

HERMANN ANSCHÜTZ-KAEMPFE (1872-1931), a German inventor and manufacturer

JEAN-BERNARD-LÉON FOUCAULT (1819-1868), a French experimental physicist and inventor of the Foucault pendulum and gyroscope

ELMER AMBROSE SPERRY (1860-1930), an American engineer and inventor who was a major contributor to American naval technical progress in World War I

Summary of Event

The invention of the gyrocompass, a major step in the development of modern navigational instrumentation, followed naturally from the revolutionary progress in transportation technology that had taken place during the nineteenth century.

Although complex in mathematical detail, the principle of gyroscopic motion is intuitively easy to grasp and to demonstrate. A spinning body such as a wheel tends to retain its orientation with respect to the two axes perpendicular to the plane of rotation, the force required to change its orientation being proportional to the angular momentum of the spinning body. Furthermore, when a disturbing torque is applied, the spinning body reacts by slowly rotating (precessing) in a direction at right angles to the direction of the torque. Gyroscopic effects were employed in the design of various objects long before the theory behind them was formally known. A classic example is a child's top, which balances, seemingly in defiance of gravity, as long as it continues to spin. Boomerangs and frisbees derive stability and accuracy from the spin imparted by the thrower. Likewise, the accuracy of rifles improved when barrels were manufactured with internal spiral grooves that caused the emerging bullet to spin. A qualitative impression of the characteristics and strength of gyroscopic force can be obtained by causing a detached bicycle wheel to rotate rapidly in the vertical plane, then, while holding it, attempting to tip it sideways and to turn in place. Both motions are surprisingly difficult, although there is no impediment to walking a direction parallel to the spinning wheel.

In 1852, the French inventor Jean-Bernard-Léon Foucault built the first gyroscope, a measuring device consisting of a rapidly spinning wheel within concentric rings that allowed the wheel to move freely about two axes. This device, like the Foucault pendulum, was used to demonstrate the rotation of the earth around its axis, since the spinning wheel, which is not fixed, retains its orientation in space while the earth turns under it. The gyroscope had a related interesting property: As it continued to spin, the force of the earth's rotation caused its axis to precess gradually until it was oriented parallel to the earth's axis, that is, in a north-south direction. It is this property that enables the gyroscope to be used as a compass. It differs from a magnetic compass in two important ways: It operates independently of the earth's magnetic field and any magnetic disturbances in the immediate surroundings, and it designates true north (the earth's axis) rather than magnetic north. Foucault also developed a mathematical analysis of gyroscopic forces. The gyroscope attracted immediate attention as an interesting physical principle and a popular children's toy.

Gyroscopic forces act to some extent in any moving object with spinning parts; with the development of motorized transportation systems, the magnitude of these forces increased dramatically and became important in vehicle engineering. In bicycles and paddle-wheel steamers, the wheels are oriented so as to act as stabilizers. In early automobiles, the flywheel had to be reoriented when it was discovered that the original orientation inhibited steering. (Unlike a train or steamship, an automobile is constantly changing direction, and gyroscopic forces that tend to keep the vehicle moving in a straight line are undesirable.) It became evident that the gyroscopic properties of spinning wheels could be used to advantage in designing vehicles that resisted tilting. In 1906, Otto Schlick patented a design for a gyroscopic stabilizer for steamships, in which the stabilizing properties of a rapidly spinning horizontal wheel were used to reduce the pitch and roll of a ship. From 1908 to 1910, Brennan attracted attention with a prototype monorail balanced by an internal gyroscopic stabilizer.

The ability of a gyroscope to resist deflection also forms the basis for sensors that detect when a moving object changes course. This property formed the basis for a simple German device (patented in 1898) for keeping a torpedo on course. When the torpedo was set in motion, a rotor was activated simultaneously; if the course of the torpedo changed sufficiently, the rotor exerted pressure on the valve controlling the compressed air propellant on that side, creating an asymetrical propulsion force sufficient to turn the torpedo toward its original course. More sophisticated gyroscopic deflection sensors were installed in the World War II era in V-1 and V-2 rockets and in early guided missiles. These sensors communicated via radio to a ground base, enabling a ground-based navigator to plot and correct the course of a missile.

From 1904 to 1905, Hermann Anschütz-Kaempfe, a German manufacturer working in the Kiel shipyards, became interested in navigation problems of submarines used in exploration under the polar ice cap. By 1905, efficient working submarines were a reality, and it was evident to all major naval powers that submarines would play an increasingly important role in naval strategy. Submarine navigation posed problems, however, that could not be solved by instrumentation designed for surface vessels. A submarine needs to orient itself under water in three dimensions; it has no automatic horizon with respect to which it can level itself. Navigation by means of stars or landmarks is impossible when the submarine is submerged. Furthermore, in an enclosed metal hull containing machinery run by electricity, a magnetic compass is worthless. To a lesser extent, increasing use of metal, massive moving parts, and electrical equipment had also rendered the magnetic compass unreliable in conventional surface battleships.

It was logical for Anschütz-Kaempfe to turn to the gyroscopic effect to produce an instrument that would both designate true north and sense changes in the submarine's orientation. Although simple in principle, production of a gyroscopic compass that would function accurately on a moving ship posed significant engineering problems; namely, the device needed to be suspended in such a way that it was free to turn in any direction with as little mechanical resistance as possible (at the same time being strong enough to withstand the inevitable pitching and rolling of a ship at sea), and it needed a continuous power supply to keep the gyro rotating at high speed.

The original Anschütz-Kaempfe gyrocompass consisted of a pair of spinning wheels connected by an electric motor. This apparatus was suspended via a shaft from a float consisting of a hollow ring immersed in mercury, which allowed free rotation around the axis of the shaft and also served as one electrical contact for the motor. Reorientation around this shaft enabled the axis of the spinning gyroscope to orient itself along the earth's axis. The shaft was connected to a compass card visible to the ship's navigator. Motor, gyroscope, and suspension system were enclosed in a set of gimbals that permitted the apparatus to retain its vertical orientation with respect to the earth despite pitch and roll of the ship. Later versions introduced an electromagnet to correct for the latitude and direction of motion of the ship. In 1912, Anschütz-Kaempfe and Max Schuler introduced a model with three rotors mounted in an equilateral triangle, which was less subject to error on a rolling ship and was the principal type used on German submarines during World War I.

In 1906, the German navy installed a prototype of the Anschütz-Kaempfe gyrocompass on the battleship *Undine* and subjected it to exhaustive tests under simulated battle conditions, sailing the ship under forced draft and suddenly reversing the engines, changing the position of heavy turrets and other mechanisms, and firing heavy guns. In conditions under which a magnetic compass would have been worthless, the gyrocompass proved a satisfactory navigational tool, and the results were impressive enough to convince the German navy to undertake installation of gyrocompasses in submarines and heavy battleships, including the battleship *Deutschland*.

Although these developments took place against a background of increasing international tension that culminated in World War I, the Anschütz Gyrocompass Company was a private commercial venture that obtained international patents. No shroud of military secrecy clouded the development of the first gyrocompass, although subsequent refinements both in Germany and in the United States were subject to wartime restrictions on strategic information.

Elmer Ambrose Sperry, a New York inventor intimately associated with pioneer electrical development, was independently working on a design for a gyroscopic compass at about the same time. In 1907, he patented a gyrocompass consisting of a single rotor mounted within two concentric shells, suspended by fine piano wire from a frame mounted on gimbals. The rotor of the Sperry compass operated in a vacuum, which enabled it to rotate more rapidly, and the suspension system eliminated some problems associated with drag and turbulence in the mercury trough used in the suspension system of the Anschütz-Kaempfe compass. The Sperry gyrocompass was in use on larger American battleships and submarines on the eve of World War I.

Impact of Event

Both the German and the American naval authorities recognized that the gyrocompass was a valuable navigational tool and were quick to undertake a program of installing it on major submarines and surface vessels. The ability to navigate submerged submarines was of critical strategic importance in World War I, when submarine warfare first played a major role in war at sea. Initially, the German navy had an advantage both in the number of submarines at its disposal and in their design and maneuverability. The German U-boat fleet declared all-out war on allied shipping, and, although their efforts to blockade England and France were ultimately unsuccessful, the tremendous toll they inflicted helped maintain the German position and prolong the war. To a submarine fleet operating throughout the Atlantic and in the Caribbean, as well as in near-shore European waters, effective long-distance navigation was critical.

Related gyroscopic devices sensing pitch (vertical movement of nose relative to tail), roll, and yaw (horizontal deviations of nose relative to tail) became important in aircraft instrumentation. A World War I-era aircraft pilot had less need of the compass function of a gyroscope than a submarine captain because the pilot could depend on landmarks for orientation, but a device giving rapid feedback on changes in orientation by the aircraft was a distinct advantage in a battle.

Gyrocompasses were standard equipment on submarines and battleships, and, increasingly, on larger commercial vessels during World War I, World War II, and the period between the wars. Although refinements were incorporated that made the compasses more accurate and easier to use, the fundamental design differed little from that invented by Anschütz-Kaempfe. Gyroscopic sensors detecting deviations from course were part of the equipment of the German V-1 and V-2 rockets and guided missiles.

Two additional devices employing gyroscopic principles are the laser gyroscope and the nuclear magnetic resonance gyroscope. The former employs two laser beams traveling in opposite directions within a rapidly spinning mirrored cavity; differences in the apparent velocity of light in the beam traveling in the direction of spin relative to the beam traveling in the opposite direction produce a change in wavelength, which in turn is used to measure changes in rotational angle. A nuclear magnetic resonance gyro functions because individual atomic nuclei spin; this spin produces the magnetic moment of the nucleus, and externally applied magnetic fields act on spinning nuclei in a manner analogous to the effects of gravity and mechanical deflection on a conventional gyroscope.

Conventional gyrocompasses are still extensively used in marine navigation, and gyroscopic deflection sensors are part of the standard instrumentation of modern aircraft, which are typically equipped with three spring-mounted gyroscopic rotors to detect the amount and speed of deflection in three dimensions. One gyroscopic rotor design suspends the spinning element electrostatically, eliminating drag from mechanical suspension elements. The technology pioneered by Anschütz-Kaempfe and Sperry has enjoyed enduring practical application.

Bibliography

Collins, A. Frederick. "The Gyroscope as a Compass." *Scientific American* 96 (April 6, 1907): 294-295. Reports the development of the Anschütz-Kaempfe gyrocompass, with detailed descriptions of the design of the device, illustrations of a preliminary and improved model, and an account of field testing performed by the German navy in 1906 and 1907.

Crabtree, Harold. *An Elementary Treatment of the Theory of Spinning Tops and Gyroscopic Motion*. London: Longmans, Green, 1914. A general reference; gives a good perspective on the increasing importance of gyroscopic motion and devices employing it during the early years of the twentieth century. Includes a simple introduction to the theory of the gyroscope and detailed descriptions of the Anschütz-Kaempfe and Sperry gyrocompasses, the Schick gyroscopic stabilizer, the Brennan monorail, and gyroscopic forces in motor vehicles.

Demel, Richard F. *Mechanics of the Gyroscope*. New York: Macmillan, 1929. An engineering text that explains the theory of the gyroscope and the theory and construction of gyroscopic devices in some detail. Covers both the Anschütz-Kaempfe and Sperry gyrocompasses and the differences between them, with more attention being given to the Sperry model. Extensive mathematical treatment of precession errors.

"The New Navy Gyroscopic Compass." *Scientific American* 106 (June 29, 1912): 588-589. A clear account, with diagrams, of the Sperry gyroscopic compass. Includes a nonmathematical description of the operation of the device and a discussion of the difficulties of using conventional compasses in early twentieth century submarines and battleships. Useful for visualizing the contemporary impact of the discovery of a practical gyroscope.

Parson, N. A. *Guided Missiles.* Cambridge, Mass.: Harvard University Press, 1958. A nontechnical introduction to guided missiles; includes a section on how a gyroscope functions as a steering device in a radio-controlled flying object.

Rawlings, A. L. *The Theory of the Gyroscopic Compass and Its Deviations.* New York: Macmillan, 1944. A textbook designed for the engineers of firms manufacturing gyrocompasses and for shipboard personnel responsible for using and maintaining them. Contains diagrams and descriptions of the Anschütz-Kaempfe gyrocompass and its modifications.

Martha Sherwood-Pike

Cross-References

The Wright Brothers Launch the First Successful Airplane (1903), p. 203; The First German U-Boat Submarine Is Launched (1906), p. 350; Ford Produces Automobiles on a Moving Assembly Line (1913), p. 542; The Fokker Aircraft Are the First Airplanes Equipped with Machine Guns (1915), p. 600; The German *Meteor* Expedition Discovers the Midatlantic Ridge (1925), p. 805; The First Jet Plane Using Whittle's Engine Is Flown (1941), p. 1187; The Germans Use the V-1 Flying Bomb and the V-2 Goes into Production (1944), p. 1235; The *Glomar Challenger* Obtains Thousands of Ocean Floor Samples (1968), p. 1876.

BARKLA DISCOVERS THE CHARACTERISTIC X RAYS OF THE ELEMENTS

Category of event: Physics
Time: 1906
Locale: University College, Liverpool, England

By studying the interaction between X rays and matter, Barkla succeeded in determining important physical characteristics of X rays and the atomic structure of matter

Principal personages:

CHARLES GLOVER BARKLA (1877-1944), an English physicist who was awarded the 1917 Nobel Prize in Physics for his work on X-ray scattering and his discovery of the characteristic Röntgen radiations of the elements

WILHELM CONRAD RÖNTGEN (1845-1923), a German physicist who discovered X rays in 1895 and was a recipient of the first Nobel Prize in Physics in 1901

GEORGE GABRIEL STOKES (1819-1903), an eminent British mathematician and physicist who theorized about the cause and nature of X rays

SIR JOSEPH JOHN THOMSON (1856-1940), a British physicist and teacher who startled scientists by announcing his experimental confirmation that cathode rays consisted of charged particles more than one thousand times lighter than the smallest atom; awarded the 1906 Nobel Prize in Physics

Summary of Event

For several decades in the nineteenth century, physicists studied the cathode rays. For an even longer time, scientists had known of the existence of atoms. Nevertheless, during the last decade of the nineteenth century, scientists still had great difficulty in comprehending the physical nature of either cathode rays or atoms. Apparently, atoms of some chemical elements were heavier, others were lighter. It was not known why. The reason could have been that atoms consisted of different materials, or because the heavier ones had more of the same materials. Chemical facts gave clues about the existence of atoms. Cathode rays appeared to be "tentacles" originating from the atom.

In December, 1895, a sequence of clues and tentacles emerged. First, Wilhelm Conrad Röntgen reported from the University of Würzburg that he had discovered X rays. In 1896, Antoine-Henri Becquerel announced to the French Academy of Sciences that uranium spontaneously emitted invisible radiations, which would blacken a photographic plate yet which seemed different from X rays. In 1898, Pierre Curie and Marie Curie detected two new elements that apparently also emitted the same types of radiations. At the same time, in England, Sir Joseph John Thomson made remarkable progress in the study of the cathode rays. He would conclude from experimental evidence that these rays consisted of charged particles. He declared that these charged particles were "matter in a new state" and that the chemical elements were made up of matter.

Therefore, by the beginning of the twentieth century, scientists were confronted with a number of intriguing questions about the nature of X rays, how one could account for radioactivity, and how one could reconcile the apparent endlessness of radioactive emanations with the conservation of energy. It was in this atmosphere of challenging scientific inquiry that Charles Glover Barkla began his scientific career. From 1899 to 1902, he conducted research with Thomson. In 1902, he attended University College in Liverpool and began his lifelong study of X rays.

The veteran mathematical physicist George Gabriel Stokes proposed the "ether pulse" theory about the nature of X rays. He hypothesized that X rays were irregular electromagnetic pulses created by the irregular accelerations of the cathode rays when they were stopped by the atoms in the target of the X-ray tube. With this theory, Thomson derived a mathematical formula expressing the scattering of X rays by electrons. Barkla's first research project was to test experimentally Stokes's and Thomson's theories.

Five years previously, Georges Sagnac had experimented in France on the absorption of X rays by solids—a phenomenon that was directly related to scattering. Sagnac found that the secondary scattered radiation was of distinctly greater absorbability. Barkla showed that the secondary radiation from light gaseous elements was of the same absorbability as that of the primary beam. He worked on air first, then extended the investigation to hydrogen, carbon dioxide, sulfur dioxide, and hydrogen sulfide. The presence of the secondary radiation was tested by an electroscope, with the assumption that the amount of ionization should be proportional to the intensity of the radiation passing through the instrument. To check the absorbability of the rays—primary and secondary—Barkla used a thin aluminum plate. He published the results in 1903. At the time, the fact that scattering did not modify the absorbability of the radiation appeared to be strong support for the ether pulse theory. From the same set of experiments, Barkla demonstrated that "this scattering is proportional to the mass of the atom." This was a highly satisfying result because it supported the theory that the atoms of different substances are different systems of similar corpuscles, where, in the atom, the number is proportional to its atomic weight. These similar corpuscles, according to most contemporary physicists, was Thomson's "matter in a new state."

In 1904, Barkla began a new series of experiments that would disclose additional physical characteristics of X rays. Because ordinary light, as the propagation of electromagnetic oscillations, is a transversal wave, it can be polarized relatively easily: When it is scattered in the direction at right angles to the incident (primary) beam, the transverse vibrations constituting the light are confined to a plane perpendicular to the primary beam. Barkla was researching the question of whether X rays could be polarized in the same way so that they could be confirmed as electromagnetic waves. This proved to be a serious challenge. It took Barkla two years to perform the difficult experiment and arrive at a clear conclusion that the scattered beam was highly polarized; consequently, X rays were most probably transversal waves, like ordinary light.

While investigating the intensity in different directions of the secondary radiation, Barkla found that light elements, such as carbon, aluminum, and sulfur, showed marked variation in intensity with direction; calcium showed much less. With iron and even heavier elements, there was practically no difference in intensity in different directions. This salient phenomenon led Barkla to investigate more closely the relation between atomic weight and absorbability. The result of his experiments showed that for light elements, the scattered radiation closely resembled the primary radiation, but for elements heavier than calcium, the scattered radiation was quite different from the primary. When Barkla examined the scattered (secondary) radiation more closely, he found that the secondary radiation from metals contained not only scattered radiation of the same character as the primary but also homogeneous radiation that was characteristic of the metallic element itself.

Meanwhile, Barkla also discovered an X-ray phenomenon that was analogous to a discovery made by Stokes: Fluorescent substances fluoresce only when exposed to light of shorter wavelength than that of the fluorescent light emitted by the substance. It is known as "Stokes' law." Barkla found that the emission of the homogeneous (secondary) radiation occurred only when the incident X-ray beam was harder than the characteristic radiation itself. Moreover, Barkla found some revealing facts about the homogeneous characteristic radiations. Beginning with calcium, and moving toward the heavier elements, the characteristic X-ray radiations form one or two series. From calcium (atomic weight 40) to rhodium (atomic weight 103), there appeared a K series; from silver (atomic weight 108) to cerium (atomic weight 140), there appeared a K series and an L series; from tungsten (atomic weight 184) to bismuth (atomic weight 208), there appeared an L series only. K radiations were softer; L radiations harder. The heavier the atom was, the harder its characteristic radiations. Such phenomena, correlating atomic weight to characteristic X rays closely, showed that the latter must have originated from the atom. In fact, Barkla's discoveries anticipated the assignment to each chemical element an atomic number, which, in general, was recognized as about one-half the atomic weight.

Impact of Event

Following these discoveries in 1906, Barkla and other physicists researched interactions between X rays and matter and achieved historic results. These achievements, accomplished from 1909 to 1923, were categorized in three stages: First, X rays interact with crystal lattices. In 1909, Max von Laue attended the University of Munich and was influenced by Röntgen and mineralogists who informed him of theories on the structure of crystal solids. Von Laue proceeded to combine the study of X rays with that of solid structures. He developed a mathematical theory based on the assumption that crystal lattices could serve as "diffraction gratings" (a type of instrument in optical experiments that demonstrates the wave character of light) for X rays. This idea was experimentally confirmed, and the confirmation has been highly praised. It opened vast potentials for studying the nature of X rays and the structure of crystal solids. Shortly after von Laue's publication of his work in 1912, William Henry Bragg and his eldest son, Lawrence Bragg, founded the science of crystallography. In particular, William Henry Bragg created the "ionization spectrometer" for measuring the exact wavelengths of X rays; Lawrence Bragg derived the influential equation now named after him. The Bragg equation tells at what angles X rays will be most efficiently diffracted by a crystal layer.

The second stage of Barkla's achievements was that X rays interact with atoms, especially heavy atoms. Henry Moseley used the Bragg spectrometer soon after its introduction to study the characteristic X rays from the atom. With a new, powerful instrument, Moseley turned to Barkla's line of investigation. Moseley could now measure Barkla's K series and L series with exactness. Significantly extending such measurements, he made wonderful discoveries that have since been called Moseley's law: the mathematical formula that relates the X-ray spectrum of an element to its atomic number. Moseley also made a series of verifiable predictions about the periodic table of elements. Tragically, Moseley was killed in World War I at the Dardanelles. Later, studies in X-ray spectroscopy and its interpretation was accomplished by Karl Manne Georg Siegbahn, winner of the Nobel Prize in 1924.

The third stage is that X rays interact with light atoms, that is, free electrons. When Barkla delivered his Nobel Prize speech in 1920 (for Physics in 1917), he declared that in the phenomena of scattering, there is strong positive evidence against any quantum theory. Three years later, in 1923, Arthur Holly Compton was to prove the folly of Barkla's statement. He followed Barkla in experimenting on the comparison of secondary X rays with primary X rays, especially when the former was scattered from light atoms. Compton experimented with the spectrometer and theoretically with the concept of photons and was awarded the Nobel Prize in 1927.

Bibliography

Allen, H. S. "Charles Glover Barkla, 1877-1944." *Obituary Notices of Fellows of the Royal Society of London* 5 (1947): 341-366. A substantial biography written by a colleague. Contains a complete list of Barkla's publications. Suitable for the nonspecialist.

Forman, Paul. "Charles Glover Barkla." In *Dictionary of Scientific Biography*, edited by C. C. Gillispie. New York: Charles Scribner's Sons, 1970. Though short, this biography is forthright in making critical comments on Barkla's scientific outlook and attitude. This essay is valuable for the nonspecialist because of its sharp and accurate comments.

Heathcote, Niels Hugh de Vaudrey. *Nobel Prize Winners in Physics, 1901-1950*. New York: Henry Schuman, 1953. Although the "physics" included in this book covers fifty years, its subject matter bears close relevance to Barkla's discoveries.

Suitable for students with a physics background.

Millikan, Robert Andrews, and Henry Gordon Gale. *A First Course in Physics.* Boston: Ginn & Co., 1906. This textbook satisfies historical interest. It contains a chapter entitled "Invisible Radiations" and includes a section entitled "Nature of X Rays."

Stephenson, Reginald J. "The Scientific Career of Charles Glover Barkla." *American Journal of Physics* 35 (February, 1967): 141-152. A comprehensive summary of Barkla's scientific career, the most influential part of which is directly related to the discoveries around 1906. The paper discusses critically Barkla's later discovery of the J series. Discusses Barkla's conflict with Compton. Useful for science students.

Stuewer, Roger H. *The Compton Effect: Turning Point in Physics.* New York: Science History Publications, 1975. A solid history of science contribution. Comprehensive and detailed, this book does not avoid formulas and diagrams. Any nonspecialist reader can follow its historical delineation.

Thomson, J. J. *Electricity and Matter.* New York: Charles Scribner's Sons, 1904. Popular lectures delivered in 1903, this book helps general readers understand the physical background of Barkla's discoveries. In 1903, Thomson published the more voluminous and famous *Conduction of Electricity Through Gases*, part of which became part of the history described above.

Wen-yuan Qian

Cross-References

Röntgen Wins the Nobel Prize for the Discovery of X Rays (1901), p. 118; Thomson Wins the Nobel Prize for the Discovery of the Electron (1906), p. 356; Bohr Writes a Trilogy on Atomic and Molecular Structure (1912), p. 507; Rutherford Presents His Theory of the Atom (1912), p. 527.

BATESON AND PUNNETT OBSERVE GENE LINKAGE

Category of event: Biology
Time: 1906
Locale: Cambridge, England

Bateson and Punnett discovered that certain hereditary characters behave as if they are physically linked, adding one of the important pieces to the puzzle of heredity and the gene theory

Principal personages:

WILLIAM BATESON (1861-1926), an English biologist who performed breeding experiments on plants and animals to study inheritance and gave the name "genetics" to this field of study

REGINALD CRUNDALL PUNNETT (1875-1967), an English biologist who collaborated with Bateson in his work with plants and farm animals, discovering the first example of sex-linkage and an explanation of sex determination

Summary of Event

In December of 1903, a six-year partnership was to begin that would lead to the discovery of genetic linkage. William Bateson, a scientist studying heredity, wrote to Reginald Crundall Punnett, a Fellow of Gonville and Caius College, inviting him to join his research team at Grantchester, Cambridge. The team consisted of a small group of volunteers. The "laboratory" was the gardens and paddocks surrounding the Bateson house, along with a plot of land on the University of Cambridge Farm. Bateson offered Punnett eighty pounds, a small sum, but all that could be afforded from the few small grants that Bateson received. The letter illustrated the two extremes of the scientific endeavor: the excitement and significance of the experimental work and the severe financial restrictions that plagued Bateson for most of his career. Punnett eagerly accepted, but refused the eighty pounds, having an income of his own. He was twenty-eight at the time, tolerant and friendly. Bateson was forty-two, forceful and stern. It was a combination that engendered mutual respect.

Bateson and Punnett concentrated their efforts on breeding experiments using sweet peas and poultry. They were searching for answers about laws of inheritance, such as how characters of inheritance are distributed among offspring and if there are predictable patterns. Before they would discover linkage, they had to establish a framework of theory concerning inheritance. Bateson had formulated a number of ideas already. He was convinced that characters are inherited as discrete units and are not "blended" in the offspring, as others were proposing. He called these unit-characters "allelomorphs," a term that survives in shortened form as "alleles," referring to individual forms of a gene. Bateson also gave the name "genetics" for the field of hereditary science.

When Bateson and Punnett started their work, the field of genetics was in its infancy. Built upon two great discoveries—Charles Robert Darwin's theory of evolution and Gregor Johann Mendel's law of segregation—the field was still awaiting the discovery of the material of inheritance. The work that would lead to linkage was based primarily on Bateson's fight to get Mendelism recognized. Mendel had been an Austrian monk, who had pursued scientific studies of peas in the gardens of the monastery. With insight and patience, he discovered that characters of inheritance, such as tall or dwarf, are carried as individual packets of hereditary information. Moreover, information for traits such as tall and dwarf could be carried simultaneously in a single plant and passed on to future generations, even though the plant itself expresses only one of these traits. The expressed trait is said to be "dominant" over the unexpressed, or "recessive" trait.

In the wake of modern molecular genetics, these packets of information can be explained as being "genes." Each trait, such as tall and dwarf, represents a different allelic form of a gene. Genes are carried on chromosomes which are housed in the nucleus of each cell. Most organisms are "diploid," carrying two complete sets of chromosomes, one set contributed by the father and the other by the mother. Alleles, therefore, exist in maternally and paternally inherited pairs. In passing the hereditary material to offspring, the organism will contribute only one allele from each pair. This is the great law of segregation. When gametes (eggs and sperm) are formed, traits (alleles) for a particular character segregate from each other. A gamete thereby receives only half of the hereditary (genetic) material of the parent. At fertilization, two gametes join, combining their genetic material to make a full complement. In this way, organisms avoid doubling their hereditary material during sexual reproduction. Mendel's work revealed this law in the absence of any knowledge of chromosomes.

Bateson's first major victory was bringing Mendel's work to the attention of the English-speaking world. Though published in 1866, the work had been buried in the obscure literature until 1900, when it was rediscovered independently by several scientists. Bateson became aware of it, as told by his wife in her memoirs, as he was riding the train to London to deliver a lecture. He had taken along a copy of Mendel's paper, reading it for the first time. Bateson always meticulously planned his lectures; in this case, however, he was so taken by Mendel's work, that he completely rewrote his lecture in order to incorporate Mendel's studies. Upon his return, he had Mendel's paper translated into English and republished with footnotes, and in the following year, he published a book called *Mendel's Principles of Heredity: A Defense* (1902). This was, in large part, in reaction to a fierce attack on Mendelism launched by a colleague, and once close friend, Walter Weldon.

Ironically, after the victory with Mendelism, Bateson and Punnett, along with E. R. Saunders, a long-time colleague of Bateson, began to accumulate data that were not in strict accordance with Mendelian principles. Mendel had demonstrated independent sorting of characters during gamete formation. If, for example, a tall plant with round seeds was also a carrier of the recessive traits, dwarf size and wrinkled seeds, the plant made gametes that carried any combination of the two traits, tall with wrinkled, dwarf with round, and the like. This independent sorting occurred only because the characters Mendel had chosen happened to be carried on separate chromosomes. It is chromosomes that sort independently during gamete formation. Characters occurring on the same chromosome, being physically linked, move together. It was this linkage behavior that Bateson, Punnett, and Saunders reported in 1906, though their interpretation did not associate heritable characters with chromosomes. In their breeding experiments with sweet peas, for example, they showed that if plants with purple flowers and long pollen grains are crossed with plants having red flowers and round pollen grains, the two characters were always passed together to the offspring. The characteristic of purple flowers was always passed with long pollen grains, red flowers with round pollen grains. Bateson and Punnett called this linkage "gametic coupling." They were unaware that it represented physical linkage of alleles by virtue of their being carried on the same chromosome. In fact, although the chromosome theory of inheritance was being formulated by others at the time, Bateson remained skeptical of it throughout his career.

The Bateson-Punnett collaboration was dominated by Bateson. His genius in experimentation was not in the novelty of his ideas but in his dogged attention to details. "Treasure your exceptions!" he had once said in a lecture at Cambridge. "Exceptions are like the rough brickwork of a growing building which tells that there is more to come and shows where the next construction is to be." It was this attention to exceptions that led to the discovery of linkage. Linkage would become part of the "rough brickwork" in the chromosome theory of inheritance. It is ironic, however, that Bateson was blind to the growing building of which his own "brickwork" was part.

Impact of Event

It is the nature of science that data can outlive the theories to which they have been applied. In this case, Bateson and Punnett's discovery of linkage was used as part of the fabric of incorrect theories. Its connection with the chromosomal theory of inheritance was not immediately understood, although it can be seen now that it lent strong support to the theory. Linkage, in fact, is the manifestation of the physical connection between genes carried on the same chromosome. It helps to verify that the chromosomes are indeed the repository of the hereditary material. At the time, however, few scientists were ready to accept this theory. Even Bateson stated that the idea of particles of chromatin conferring all properties of life surpassed the rational. It was this inability to imagine hereditary material solely as particulate material that prevented the full appreciation of the implications of linkage.

The international community of scientists, in general, was divided on theories of inheritance. In France, for example, there was a strong preference for the Lamarckian view that acquired characteristics could be passed on to offspring. In Germany, scientific sentiment resisted acceptance of the chromosome theory of heredity, preferring the claim that the cytoplasm, which lies outside the nucleus, played the vital role in inheritance. In Denmark, the famous botanist Wilhelm Johannsen, who rejected Lamarckian views and coined the term "gene," was unwilling to consider that the gene might correspond to particulate matter in the chromosome. An American scientist, Thomas Hunt Morgan, was the major proponent of the chromosome theory of inheritance, showing in 1910 that the behavior of chromosomes obeyed the law of segregation and could explain linkage. Yet, Morgan was slow to accept Mendelism. Bateson did not accept Morgan's chromosome theory until 1921, when, after visiting Morgan's laboratory, Bateson gave it tentative acceptance, only to return to his criticism in his last paper in 1926. The certainty of the connection between Mendel's principles, linkage, and the chromosomes would not come until 1953, when James D. Watson and Francis Crick finally broke the genetic code.

Bateson was fully cognizant of the advantages his work could have for agriculture. If characters within a population could be manipulated through planned breeding, then superior hybrids of cash crops could be produced. Bateson worked hard to convince the English government that this type of research should be supported with government funds. He even developed a strain of sugar beet that would not bolt early and a strain of flax that contained especially long fibers. Government policies did shift, and this marked the beginning of the government taking an active role in support of scientific research.

The unraveling of genetic theory also helped to establish the study of eugenics, the manipulation of genes in human populations. Even Bateson, though he never advocated applying eugenics, wondered what biologists would do with the knowledge. He wondered if it could help to produce "healthier, wiser, or more worthy" people. It took the application of eugenics in the horrific experiment of Nazi Germany to still the enthusiasm for eugenics. It was ample warning that the field of genetics—the science named by Bateson—must never be used to manipulate the genes within a human population through selective breeding.

Bibliography

Bateson, Beatrice. *William Bateson, F. R. S. Naturalist*. Cambridge, England: Cambridge University Press, 1928. This is a memoir by William Bateson's wife, written after his death. It contains a number of Bateson's letters and lectures. The letters are compiled with commentary by Beatrice Bateson.

Bateson, W., E. R. Saunders, and R. C. Punnett. "Experimental Studies in the Physiology of Heredity." *Reports to the Evolution Committee of the Royal Society*, Report III (1906): 1-53. This is the original paper announcing the discovery of linkage. Its long columns of data are testimony to the changes that have occurred in science publishing. As of the 1990's, journal articles normally include only summarizing graphs and charts of the original data. Also, a "Miss" or "Mrs." was required to identify any woman author; therefore, the second author is identified as "Miss" E. R. Saunders in this paper. Interesting reading for those who seek information on the history of genetics.

Bowler, Peter J. *The Mendelian Revolution*. Baltimore: The Johns Hopkins University Press, 1989. The discussion on the development of the modern concepts of genetics is geared for the nonspecialist. Several chapters cover Bateson's discoveries. The work is quite sophisticated, and includes an extensive bibliography.

Cock, A. G. "William Bateson, Mendelism and Biometry." *Journal of the History of Biology* 6 (1973): 1-36. The main body of this text is fairly technical, however, a section is appended describing the experimental notebooks of Bateson and Punnett, with photographs of selected pages, including several where Bateson and Punnett had scribbled in the margins wagers they had made on the outcome of certain experiments.

Crowther, J. G. *British Scientists of the Twentieth Century.* London: Routledge & Kegan Paul, 1952. This book includes a very readable, sixty-two-page chapter on Bateson's life. In nontechnical terms, it discusses Bateson's scientific work, including the discovery with Punnett of linkage. A line-drawing of Bateson is included, along with a short bibliography.

Dunn, L. C. *A Short History of Genetics.* New York: McGraw-Hill, 1965. Includes an eleven-page chapter on Bateson's work. Although fairly nontechnical, it is a college-level text, and includes a short glossary and an extensive bibliography.

Kevles, Daniel. "Genetics in the United States and Great Britain, 1890-1930." *Isis* 71 (1980): 441-455. This is an excellent summary of Bateson's professional career and of his influence on scientific thought in England and the United States.

Mayr, Ernst. *The Growth of Biological Thought: Diversity, Evolution, and Inheritance*. Cambridge, Mass.: Harvard University Press, 1982. This is a monumental work on the history of genetics and evolution. Mayr is a world-famous expert in the field, and is able to write thought-provoking prose that is as stimulating to the professional in the field as it is accessible to the average reader. The book includes a full set of references and an excellent index. Considered a classic in biology.

Punnett, Reginald Crundall. "Early Days of Genetics." *Heredity* 4 (April, 1950): 1-10. This was an address delivered at the Hundredth Meeting of The Genetical Society, in Cambridge, on June 30, 1949. It is a delightful account of Punnett's collaborative work with Bateson, with details that reveal the personalities of the two men and the daily routines of their work.

Rothwell, Norman V. *Understanding Genetics.* 4th ed. New York: Oxford University Press, 1988. A well-written college text that discusses genetics in general, allowing the reader to put linkage into context. It includes an extensive glossary, problem sets with answers, numerous diagrams and photographs, and a bibliography after each chapter.

Mary S. Tyler

Cross-References

De Vries and Associates Discover Mendel's Ignored Studies of Inheritance (1900), p. 61; Sutton States That Chromosomes Are Paired and Could Be Carriers of Hereditary Traits (1902), p. 153; Punnett's *Mendelism* Contains a Diagram for Showing

Heredity (1905), p. 270; Hardy and Weinberg Present a Model of Population Genetics (1908), p. 390; Morgan Develops the Gene-Chromosome Theory (1908), p. 407; Johannsen Coins the Terms "Gene," "Genotype," and "Phenotype" (1909), p. 433; Sturtevant Produces the First Chromosome Map (1911), p. 486; Avery, MacLeod, and McCarty Determine That DNA Carries Hereditary Information (1943), p. 1203; Watson and Crick Develop the Double-Helix Model for DNA (1951), p. 1406.

COTTRELL INVENTS THE ELECTROSTATIC PRECIPITATION PROCESS

Category of event: Applied science
Time: 1906
Locale: University of California, Berkeley

Cottrell developed the electrostatic precipitator, which is used to remove particulates and fumes from atmospheric emissions for pollution control

Principal personages:

FREDERICK GARDNER COTTRELL (1877-1948), an American educator and scientist who was the inventor of the electrostatic precipitator

E. S. HELLER, a prominent San Francisco attorney and business partner of Cottrell

EDMOND O'NEILL (1859-1933), a professor of chemistry and business partner of Cottrell

HENRY EAST MILLER, a consulting chemist who, along with Cottrell, Heller, and O'Neill, formed the International Precipitation Co. and the Western Precipitation Co.

WALTER A. SCHMIDT (1883-1962), a chemical engineer who developed the bare wire discharge electrode used in today's electrostatic precipitators

Summary of Event

The advent of the Industrial Revolution led to the development of severe air pollution problems. Prior to this time, there were periods when coal burning was forbidden. These earlier complaints, however, were based mainly on aesthetic reasons. Large-scale industrialization resulted in much more serious environmental problems.

Various methods of pollution control, including the use of "tall stacks" to dissipate the pollutants and baghouses, to contain the particulates by a filtering mechanism very similar to that used in a vacuum cleaner, were available by the early 1900's. Though some success was achieved by these techniques, many of the pollutant streams could not be treated successfully by these methods. It was evident that new techniques were needed.

In 1824, M. Hohlfeld, a mathematics teacher from Leipzig, Germany, first suggested the precipitation of smoke particles by electricity. In the 1880's, Sir Oliver Lodge attempted to commercialize this phenomenon. In tests along the Mersey River, where the fogs periodically disrupted Liverpool shipping, Lodge employed a single electrode process to dissipate these fogs. The resulting electrical discharge was sufficient to clear the air within a radius of 15 meters around the electrode, the moisture coalescing into small flakes of snow. In 1884, however, when Lodge attempted to apply his ideas at the Dee Bank Lead Works in Willington Quay-on-Tyne, Wales, the process failed.

As a youth, Frederick Gardner Cottrell exhibited a wide range of interests. During the early 1890's, when he was in his young teens, he advertised himself as a landscape photographer, printer, electrician, chemist, and telegraph operator. These early enthusiasms, particularly his interest in electricity and electrical apparatus, provided Cottrell with the seed of an idea for one of his most important contributions to science and technology, electrostatic precipitation. The Christmas, 1890, issue of the *Boys' Workshop* (a weekly publication that Cottrell edited) contained an article, "The Progress of Science," which referred to "two very important applications of electricity." The first application was welding and the second was a very brief allusion to "the deposition of smoke and dust by electrical aid."

Cottrell's later interest in air pollution control was sparked in 1905 by the difficulties experienced by the DuPont plant near Pinole, California, which produced explosives and acids. Sulfuric acid was produced by the newly developed "Mannheim process," a contact method in which gaseous sulfur dioxide and oxygen are passed through an iron oxide catalyst to produce sulfur trioxide, which is converted to concentrated sulfuric acid. An undesired by-product was arsenic, which poisoned the catalyst. Cottrell's first solution was the development of a centrifuge, a 76-centimeter-long, 3.8-centimeter-diameter tube, which was rotated at high speed to separate the arsenic from an acid mist. In spite of the success attained in the laboratory, the pilot plant at Pinole was a dismal failure.

Working in his laboratory at the University of California, Berkeley, Cottrell pursued another idea. In early 1906, he applied electricity to the problem of sulfuric acid mists. Cottrell's efforts differed from Lodge's previous attempts in three major ways: First, the use of rectified alternating (AC) current, the introduction of a "pubescent" discharge electrode, and the application of a negative polarity to the discharge electrode. Cottrell's first tests required that one of his students learn to smoke a pipe. The pipe smoke was blown into a glass jar containing a pubescent discharge electrode, a cylindrical wire screen around which were wrapped several turns of asbestos sewing twine. The walls of the jar served as the collecting electrode. Subsequently, Cottrell made a miniature sulfuric acid plant from odds and ends around his laboratory and tested the process on acid mist.

When a high voltage was applied across the electrode and the jar sides, electrons were emitted from the highly charged discharge electrode. These electrons then collided with the surrounding gas molecules, ionizing them. The negative gas molecules migrated to the relatively positive collection electrode and, in doing so, collided with and transferred their negative charge to the entrained smoke or sulfuric acid mist particles. The negative smoke, or mist particles, then migrated to the collection electrode, where they were neutralized and collected at sides and bottom of the jar.

Cottrell soon discovered that to handle large volumes of moving gases, it was necessary to use a direct (not alternating) current (DC). To attain the high voltages necessary for these tests, he transformed the 120 volt lighting system in the chemistry building to 10,000 volts, then commutated the AC current to intermittent DC by a homemade rotating contact maker.

To develop and commercialize the process, Cottrell, along with Henry East Miller, E. S. Heller, and Edmond O'Neill, formed two companies, International Precipitation Co. and Western Precipitation Co. The former company obtained and held all the patents, whereas the latter was the operating unit. Cottrell then returned to the DuPont powder works at Pinole for pilot tests, to redeem his failure with the centrifuge. The new precipitator was able to treat successfully 30-60 cubic meters per minute of gas.

Before negotiations regarding licensing of the process could be concluded with the DuPont plant, Selby Smelting and Lead Co. contacted Cottrell. The smelter was in deep environmental trouble. Gases from smelter operations can contain substances such as sulfur dioxide, sulfur trioxide, and arsenic, zinc, lead, copper, and antimony salts. Since 1905, there had been complaints about odor, decreased grain production, and increased corrosion of window screens and wire fences in the vicinity. The company was taken to court. The Selby management heard of the work at nearby Pinole and were anxious to obtain the process.

The Selby smelter had three separate stacks with three separate problems. Each had to be handled individually. The first stack emitted lead fumes from the lead blast furnace. These emissions were collected successfully by a known process, a baghouse. The second stack emitted sulfuric acid mists from the refinery, where the acid was used to dissolve silver from the gold/silver alloy extracted from the lead. Cottrell developed a 1 meter by 1 meter lead flue, in which several rows of vertical lead plates were suspended. Each was 10 centimeters wide and 1 meter long, and they were placed 10 centimeters apart. Between each was a lead-covered iron bar with pointed projections of mica, which served as the negative discharge electrode. By October, 1907, the precipitator was declared a success. A heavy white plume was converted into an occasional, almost imperceptible, thin white puff. The third stack discharged a 15,240 cubic meters per minute dense white mixture of sulfur dioxide gas, sulfuric acid fumes, and arsenic and lead salts from the roasters. Cottrell and his colleagues spent more than two years attempting to solve this problem with no real success. Eventually, the smelter had to install new roasters.

By 1910, Cottrell discarded the pubescent electrode idea. The Riverside Portland Cement Co., emitting 100 metric tons per day of high-temperature, partially processed cement dust, was under an injunction for destroying the nearby orange groves. In order to handle these dusts, the pubescent electrode was replaced by a bare wire of moderate diameter developed by Walter A. Schmidt, one of Cottrell's former pupils. In 1912, Cottrell founded the Research Corporation, a nonprofit organization that supported basic research at colleges and universities. He assigned his precipitator patents to the corporation as an endowment.

Impact of Event

The electrostatic precipitator had an immediate impact on many industries, not only regarding their emissions but also in a completely unexpected way. Prior to World War I, the United States imported from Germany about 1 million tons of potash annually for fertilizer. In 1915, three years after the Riverside installation was completed, Germany embargoed these exports. Until the U.S. potash industry could be developed, the potash in cement dust proved to be an exceedingly valuable by-product. The dust, collected by precipitator, was processed to extract the potash, which, in turn, was sold to area farmers at a cost which, at that time, exceeded the value of the cement being produced.

Over the years, the value of the by-products in the materials collected has proved to be a major incentive for the use of precipitators. Though other collection devices, such as wet scrubbers, can collect more than 99 percent of the particulates or fumes emitted from a stack, for many applications, a dry process is preferred. In the cement industry, for example, on the average one-third of the raw materials are converted to dust. Without recycling, this would result in losses of approximately 150 metric tons per day per kiln of partially processed materials. Even at the Riverside plant, it is estimated that Cottrell's original precipitator, during its thirty-eight-year lifetime, collected more than 1 million tons of dust. A dry process, such as a precipitator, produces a dust product that is eminently suitable for recycling directly back to the kiln.

The same situation exists for many other industries. Many times, the emissions contain substances that have significant economic value and can be recovered. For example, many years after an older smelting plant in Montana closed, the soil surrounding it was stripped off and processed at a nearby newer Anaconda Smelting and Refining Co. plant. The topsoil was sent through the concentrator and smelted at the new plant with a reported recovery of more than $1 million in copper and other metals. Today, many of the metals and sulfur oxides in smelter emissions can be collected on-site and reprocessed immediately.

The economic impact of precipitators is not restricted to the value of the collected particulates. Reducing the quantity of air emissions not only improves the environment aesthetically but also has had a major economic impact on society. It has been estimated that for every $90 spent on air pollution control, $240 is saved in damages to health and property. Further, if air pollution could be reduced to safe levels in all factories and power plants, it would produce a 75 percent decline in illness and, consequently, save $16 billion in health costs annually.

Precipitators are the "method of choice" in many industries. They are versatile and efficient, often approaching 99.9 percent efficiency. They permit collection of very fine, hot, and/or corrosive particles. Operating and maintenance costs are low, compared to other high-efficiency collectors.

The 1970 Clean Air Act forced many U.S. industries to reduce their air emissions. Nevertheless, even with the increased emphasis on developing new, efficient, and economical techniques of pollution control, there have been no significant innovations in air pollution control methods in the years since Cottrell's work. Iron and steel companies, non-ferrous smelters, foundries, power plants, cement and lime kilns, and pulp and paper mills are merely a few of the industries that often employ electrostatic precipitators for collection of their dusts, fumes, and/or mists.

Bibliography

Cameron, Frank. *Cottrell: Samaritan of Science*. Garden City, N.Y.: Doubleday, 1952. Biography of Cottrell that gives insights into his life, his work, and his personality. Well-researched, interesting reading for all who enjoy science biographies. Written for the nonscientist. Indexed.

Cottrell, Frederick. "Electrical Precipitation: Historical Sketch." *Transactions of the American Institute of Electrical Engineers* 34 (January, 1915): 387-396. Focuses on early history and development of precipitation process, including the work of Hohlfeld and Lodge and on smoke abatement work at the Mellon Institute. Cottrell discusses his preliminary ideas of using precipitation for fog abatement.

Sell, Nancy J. *Industrial Pollution Control: Issues and Techniques.* New York: D. Van Nostrand Reinhold, 1981. Part textbook, part reference book that explains in relatively nontechnical language not only the advantages, limitations, and operating principles of precipitators but also their uses in a variety of industries. Suitable for those with some minimal physical science background. Illustrated, with a bibliography.

Vesilind, P. Aarne, and J. Jeffrey Peirce. *Environmental Engineering*. Ann Arbor, Mich.: Ann Arbor Science, 1982. Textbook designed for those with some physical science or engineering background; includes a good, although short, section on electrostatic precipitator design. Contains an empirical efficiency equation and a useful chart comparing the removal efficiencies of various collection techniques. Describes environmental effects of specific pollutants. Illustrated, with bibliography.

Yost, Edna. *Modern Americans in Science and Technology.* New York: Dodd, Mead, 1962. Cottrell is one of thirteen American scientists discussed. Good summary of the early applications of the precipitation process, the savings attained by dust recovery, and the establishment of Cottrell's Research Corporation. Appropriate for young readers; nontechnical. Indexed.

Nancy J. Sell

Cross-References

Fleming Files a Patent for the First Vacuum Tube (1904), p. 255; Steinmetz Warns of Pollution in *The Future of Electricity* (1908), p. 401; Carson Publishes *Silent Spring* (1962), p. 1740.

FRÉCHET INTRODUCES THE CONCEPT OF ABSTRACT SPACE

Category of event: Mathematics
Time: 1906
Locale: Paris, France

Fréchet allowed points in abstract sets to be lines or curves, rather than geometric points, and defined a notion of distance between such objects, thereby generalizing the notion of distance in Euclidean three-dimensional space

Principal personages:

MAURICE-RENÉ FRÉCHET (1878-1973), a French mathematician who was one of the originators of the mathematical fields of functional analysis and topology

FRIGYES RIESZ (1880-1956), a Hungarian mathematician who refined and extended Fréchet's work in topology and one of the founding fathers of functional analysis and operator theory

JACQUES-SALOMON HADAMARD (1865-1963), a French mathematician who was known for his work in partial differential equations

VITO VOLTERRA (1860-1940), an Italian mathematician whose influential work spanned areas including elasticity, hydrodynamics, differential and integral equations, and the beginnings of functional analysis

CESARE ARZELÀ (1847-1912), an Italian mathematician who originated the notion of sequential compactness, providing Fréchet with motivation for one of his main abstract concepts

Summary of Event

By the late nineteenth century, abstraction and generalization had become significant trends in mathematics research. In his 1906 doctoral thesis, Maurice-René Fréchet considered abstract sets containing elements of a "general nature." Prior to this, sets were regarded as collections of specific objects such as numbers or points in the plane or collections of curves. In addition, Fréchet introduced a generalized notion of distance on his sets, which allowed him to use ideas from geometry in new ways. The combination of an abstract set and a distance on it (that is, an abstract metric space) was the starting point for two major areas of twentieth century mathematics: functional analysis and topology.

The term "space" as used in mathematics usually refers to a set having some geometric structure. (Fréchet continued to use the term *ensemble abstraite*, abstract set, rather than *espace abstraite*, abstract space, until the mid-1920's.) Abstract space can be motivated most easily by considering a sequence of problems. A single algebraic equation in one unknown usually has a solution consisting of a single number. When solving two equations in two unknowns, a solution may consist of a pair of numbers, which is considered as a point in the plane. Similarly, with three equations in three unknowns, a solution will often be three numbers and can be considered as a point in Euclidean three-dimensional space. If there are n equations in n unknowns, it is reasonable to look for n numbers as a solution and call the result a point in n-dimensional space. By the time of Fréchet, a generalization of this problem known as an integral equation was receiving attention. A solution to an integral equation is a function. (A function is a rule that takes input from one set and produces output in a possibly different set. A common example is the function that takes a number and produces the square of the number.) The set of functions that are solutions to integral equations can be thought of as an infinite dimensional space.

Fréchet's ultimate goal was to study functionals; that is, functions defined on spaces of functions. (An example is the functional that takes a curve and produces its length.) His novel approach was to ignore initially the individual aspects of the classes of functions and work instead on abstract sets. His first task was to impose some geometric structure on an abstract set. Geometry is involved because functions can be represented often by lines or curves in the plane or as surfaces in three dimensions. Although the idea of giving a class of functions a "geometrical" structure can be traced back to George Bernhard Riemann's doctoral thesis in 1851, Fréchet's contribution was to use ideas from ordinary three-dimensional geometry as motivation to ask questions about spaces of larger (and even infinite) dimension.

As with most new ideas in mathematics, Fréchet's ideas were motivated by the problems of the day. One such source was the work of Vito Volterra in the calculus of variations. An early problem in this area was to find the curve in the plane having a fixed length that encloses the largest area. The answer was a circle, but the rigorous verification of this (and the solution of related problems) was far from obvious. It was necessary to take an arbitrary closed curve (with a fixed perimeter) and compute the enclosed area. This process described a function: For each curve, produce its area. This is an example of what Volterra called a "function of lines" (now called a functional) and is a typical example of a functional acting on an abstract set.

Fréchet's thesis, *Sur quelques points du calcul fonctionnel* (1906), was written in Paris at the École Normale Supérieure under the direction of Jacques-Salomon Hadamard. The first of two parts consisted of the preliminary steps toward what is now called point-set topology. The goal was to be able to define the limit of a sequence of elements in an abstract set. He did this in three ways of varying generality. His third approach—and the only one that has flourished—was through what Fréchet called an *écart*. This construction is now called a metric, so named by Felix Hausdorff in his famous book on set theory and topology, *Grundzüge der Mengenlehre* (1914). For each pair of points x and y in a set, a metric assigns a nonnegative number to represent the distance between the points. This number (that is, distance) is denoted by d(x,y). Fréchet's notation is simply (x,y). It satisfies the following simple properties: The distance from x to y is the same as that from y to x; d(x,y) is equal to zero if and only if x equals y; and the triangle inequality: The distance from x to y is less than (or equal to) the sum of the distances from x and y to another point z. Using this idea of distance, Fréchet was able to define continuity in abstract sets which enabled him to apply his new ideas to some classical spaces of functions. (A function is continuous if input points that are "close together" produce output that is "close." Geometrical properties of the input are transferred to the output by continuous functions.)

The topological concepts of compactness, completeness, and separability were important theories from Fréchet's thesis. The first of these is especially important for solutions of equations in infinitely many unknowns. If a sequence of elements in a set is converging, the limiting element is guaranteed to exist in the set if the set is known to be compact. This allows equations to be solved by producing a sequence of approximate solutions; the actual solution is then the limit of the approximations. This idea was used long before Fréchet and several conditions were known to guarantee the existence of the limit solution. The works of Cesare Arzelà and Giulio Ascoli on equicontinuity had the most influence on Fréchet's thesis in this regard. Yet, it was Fréchet who was able to isolate the essential ideas and unite them in an abstract framework. Fréchet also applied the abstract concepts of compactness, completeness, and separability to concrete examples of spaces of functions.

Another important mathematician who was influential in creating the notion of abstract space was Frigyes Riesz. His 1907 paper on the origins of the concepts of space was in large measure a quasi-philosophical paper on a mathematical model for the geometry of space as needed in physics. Riesz, however, introduced the concepts of derived set, neighborhood, and connectedness which came to be standard ideas in Hausdorff's topology. Fréchet's work was an important stimulus for future work of Riesz in his development of the notion of the function space consisting of square integrable functions and a metric on this space of functions. Both Fréchet and Reisz published theorems providing concrete representations of certain kinds of abstract linear functionals. This early work of Riesz and Fréchet, motivated by Hadamard, was the beginning of a fruitful area of modern mathematics known as functional analysis.

Impact of Event

The immediate effect of Fréchet's thesis was to initiate the study of point-set topology and provide tools for early work in functional analysis. His definition of metric space is the one used today, and the study of metric spaces is still ongoing. Not all of the concepts that Fréchet defined were sufficiently powerful to solve the relevant problems of the day; they needed further refinement by others. For example, his notion of compactness was not powerful enough to treat convergence problems outside of metric spaces and other mathematicians (for example, Hausdorff) needed to generalize the concept. It was also found that basing the concept of convergence on sequences was too restrictive and the idea of neighborhood proved more fruitful. (Fréchet, Riesz, Hausdorff, and others made advances in this direction between

1906 and 1920.) Rather than being remembered for his theorems and their applications, it is instead the objects that Fréchet defined and his approach to abstract space for which he is best known. Although he continued to work in the field of topology and analysis until about 1930 (at which time he moved into the study of probability), Fréchet's thesis was his most influential work.

The abstract, axiomatic approach to mathematics that is evident in Fréchet's thesis continued and reinforced a growing tendency in mathematics. Axiomatics refers to developing mathematics by precise statements of axioms and definitions followed by the logical consequences of the axioms—careful statements and proofs of theorems. Classical Euclidean geometry has been presented axiomatically for centuries, but the rest of mathematics had not begun to develop this way until the later part of the nineteenth century. Beginning in earnest with the work of David Hilbert at the beginning of the twentieth century, mathematics has developed axiomatically. The axiomatic aspect of Fréchet's work was especially important for the future of functional analysis. This area of mathematics, originating in the early 1900's, is characterized by applying the abstract approaches of algebra and topology to problems in classical mathematics, such as differential and integral equations, as well as to parts of physics such as quantum mechanics. It is important for the development of functional analysis to have an abstract view of the problems so as to extract their essential and common elements. It is precisely this point of view that Fréchet helped popularize.

Bibliography

Dieudonné, Jean. *History of Functional Analysis.* New York: North-Holland, 1981. The author is one of the important mathematicians of the twentieth century. Several of the later chapters in this book contain technical material, but the other chapters can still be profitably read, occasionally skipping over some technical details.

Fréchet, Maurice. "From Three-Dimensional Space to the Abstract Spaces." In *Great Currents of Mathematical Thought*, edited by François Le Lionnais. New York: Dover, 1971. This is one of the few papers of Fréchet in English. It provides basic background and motivation for considering abstract space. The article also includes a discussion of metric spaces, topological vector spaces, and applications to functional analysis.

Kline, Morris. *Mathematical Thought from Ancient to Modern Times.* New York: Oxford University Press, 1972. This monumental, well-written book contains two chapters relevant to the work of Fréchet. Chapter 46 gives background and an outline of functional analysis; chapter 50 treats the beginnings of topology. The author is a master of exposition and has written so that both the general reader as well as the professional mathematician can profit.

Taylor, Angus E. "A Study of Maurice Fréchet: I. His Early Work on Point Set Theory and the Theory of Functionals." *Archive for the History of Exact Sciences* 27 (1982): 233-295. This article contains a wealth of information about the life and work of Fréchet as well as the mathematicians of his time. There are some technical passages, but the casual reader may still glean much information about Fréchet and the significance of his work. Two other articles from the same journal, 34 (1985): 270-380 and 37 (1987): 25-76, give a well-rounded view of the work of Fréchet up through 1930.

Temple, George. *One Hundred Years of Mathematics: A Personal Viewpoint*. New York: Springer-Verlag, 1981. The first few sections of chapter 9 discuss the origins of topology and some later sections of the chapter treat the foundations of functional analysis. More technical than biographical, the book is still accessible to the interested nonspecialist.

Michael McAsey

Cross-References

Hilbert Develops a Model for Euclidean Geometry in Arithmetic (1898), p. 31; Mandelbrot Develops Non-Euclidean Fractal Measures (1967), p. 1845.

HOPKINS SUGGESTS THAT FOOD CONTAINS VITAMINS ESSENTIAL TO LIFE

Category of event: Biology
Time: 1906
Locale: Cambridge, England

Hopkins suggested that trace ingredients other than carbohydrates, proteins, and fat are essential to the diet in humans, leading to the concept of vitamins

Principal personages:

SIR FREDERICK HOPKINS (1861-1947), an English biochemist who was the father of biochemistry in England and a cowinner of the 1929 Nobel Prize in Physiology or Medicine

CHRISTIAAN EIJKMAN (1858-1930), a Dutch physician who solved the problem of beriberi, a vitamin deficiency, and shared the 1929 Nobel Prize with Hopkins

CASIMIR FUNK (1884-1967), a Polish biochemist who coined the term "vitamine" after isolating the vitamin nicotinic acid

Summary of Event

Sir Frederick Hopkins did pioneering work in the field of diet and nutrition that led to the discovery of vitamins. Vitamins are organic molecules which are needed in the body in minute quantities: thus, they are called micronutrients. Vitamins play a vital role in the biochemical reactions taking place in the cells of the body. They aid enzymes, which catalyze, or speed up, biochemical reactions. They are essential to the diet of humans.

Thirteen vitamins are known to be required in the diet of human beings for normal growth and development. Vitamins are divided into two classes: water-soluble and fat-soluble. The water-soluble vitamins include vitamin B (thiamine, riboflavin, pantothenic acid) and vitamin C (or ascorbic acid). The fat-soluble vitamins include vitamins A, D, E, and K, which are not soluble in water.

Hopkins was an English biochemist at the forefront of biochemical thought and the combining of chemistry and metabolism of the body. His research brought together these two diverse areas. Hopkins was trained early as an analytical chemist. He entered biochemistry relatively late in his career but in the infancy of the biochemistry field. Hopkins used his analytical chemistry skills and had an enormous impact on the field of experimental biochemistry. Hopkins was invited to the University of Cambridge in September, 1898. By 1914, he was appointed as the first professor of biochemistry at Cambridge. His early research was in uric acid. Hopkins' ability as an analytical chemist allowed him to develop and improve the method for uric acid determination in the urine. His interest in diet and uric acid excretion led to his interest in proteins. In the early 1900's, Hopkins and Sydney W. Cole, a student, described in a series of articles the isolation of the first essential amino acid, tryptophan. After several more papers on proteins were published, this investigation led him to studies on diet and protein.

Hopkins then began studying diet and metabolic function. In his 1906 address to the Society of Public Analysts titled "The Analyst and the Medical Man," Hopkins alluded to some trace substance that might be required in the diet in addition to proteins, fats, and carbohydrates. Hopkins stated: " . . . no animal can live upon a mixture of pure protein, fat, and carbohydrate, and even when the necessary inorganic material is carefully supplied, the animal still cannot flourish." Hopkins believed that rickets and scurvy may be caused by this missing dietary factor. Christiaan Eijkman had done pioneering work on beriberi, and this was interpreted in the same manner. In 1912, Hopkins published a paper in the *Journal of Physiology* entitled "Feeding Experiments Illustrating the Importance of Accessory Factors in Normal Dietaries." These experiments showed that animals cannot grow when fed a synthetic diet of pure proteins, fats, carbohydrates, and salts. Hopkins added milk in minute quantities to the diet of the animals. This enabled the animals to utilize the synthetic food mixture for growth. Hopkins concluded that some substance in trace quantities was present in normal foodstuffs and only small quantities were required for growth. Looking back at his research in a 1922 Chandler Medal Address at Columbia University, Hopkins talked about his pioneering work on what are now called vitamins. He at first expressed disbelief that in a synthetic diet of pure proteins, fats, carbohydrates, and salts, there was some other substance required for growth of the animal. "So much careful scientific work upon nutrition had been carried on for half a century and more—how could fundamentals have been missed?" Hopkins realized that the pure foodstuffs might not be as pure as scientists had thought.

Even after Hopkins published his results, vitamins were not universally accepted. Other researchers were utilizing purified substances and found that rats could grow quite well on this diet. The synthetic diets, however, may not have been quite as pure.

Hopkins' pioneering research laid the groundwork for other researchers to study the vitamin hypothesis. Research in the United States during World War I by Thomas Osborne and Lafayette Benedict Mendel, and E. V. McCollum and his colleagues led to the distinction of water-soluble and fat-soluble vitamins. By 1920, there was still skepticism about the vitamin hypothesis. By 1923, even the staunchest disbelievers had come around finally to the vitamin hypothesis. Eventually, in 1929, Hopkins and Eijkman were awarded the Nobel Prize in Physiology or Medicine for their work on vitamins and the vitamin hypothesis.

A large amount of Hopkins' work on diet and metabolic function was initiated because of the outbreak of World War I. During the war years, Hopkins served on the Royal Society Food Committee. Food rationing and nutrition were his primary concern. Butter was a scarce and an expensive food commodity, while margarine was cheap and readily available. There was considerable discussion about the nutritional quality of margarine compared to butter. This discussion evolved because of

Hopkins' work on accessory food factors (or vitamins). In 1917, Hopkins initiated work on behalf of the margarine industry to study the nutritional quality of margarine. His results showed clearly that margarine was nutritionally inferior to butter because it lacked vitamin A (later, it became clear that margarine lacked both vitamin A and D).

After the war, Hopkins acted as a consultant to the margarine industry. Research continued to investigate the possibilities of introducing vitamins to margarine. J. C. Drummond surveyed natural sources for vitamin A and D. By 1926-1927, vitamins A and D were introduced into margarine as a commercial product. In 1928, they received certified approval from the Pharmaceutical Society. Vitamin-enriched margarine is now a standard in the supermarket and is most likely little or not at all inferior to butter in caloric value and vitamin quality. Hopkins then turned to other areas of research. By then, biochemistry had begun to be taught at the entry level in college throughout the English university system. In 1925, Hopkins was knighted and was awarded the Copley Medal of the Royal Society in 1926. His most impressive distinction, however, came in 1929 when he shared the Nobel Prize with Eijkman.

Hopkins was actively conducting research in many other fields of biochemistry and physiology. Research with Walter Fletcher showed that muscle activity and lactic acid production are correlated with each other. Hopkins' later studies were on sulfur-containing vitamins, and he discovered a tripeptide (three amino acids linked together). He published a series of papers on the structure and function of the molecule. Hopkins' impact on the field of biochemistry was far-reaching.

Impact of Event

The existence of vitamins has been known since the time of the ancient Greeks. Aristotle knew that raw liver contained an ingredient that could cure night blindness, a deficiency of vitamin A. An early account on the cure of scurvy, a vitamin C deficiency, was recorded in 1535 during the voyages of Jacques Cartier to Newfoundland. Captain Cook was also aware of the antiscurvy properties of lime juice in the 1700's.

Eijkman entered the field of vitamin research in 1886 working in the Dutch East Indies. He studied the problem of beriberi, a deficiency of the B vitamin, thiamine. In 1896, a disease broke out among chickens being used for research. Eijkman, trained in microbiology, tried to find the germ responsible for the outbreak but was unsuccessful. He traced the symptoms finally to a diet of highly polished rice that the chickens had been eating. He switched their diet to whole rice or the highly polished rice and the husks and the symptoms cleared up. Thus, he was the first to notice a dietary deficiency disease. Eijkman, however, did not understand the significance of his findings; it was through the work of Hopkins that the existence of vitamins became established.

After Hopkins' address in 1906 about the existence of vitamins, many researchers worked to isolate these trace substances. In 1912, Casimir Funk, a Polish biochemist, purified a substance he thought cured beriberi. He coined the term vitamine (vital amine). It turned out that he actually isolated nicotinic acid (niacin), a cure for pellagra, which was contaminated with thiamine. In later years, the "e" was dropped from vitamine.

After years of dietary research, scientists are now quite certain that all the required vitamins have been identified for humans. Much of the research after Hopkins was concerned with identifying their structures and functions within the cell. Other areas of research are concerned with vitamin requirements of humans. The requirements of individuals vary depending on the diet of the individual. Nevertheless, there are recommended daily dosages for adequate growth. Vitamin dosage has aroused much debate. Vitamin deficiencies still occur regularly in the United States. The most common deficiencies are of vitamin B (thiamine, niacin, riboflavin, and folic acid) and vitamin C (ascorbic acid). These deficiencies are usually caused by an inadequate diet that lacks the essential vitamin. Alcoholics usually have a thiamine deficiency because they obtain a large amount of their calories from alcoholic beverages. The most common worldwide deficiency is of folic acid, which affects the absorption of nutrients from the small intestine. These vitamins are water-soluble and must be obtained in the diet on a regular basis. Excess amounts of the water-soluble vitamins are excreted in the urine and are not stored in the body.

Deficiencies of the fat-soluble vitamins (A, D, E, K) are rare in the United States. These vitamins may be stored in the body and may give an individual an adequate supply for up to six months. People with intestinal or pancreatic ailments which hinders fat absorption may show deficiencies in the fat-soluble vitamins. Another problem associated with fat-soluble vitamins is that large doses are toxic to humans because they are not readily excreted in the urine.

The knowledge of vitamins and diet has increased tremendously during the twentieth century. Hopkins' work has greatly aided an understanding of diet and nutrition, not only with his vitamin hypothesis but also with his work with essential amino acids.

Bibliography

Asimov, Isaac. "Sir Frederick Gowland Hopkins." In *Asimov's Biographical Encyclopedia of Science and Technology.* 2d ed. Garden City, N.Y.: Doubleday, 1982. This book contains the biography of 1,510 great scientists from the ancient times to the present in chronological order. It is an easy book and the biographies are short. A nice feature is that some of the biographies are cross-referenced for major advances in the field.

Baldwin, Ernest. "Frederick Gowland Hopkins." In *Dictionary of Scientific Biography,* edited by Charles Coulston Gillispie. Vol. 6. New York: Charles Scribner's Sons, 1972. A good reference volume on famous scientists. The biographies are well written and in much greater detail than Asimov's book (cited above). They also have references at the end of each biography. For the nontechnical reader.

Lehninger, Albert L. "Human Nutrition." In *Principles of Biochemistry.* New York:

Worth, 1982. A well-written chapter in an elementary biochemistry college textbook. Written for the beginner who has little understanding of biochemistry. Includes references at the end of the chapter.

Needham, Joseph, and Ernest Baldwin, eds. *Hopkins and Biochemistry.* Cambridge, England: W. Heffer and Sons, 1949. A book containing Hopkins' uncompleted autobiography, excerpts from a paper with commentaries, and selected addresses. This interesting collection is for the nontechnical reader. Other sections can be detailed for the beginner. The book gives one a feel for the accomplishments of Hopkins.

Williams, Trevor I., ed. "Sir Frederick Gowland Hopkins." In *A Biographical Dictionary of Scientists.* 3d ed. New York: John Wiley & Sons, 1982. For the interested layperson. The biographies are short and easy to understand but do not contain cross-references to other biographies. Includes references at the end of each biography.

Lonnie J. Guralnick

Cross-References

Hopkins Discovers Tryptophan, an Essential Amino Acid (1900), p. 46; Grijns Proposes That Beriberi Is Caused by a Nutritional Deficiency (1901), p. 103; McCollum Names Vitamin D and Pioneers Its Use Against Rickets (1922), p. 725; Steenbock Discovers That Sunlight Increases Vitamin D in Food (1924), p. 771; Szent-Györgyi Discovers Vitamin C (1928), p. 857.

MARKOV DISCOVERS THE THEORY OF LINKED PROBABILITIES

Category of event: Mathematics
Time: 1906
Locale: St. Petersburg (Leningrad), Soviet Union

Markov's formal development of linked probabilities (Markov chains) provided new computational models for a wide variety of random processes

Principal personage:
ANDREY ANDREYEVICH MARKOV (1856-1922), a Soviet mathematician

Summary of Event

Before Pafnuty Chebyshev, Andrey Andreyevich Markov, and A. N. Kolmogorov, a number of oblique working definitions of key statistical properties had been offered, mainly by natural scientists applying statistical methods. These definitions, however, frequently did not coincide or in some cases were even contradictory. This was also true for basic statistical methods such as the use of histograms in count data analysis. A histogram is constructed by dividing the horizontal axis into segments or classes of a certain data value (such as size, temperature, and the like), which cover the entire range of data values. Over each segment, a rectangle is constructed whose height is proportional to the number of data points in each class segment. This histogram shows the relative class frequencies, which in the limits of smaller subdivisions and larger numbers of samples approaches the underlying probability density distribution for the population. A key example of functional, yet vague, statistical concepts awaiting further development was James Bernouli's so-called law of large numbers. Formally stated, Bernouli's theorem asserts that frequencies of occurrence for independent chance events must converge eventually in the limit of large numbers of observation to the underlying probability law. In other words, this theorem stated basically that the experimental histogram of samples from a statistical population would match the theoretical probability distribution function more closely, defining a given population as the number of selected samples increases without bound.

Even by the late 1880's, satisfactory proofs under general assumptions had not been found for the large number law, nor had its limits of applicability been well determined. This was particularly true for the case of events whose statistical independence or relation was not known or determinable beforehand. Thus, for example, the German statistician Wilhelm Lexis, although arguing for the conceptual modeling of statistical interpretation of natural events on the results of a notional urn drawing, did not believe that evolutionary or otherwise linked or connected series were susceptible to formulation in terms of quantitative probability theory. As numerous investigators pointed out near the beginning of the twentieth century, in its original (chiefly heuristic) form, the large number law gave nothing useful about the precise manner or function form in which statistical averages approach their limit. Many statisticians believed that the law of large numbers is true only for statistically independent events or trials. Nevertheless, without a more rigorous law of large numbers, probability theory would lose its common intuitive foundations and so the issue could not be dismissed simply.

In contrast to the German and French schools of statistics, Soviet statisticians (almost exclusively at the University of St. Petersburg) were strongly theoretical in focus, more interested in formal existence and consistency conditions than in applications of statistical laws. It was precisely this group which, from their own perspective, addressed the problems of consistently defining the interpretation and applications scope of the law of large numbers. In 1867, Chebyshev found the first elementary proof of the law of large numbers (LLN) as well as the "central limit theorem" (CLT). As an advanced graduate student of Chebyshev, Markov in 1884 published a shorter and more perspicious reformulation of Chebyshev's proofs, which began a series of exchanges and further clarifications of the basic concepts underlying, and scope of application, of the large number and central limit theorems to linked, as well as totally independent, events. From a purely theoretical perspective, Markov approached these questions indirectly by considering whether the LLN applied equally to dependent as well as to independent random variables. Then, using the Markov model of (weaked/linked/conditional) dependence to verify the LLN, he examined whether sums of dependent variables satisfy the CLT.

Specifically, in Markov's subsequent work, he focused closely on the theoretical laws governing what is in effect the convergence of empirical histograms to theoretical "probability distributions functions" (pdf's) for a variety of conditions of weak statistical dependency or sample interlinking, as well as independence. The general classes of weakly linked statistical events considered originally by Markov can be considered a generic mathematical model of a random process with only specifically delimited after-effects or with a very short-term memory. This model describes any physical or other system in which the probability of change or transition from one state to another state depends only on the state of the system at the present or immediately preceding time and not on the longer prior history of the process.

A perennial problem by the end of the nineteenth century, related to the issue of independent versus dependent statistical data, was that of estimating the statistical variance of time or space averages of ocean, weather, geologic, or other statistically measured data that could not, strictly speaking, be assumed as totally independent random or unconnected by virtue of their (causal) adjacency. One of many practical examples of events that both common-sense and science describe as linked are meteorologic observations. In 1852, French mathematician Adolphe Quetelet proposed what is probably the first probabilistic model to fit weather observations of consecutive rainy (or cold) and dry (or warm) days. To account for the frequent persistence, or temporal linkedness of rainy (or dry) weather, Quetelet devised for the occasion a simple mathematical formula that explicitly captures an apparently random element but that also exhibits the influence or effect of one or more previous elements on subsequent events. Neither Quetelet nor other largely empirical science-oriented statisticians of the era pursued further the more general possibilities of accounting for linked events or processes.

The formal properties, and applications potential, of a general method for describing events with linked probabilities were stumbled upon by Markov at the beginning of the twentieth century. In his 1906 paper, "Extension of the Law of Large Numbers," Markov proved for the first time that both the number of occurrences of a studied event and the sequence of its associated random variables obey the law of large numbers. Assuming a simple one-link only (first-order chaining) of events, and an unconditional normalized total probability for the event of p, the m-th order transition probabilities, or probability of a change of binary-determined states between temporally-adjacent events, was first derived by Markov to be given by the relation $R_m = p + (1\text{-}p)\ (p_1\text{-}p_2)^m$. This relation can be interpreted as linking the general conditions under which a system of equations has a unique solution with the specific parameters defining the chaining of events. As stated in his 1907 paper, "Extension of the Limit Theorems of Probability Theory to a Sum of Variables Connected in a Chain," these weakly dependent, or linked, sequences are definable as "numbers connected into a chain in such a manner that, when the value of one of them becomes known, subsequent numbers become independent of the preceding ones." Likewise, Markov, in publications in 1908 and 1910 to 1912, considered more complex (higher-order) linkings or chainings of events whose probabilities of occurrence depend upon the outcomes of two or more prior trials. In modern probability theory, a sequence is considered to be a Markov chain if the pdf governing its underlying process can be said to have connection, link, or memory extending to one or more prior events. Alternately expressed, in the Markov model, given the present, the future is independent of the past. More generally, random processes having a dependence between successive terms as an intrinsic property of the underlying process are defined as Markov processes.

The classical example of a (multiple) Markov chain, as a semiserious test of the efficiency of information-transfer originally given by Markov himself, is that of the written (Soviet) language in Alexander Pushkin's novel *Evgeny Onegin* (1825-1833; *Eugene Onegin*, 1881). Here, letters of the alphabet are subdivided into type states, denoted by 0 if a vowel and 1 if a consonant, so that a page of written text appears as a sequence of the occurrence of 0's and 1's. The vowels and consonants form a first-order Markov chain if, given any string of letters, the probability for the next letter to be a vowel or consonant (0 or 1) is the same as the probability that the next letter will be a vowel or consonant if one knows only the last letter of the entire story.

Although there is no direct evidence that the St. Petersburg school's original work to theoretically validate probability theorems to the case of dependent variables was motivated by prior publications by Lexis or his contemporaries between 1907 and 1911, Markov and a colleague repeatedly referred to several examples and applied problems from their publications. In addition, Markov and his colleagues generally indicated several other examples of physical phenomena exhibiting linked probabilistic behavior. These included the theories of molecular random ("Brownian") motion of Albert Einstein and Paul Langevin, biologic theories of the extinction of genetic families, and the so-called random walk problem. (The random walk, paradigmatic for many probability applications, is defined generically by the motion along a straight line which, at each unit time, can move one unit left or right or not at all, whose probabilities depend only on position.)

Impact of Event

Almost since their discovery, Markov chains have proven to be a powerful method for modeling a wide variety of physical, chemical, and biologic phenomena involving time-dependent/transient and long-run/steady-state behaviors. As early as 1908, A. K. Erlang carried out studies of the steady-state behavior of commercial telephone exchange traffic (theory of queues), deriving what is now known as the Kolmogorov equations for a finite Markov process. In addition to stock exchange speculation, and particularly telephone, mail, and road traffic, similar Markov queue models have been applied widely to landing of aircraft and ships, assembly-line component breakdown, scheduling and checkout for clinics and supermarkets, and inventory maintenance. In the first three decades of the twentieth century, problems of quantitative epidemic spreading and population growth using a second order Markov model were considered. In addition to the theory of elementary particle collisions and cascades, the theory of statistical mechanics describing the physics of molecules—an important Markov chain application—is the so-called nearest neighbor system used, for example, as models for crystal lattices; likewise, in chemistry for treating chemical kinetics and diffusion-controlled reactions. During the past decades, following the work of Ludwig von Mises, a number of researchers in operations research have combined the methodologies of Markov models and Bayesian decision analysis to facilitate quantitative solution of complex problems in economic and military equipment procurement and deployment.

As exhaustively detailed in Albert T. Bharucha-Reid's advanced monograph, thousands of papers have been published on specialized applications of temporal and spatial Markov series. Theoretically, the mathematical methods initiated by Markov were extended later and formalized by Aleksandr Khinchiny, Norbert Wiener, and notably Kolmogorov, establishing probability theory as an identifiable and rigorously founded subdiscipline.

Bibliography

Bartos, Otomar J. *Simple Models of Group Behavior.* New York: Columbia University Press, 1967. Examines several applications of Markov process models to predicting the persistence of specific behaviors in small group settings studied by sociology.

Bharucha-Reid, Albert T. *Elements of the Theory of Markov Processes and Their Applications.* New York: McGraw-Hill, 1960. Offers an encyclopedia of basic and advanced Markov theory and applications to physical, chemical, astronomical, biological, engineering, and other sciences.

Damerau, Frederick J. *Markov Models and Linguistic Theory.* The Hague: Mouton, 1971. An important theoretical study concerning the degrees of approximation to the English language by Markov models of progressively higher order.

Gani, Joseph M., ed. *The Craft of Probabilistic Modeling*. New York: Springer-Verlag, 1986. Includes sufficient theory and applications. Remains within the scope of general undergraduate readers. Motivates the developments in a conceptual discussion of probability theory as a whole.

Gigerenzer, Gerd, et al. *The Empire of Chance: How Probability Theory Changed Science and Everyday Life*. New York: Cambridge University Press, 1989. Extensive discussions of empirical probability rules and applications in the natural sciences throughout the nineteenth and twentieth centuries.

Gray, Robert M., and Lee D. Davisson. *Random Processes: A Mathematical Approach for Engineers.* Englewood Cliffs, N.J.: Prentice-Hall, 1986. Provides a clear and concise introduction for discrete random processes.

Gerardo G. Tango

Cross-References

Lebesgue Develops a New Integration Theory (1899), p. 36; Fréchet Introduces the Concept of Abstract Space (1906), p. 325; Mises Develops the Frequency Theory of Probability (1919), p. 664.

OLDHAM AND MOHOROVIČIĆ DETERMINE THE STRUCTURE OF THE EARTH'S INTERIOR

Category of event: Earth science
Time: 1906 and 1910
Locale: England and Croatia (now Yugoslavia)

Oldham, Mohorovičić, and other seismologists, using data from earthquakes, revealed the layered internal structure of the earth

Principal personages:

RICHARD DIXON OLDHAM (1858-1936), the Irish seismologist who elucidated the triple-phase motion of earthquakes and showed that the earth has a core

ERNST VON REBEUR-PASCHWITZ (1861-1895), a German inventor of the first accurate horizontal-pendulum seismograph and the first to detect earthquake vibrations that had passed through the body of the earth

SIMÉON-DENIS POISSON (1781-1840), a French mathematician known for his work in the development of statistics

BENO GUTENBERG (1889-1960), the German-American seismologist who determined the mantle/core boundary

ANDRIJA MOHOROVIČIĆ (1857-1936), a Croatian (Yugoslav) meteorologist who achieved fame for his seismological studies and showed the discontinuity, which now bears his name, between the earth's mantle and crust

INGE LEHMANN (1888-?), a Danish seismologist who hypothesized, in 1936, the existence of the inner core

Summary of Event

The deep interior of the earth is as inaccessible to humans as the next galaxy and has provided a convenient location for religions to place the fire and brimstone of hell, for science-fiction writers to send intrepid heroes on voyages of discovery, and for crackpots to cite as the home of advanced civilizations and the source of UFOs. The scientific picture of the subterranean structure must rely upon indirect evidence; the picture has thus changed over time as new evidence became available. Only in the twentieth century, with the development of new measurement techniques, has a clear picture begun to emerge.

The preeminent geological text of the latter half of the seventeenth century, Athanasius Kircher's *Mundus subterraneus* (1664), described the earth with a fiery central core from which emerged a web of channels that carried molten material into fire chambers or "glory holes" throughout the "bowels" of the earth. Some details of Kircher's picture were speculative, but the basic idea of a hot, liquid interior was a natural inference from observations of volcanic eruptions and of igneous rocks at the surface that appeared to have been formed by melting and recrystallization at high temperatures. The "contraction" theory, popular through the first half of the nineteenth century, held that the earth had formed from material from the sun and had since been slowly cooling, creating a solid crust. As the fluid cooled, it would contract and the crust would collapse around the now smaller core, causing earthquakes and producing wrinkles that formed the surface features of mountains and valleys. Given what little data scientists had, this was a plausible theory. Physicists in the 1860's, however, pointed out several physical consequences of such a model that appeared to be violated. For example, the gravitational force of the moon, which creates tides in the ocean, should produce tides also in a fluid interior that should either cause the postulated thin crust to crack and produce earthquakes whenever the moon was overhead or, if the crust were sufficiently elastic, cause tides in the crust as well. Instruments were developed to detect crustal tides, but the results were all negative. By the beginning of the twentieth century, geologists were forced to abandon the contraction theory and conclude that the earth probably was completely solid and "as rigid as steel." Thus, when Alfred Wegener suggested the idea of continental drift in 1912, it was dismissed on this basis as being physically impossible. Ironically, the very measuring instruments that overturned the simple liquid interior view of the contraction theory would, in turn, disprove the solid earth view that had challenged and replaced it; seismographs—such as the inverted pendulum seismograph invented in 1900 by Emil Wiechert—would provide a completely new source of information about the internal structure of the earth.

In 1900, Richard Dixon Oldham published a paper entitled "On the Propagation of Earthquake Motion to Great Distances" in *Philosophical Transactions of the Royal Society of London*, which established that earthquakes give rise to three separate forms of wave motion that travel through the earth at different rates and along different paths. This tripartite division of disturbances or "phases" had been observed earlier, most notably in 1894 by Ernst von Rebeur-Paschwitz, a German scientist who had developed the first accurate horizontal-pendulum seismograph. He had been the first person to detect, in 1889, earthquake vibrations that had passed through the body of the earth and had promoted the idea of an international network of seismological stations to gather seismic data on standardized instruments. It was Oldham, however, who correctly analyzed the nature of the phases and realized how they could be used to reveal the earth's internal structure.

When an earthquake occurs, it causes waves throughout the earth that fan out from the quake's point of origin or "focus." Primary (P) waves are longitudinal, and emanate like sound waves from the focus, successively compressing and expanding the surrounding material along the direction of travel. Secondary (S) waves are transverse, vibrating at right angles to the directon of travel without causing any change in volume. While P waves can travel through gases, liquids, and solids, S waves can travel only through solids. S waves travel at about two-thirds the speed of P waves. Surface waves, the third type, travel as undulations only near the earth's surface. It is interesting to note that the mathematical theory of wave propagation was already highly developed well before Oldham's time; indeed, the equations of motion for a disturbance in an elastic medium had been determined some seventy years earlier. Siméon-Denis Poisson, for example, had showed that there could be two types of waves (the P and S waves) that would be transmitted at different speeds through such a medium. Other mathematicians had determined the nature of the waves as well as their properties of reflection and refraction. Once it was established that earthquakes produced waves that could be analyzed according to these mathematical laws, seismic data became a rich new source of information about the earth's interior; seismologists now speak of seismic data as being able to provide an X ray of the earth. In his groundbreaking 1906 article, "The Earth's Interior as Revealed by Earthquakes," Oldham analyzed worldwide data on fourteen earthquakes and revealed zones with fewer than expected P waves in a pattern such as could be caused by the differential bending of the waves upon entering and leaving a spherical core. In essence, Oldham had detected a seismic wave "shadow" of the earth's core. In recognition of his work, Oldham was awarded the Lyell Medal in 1908 and was elected a Fellow of the Royal Society in 1911.

Seismologists now acknowledge that Oldham's techniques were crude, but the first step had been taken and improved analysis came quickly. In 1912, Beno Gutenberg was able to establish a sharp change in wave velocity at a depth of about 2,900 kilometers (about half the earth's radius). His figure still stands as the accepted boundary of the core and the layer above (the mantle), a boundary now called the Gutenberg Discontinuity.

Oldham had also recognized in his data the suggestion of a thin outer crust, but his data was insufficient to determine its depth. This estimation was to be made by Andrija Mohorovičić, a professor at the University of Zagreb, from an analysis of an earthquake that hit Croatia's Kulpa Valley in late 1909. He published his findings the next year in an article entitled "Das Beben vom 8.X. 1909" (the earthquake of October 8, 1909). He had noticed that at any one seismic station, both P and S waves from the earthquake appeared in two sets, but the time between the sets varied with the distance of the station from the earthquake's focus. Based upon these differential arrival times, Mohorovičić reasoned that waves from a single shock were taking two paths, one of which traveled for a time through a "faster" type of rock; he calculated the depth of this change of material. His estimate fell within the now-accepted figures of 20 to 70 kilometers (the crustal thickness varies under the continents and shrinks to between 6 and 8 kilometers under the ocean). This boundary between the crust and mantle is now called the Mohorovičić Discontinuity, or the Moho.

Impact of Event

The use of seismic data to plumb the earth's interior produced an outpouring of research, both theoretical and experimental. By the mid-1920's, seismologists had learned to make subtler interpretations of their data and realized that no S waves had been observed to penetrate through the core. Since S waves cannot pass through liquid, this was reason to think that the core was liquid. Independent support for this hypothesis came from rigidity studies. Seismic waves could be used to determine the rigidity of the mantle; it was revealed to be much greater than the average rigidity of the earth as a whole (the existence of a low-density fluid region would account for this discrepancy). In 1936, however, Inge Lehmann was able to show that P waves passing close to the earth's center changed velocity slightly, and she correctly inferred the existence of an inner core to account for this phenomenon. The current picture of the earth now includes a liquid outer core surrounding a solid inner core about 1,200 kilometers in radius.

The structure of the core plays a key role in theories of the earth's magnetic field. The rotation of the earth and the presence of the solid inner core are thought to influence the flow of the electrically conducting material in the outer core, which acts as a dynamo to generate the field. Given this conclusion, analysis of the historical changes in geomagnetism (as recorded and preserved in magnetized rocks) can, in turn, provide information about the core's history. The data suggests that the basic structure of the core has persisted for at least three-fourths of the earth's history.

Research into the chemical composition of the various layers has progressed slowly. Scientists now generally accept that the core consists mainly of iron, and the mantle and crust of silicates, but details are still a matter of controversy. This investigation is closely tied with speculation about the formation of the earth; it is recognized that conclusions about internal chemical composition must fit within a plausible story about how the material could have gotten there, and this has given rise to the development of new theories about planet formation. Knowledge of mantle structure and composition is essential also in theories of plate tectonics and continental drift.

On the experimental side, the usefulness of seismic data spurred development of specialized and more accurate measuring instruments. For example, different types of seismographs now are used to capture the different character of movements near an earthquake's focus (within 10 degrees) and movements at a distance. Some seismographs are designed to be sensitive to particular wave periods, enabling specific parts of the seismic spectrum to be studied closely. Use of underground nuclear bomb detonations to set up seismic disturbances with some degree of control of intensity was tried successfully starting in the late 1940's. In the 1960's, an attempt was made to gather information about the crust and mantle. Scientists tried to drill through a thin part of the crust under the ocean to reach the Mohorovičić Discontinuity, but the oil-drilling technology they had at their disposal was inadequate to the task, and the "Mohole Project" was eventually abandoned.

Bibliography

Brush, Stephen G. "Inside the Earth." *Natural History*, February, 1984, 26-34. A good introduction for the lay reader, this historical overview article is written in a clear manner that should be accessible to anyone. Without going into technical details, Brush does a superb job of explaining the various theories about the nature of the interior of the earth that have been proposed since the seventeenth century. The article does not contain any references or bibliography, but there are photographs of some of the scientists whose work was most important.

Bullen, K. E. "The Interior of the Earth." *Scientific American* 193 (September, 1955): 56-61. A good source for details of the layered structure of the earth, as revealed by seismic data. The information is presented chronologically, and the article notes persons who were most important in the development of the field, from those involved in the development of seismographs to the seismologists who interpreted the data. Bullen's explanations of the use of seismic data are authoritative. Contains photographs of early seismometers, graphs, and diagrams to illustrate the main points of the article. Reprinted as chapter 27 of *Adventures in Earth History*, edited by Preston Cloud. San Francisco: W. H. Freeman, 1970.

Davison, Charles. "Richard Dixon Oldham: 1858-1936." *Obituary Notices of the Fellows of the Royal Society* 2, no. 5 (1936): 110-113. No books or articles have been devoted to Oldham or his work, and published biographical data is limited to three obituaries. This obituary notice covers the major events of Oldham's life, briefly describes his most important research, and includes a photograph of Oldham.

Jeanloz, Raymond. "The Earth's Core." *Scentific American* 249 (September, 1983): 56-65. Discusses theoretical and experimental work on the nature of the earth's core up to the early 1980's. Jeanloz does an excellent job of highlighting the lines of evidence for and against current alternative hypotheses regarding the composition and development of the core. Primarily explains views on how the core generates the earth's magnetic field and how data on the field can provide information about the core. Clear diagrams highlight important concepts.

Mather, Kirtley F., ed. *Source Book in Geology, 1900-1950*. Cambridge, Mass.: Harvard University Press, 1967. This compilation of historically important original articles on geological subjects from the first half of the twentieth century includes major portions of Oldham's article "The Constitution of the Interior of the Earth as Revealed by Earthquakes" (omitting only the data sections), and a very brief biographical sketch. Gutenberg's article "Mobility of the Earth's Interior" (1941) is included.

Robert T. Pennock

Cross-References

Weichert Invents the Inverted Pendulum Seismograph (1900), p. 51; Wegener Proposes the Theory of Continental Drift (1912), p. 522; Gutenberg Discovers the Earth's Mantle-Outer Core Boundary (1913), p. 552; Lehmann Discovers the Earth's Inner Core (1936), p. 1065.

WILLSTÄTTER DISCOVERS THE COMPOSITION OF CHLOROPHYLL

Category of event: Chemistry
Time: 1906-1913
Locale: Federal Institute of Technology, Zurich, Switzerland

Willstätter unified thought concerning the material responsible for photosynthesis in plants and showed that this activity was the result of two basic chlorophyll molecules

Principal personages:
RICHARD WILLSTÄTTER (1872-1942), a German professor of organic chemistry and winner of the 1915 Nobel Prize in Chemistry
ARTHUR STOLL (1887-1971), a Swiss chemist who, beginning as a student, collaborated with Willstätter on the chlorophyll research

Summary of Event

When Richard Willstätter fled from the Gestapo in Nazi Germany to Switzerland in 1939, one of the few items that he took with him was his highly prized 1915 Nobel Prize certificate honoring him for chemical research. This work had resulted in establishing the composition and partial structure of chlorophyll and the certificate was, very appropriately, decorated around its border with green leaves, blue cornflowers, scarlet geranium blossoms, and cherry-red berries. It was the presence of these and other colors in the plant kingdom that had provided the initial observations leading to this very productive phase of his research. In hindsight, it can be seen that Willstätter's life as a chemist prior to this work was a very direct preparation for the success that he enjoyed. As a doctoral student under Adolf von Baeyer and Alfred Einhorn, Willstätter learned and practiced the techniques then available for structural characterization by working on cocaine, tropine, and other molecules extracted from plants. The general method available at the time involved a series of chemical reactions in which compounds of known composition were reacted with the original material (or a material derived previously from the original), with the goal of degrading the original into smaller fragments. These reactions were followed by tedious purification of the products, often formed in very low yields, until a simple molecule of known structure resulted. Once these derivatives were identified, the series of reactions was logically followed backward to determine the structure of the starting material that could be responsible for the observed results. For the most part, the reconstructed trail was neither direct nor clear, and many sideline possibilities had to be ruled out by performing further reactions before a definitive answer could be found. The answers were also very open to challenge and, unless the utmost care was exercised at each step of the way, could be easily refuted. The heart of Willstätter's research for the remainder of his active career was modifying, adapting, and refining these basic tools in order to establish a clearer trail leading to the structure of molecules.

After completing research for his doctoral degree, Willstätter continued in the area of organic structure research but shifted to applying the degradation and derivatization techniques to synthetic quinones, which formed the chemical basis of the German dye industry. Upon accepting a position as full professor at the Federal Institute of Technology in Zurich in 1905, Willstätter inaugurated a new research effort directed at understanding how photosynthesis (the conversion of light energy and inorganic material into plant tissue) occurred. The first step in this major research effort was the study of chlorophyll. Willstätter produced twenty-four journal articles in conjunction with his students and a classic book, *Investigations on Chlorophyll: Methods and Results*, written with the collaboration of Arthur Stoll, another student.

At the time Willstätter began studying photosynthesis, very little was known about the source of color in plants; however, the generally accepted view was that each variation of green observed was the expression of a unique chemical molecule. Therefore, it was thought that there must be a vast multitude of molecules responsible. These were, as a group, referred to as chlorophylls and were postulated to be the sites at which photosynthesis occurred.

Willstätter worked in an area where there was scant guidance from earlier research. Building on the meager knowledge available, he created techniques to extract plant material and to recover, unchanged, the constituent chemical substances. The raw material for developing these processes was dried leaves of plants. With the raw extract in hand, he worked to invent, improve, and characterize methods, which enabled him to separate and to purify the chemical substances resulting from the extractions. Some of these methods involved the use of enzymes, a then poorly understood class of biological catalysts; others used the technique of adsorption chromatography, which was just developing.

Of particular interest is his finding that there are but two chlorophylls: a blue-green, or alpha form, and a yellow-green, or beta form, rather than the myriad forms previously projected. All the hues of green observed resulted from only these two molecules. Willstätter demonstrated this simplicity in the presence of diversity by extracting and analyzing the green parts of more than two hundred species of plants. Only the two types of chlorophyll were found. Further, he was able to show that the two forms always existed in a near constant ratio to each other, regardless of the plant source.

The extractive fractionation methods used by Willstätter can be described as depending on the selective solubility of the various plant components in different solvents at different acidities. For illustration, consider the final step in the separation of chlorophyll. This separation resulted when an alcohol solution containing the refined plant extract was treated with petroleum ether. Pure chlorophyll is not soluble in the petroleum ether, and it crystallized from the solution in good yield. Other components remained in either the alcohol or petroleum ether solutions. The intensity of color in the various solutions was used as a means of following the course of the separations and also as a means of assessing the purity of the fractions. At this point in the purification, the chlorophyll was not a single substance yet, but rather a mixture of the two chlorophyll forms. One investigation began with 2 kilograms of dried leaves and, following nearly a dozen extractive steps using a variety of solvents, produced slightly less than a gram of a mixture of the two chlorophylls. Further extraction using methanol and a low-molecular-weight hydrocarbon produced the separation of the two, with the alpha form being more soluble in the hydrocarbon and the beta in the alcohol.

The culmination of Willstätter's work was the establishment of the chemical composition of these two forms of chlorophyll. Magnesium, previously postulated by others to be an impurity, was shown to be an integral part of the molecule. Phosphorus, thought by others to be a part of the molecule, was shown to be an impurity. The two chlorophylls were shown to be magnesium complexes of dicarboxylic acids esterified with methanol and a new unsaturated alcohol, named phytol. Each chlorophyll form is a very large molecule. The alpha form contains fifty-five carbon atoms, seventy-two hydrogen atoms, four nitrogen atoms, five oxygen atoms, and a magnesium atom. The beta form differs by having one less oxygen atom and two less hydrogen atoms.

Willstätter also established the similarity of these molecules to hemoglobin with the magnesium of chlorophyll and the iron of hemoglobin playing similar structural roles. This similarity had been intuitively conceived as early as 1851, but Willstätter's work was the earliest establishing the idea as fact.

Impact of Event

Willstätter's research may be seen as the beginning of the understanding of the process by which simple sugars are synthesized from carbon dioxide and water by utilizing the energy of sunlight. This is the process known as photosynthesis. Willstätter's contribution also marks the point from which knowledge of the structure of large, biologically important, molecules stems. The importance of this series of experiments can be noted in the fact that several Nobel Prizes have been awarded for contributions depending directly on Willstätter's earlier work. The 1930 award was given to Hans Fischer for work toward a further understanding of the composition of chlorophyll and hemin, the oxygen-carrying component of blood. In 1961, the award was granted to Melvin Calvin for research leading to an understanding, at the molecular level, of the process of photosynthesis. Robert Burns Woodward received the award in 1965 for the complete synthesis of chlorophyll. Less directly, the award to Sir Robert Robinson in 1947 for investigation of materials of biological importance derived from plants is also related. This sequence makes clear the idea, often expressed in science, of modern workers "standing on the shoulders of giants" who preceded them. Others have used methods contributed by Willstätter to delve into the composition of many biologically important molecules.

Recognizing the uniqueness of the chlorophyll molecule, its composition, and the role that it plays in the photosynthetic process has had significant impact on the human condition. Willstätter's work, and the extensions made from it by others, has allowed an understanding of plant growth that has guided agriculturists in their search for greater plant productivity. Of most direct impact was the understanding of the important role of magnesium as a plant nutrient. As an integral part of the chlorophyll molecule, magnesium must be readily available to any growing plant. This means that this essential element, along with others, must be supplied by fertilization in soils which show a magnesium deficiency. Understanding this connection has allowed successful crop production on land previously thought to be too poor for farming and, in consequence, has led to an increase in food available for the world's increasing population.

Although Willstätter is most recognized for research on chlorophyll, his contribution to the definition of the role of teacher and investigator should not go unmentioned. Each of his writings on chlorophyll was accomplished with a student as a collaborator. Willstätter was a member of a small group of chemists who pioneered the field of organic chemistry by breaking new ground and by devising new methods of experimentation. For this effort, Willstätter is often referred to as one of the fathers of organic chemistry.

Bibliography

Armstrong, Henry E. "Scientific Worthies: XLV. Richard Willstätter." *Nature* 120 (July 2, 1927): 1-5. One of a series of articles on the contributions and lives of scientists written by one who knew Willstätter personally and who conducted research in the same field. Willstätter's work on chlorophyll and other pigments is summarized, as is his life and scientific background. Most important, this reference provides a philosophical look at the impact of Willstätter and his work on the chemistry of the day.

Huisgen, Rolf. "Richard Willstätter." *Journal of Chemical Education* 38 (January, 1961): 10-15. A biography emphasizing the chemical career of Willstätter. Sections deal with his graduate school research on alkaloids; his early work on quinones; the work with a variety of plant pigments, including chlorophyll, and with photosynthesis; and his late work on enzymes. The chemistry presented is written at a level that is accessible to an interested layperson.

Rabinowitch, Eugene, and Govindjee. *Photosynthesis.* New York: John Wiley & Sons, 1969. Provides a broad introduction to the physicochemical mechanism of the primary process in photosynthesis and the enzymatic mechanisms closely related to it. Also discussed is the role of photosynthesis in nature. This survey shows the place that Willstätter's work has in the modern understanding of the subject. The book is suitable for interested readers with a high school background in science.

Robinson, Robert. "Willstätter Memorial Lecture." *Journal of the Chemical Society,* 1953, 999-1026. This address was presented to the Chemical Society in London on the tenth anniversary of Willstätter's death. The chemistry included may be too complex for a layperson, but the address includes much more than chemistry.

Included are Willstätter's connections to others in the scientific community and several reminiscences that allow an understanding of the style of the man as a citizen and as a scientist.

Willstätter, Richard. *From My Life: The Memoirs of Richard Willstätter.* Translated by Lilli S. Hornig. New York: W. A. Benjamin, 1965. Arthur Stoll collected partially completed notes that Willstätter had prepared for a biography and posthumously published them. Some areas lack precision, because the notes were written in exile and without reference to any of the original sources. Presents a clear look at Willstätter's scientific life and his life as a Jew in Germany during the early part of the twentieth century.

Kenneth H. Brown

Cross-References

Krebs Describes the Citric Acid Cycle (1937), p. 1107; Hodgkin Solves the Structure of Penicillin (1944), p. 1240; Sanger Wins the Nobel Prize for the Discovery of the Structure of Insulin (1958), p. 1567; Calvin Wins the Nobel Prize for His Work on Photosynthesis (1961), p. 1703; Barton and Hassel Share the Nobel Prize for Determining the Three-Dimensional Shapes of Organic Compounds (1969), p. 1918.

THE FIRST GERMAN U-BOAT SUBMARINE IS LAUNCHED

Category of event: Applied science
Time: August 4, 1906
Locale: Danzig, Germany

The German U-boat fleet proved to be the catalyst that brought the United States into World War I and assured the ultimate defeat of Germany

Principal personages:

RAYMOND LORENZO D'EQUEVILLEY-MONTJUSTIN (1860-1931), a Spanish engineer who offered the plans for an experimental submarine to the Krupp firm, which eventually evolved into the first U-boat

MAXIME LAUBEUF (1864-1939), a French engineer instrumental in developing the plans for what became the first German U-boat

ALFRED VON TIRPITZ (1849-1930), a German admiral and architect of the German high-seas fleet

WILLIAM II (1859-1941), the German emperor who formulated the "world policy" that led Germany into competition with other European states in building a high-seas fleet

GUSTAV KRUPP (1870-1950), a German industrialist who built the prototypes of the German U-boat

Summary of Event

The launching of *Unterseeboot-eins* (underwater boat number 1) at Danzig on August 4, 1906, hardly appeared to any observer as an epochal event. The German navy was a relative latecomer into the new era of submarine warfare. The English, the French, the Russians, and the Americans already possessed operational submarine fleets before the architect of the German high-seas fleet, Grandadmiral Alfred von Tirpitz, finally consented to divert some of the resources of the German navy to the building of submarines. Despite its unprepossessing beginnings, however, the U-boat has had an enormous influence on the course of world events.

Submarines were not new in 1906. Cornelis Drebbel, a Dutch inventor, conducted the first recorded successful test of a submarine vessel in the Thames River in England in 1620. Over the next four years, he successfully navigated his craft 4 to 5 meters below the surface of the river, once (according to tradition) with King James I as a passenger. During the eighteenth century, inventors patented fourteen designs for submersible vessels in England alone.

The first recorded attempt to use a submarine as a weapon of war occurred during the opening phases of the American Revolution. David Bushnell, a student at Yale University, built an oval-shaped, submersible vessel named the *Turtle*, in which he proposed to approach an English ship in Boston Harbor, attach an explosive charge to its hull, and sink it. The *Turtle* succeeded in approaching the English ship unobserved, but its one-man crew was unable to attach the charge to the copper bottom of the English warship. The first successful submarine attack on a surface ship also occurred in the United States, during the Civil War. On February 17, 1864, a Confederate submarine named *Hunley* (after its inventor, H. L. Hunley) attacked and sank the Union vessel *Housatonic*. The *Hunley*'s crew used a device called a spar torpedo to dispatch the Union craft. A spar torpedo consisted of a powerful explosive charge attached to a long pole. In order to sink the *Housatonic*, the *Hunley* had to approach so near to the enemy vessel that it, too, sank in the ensuing explosion.

Interest in submersible vessels grew in Germany during the mid-nineteenth century, spurred by the heroics of Sebastian Wilhelm Valentin Bauer. In 1850, Bauer built for the government of Schleswig-Holstein a submarine vessel incorporating all the essential elements of later military submarines. Bauer's vessel, named *Brandtaucher* (sea diver), sank during its first test in the Baltic Sea in February, 1851. It remained on the sea bottom until 1887, when the Imperial German government had it raised and, after a restoration program that took nineteen years, placed on display in the Berlin Naval Museum. Bauer later emigrated to England, where he was instrumental in initiating the English submarine development program. He then went to Russia, where he built and successfully tested an enlarged version of his original *Brandtaucher* for the Russian navy. The Russians, however, never deployed the vessel in a combat role.

In the 1870's, the locus of research into development of submersible vessels adaptable to military uses shifted once again to the United States. John P. Holland, generally recognized as the father of modern military submarines, made several important innovations in submarine design from 1875 to 1900. These innovations included a workable electric engine for driving submarines while under water, water ballast to facilitate diving, and horizontal rudders to cause the boat to dive. In 1900, Holland delivered a vessel named the *Plunger* to the United States Navy incorporating all of his innovations. The English navy ordered several submarines of the *Plunger* class, some of which saw action in World War I.

After the unification of the German states into the German empire in 1871 under the leadership of Otto von Bismarck, German military planners assigned a secondary role to the nation's naval forces. Primarily responsible for coastal defense, the German navy had only limited offensive capability. That situation began to change in the 1880's with the acquisition by Germany of overseas colonies. German industrialists, in need of raw materials and new markets for their manufactured products, began to pressure the German government to build a high-seas fleet capable of protecting German commerce on any of the world's seas. Industrialists organized groups such as the Naval League and the Colonial League to propagandize in favor of a massive naval building program. Bismarck opposed these groups with some success as long as he was chancellor of the empire.

In 1888, William II ascended the German throne as Kaiser. Whereas his grandfather, William I, had been content to delegate the responsibility for governing the

German empire to Bismarck, William II was determined to rule as well as reign. Naïvely enthusiastic about colonial expansion, William clashed with Bismarck on that and other issues, ultimately resulting in Bismarck's dismissal in 1890. After the old chancellor's departure, William immediately abandoned Bismarck's diplomatic strategy, which had virtually assured that there could be no general European war. The new Kaiser enthusiastically began to cooperate with industrialists who advocated the creation of a German high-seas fleet and the expansion of German colonial holdings into a worldwide empire rivaling that of England. The naval rivalry with England and William's new "world policy" were important links in the chain of events that led to World War I.

By 1905, the German naval building program reached sufficient levels to represent a legitimate threat to English naval hegemony. The English admiralty's position at the time was that the English fleet must be at least as large as the fleets of the next three largest naval powers in Europe combined in order to assure the security of the home islands and English commerce. Because of the expanding German navy, English government officials were compelled to increase their own naval building program in order to comply with the admiralty's position. Faced with the massive expansion of the English navy, Admiral von Tirpitz agreed finally in 1905 to support the development of submarines capable of offensive operations. Because of Gustav Krupp's efforts, very little time expired before Germany had an operational submarine fleet.

Raymond Lorenzo d'Equevilley-Montjustin had approached the Krupp firm in 1901 with plans for a double-hulled submarine capable of long-range offensive operations. D'Equevilley-Montjustin actually had offered his designs to the French government first, but the French navy bought and developed an almost identical design offered them by a Frenchman named Maxime Laubeuf, who had worked closely with d'Equevilley-Montjustin in Paris. Which of the two men actually designed the submarine is uncertain. The Krupp firm approached Tirpitz for financial support to develop the submarine, but Tirpitz declined, voicing the opinion that submarines could never compete in offensive operations against surface vessels. Consequently, the Krupp firm began development of d'Equevilley-Montjustin's design using its own resources in February, 1902. Krupp completed the prototype of d'Equevilley-Montjustin's vessel on June 8, 1903, but financial constraints had forced abandonment of many of the original features of the submarine. Krupp's boat measured 13 meters. She had an operational range of 40 nautical kilometers at a surface speed of four knots. Driven by a 65-horsepower electric motor, the boat could make 5.5 knots submerged. Krupp invited the Kaiser to observe tests of the vessel in 1904. Both William II and his sons displayed considerable interest in the vessel, but Tirpitz remained opposed to spending scarce naval funds for further development of a vessel unproven in combat.

With the outbreak of the Russo-Japanese War in 1904, Krupp managed to convince the Russian government to buy not only his prototype but also three more boats of the same design (which came to be called the *Karp* class) as well. During the building of the submarines for the Russian navy, Krupp's engineers greatly improved on the original model. The improved submersibles finally won over Tirpitz. In 1905, the admiral agreed to appropriate 1.5 million marks for submarine development and agreed to purchase several of the vessels for the Germany navy.

Tirpitz' decision resulted in the launching of the *U-1* on August 4, 1906. The *U-1* measured 42 meters and displaced 234 tons with a crew of twenty men and officers. Armed with a 46-centimeter bow torpedo tube with three self-propelled torpedos and an 88-millimeter deck gun, the boat would make 10.8 knots on the surface and 8.7 knots submerged. With a cruising range of 2,414 kilometers, the *U-1* had definite offensive capabilities. The German navy had acquired a fateful new weapon.

Impact of Event

The launching of *U-1* made no particular impression in 1906, even on the other naval powers of the world. The Germans entered the race later, to develop submarine technology for military purposes, and progressed slowly. When World War I began, Germany was well behind the other great powers, both in numbers of submarines and in submarine technology. (France had 123 submarines, England had 72, Russia had 41, the United States had 34, and Germany had only 26.) Nevertheless, the German submarine fleet quickly became one of the most important weapons and the most controversial.

On September 22, 1914, *U-9*, commanded by Kapitanleutnant Otto Weddigen, sank three English armored cruisers in a naval engagement lasting less than one hour. Weddigen's demonstration of the offensive capabilities of submarines and their effectiveness against surface vessels revolutionized naval strategy. The engagement forced the English admiralty not only to acknowledge the vulnerability of its surface warships and change the way they were deployed but also to face the inevitability of submarine attacks on English commercial vessels. Since England had been a net food-importing nation for more than a century, the specter of certain defeat loomed if the German submarines could sufficiently curtail the number of merchant ships reaching the English isles.

On October 14, 1914, Kapitanleutnant Feldkirchner, commanding the *U-17*, sank an English steamer. Despite the agreement reached between the great powers at Geneva in 1907, regulating submarine warfare and prohibiting surprise attacks on civilian ships, the German high command declared a submarine blockade of the English isles and proceeded to sink ever-increasing numbers of commercial vessels into 1915. Desperate to combat the submarine campaign that threatened to take England out of the war, the English admiralty launched a propaganda campaign in the United States against the German actions. The German government, anxious to avoid America's entry into the war on the side of its enemies, tried to avoid anything that might antagonize the United States.

On May 7, 1915, a German U-boat sank the English passenger liner *Lusitania* with a loss of 1,198 civilian lives. Among the casualties were 124 Americans, despite the fact that the German embassy in the United States had taken out a full-page advertisement in *The New York Times* prior to the *Lusitania*'s departure warning potential passengers that the *Lusitania* was carrying matériels of war into a zone of war and was liable to be attacked by German submarines. English propagandists in the United States made much of the incident, even insisting that the U-boat had deliberately launched a second torpedo into an already sinking vessel while the civilians were trying desperately to abandon ship. In actuality, the second explosion resulted from the ignition of the ammunition being carried illegally by the *Lusitania*. The vehement American protest about the *Lusitania* incident and several similar occurrences during 1915 ultimately resulted in the German abandonment of unrestricted submarine warfare out of fear of antagonizing the United States.

By 1917, Germany's military situation became desperate. General Erich Ludendorff and Field Marshall Paul von Hindenburg, the virtual military dictators of Germany by that time, decided to chance American entry into the war in a risky gamble to force England to capitulate. They hoped that a resumption of unrestricted submarine warfare could knock out England before the United States could be mobilized sufficiently to affect the outcome of the war. Accordingly, the German navy resumed unrestricted attacks on all vessels sailing into proximity to the English isles in February, 1917. After the sinking of three U.S. merchant vessels, the United States government broke off diplomatic relations with Germany. A month later, when President Woodrow Wilson asked the U.S. Congress for a declaration of war against Germany, he cited the resumption of unrestricted submarine warfare as the primary cause. The entry of the United States into the war ensured Germany's ultimate defeat. The lasting antagonism between the two nations engendered by wartime propaganda was a major factor influencing America's entry into World War II, again ensuring defeat for Germany. Thus, the seemingly innocuous launching of *U-1* in 1906 was an integral factor in the outcomes of the two great conflicts that shaped the political and social organization of the contemporary world.

Bibliography

Archer, William. *The Pirate's Progress: A Short History of the U-Boat*. New York: Harper & Brothers, 1918. Archer's book is a very critical and unflattering account of German submarine activities during World War I. Obviously convinced by wartime atrocity propaganda that submarine warfare is somehow more barbaric than other forms of combat, he condemns the Germans for adopting it. Should be read with an understanding that Archer is extremely biased.

Botting, Douglas. *The U-Boats*. Alexandria, Va.: Time-Life Books, 1979. A popular and entertaining account of U-boat development from its beginning to the end of World War II. Relatively free of bias, the book contains numerous photographs and tables containing the dimensions and specifications for most U-boats. Contains little information about the *U-1*, but does put it in the perspective of German U-boat development.

Friedman, Norman. *Submarine: Design and Development*. Annapolis, Md.: Naval Institute Press, 1984. Places German submarine development into perspective with the other nations of the world. The section on pre-World War I U-boat evolution is short but informative.

Gibson, R. H., and Maurice Prendergast. *The German Submarine War, 1914-1918*. New York: Richard R. Smith, 1931. A surprisingly objective account of the U-boats in World War I, including an estimate of their effectiveness in the conflict. Contains a lengthy section on the development and evolution of the U-boat before World War I and speculation about the effectiveness of submarines in future conflicts.

Keatts, Henry, and George Farr. *U-Boats*. Kings Point, N.Y.: American Merchant Marine Museum Press, 1986. Replete with rare photographs and specifications for most German submarines launched between 1906 and 1945, this book will be enjoyable reading for everyone interested in the development of submarine warfare. The authors also pay particular attention to German submarines sunk in U.S. waters, including accounts of amateur and U.S. Navy efforts to explore their wreckage.

Manchester, William. *The Arms of Krupp, 1587-1968*. Boston: Little, Brown, 1968. A scholarly chronicle of the evolution of the Krupp firm into the world's largest producer of arms. Contains a short but illuminating section on Krupp's role in the development of the U-boat and places it in perspective with the other activities of the firm.

Rössler, Eberhard. *The U-Boat: The Evolution and Technical History of German Submarines*. Translated by Harold Erenberg. Annapolis, Md.: Naval Institute Press, 1981. Only those with an interest in very specific technical details of German U-boats should attempt this book. It is laden with figures concerning cruising speed, thickness of armor, engine specifications, number and explosive power of torpedos, and the like, to the point that the average reader is overwhelmed by a sea of details. Many rare photographs.

Paul Madden

Cross-References

Anschütz-Kaempfe Installs a Gyrocompass onto a German Battleship (1906), p. 303; Langevin Develops Active Sonar for Submarine Detection and Fathometry (1915), p. 620.

THOMSON WINS THE NOBEL PRIZE FOR THE DISCOVERY OF THE ELECTRON

Category of event: Physics
Time: December, 1906
Locale: Stockholm, Sweden

Thomson's discovery of the electron explained the nature of the cathode rays, provided explanations for problems with currents in gases, and paved the way for advances in understanding atomic structure

Principal personages:

SIR JOSEPH JOHN THOMSON (1856-1940), an English physicist, professor and director of the Cavendish Laboratory at the University of Cambridge in England, and winner of the 1906 Nobel Prize in Physics for the discovery of the electron

CHARLES THOMSON REES WILSON (1869-1959), a Scottish physicist who developed the famed "Wilson's cloud chamber," an apparatus that enabled Thomson to determine the charge of the electron; shared the 1927 Nobel Prize in Physics with Arthur Holly Compton

WILHELM CONRAD RÖNTGEN (1845-1923), a German physicist who discovered X rays, for which he was awarded the 1901 Nobel Prize in Physics

PHILIPP LENARD (1862-1947), a Hungarian-born, German-educated professor of physics at the University of Kiel who won the 1905 Nobel Prize in Physics for his work on cathode rays

Summary of Event

In his celebrated work, *Treatise on Electricity and Magnetism*, published in 1873, James Clerk Maxwell stressed the need to study the complex processes involved in electric discharge in gases in order to understand the nature of the charge and the medium. In 1879, an English chemist, Sir William Crookes, who had invented the "Crookes's tube" and was the first to observe radiations emitted from a cathode in an evacuated glass tube through which electric discharges occurred, published an extensive series of attributes of these rays. Among other properties, Crookes noted that cathode rays cast shadows and were bent by a magnetic field, and concluded that they were made up of particles. Upon the suggestion by Hermann von Helmholtz, Eugen Goldstein of Berlin studied the cathode rays exhaustively and published an impressive paper in the English *Philosophical Magazine* in 1880, firmly convinced that these rays were a form of waves. Thus, in 1880, the divergence of opinion regarding the nature of cathode rays became the central problem. Crookes and the leading English physicists believed that the cathode rays consisted of electrified particles, while the German physicists, led by Heinrich Hertz, were certain that these rays were waves.

In 1883, Hertz found that the cathode rays could be bent if one applied a magnetic field outside a discharge tube. His attempt to measure the magnetic field because of the discharge inside the tube (between two parallel glass plates, which enabled him to determine the current distribution therein) gave no significant correlation to the direction of the cathode rays. Further, Hertz applied static electric fields both inside and outside the tube via parallel conducting plates connected to batteries up to 240 volts. This would presumably produce a force perpendicular to the direction of the rays, deflecting them if they were composed of charged particles. In both situations, he obtained a null result. His long series of experiments seemed to confirm the basic premise with which he had started, namely, cathode rays were waves. His pupil, Philipp Lenard, continued the study of cathode rays, concentrating upon their properties outside the tube, which made them easier to handle. He showed that once out of the tube, the rays rendered the air a conducting medium and blackened the photographic plates; moreover, the distance they traveled depended upon the weight-per-unit area of matter, not upon its chemical properties, and the magnetic deflection was independent of the gas inside the tube. Like Hertz, Lenard believed that he was dealing with a wave phenomenon.

In 1895, Jean-Baptiste Perrin, repeating Crookes's experiment with improved equipment, succeeded in collecting from cathode rays negatively charged particles in an insulated metal cup. This appeared to cast doubt about their wave aspects and consequently, by late 1895, two divergent views prevailed among the leading physicists as to the nature of electric charges. One group thought of them as portions of fluids consisting of large numbers of "molecules of electricity," or electrons. The other group regarded the "charge" as a result of an unknown form of stress in ether, attached to matter being rendered visible. Therefore, the nature of the cathode rays still remained to be resolved.

In 1895, at the University of Würzburg, while studying discharges produced by an induction coil in an evacuated Crookes's tube, Röntgen accidentally discovered X rays produced by cathode rays as they impacted a platinum target. Among other properties, such as penetrability through matter, X rays were found to ionize gaseous media, making it conduct, which accelerated the study of conductivity in gases. Sir Joseph John Thomson's use of the property of X rays proved to be the pivotal point in guiding him toward the discovery of the electron.

In the hope of resolving the controversy on the nature of cathode rays, Thomson repeated Perrin's experiment with minor modifications in the collection and measurement of the charges. Using a magnetic field to bend the rays, he collected them in a metal cup placed away from the direct line. He found that the charge in the cup reached a steady state after attaining a maximum value, which he correctly explained as caused by leakage into surrounding space. Hertz had failed to observe electric deflection of the cathode rays. Thomson, using two conducting plates between the cathode rays within the tube, applying an electrostatic field between the plates, and utilizing a better vacuum technique compared to that available to Hertz, was able to observe deflection of the rays, showing that they were composed of negatively charged particles. He correctly explained that Hertz's failure to observe electric deflection of the cathode rays was caused by their ionizing property in excess amount of gas in the tube, thus shielding them from the very field meant to deflect them. By the simultaneous application of the electric and magnetic fields to the cathode rays, Thomson obtained the velocity v of the cathode ray particles. On the assumption that the particles carried a charge e and had mass m, Thomson succeeded in obtaining the crucial ratio of charge to mass, that is, e/m, showing that it was seventeen hundred times the corresponding value for hydrogen atoms. He further showed that the constant e/m was independent of velocity v, the kind of electrodes used, and the type of gas inside the cathode ray tube.

Using Charles Thomson Rees Wilson's newly developed "cloud chamber," Thomson was able to obtain the value of the charge e; from the ratio e/m, it was simple to compute the numerical value of m. Hence, the smallness of the mass, combined with the relatively large velocity of the cathode ray particles, also explained Hertz's observation, namely, that the rays penetrated thin sheets of metals. Obviously, massive particles could not do so. From Lenard's result of constancy of magnetic deflection of the rays and independence of chemical properties, Thomson soon realized that he had discovered a universal module of atoms found in radioactive substances, alkali metals bombarded by ultraviolet light, and in a variety of gaseous discharge phenomena. Thomson's discovery that the cathode ray particle is universal and fundamental to the understanding of the structure of all matter unraveled the puzzling aspect of conductivity of gases, the nature of electricity, and the wave-particle controversy.

Impact of Event

Thomson's discovery of the electron and the recognition of the fact that it carried a natural unit of charge and was the universal component of all atoms, marked the beginning of a new and exciting period in atomic research. The confirmation concerning the particle nature of cathode rays also came from other quarters. For example, Pieter Zeeman's observation of the widening D lines in the spectrum of sodium, explained by Hendrik Antoon Lorentz's theory, caused by changed electron configurations of the atoms in the presence of a magnetic field, gave a value of e/m, comparable to those obtained by Thomson.

Based on his discovery and a study of mechanical stability of the electrons under the influence of the electrostatic force, Thomson showed that the electron must circulate about the atom's center, constrained to move in concentric circles. Calling the cathode ray particles "corpuscles" and speculating that their number increased proportionally to the atomic weights, Thomson attempted to explain the structure of the chemical elements and their properties. From this early model, he drew several important conclusions. First, since the electrons must accelerate as they move in circles around the atomic center, they will radiate. Therefore, such an arrangement of electrons cannot be stable since n, the number of electrons, was assumed to be of the order of one thousand times the atomic weight A. Since experimental work on alpha, beta, and gamma scattering, performed under the supervision of Thomson at Cavendish to verify his theory and obtain n of chemical elements, had led to negative conclusions, this served as a basis for Ernest Rutherford's model of the nuclear atom. Additionally, Thomson's discovery of the order of n fostered the development of the scattering theory, which was to play an important role in the evolving research work in atomic and nuclear physics. The instability of his model atom was instrumental in the formulation of the quantum hypothesis and the discovery of the atomic quantum levels. Finally, Thomson's electron distribution in atoms provided analogies to the behavior of the chemical elements and in particular the population of electrons in the atoms of contiguous elements in the periodic table that differed by unity.

Bibliography

Crowther, J. G. *The Cavendish Laboratory, 1874-1974*. New York: Science History Publications, 1974. In this volume, the author leads the reader from the origin of the Cavendish Laboratory through its period of rapid growth to its present-day organization, a crucial role in the rapid growth and accomplishment of the personnel of this unique research laboratory. This volume, in addition to providing a complete account of the researchers at Cavendish and their achievements, contains an exhaustive list of references.

Nobelstiftelsen. *Physics: 1901-1921*. New York: Elsevier, 1967. Sir Joseph John Thomson's Nobel lecture of 1906, detailing his discovery of the electron, is relatively short, easy to follow, and by far the best source of information on the subject. His biographical sketch, although brief, contains a complete list of honors bestowed upon him.

Strutt, Robert John, Fourth Baron Rayleigh. *The Life of Sir J. J. Thomson*. Cambridge, England: Cambridge University Press, 1942. At Cavendish in 1899, Robert John Strutt, the Fourth Baron Rayleigh, began his research work on ionization produced by radiations from various radioactive substances and was close to Thomson's own research activities. This biography of Thomson is the authoritative text that is often quoted, for good reason. One begins to appreciate the contribution of Thomson and the people he inspired in the field of science after reading this work.

Thomson, George Paget. *J. J. Thomson and the Cavendish Laboratory in His Day*. London: Thomas Nelson, 1964. George Paget Thomson, the son of Sir Joseph John Thomson, an eminent physicist, who received the Nobel Prize for his work on the diffraction of electrons in crystals in 1937, narrates the life and achievements of his famous father and those who worked with him at the University of Cambridge. This volume contains not only the story of Thomson's life and work but also a lucid account of the history of the development of physics at the beginning of the twentieth century in England and Europe.

Thomson, J. J. *Cathode Rays*. Stanford, Calif.: Academic Reprints, 1954. This classical scientific paper, originally published in the *Philosophical Magazine* (1897), in which Thomson methodically details his findings concerning the cathode rays and his discovery of the electron provides a proper historical perspective and an understanding of the working of a powerful mind. Although the subject matter is technical, the reader should find it highly rewarding.

____________. *Recollections and Reflections*. London: G. Bell & Sons, 1936. An autobiography is necessarily "recollections and reflections" in the true sense of the words. Thomson was an accomplished writer, scholar, and teacher accustomed to scrupulously accurate recording of his thoughts and observations. In this last work, Thomson gives a candid account of his many achievements, preserving a balanced historical perspective. This volume spans the history of physics at its most exciting period and is a pleasure to read.

V. L. Madhyastha

Cross-References

Röntgen Wins the Nobel Prize for the Discovery of X Rays (1901), p. 118; Bohr Writes a Trilogy on Atomic and Molecular Structure (1912), p. 507; Rutherford Presents His Theory of the Atom (1912), p. 527.

FESSENDEN PERFECTS RADIO BY TRANSMITTING MUSIC AND VOICE

Category of event: Applied science
Time: December 24, 1906
Locale: Brant Rock, Massachusetts

Fessenden revolutionized radio broadcasting by transmitting music and voice for the first time

Principal personages:

REGINALD AUBREY FESSENDEN (1866-1932), the first major experimenter with wireless radio, pioneering radio broadcasting as it is known today

GUGLIELMO MARCONI (1874-1937), an Italian physicist and inventor who is usually credited as the inventor of wireless radio

Summary of Event

The first major experimenter in the United States to work with wireless radio was Reginald Aubrey Fessenden. This transplanted Canadian was a skilled, self-made scientist, but like Thomas Alva Edison, he lacked the business acumen to gain full credit and wealth due such path-breaking work. Guglielmo Marconi is most often remembered as the person who invented wireless (as opposed to telegraphic) radio. There is a fallacy in the Marconi claim, however. There was a great difference between the contributions of Marconi and Fessenden. Marconi limited himself to experiments with radio telegraphy; that is, he sought to send through the air messages that were currently being sent by wire—dots and dashes. Fessenden sought to perfect radio telephony: voice communication by wireless. Fessenden pioneered the essential precursor to modern radio broadcasting. At the beginning of the twentieth century, Fessenden spent much time and energy publicizing his experiments, thus promoting interest in the new science of radio broadcasting.

Fessenden had a varied background. He worked with Edison, Westinghouse Electric Corporation, and the United States Weather Bureau of the Department of Agriculture. The three years he spent working with Edison shaped his values about the process of invention. He sought to make a name for himself, aspiring to become as famous for the miracle of radio as Edison had become for the phonograph, electricity, and the motion picture.

Fessenden began his career as an inventor while working for the Weather Bureau. He set out to invent and innovate a system by which to disseminate weather forecasts to users on land and on the sea—by radio. Fessenden believed that his technique of continuous waves in the radio frequency range would provide the power necessary to carry Morse telegraph code yet be effective enough to handle voice communication. He would turn out to be correct. He conducted experiments as early as 1900 at Rock Point, Maryland, about 80 kilometers south of Washington, D.C., and registered his first patent in the area of radio research in 1902.

In 1900, Fessenden asked the General Electric Corporation to produce a high-speed generator of alternating currents—or alternator—to use as the basis of his radio transmitter. This proved to be the first major request for wireless radio apparatus useful to project voices and music. It took the engineers three years to design and deliver the alternator. Meanwhile, Fessenden worked on an improved radio receiver. To fund his experiments, Fessenden piqued the interest of financial backers, who put up one million dollars to create the National Electric Signaling Company in 1902.

He and a small group of hand-picked scientists worked at Brant Rock on the Massachusetts coast south of Boston. Fessenden followed the methods he had learned under Edison, creating a small, crude shop and learning by trial and error. He would work outside the corporate system and then seek fame and glory based on his own work, not something owned by a corporate patron.

Fessenden's moment of glory came on December 24, 1906, with the first announced broadcast of his radio telephone. Using an ordinary telephone microphone and his special alternator to generate the necessary radio energy, and following a private demonstration a month earlier, Fessenden alerted ships up and down the Atlantic coast with his wireless telegraph and arranged for newspaper reporters to listen in from New York City. He "broadcast" from Brant Rock. Fessenden made himself the center of the show. He played the violin, sang, and read from the Bible. Anticipating what would become standard practice fifty years later, Fessenden also transmitted the sounds of a phonograph recording. He ended this first broadcast by wishing those listening a merry Christmas. A similar, equally well-publicized demonstration came on December 31.

If one considers the audience of telegraph operators on the ships of the day and newspapers as the means of mass dissemination, then this public display in late 1906 represented the first radio broadcast. It was scheduled, and listening required no special knowledge of code because the sounds consisted of voice and music. Newspaper reporters told of shipboard radio operators "hearing" the voices of angels rather than the usual static and Morse code.

Although Fessenden was skilled at drawing attention to his inventions and must be credited, among others, as one of the engineering founders of modern principles of radio, he was far less skilled at making money with his experiments, and thus his long-term impact was limited. National Electric Signaling Company had a fine beginning and for a time had been a supplier to the boats of United Fruit Company. The financial panic of 1907, however, wiped out an opportunity to sell the Fessenden patents—at a vast profit—to a corporate giant, the American Telephone and Telegraph Corporation. Fessenden found himself deserted by his original backers. As had been the case with Edison, backers loved the promise of his experiments but avoided failures. Fessenden was not a rich man; once his financial support dried up, he was isolated. Thereafter, Fessenden's career went downhill. Much like Edison in the twentieth century, Fessenden spent more and more of his time in court trying to protect his inventions. His laboratory career was over. Eventually, this led to a series of extended lawsuits and, in the end, the sale of patents to Westinghouse Electric Corporation following World War I. Fessenden's experience again proved that simple invention (and skilled promotion) was not enough. One had also to be an innovator, skillfully developing a product that could sell in the economic marketplace and then have the skills to market it. Fessenden's invention, spectacular in its day, required massive apparatus that was too cumbersome to use permanently. Thus, he, along with a handful of others, had to be satisfied with knowing he helped to spawn the modern age of radio and television communications.

Impact of Event

In many countries, several scientists and inventors claimed to be the first radio broadcaster. How can one know which was the first radio broadcast? As an industry, radio broadcasting as it is currently known did not commence until late in 1920, when the Westinghouse Electric Corporation of Pittsburgh, Pennsylvania, established its station, KDKA. Many experiments and demonstrations preceded KDKA. Of these experiments, Fessenden's broadcast of December 24, 1906, to ships along the Atlantic seaboard ranks as a pioneering effort. Fessenden's transmission was intended for the audience of the day.

Had there been more receiving equipment available and in place, a massive audience could have heard his voice and music. Fessenden had the correct idea, even to the point of playing a crude phonograph record. Yet, neither Fessenden, Marconi, nor their many other rivals were able to establish permanently a regular series of broadcasts. Their "stations" were experimental and promotional. At best, they performed a series of demonstrations. Fessenden, however, became better known than most through his skills in generating favorable publicity, and his equipment was more powerful. Fessenden was what one might consider today an amateur radio operator. Indeed, in the years immediately preceding World War I, the creation of a growing community of amateur radio hobbyists across the nation must be judged the principal effect of the Fessenden experiments. These early radio enthusiasts rarely ventured into voice or musical broadcasting but continued to use point-to-point communication through Morse code. Nevertheless, this development marked the beginning of modern mass communication.

It took the needs of World War I to encourage broader use of wireless radio, based on Fessenden's experiments. Suddenly, communicating from ship-to-ship or ship-to-shore became a matter of life or death. Generating publicity was no longer necessary. Governments fought over crucial patent rights. The Radio Corporation of America pooled vital knowledge. Ultimately, RCA came to acquire the Fessenden patents. Radio broadcasting commenced, and the radio industry, with its multiple uses for mass communication, was off and running. Pioneers like Fessenden were left for the history books.

Fessenden's is another case of an independent inventor trying to go against the trend that modern science is created within corporate laboratories. One could not work outside the given institutional relations and hope to succeed. Fessenden would prove that Edison ended an era, the age of the inventor as lone wolf.

Bibliography

Aitken, Hugh G. J. *The Continuous Wave: Technology and American Radio, 1900-1932*. Princeton, N.J.: Princeton University Press, 1985. The author, a long-time student of radio history, analyzes the technical shifts that led to the configuration of modern radio. This is a highly specialized and difficult work but is the standard for technological history.

Barnouw, Erik. *A History of Broadcasting in the United States.* Vol. 1, *A Tower of Babel, to 1933*. New York: Oxford University Press, 1966. Barnouw's work stands as the standard multivolume history of broadcasting in the United States. He covers the coming of radio in this first volume.

Douglas, Susan J. *Inventing American Broadcasting, 1899-1922*. Baltimore: The Johns Hopkins University Press, 1987. This scholarly book offers a richly illustrated account of an important era in media history. She identifies the use of newspaper and magazine publicity as crucial for the cultural evolution of radio since the press legitimated the new technology.

Fessenden, Helen M. *Fessenden: Builder of Tomorrow.* New York: Coward-McCann, 1940. This is the inventor/promoter's wife arguing after his death for his proper role in the creation of modern radio. In 1974, Arno Press reprinted this volume with a new index.

Head, Sydney W., and Christopher H. Sterling. *Broadcasting in America*. 5th ed. Boston: Houghton Mifflin, 1987. The standard introduction to the institutions of radio and television in the United States. The book begins with an analysis of the invention of wireless radio broadcasting.

Lichty, Lawrence W., and Malachi C. Topping. *American Broadcasting: A Source Book on the History of Radio and Television*. New York: Hastings House, 1975. This book contains articles and documents concerning the history of radio and television. The invention of radio transmission is treated in some detail.

Sterling, Christopher H., and John M. Kittross. *Stay Tuned: A Concise History of American Broadcasting*. Belmont, Calif.: Wadsworth, 1978. This is the standard, one-volume history of radio and television in the United States. A good beginning point.

Douglas Gomery

Cross-References

Marconi Receives the First Transatlantic Telegraphic Radio Transmission (1901), p. 128; Transatlantic Radiotelephony Is First Demonstrated (1915), p. 615; The Principles of Shortwave Radio Communication Are Discovered (1919), p. 669; Armstrong Perfects FM Radio (1930), p. 939; The First Color Television Broadcast Takes Place (1940), p. 1166.

HALDANE DEVELOPS STAGE DECOMPRESSION FOR DEEP-SEA DIVERS

Category of event: Medicine
Time: 1907
Locale: Lister Institute of Preventive Medicine, London, England

Haldane developed a method for deep-sea divers to ascend to the surface without suffering decompression sickness

Principal personages:

JOHN SCOTT HALDANE (1860-1936), an English physiologist and science philosopher who was a pioneer in the study of gas exchange during respiration

ARTHUR EDWIN BOYCOTT (1877-1938), an English pathologist who assisted Haldane in his first decompression experiments and coauthored the resulting paper on the prevention of compressed air disease

G. C. C. DAMANT, a lieutenant in the Royal Navy who was the first human subject for Haldane's decompression experiments

Summary of Event

It has long been known that individuals who have been subjected to high atmospheric pressure cannot be returned rapidly to normal pressure without risking painful symptoms. During the mid-1840's, in coal mines pressurized to keep out water, the miners frequently suffered muscle pains upon ascending to the surface. These symptoms were first studied in 1854, when it was noted further that a return to compressed air alleviated the pain.

In 1878, the French physiologist Paul Bert wrote a landmark book called *La Pression barométrique* (*Barometric Pressure*, 1943), in which he presented evidence that the so-called decompression sickness resulted from the formation of nitrogen gas bubbles, which blocked the circulation.

Deep-sea diving had been plagued by decompression sickness (also known as compressed air disease, or "the bends," because of the slightly bent, limping posture of its sufferers) since individual diving attire had been developed in the early 1800's. By the beginning of the twentieth century, there was still no safe method for retrieving a person from deep dives.

England's Royal Navy used divers in its routine operations. Intent on finding a way to spare these men the risks of pain, paralysis, or even death as a result of decompression sickness, it commissioned John Scott Haldane—a physician and man of letters, famous for his studies of respiratory physiology—to develop a decompression schedule that could be written down in tabular form and distributed to its fleet divers. Haldane was a scientist with a strongly philosophical nature. He disapproved of the generally accepted line between "pure" science—that carried out for its own sake—and applied science—that employed as useful technology. He was a laboratory scientist who envisioned the results of his labors as being linked intimately to the general welfare. As such, he was eager to apply his theories and insights to the human working environment. Already famous for his studies of the composition of the air in schools and dwellings and for his identification of carbon monoxide as the cause of death in coal mining disasters, he accepted the Royal Navy's commission with alacrity and began his studies of what came to be known as stage decompression.

It was standard knowledge that as one descends through a body of water, every 33-foot increment of depth exerts an additional pressure of one atmosphere. Thus, a diver at 33 feet experiences two atmospheres of pressure (the pressure of the air plus the extra atmosphere of the 33 feet of water), a diver at 66 feet is subjected to three atmospheres, and so on. Haldane was aware that cases of decompression sickness never occurred, even with rapid decompression, when a diver suddenly ascended from two atmospheres to one. He reasoned that it must, therefore, be equally safe to decompress suddenly from four atmospheres to two, or from six to three, as long as the two-to-one ratio is maintained. Thereafter, in order to avoid the bends, the ascent would have to proceed more slowly to give nitrogen that had been dissolved in the tissues under pressure the necessary time to emerge from the blood and be vented by the lungs. Haldane's method therefore consisted in an initial reduction of pressure to one-half as rapidly as possible, followed by continued, slower decompression in a series of stages designed to prevent the rapid bubbling of nitrogen out of the blood.

Haldane carried out his first live experiments on goats in a steel compression chamber at the Lister Institute of Preventive Medicine in London. He was assisted in this work by Arthur Edwin Boycott and by Lieutenant G. C. C. Damant of the Royal Navy. They exposed several of the animals for long periods to 6 atmospheres of pressure (equivalent to a dive of 165 feet) and then decompressed them suddenly to 2.6 atmospheres with no ill effects. When the animals were dropped from 4.4 to 1 atmosphere (far in excess of the two-to-one proportion), 20 percent of the animals died and 30 percent experienced the bends.

Immediately after the animal experiments, Lieutenant Damant volunteered as the first human subject. He crawled into the chamber, experienced the rapid pressure differentials, and emerged with no symptoms. Haldane had proved conclusively that it was quite safe to halve the absolute pressure rapidly, regardless of depth of dive.

Haldane realized, however, that he had only part of the solution. The danger of decompression sickness lay in what remained of the ascent after the pressure on the diver's body had been rapidly halved. This required the painstaking working out of decompression tables that took into account not only the depths to which a diver had descended but also the amount of time the diver had remained underwater. These factors taken together actually determined the degree of saturation of body tissues with nitrogen.

The complexity of the task of formulating accurate decompression tables can be appreciated when one considers that the various tissues of the body absorb nitrogen at different rates. The rapidity with which a given tissue becomes saturated with nitrogen is dependent upon two factors: the solubility of nitrogen in that tissue and the rate at which nitrogen is transported to the tissue by the blood. Nitrogen is very soluble in fatty tissues, for example, but much less so in nonfatty, or aqueous, tissues. Conversely, fat is served poorly by the circulation, so the rate of nitrogen saturation of that tissue is slow, whereas aqueous tissues quickly reach their nitrogen saturation points because they are well supplied with blood vessels. Haldane, therefore, divided human tissues into two types: slow—those, such as fat, that are poorly vascularized but have a high capacity for storing nitrogen; and fast—those parts of the body, such as the brain, that are well vascularized and quickly become saturated with nitrogen.

In drawing up his decompression tables, Haldane realized that saturation rates for the various tissues varied from seconds to hours. Because the rate of saturation also changed as the individual tissues took up nitrogen, it was impossible to determine on paper when exactly a given tissue would be fully saturated. Haldane struck upon the idea of "half-time tissues." A tissue half-time is the time required for nitrogen to reach half its saturation value. Because it was not possible to assign precise half-times to real tissues, he arbitrarily chose to designate five-, ten-, twenty-, forty-, and seventy-five-minute half-time tissues for mathematical convenience in predicting the uptake and elimination of nitrogen. Although he realized that the longest half-time tissue in the body was sixty minutes, it is typical of Haldane's conservatism and his emphasis on safety that seventy-five minutes should represent the longest half-time tissue in his decompression schedules.

It was these carefully worked-out calculations of the rates at which fast and slow tissues eliminated their nitrogen that made Haldane's tables so valuable. With them, a diver was able to regulate the rate of decompression so that no part of the body was at any point so supersaturated with nitrogen that bubbles would begin to form in the circulation. Haldane's method of ascent by stages replaced the previous method of continuous decompression, which took much longer and seldom left the diver fully decompressed by the time the surface was reached. Haldane, Boycott, and Damant coauthored these findings, and the Haldanian decompression tables were adopted immediately by the Royal Navy.

Impact of Event

After Haldane's experiments at the Lister Institute, Lieutenant Damant made numerous deep dives at sea with the use of the Haldanian decompression tables. He achieved a maximum depth of 35 fathoms (210 feet). Following stage decompression, he showed no symptoms of the bends.

Haldane's method of stage decompression immediately altered diving practice, and his schedules were universally adopted. After the Royal Navy published his tables, decompression sickness virtually disappeared among its divers. One of the most notable tests to which the tables were put involved the recovery of a United States submarine off Honolulu in 1915. The diving crew descended to a depth of 50 fathoms (15 fathoms more than Lieutenant Damant's deepest test dive) and completed the salvage operation without any of the divers experiencing decompression sickness.

Stage decompression can be a lengthy process, a slow ascent in stages after an initial rapid decompression being its underlying principle. Bodily nitrogen that has been subjected to high pressure must be given adequate time to leave the blood and be exhaled from the lungs as a diver ascends to the surface. Even after it is gone from most areas of the body, nitrogen tends to linger in the fatty tissues, where it is especially soluble. A diver who has been at a depth of 190 feet for sixty minutes, for example, must be decompressed for about three hours before being brought to the surface. Although there existed primitive decompression chambers in Haldane's time, they were often difficult to control, so Haldane favored resubmergence to appropriate depth in those cases where divers were experiencing symptoms of the bends.

Because of the length of decompression, questions were raised about ways to cut down on the times involved. "It appears that the times cannot be cut down without risk of trouble," wrote Haldane in his monumental book *Respiration* (1922), "unless the divers are placed in the [decompression] chamber as a matter of routine after each dive." Although Haldane never actively promoted this technique as a method for shortening in-water decompression times, it is in common use today if a diver needs to be removed as quickly as possible from badly polluted waters, or when the weather threatens a normal decompression. This method involves bringing divers to the surface immediately and placing them in a decompression chamber within five minutes. The pressure is then reapplied and the divers are decompressed. The actual length of decompression remains about the same.

Haldane's original schedules were found to be highly accurate over their middle range but divers soon learned that they could cut corners on short, shallow dives without risking the bends. Conversely, they found that Haldane's schedules were not conservative enough on longer, deeper dives. They have, therefore, been modified through experience over the years to resolve these difficulties.

Haldane had a spirited commitment to the constructive application of scientific principles for society's well-being. This code characterized his investigations not only in deep-sea diving but also in mining and tunneling. The fact that his method of stage decompression has allowed humans to persist in these environments has earned for Haldane a reputation for being the prime mover of modern times in respiratory physiology.

Bibliography

Douglas, C. G. "John Scott Haldane." *Obituary Notices of Fellows of the Royal Society of London* 2 (December, 1936): 115-139. It is surprising that, given Haldane's accomplishments and his stature as one of the foremost scientific movers of his time, this is the only detailed biographical information on him. A fitting and, in the space of a chapter, comprehensive tribute to Haldane, the scientist and thinker.

Guyton, Arthur C. *Textbook of Medical Physiology.* 6th ed. Philadelphia: W. B. Saunders, 1980. The classic text for students of animal physiology. Contains a detailed chapter on aviation, space, and deep-sea diving physiology. Clearly written and detailed, and with abundant references, it is an invaluable source of information for the reader desiring a grounding in the principles of diving physiology.

Haldane, John Burdon Sanderson. *Adventures of a Biologist.* New York: Harper & Brothers, 1940. Son of John Scott Haldane, J. B. S. Haldane was an accomplished man, having made major contributions in widely disparate areas of knowledge. Provides the most personal portrait in existence of his father, including his routine participation as a human subject in his own experiments.

Haldane, John Scott. *Respiration.* New Haven, Conn.: Yale University Press, 1922. Haldane's monumental work on the physiology of respiration. A compilation of lectures Haldane gave at Yale in 1916. Provides a detailed account not only of what was known about respiration in Haldane's time but also of the author's pioneering experiments. Still considered a classic of respiratory physiology.

Strauss, Richard H. *Diving Medicine.* New York: Grune & Stratton, 1976. A highly readable textbook, valuable for its clear presentation of both the history and physics of diving, including Haldane's contributions. Although intended as an introduction to the medical aspects of diving for physicians, it assumes little prior knowledge of human biology and is accessible to the interested reader.

Robert T. Klose

Cross-References

Beebe and Barton Set a Diving Record in a Bathysphere (1934), p. 1018; Cousteau and Gagnan Develop the Aqualung (1943), p. 1219.

HERTZSPRUNG DESCRIBES GIANT AND DWARF STELLAR DIVISIONS

Category of event: Astronomy
Time: 1907
Locale: Denmark

Hertzsprung described a classification system for stars in which stellar objects were categorized by size versus stellar spectral type and in which giant and dwarf stars were described

Principal personages:

EJNAR HERTZSPRUNG (1873-1967), a Danish astronomer who devised a classification system for stars based on luminosity and spectral class

KARL SCHWARZSCHILD (1873-1916), a German astronomer who was Hertzsprung's mentor

HENRY NORRIS RUSSELL (1877-1957), an American astronomer who developed a stellar classification scheme similar to Hertzsprung's

H. E. LAU, an astronomer who taught Hertzsprung the fundamentals of observational astronomy

Summary of Event

By the beginning of the twentieth century, astronomers had turned their telescopes on the stars and had begun the exceptionally arduous task of cataloging them. This work was an extension of the work of Hipparchus, a Greek astronomer who cataloged the stars (one thousand entries) in 130 B.C. Hipparchus' catalog contained the brightest of the few thousand stars he could make out on any clear night with his unaided eye. When astronomers of the early 1900's began their tasks, they used powerful telescopes and recorded their observations on sensitive photographic films. By combining these methods, suddenly millions of stars became visible, making the task of cataloging one of the most challenging and demanding tasks of science.

The first detailed stellar catalogs of the nineteenth century included star identifications (names or numbers), positions, and star motions. By 1875, new scientific techniques had been developed and added even more information about stars. The technique of spectral photography and analysis enabled astronomers to tell much about the composition of stars by an analysis of the spectral shifts, and even the motion and velocity of stars in the line of sight could be determined.

Astronomers Angelo Secchi and William Huggins determined that there were only a handful of basic stellar spectral types in a broad series that appeared to flow from one type to another, so Secchi proposed a spectral classification scheme that included four main types. In 1901, Antonia C. Maury and Annie J. Cannon of the Harvard College Observatory proposed that the classification system of stars include seven spectral types indexed and classified by letters. Maury proposed a classification system different from Cannon's. Cannon's scheme is still used today, classifying stars by a letter scheme (O, B, A, F, G, K, and M). The Harvard College catalog became known as the *Henry Draper Catalog*, named for Henry Draper, a pioneer in stellar spectrophotometry. The Draper catalog contained information on some 225,320 stars.

Unable to make a good living in astronomy, Severin Hertzsprung encouraged his son, Ejnar Hertzsprung, to pursue a degree in chemical engineering. Ejnar Hertzsprung was graduated from the Polytechnical Institute in Copenhagen in 1898 without any formal education in astronomy. His father died four years later, and upon his death, all of his astronomy books were sold. Ejnar would later state with some irony, "Nobody imagined that I should become an astronomer." Hertzsprung became a chemist in St. Petersburg, then in 1901 seized the opportunity to follow an interest he held in photochemistry. Hertzsprung traveled to Friedrich Wilhelm Ostwald's laboratories in Leipzig. During his stay in Germany, Hertzsprung had decided he wanted to pursue a career in astronomy. Because of his engineering and photochemistry background, Hertzsprung brought to his astronomy career an interesting variety of useful talents and knowledge, but he had no idea of what the duties of an astronomer were. Hertzsprung began his apprenticeship at the Copenhagen University's observatory under the tutelage of H. E. Lau, as well as doing some work at the Urania Observatory in Frederiksberg.

With perhaps much better training in photographic techniques and use of specific emulsions than most other astronomers, Hertzsprung brought a unique variety of talents to bear on photographing stellar fields at the observatories where he worked. Hertzsprung began to collect and study the distribution of stellar images from these plates. At this time, Hertzsprung was considered an amateur astronomer, an apprentice at best. Yet, it was during this period that he would make the pivotal contribution of his life. From a collection of observations, Hertzsprung published two papers in *Zeitschrift für wissenschaftliche Photographie*, a photophysics and photochemical journal. The Hertzsprung papers were published in 1905 and 1907, respectively.

The papers he titled, "Zur Strahlung der Sterne" (on the radiation of the stars), used the star catalog of Maury and her spectral classification techniques. Hertzsprung demonstrated that a certain class of stars (which she called the c class) were more luminous than the remainder. Later, this idea would be developed to form the concept of "spectroscopic parallax," which has become one of the primary means by which the distances to stars and galaxies are determined.

Most significant, however, these papers also contained the profound discovery of giant and dwarf stars among the dense masses of star points on his photographic plates. Hertzsprung made the fundamental discovery that all stars could be categorized into two series of stars, one series that has become known as the "main sequence," and the other containing the high-luminosity stars, or giant stars. Later, a diagram would be made of Hertzsprung's discovery and would bear his name. This diagram has become one of the fundamental tools of modern astronomy.

Hertzsprung began a correspondence with Germany's most famous astronomer,

Karl Schwarzschild, and sent him copies of his papers. Schwarzschild immediately recognized the importance of Hertzsprung's findings. He invited him to visit with him at the University of Göttingen in 1909. Later that year, Hertzsprung was appointed lecturer at Göttingen. He resigned, however, after only a few months to join Schwarzschild, who had become director of the Astrophysical Observatory at Potsdam. Hertzsprung was hired as the senior staff astronomer. Largely because of Schwarzschild's mentorship, Hertzsprung's professional career in astronomy was established.

In 1913, American astronomer Henry Norris Russell presented a paper to the Royal Astronomical Society, in which he presented his ideas on his own independent findings on giant and dwarf stars. The work done earlier by Hertzsprung was unknown to him. Later, the detailed findings of both astronomers was graphically depicted on what has become known as the Hertzsprung-Russell diagram. The modern Hertzsprung-Russell diagram (also known as the H-R diagram) plots star types on a graph showing spectral classes (such as those by Cannon), absolute magnitude, luminosity, and temperature.

Hertzsprung's work (in 1907) on the relationship of the radiation of a star and its color based on stars located within a cluster is often overlooked. This information would later establish the age of the stars of the Pleiades and stars in the neighborhood of the sun. With his method, Hertzsprung would later publish the first color-magnitude diagram ever released. Using the information on spectroscopic parallax, Hertzsprung would determine the distance to the Small Magellanic Cloud, for which he would be awarded the gold medal of the Royal Astronomical Society in 1929.

Impact of Event

Ejnar Hertzsprung established an important link between observational and theoretical astronomy. Through his careful analysis of the data he was collecting (largely in a cataloging effort), Hertzsprung forged a link between the color of a given star and many other key properties, including its intrinsic (natural) brightness, its temperature, and even its size.

As Hertzsprung discovered through the cataloging efforts of the staff at the Harvard College Observatory, in the relative handful of star spectral classes, there were other key features of stars that could be linked and plotted on a relatively simple graph. At first glance, it seemed probable that the kinds of stars in the universe were perhaps as infinitely diverse as were their profuse numbers. Yet, by establishing the links between the star characteristics, Hertzsprung also organized and constrained the very boundaries of the science of stellar astronomy.

Hertzsprung discovered that there was a significant population of stars outside the "main sequence" of stars (of which the sun is one). He called these "giant" and "dwarf stars" because they were different significantly in characteristics from the main sequence stars. By establishing the main sequence, the unique nature of the giant and dwarf stars became strikingly clear.

Such a sequencing method would point later to important evolutionary aspects of stellar development and also the amount of time each star could exist. Such an incredible diversity of science has emanated from the H-R diagrams that there are few areas of modern astronomy that do not rely on the diagram as a fundamental cornerstone.

One of the vital aspects of theoretical astronomy that emerged as a corollary to the H-R diagram is that since the kinds of stars depicted on and classified by the diagram are apparently universal, there must be a ubiquitous process of stellar development and evolution occurring everywhere. This has, in turn, led to speculations of how the sun will behave and what the duration of the sun's lifetime will be. Looking at other stars on the H-R diagram, one can follow the sun's evolutionary position from birth to death as a red giant star.

In another important aspect, the diagram allows astronomers and other scientists to understand much about the environment of any star and near-space, so that the conditions on any theoretical planet in orbit about any star can be theorized. From this speculation and knowing the population distribution of each type of star on the diagram, complex theories have been developed, such as how many of each class of star could support conditions for the development of life in the universe.

Bibliography

Asimov, Isaac. *Extraterrestrial Civilizations*. New York: Crown, 1979. In this work, use is made of stellar types and classes to detail how life could evolve and survive around different types of stars (including giant and dwarf) found on the H-R diagram. It is a good example of how deeply the H-R diagram had woven its way into all aspects of astronomy. The book is easy to understand and is an important book on the topic, although it is not illustrated.

Bok, Bart J., and Priscilla F. Bok. *The Milky Way*. 5th ed. Cambridge, Mass.: Harvard University Press, 1981. This is the definitive book on the Milky Way galaxy, written by two of the world's experts on the subject. Well written, easy to understand, and includes a detailed deliberation of the H-R diagram and its relevancy to the discussion of the nearer stars in the vicinity of the sun. Illustrated.

Cornell, James, and Alan P. Lightman, eds. *Revealing the Universe*. Cambridge, Mass.: MIT Press, 1982. This book discusses in some detail specific areas of observational and theoretical astronomy. The H-R diagram is included in some of the discussions, specifically on how the shape of stellar interiors predicts the shape of the H-R diagram. The book is somewhat technical and is best approached by those with a scientific background. Illustrated.

Hartman, William K. *Astronomy: The Cosmic Journey*. Belmont, Calif.: Wadsworth, 1978. Heavily illustrated, and approaches astronomy in all its basic features. Offers a detailed discussion of not only the H-R diagram but also discusses giant and dwarf stars. For the general reader, especially those with a minimal grounding in science.

Harwit, Martin. *Cosmic Discovery*. New York: Basic Books, 1981. This book covers "the search, scope and heritage of astronomy." Discusses some of the history of

Hertzsprung's development of the H-R diagram and his use of the classification data from the Harvard College Observatory. The book is a fascinating expedition through the history and foundation of astronomy; for a wide audience. Illustrated.

Kaler, James B. "Journeys on the H-R Diagram." *Sky and Telescope*, May, 1988, 483-485. Offers a concise view of the H-R diagram and its many uses. Gives a complete discussion of the giant and supergiant classes and why they lie outside the main sequence. Illustrated.

Dennis Chamberland

Cross-References

Hertzsprung Notes the Relationship Between Color and Luminosity of Stars (1905), p. 265; Leavitt's Study of Variable Stars Unlocks Galactic Distances (1912), p. 496; Slipher Obtains the Spectrum of a Distant Galaxy (1912), p. 502; Russell Announces His Theory of Stellar Evolution (1913), p. 585; Eddington Formulates the Mass-Luminosity Law for Stars (1924), p. 785; Chandrasekhar Calculates the Upper Limit of a White Dwarf Star's Mass (1931), p. 948.

LOUIS AND AUGUSTE LUMIÈRE DEVELOP COLOR PHOTOGRAPHY

Categories of event: Applied science and chemistry
Time: 1907
Locale: Lyons, France

The Lumière brothers introduced the autochrome plate, the first commercially successful process in which a single exposure in a regular camera produced a color image

Principal personages:
LOUIS LUMIÈRE (1864-1948), a French inventor and scientist
AUGUSTE LUMIÈRE (1862-1954), an inventor, physician, physicist, chemist, and botanist
ALPHONSE SEYEWETZ, a skilled scientist and assistant of the Lumière brothers

Summary of Event

In 1882, Antoine Lumière, painter, pioneer photographer, and father of Auguste and Louis, founded a factory to manufacture photographic gelatin dry-plates. After the Lumière brothers took over the factory's management, they expanded production to include roll film and printing papers in 1887 and also carried out joint research that led to fundamental discoveries and improvements in development and other aspects of photographic chemistry.

While recording and reproducing the actual colors of a subject was not possible at the time of photography's inception (about 1822), the first practical photographic process, the daguerreotype (silver halide process invented by Jacques Daguerre in 1837), was able to render both striking detail and good tonal quality. Thus, the desire to produce full-color images, or some approximation to realistic color, occupied the minds of many photographers and inventors—including Louis and Auguste Lumière—throughout the nineteenth century.

In the course of their research into the area of color photography, Auguste and Louis attempted to make the Lippmann interference process (a method of recording full-color images directly by means of interference between wavelengths, without the use of dyes or colorants, which was invented by Gabriel Lippmann in 1891) commercially possible as early as 1892. While the Lumière brothers became the most successful investigators and developers of the process, it never became a practical technique, since the emulsion was at least 100,000 times slower than that of ordinary films.

Reproducing the colors of nature by indirect means, however, was found to be possible in two ways: by the addition of colored lights or by the mixture of various pigments. Although both were proposed at the same time, the first process that met with any practical success was based on the theory, expounded by James Clerk Maxwell in 1861, that any color can be created by adding together red, green, and blue light in definite proportions. This theory of additive color holds true only for colored light; the subtractive color process, on the other hand, is based on the mechanical mixture of pigments. While white light is produced by reflecting or adding all the light rays that fall upon it (namely, the additive primary colors: red, green, and blue), black absorbs or subtracts all the rays falling on it (namely, the subtractive primary colors: cyan, magenta, and yellow).

Maxwell, taking three negatives through screens or filters of these additive primary colors, and making slides made from them projected through the same filters onto a screen so that their images were superimposed, found that it was possible to reproduce the exact colors as well as the form of nature. Unfortunately, since colors could not be recorded in their tonal relationships before the end of the century, Maxwell's experiment was unsuccessful. Although Frederick E. Ives of Philadelphia, in 1892, optically united three transparencies so that they could be viewed in proper register (exact matching in position of the filter and the plate) by looking through a peephole, viewing the transparencies was still not as simple as looking at a black-and-white photograph.

The first practical method of making a single photograph that could be viewed without any apparatus was devised by John Joly of Dublin in 1893. Instead of taking three separate pictures through three colored filters, he took one negative through one filter minutely checkered with microscopic areas colored red, green, and blue. The filter and the plate were exactly the same size and were placed in contact with each other in the camera. After the plate was developed, a transparency was made, and the filter was permanently attached to it. The black and white areas of the picture allowed more or less light to shine through the filters; if viewed from a proper distance, the colored lights blended to form the various colors of nature. Thus, while the principles of additive and subtractive color and their potential applications in photography had been discovered and even experimentally demonstrated by 1880, a practical process of color photography utilizing these principles could not be produced until a truly panchromatic emulsion was available, since all methods depend upon making a record of the primary color content of light from the subject. This necessary improvement in emulsion response was achieved from 1903 to 1906 and was immediately applied to color photography by the Lumière brothers.

Louis and Auguste Lumière, along with their research associate Alphonse Seyewetz, succeeded in creating a single-plate color process in 1903. It was introduced commercially as the autochrome plate in 1907 and was soon in use throughout the world. Autochrome was a random-dot screen process that created its effect by the additive color synthesis. A glass plate was coated with a double layer of starch grains dyed to act as primary additive color filters; spaces between grains were filled with carbon dust so that no unfiltered light would degrade the image. After the filter layer was coated with a panchromatic emulsion, the plate was exposed through the base (glass) side so that the subject colors were analyzed by the filter layer before exposing the emulsion. Reversal processing developed the emulsion to a black-and-white negative image and then converted it to a positive transparency. When viewed from the emulsion side (to correct for the left-right reversal caused by the camera lens), the filter layer colored the transmitted light, while the positive image modulated the intensities; the eye blended the additive color sensations thus received to produce a full-color image. Later versions of the plate used resin globules instead of the starch and carbon particles.

In order for details of the image to be analyzed in red, green, and blue, the filter elements of the screen were required to be extremely small. They were, in fact, dyed particles that averaged 0.015 millimeter in size. Various particle materials were used at different times, including yeasts, dried ferments, bacilli, and powdered enamels; however, potato starch grains were the most successful in early versions of the process.

Separate batches of particles were dyed red, green, and blue and were blended to produce a mixture with a uniformly neutral (gray) appearance. A glass plate was covered with a thin layer of transparent varnish, and the particle mixture dusted on. When dry, this layer was coated with varnish and a second layer was dusted on. Finally, carbon black (powdered charcoal) was dusted on to fill any remaining spaces between the colored particles of the second layer. The carbon particles blocked any light from passing completely through the screen in areas where spaces in both layers coincided, or through only the grains of the lower layer below spaces in the upper layer; either case would permit unwanted light to degrade the final image.

When the screen was completed, it was coated with a panchromatic emulsion. Exposure was through the base side of the plate so that the screen could analyze the light before it affected the emulsion. The processed image was viewed from the emulsion side to counteract the left-right reversal produced by the camera lens.

While this method entails color separation through the three filters, each separation "negative" is the size of one tiny starch grain. Employing special chemical treatment after development, the image is transformed from negative to positive, thus becoming a transparency of full photographic values, plus the colors supplied by the starch grains.

This process is one of the many that takes advantage of the limited resolving power of the eye. Grains or dots too small to be recognized as separate units are accepted in their entirety and, to the sense of vision, appear as tones and continuous color. Every grain of filter color in the autochrome screen serves two purposes: First, it separates the colors of the subject for photographic purposes, and then it permits itself to be seen in its allotted space.

One drawback to the autochrome plate was that it required forty to sixty times more exposure than the best black-and-white plates. In addition, although it was possible theoretically to contact-print one plate onto another, or to copy it with a camera, in practice, the results generally were unacceptable. Sharpness suffered because the plates could not be placed emulsion-to-emulsion (if they were, light exposing the second plate would not be screened), the grain factors of the two screens multiplied to produce a grainier image, and colors changed because the transmission characteristics of the screens were not perfect and tended to differ from batch to batch. Therefore, each autochrome plate was unique.

Impact of Event

The Lumières' autochrome plate paved the way for the commercial success of color photography. Prior to their discovery, color photography involved the difficult process of producing three negatives to form one color image. The Lumière brothers, by obtaining the three necessary color exposures on one plate, made color photography accessible to everyone with an ordinary camera.

The autochrome plate was rivaled eventually by other processes, especially Agfa, which has a granular screen quite similar to the autochrome; Dufaycolor, which is a flexible film, but one in which the color screen is a very fine mosaic instead of an unsymmetrical arrangement of grains; and the Finlaycolor process, in which the color screen is not integral with the color plate, but is located on a separate glass. It is placed in contact with the sensitive plate in the plate holder, and the exposure made through it results in the usual minute color separations; however, the plate is developed and fixed as a negative after the color screen has been removed. The Finlay screen is also a symmetrical mosaic of color elements; when transparent positives are made from the negative, the mosaic image on the negative becomes symmetrically duplicated on the positive. Color is supplied by the use of a viewing screen, which has identical color arrangement and can be registered with the transparency. Since these viewing screens can be purchased in quantities and matched up with additional positives, duplicate color transparencies can be obtained from one negative. Aligning a Finlay screen over a transparency requires great care; misalignment causes either a moiré (wavy) pattern or inaccurate colors.

While the autochrome plate remained one of the most popular color processes until the 1930's, soon this process (as well as other additive color processes, including Agfa, Dufay, and Finlay) was superseded by subtractive color processes. Leopold Mannes and Leopold Godowsky, both musicians and amateur photographic researchers who eventually joined forces with Eastman Kodak research scientists, did the most to perfect the Lumière brothers' advances in making color photography practical. Their collaboration led to the introduction in 1935 of Kodachrome, a subtractive process in which a single sheet of film is coated with three layers of emulsion, each sensitive to one primary color. A single exposure produces a color image.

Color photography is now commonplace. The amateur market is enormous, and the snapshot is nearly always taken in color. Commercial and publishing markets use color extensively. Even photography as an art form, which has been in black and white for most of its history, has turned increasingly to color.

Bibliography

Eder, Josef Maria. *History of Photography.* Translated by Edward Epstean. Reprint.

New York: Dover, 1972. In this lengthy and excellent work, the author discusses the Lumières' autochrome process in the context of the historical development of color photography. The work contains a brief biography of the author, extensive notes, and numerous reproductions.

Mees, C. E. Kenneth. *From Dry Plates to Ektachrome Film: A Story of Photographic Research*. New York: Ziff-Davis, 1961. In this work, Mees, a scientist at Kodak, discusses the Lumière brothers' contribution to photographic science in a chapter devoted to color photography. Contains numerous illustrations and photographs.

Newhall, Beaumont. *The History of Photography: From 1839 to the Present*. 5th rev. ed. New York: Museum of Modern Art, 1982. This standard work on the historical development of photography offers a brief discussion of the development of the autochrome process. Newhall offers the history of a medium rather than a technique. Work contains many black-and-white illustrations.

Ostroff, Eugene, ed. *Pioneers of Photography.* Springfield, Va.: SPSE, 1987. The Lumière brothers' work is described in several essays in this excellent collection. The autochrome process is discussed in its historical context rather than in extensive technical detail. Work contains bibliographic entries after each essay and numerous illustrations and photographs.

Sipley, Louis Walton. *Photography's Great Inventors*. Philadelphia: American Museum of Photography, 1965. This slim but informative volume devotes one section to the work of the Lumière brothers. The sketch contains biographical information as well as a brief discussion of the autochrome process in the context of the brothers' contribution to phototechnology. Short bibliography follows each entry.

Genevieve Slomski

Cross-References

The First Color Television Broadcast Takes Place (1940), p. 1166; Land Invents a Camera/Film System That Develops Instant Pictures (1948), p. 1331.

HARRISON OBSERVES THE DEVELOPMENT OF NERVE FIBERS IN THE LABORATORY

Categories of event: Biology and medicine
Time: Spring, 1907
Locale: Yale University, New Haven, Connecticut

Using the new technique of tissue culture, Harrison observed the development of nerve fibers from neural tissue removed from a frog embryo

Principal personages:

ROSS GRANVILLE HARRISON (1870-1959), a Yale University zoologist and longtime managing editor of *Journal of Experimental Zoology*

SANTIAGO RAMÓN Y CAJAL (1852-1934), a distinguished Spanish histologist who was awarded the 1906 Nobel Prize in Physiology or Medicine for his work on neuroanatomy

WILHELM HIS (1831-1904), a Swiss anatomist known for his work on the origin of nerve tissue in the embryo

VICTOR HENSEN (1835-1924), a German physiologist who developed early theories on nerve fiber development

THEODOR SCHWANN (1810-1882), a German biologist who was the developer of the cell theory of nucleated cells and an early investigator of the origin of tissue in embryos

Summary of Event

With the formulation of the cell theory by Theodor Schwann in 1839, the cellular nature of differentiated tissues such as those found in the nervous system had become apparent. During the ensuing decades, the nature of the nervous system had undergone extensive investigation, and an overview of its embryonic development had become firmly established as the nineteenth century drew to a close. It was clear that during early stages of embryonic development, the neural cleft, or tube, consisted of histologically identical cells. As differentiation proceeded, peripheral nerve fibers began to extend from the neural tube in the form of elongated structures called axons, the ends of which were characterized by an extensive network of fine processes. The nature of the formation of these nerve fibers, the development of which resulted in the formation of cell connections within the nervous system, was the source of considerable controversy among those scientists who studied neural biology.

Three major theories had been advanced that attempted to account for their formation. Two of these hypotheses were based primarily on observations during the embryonic stage of development and were limited by the extent of nineteenth century experimental technology. During the latter part of the 1830's, Schwann discovered the presence of a membranous sheath that surrounded certain forms of nerve fibers, the source of which is named the Schwann cell. It was suggested that the nerve fiber was, in fact, derived from the Schwann cell. Further observations eventually demonstrated the separate nature of the fiber and its surrounding sheath, but the theory did hold sway among a number of investigators. A second, more credible theory suggested that nerve fibers resulted from the differentiation of a preformed set of protoplasmic connections, or bridges, the nature of which was somewhat nebulous. The major proponent of this theory, Victor Hensen, was himself a prominent neurophysiologist. While Hensen acknowledged a role for protoplasmic movement in the formation of the nervous system, he argued that there was insufficient evidence to suggest that such movement resulted in the outgrowth of nerve processes. In 1906, Hans Held published a modification of Hensen's theory, in which he suggested that the protoplasmic bridges observed by Hensen served as a type of scaffold, or "substratum," into which the protoplasmic material from the neural cells extended. As Ross Granville Harrison pointed out in 1910 in his classic paper on the subject, Held's treatment of Hensen's theory was in reality a refutation.

The third theory of fiber development, initially the work of Wilhelm His, but more extensively developed by Santiago Ramón y Cajal, suggested that the formation of the fibers resulted from protoplasmic outgrowths of preexisting cells within the nerve ganglia. Although the latter two schools of thought on fiber development were, to some extent, mutually exclusive, they represented differences of opinions based on observations of minute histological detail, often using material prepared in an identical manner. It was clear that a correct understanding of the process was dependent upon a new experimental approach.

In 1901, while at Johns Hopkins University, Harrison began a series of experiments in which he observed the formation of peripheral nerves during the early stages of frog embryo development. First, he was able to establish that only nerve cells were involved in axon formation and that the Schwann cell was clearly extraneous. In a series of experiments that he continued upon joining the faculty at Yale University in 1907, Harrison also demonstrated that nerve ganglia were the "one essential element in the formation of the nerve fiber." Axons never developed following removal of the nervous system from embryos; when nerves in the tadpole fin were severed, additional fibers extending from the severed nerves could be observed. Further, when ganglia were transplanted into other areas of the embryo, the development of nerve fibers could be observed in those areas. These experiments persuaded Harrison that the theories promoted by His and by Ramón y Cajal were probably correct.

The primary difficulty in establishing the validity of their view was the inability to study fiber development in the absence of extraneous tissue. It was necesary to test the ability of the undifferentiated nerve cell, the neuroblast, to develop nerve fibers outside local influences, particularly other types of tissue. If the neuroblast was the source of the axon, it should still be capable of forming a nerve fiber even when in the presence of a foreign medium. By 1907, Harrison had solved this problem with his development of a method for growing animal cells or tissues in culture outside the organism. Initially, Harrison placed sections of tissue, extracted from frog embryos at a stage prior to nerve fiber development, in drops of nutrient fluid at physiological concentrations on the underside of a microscope slide. When the material was sealed with paraffin, the preparation could be observed over a period of several days. Under these conditions, no differentiation of nerve tissue could be observed. Harrison achieved better results when he introduced a solid gelatinous substrate suitable for attachment and cell movement. The rationale was that clotted serum would best resemble chemically the natural embryonic environment. Harrison was aware also of the earlier experiments carried out by Leo Loeb, who in 1902 transplanted within the bodies of animals small fragments of epidermal tissue that had been embedded in agar or clots of blood. Initially, Harrison tried clotted chicken serum as a substrate, but found frog lymph to be more useful. Using careful, aseptic techniques, Harrison found he was able to maintain his specimens as long as four weeks in culture.

After considerable manipulation of experimental techniques, Harrison was able finally to devise a procedure that allowed him to carry out the necessary observations, ultimately involving more than two hundred preparations. Within isolated pieces of undifferentiated nerve tissue, Harrison observed protoplasmic (amoeboid) movement that resolved itself into numerous fibers extending into the surrounding clot of frog lymph. He noted the enlarged, swollen ends of the fibers, which closely resembled those found in sections from normal embryos at an analogous stage of development. The rate of growth of the axons was particularly striking. In his original paper of 1907, Harrison observed the formation of one fiber, 20 microns in length, in less than thirty minutes. The largest fibers measured at that time were 0.2 millimeter in length. Repeated observations refined these calculations further. He found the rate of lengthening of the nerve fiber to vary considerably, ranging anywhere from 16 microns per hour to a maximum rate of 56 microns per hour. The largest fiber measured reached a length of 1.15 millimeters; Harrison noted that development was followed over its entire period of growth, a total of fifty-three hours.

Harrison was concerned that it be firmly established that what he observed in culture was a true analogy of the situation within the embryo. Consequently, he followed the activities in culture of several types of embryonic tissue. In all cases, he found that each type of isolated tissue underwent development in a manner similar to what was known to occur in the animal. Also, it was readily apparent that similar procedures could be applied to study the influences of various experimental conditions on the development of the tissue. Harrison reported his results in 1907, and the work was immediately the recipient of considerable public acclaim.

Impact of Event

The publication, in 1907, of Harrison's research into cultivation of animal tissues in culture quickly made an impact on the work of other cell biologists. The popular media proclaimed it to be a notable scientific discovery. Originally, Harrison had chosen the frog, a cold-blooded animal, because of the lower incubation temperature necessary to maintain the cells. Interest, however, began to develop into research on tissues from warm-blooded animals. Harrison's student, Montrose Burrows, attempted to grow explants from chick embryos in blood plasma, with limited success. By 1912, Burrows had joined the laboratory of Alexis Carrel, where it was found that explants could be made to grow indefinitely by periodic passage into fresh media. Carrel also developed a glass vessel, the Carrel flask, in which the cells could be maintained.

For some years, Harrison continued his research in tissue culture. Generally, his experiments involved modification of various factors, such as the forms of solid support or type of fluid medium. Harrison published this work in 1914, marking an end to his studies on the subject.

In 1917, the Nobel Committee recommended that Harrison be presented the award in Physiology or Medicine "for his discovery of the development of the nerve fibers by independent growth from cells outside the organism." Unfortunately, because of World War I, the Committee decided it would not award a prize. Harrison was nominated again in 1933, but the Committee refused to recommend an award for his work, ironically citing the justification that his experimental methods were by then of "limited value."

Adaptation of Harrison's methods continued. In 1943, Wilton Earle developed a mouse cell line that had the ability to grow indefinitely in the laboratory. Katherine Sanford, Gwendolyn Likely, and Earle demonstrated the possibility of growing single cells in laboratory vessels when they cloned Earle's mouse cells in 1948. Even human cells could be shown to grow in culture, when in 1952, George Gey established a cell line developed from an explant of cervical carcinoma. Normal human cells could be maintained in culture also. In 1961, Leonard Hayflick described the characteristics of human cells derived from fetal tissue and provided an extensive comparison of these cells with their counterparts in "immortal," often cancerous, cell lines.

The impact of this technology on research in the biological sciences cannot be overestimated. The effects of chemical or environmental changes on cells can be monitored easily. Genetic defects can be determined through the growth of cells extracted by amniocentesis, and even research at the molecular level can be carried out. Perhaps the greatest impact has been in the field of virology. Since viruses are intracellular parasites, they could be grown only in living animals. With the development of cell and tissue culture, however, growth of viruses could be carried out in laboratory vessels. In 1949, John Enders and his coworkers were able to propagate poliovirus in culture, an event that culminated in the development of the polio vaccine within a decade. By 1990, vaccines had been developed in a similar manner against most major viral illnesses.

Bibliography

Harrison, Ross Granville. "The Cultivation of Tissues in Extraneous Media as a Method of Morphogenetic Study." *Anatomical Record* 6 (1912): 181-193. By

1912, Harrison had completed most of his work on growth and differentiation of tissue in culture. This work, originally presented as a lecture for a symposium on tissue culture held in 1911, was a description of the theoretical basis for his earlier experiments and their relationship to morphogenesis.

__________. *Organization and Development of the Embryo*. Edited by Sally Wilens. New Haven, Conn.: Yale University Press, 1969. Includes a thorough description of Harrison's work on embryonic development, especially the importance of biological symmetry. A chapter on the role of tissue culture provides background to his early work.

__________. "The Outgrowth of the Nerve Fiber as a Mode of Protoplasmic Movement." *Journal of Experimental Zoology* 9 (1910): 787-846. Harrison's classic work on the subject. Provides a complete description of Harrison's work. Includes a description of the procedures followed by Harrison as he became the first person to grow animal tissues outside the body.

Nicholas, J. S. "Ross Granville Harrison." In *Biographical Memoirs*. The National Academy of Sciences of the United States. Vol. 35. New York: Columbia University Press, 1961. Nicholas has produced an outstanding biography of Harrison. A concise, yet thorough, synopsis of Harrison's career. Includes a discussion of the controversy that arose over the failure of the Nobel Committee to award a prize to Harrison for his work.

Pollack, Robert, ed. *Readings in Mammalian Cell Culture*. 2d ed. Cold Spring Harbor, N.Y.: Cold Spring Harbor Laboratory, 1981. A compilation of articles covering the history of mammalian cell culture. Included is a copy of Harrison's 1907 article, "Observations on the Living Developing Nerve Fiber," the work that established firmly that nerve fibers develop from central nerve cells. A brief synopsis of each paper is provided.

Richard Adler

Cross-References

Ramón y Cajal Establishes the Neuron as the Functional Unit of the Nervous System (1888), p. 1; Landsteiner Discovers Human Blood Groups (1900), p. 56; Salk Develops a Polio Vaccine (1952), p. 1444; Sabin Develops an Oral Polio Vaccine (1957), p. 1522.

HABER DEVELOPS A PROCESS FOR EXTRACTING NITROGEN FROM THE AIR

Category of event: Chemistry
Time: 1908
Locale: Germany

Haber developed the first successful method for converting nitrogen from the atmosphere and combining it with hydrogen to synthesize ammonia, a compound used as a fertilizer for agriculture

Principal personages:

CARL BOSCH (1874-1940), a German chemist who was the cowinner of the 1931 Nobel Prize in Chemistry and inventor of methods for synthesis of gasoline

CARL ENGLER (1842-1925), a German chemist who originated a theory for the formation of petroleum and natural gas by decay of prehistoric animals

FRITZ HABER (1868-1934), a German chemist who was awarded the 1918 Nobel Prize in Chemistry and invented poison gas

WALTHER HERMANN NERNST (1864-1941), a German physicist and chemist who developed the third law of thermodynamics

WILHELM OSTWALD (1853-1932), a German chemist and 1909 Nobel laureate in chemistry who is sometimes called the father of physical chemistry

Summary of Event

The nitrogen content of the soil, essential to plant growth, is maintained normally by the deposition and decay of old vegetation and by nitrates in the rainfall. If, however, the soil is used extensively for agricultural purposes, more intensive methods must be used to maintain soil nutrients, such as nitrogen. One such method used by farmers is crop rotation, in which successive divisions of a farm are planted in rotation with clover, corn, or wheat, for example, or allowed to lie fallow for a year or so. The clover is able to absorb nitrogen from the air and deposit it in the soil from its root nodules. As population has increased, however, farming has become more intensive, and the use of artificial fertilizers—some containing nitrogen—has become almost universal.

Nitrogen-bearing compounds, such as potassium nitrate and ammonium chloride, have been used for many years as artificial fertilizers. Much of the nitrate used, mainly potassium nitrate, came from Chilean saltpeter, of which a yearly amount of half a million tons was imported at the beginning of the twentieth century into Europe and the United States for use in agriculture. Ammonia was produced by dry distillation of bituminous coal and other low-grade fuel materials. Originally, coke ovens discharged this valuable material into the atmosphere, but more economical methods were found later to collect and condense these ammonia-bearing vapors.

In the first decade of the twentieth century, the cyanamide process for using atmospheric nitrogen was developed by Adolf Frank (1834-1916) and Heinrich Caro (1834-1910). The Frank-Caro process produced calcium cyanamide through the reaction of nitrogen from the air on calcium carbide, obtained from limestone and coal in an electric arc. Cyanamide can be converted readily into ammonia by reacting it with water. Until 1904, however, the direct combination of nitrogen with hydrogen to form ammonia had not been accomplished on a commercial scale. William Ramsay (1852-1916) and Sydney Young (1857-1937) had conducted experiments in England in 1884 to do this, but with unsatisfactory results.

In 1904, Fritz Haber and his coworkers began experiments on the conversion of nitrogen and hydrogen into ammonia by passing various mixtures over catalytic agents, such as iron filings at temperatures of up to 1,000 degrees Celsius and at various pressures. (Catalytic agents speed up a reaction without affecting it otherwise.) In the process, they studied the equilibrium conditions in the reaction and found that, at pressures of about two hundred atmospheres and temperatures not above 500 degrees Celsius, the production of ammonia in practical yields from its elements could be carried out in a continuous process. Other catalysts, such as uranium and osmium, produced even higher yields but were considerably more costly than iron.

In 1910, a factory for the production of ammonia from atmospheric nitrogen by the Haber process was erected near Frankfurt am Main, Germany; it had an annual production capacity of 10,000 tons of ammonia. Wilhelm Ostwald had developed a method for converting ammonia into nitric acid, which, in turn, could be converted readily into calcium nitrate by reacting it with limestone. A hydroelectric method of conversion was developed in Norway, but it required a cheap source of electric power. In areas where no such cheap source was readily available, including Germany, the Haber process for making ammonia was the preferred one.

Germany had practically no source of fertilizer-grade nitrogen at the beginning of the twentieth century. Almost all its supply came from the deserts of northern Chile. As demands for nitrates increased, it became apparent that the calculated supply from these vast deposits would be exceeded. Other sources needed to be found, and the almost unlimited supply of nitrogen in the atmosphere (80 percent nitrogen) was an obvious source.

The combination of nitrogen and hydrogen to form ammonia could be induced by electrical discharge, but this required enormous amounts of electrical energy, then available only where there was hydroelectric power. When Haber began his experiments on ammonia production in 1904, there was no known method for the spontaneous conversion of nitrogen and hydrogen into ammonia.

Nitrogen and oxygen will combine in an electric arc to form oxides of nitrogen, as Walther Hermann Nernst had discovered. Because oxides of nitrogen can be converted readily into nitric acid and, in turn, nitrates, this possibility aroused considerable interest as a process for "nitrogen fixation," as the process of direct combination of atmospheric nitrogen into soluble compounds became known. The large amounts of electrical energy necessary to maintain the electric arc, and to heat large masses of air to high temperatures, imposed practical limits on this method. Haber spent many years studying the synthesis of nitric oxide by electrical discharge.

Haber gave up the idea of fixing nitrogen from the atmosphere by oxidizing it to nitric oxide. He turned, instead, to reducing it with hydrogen to ammonia. First, he looked into combining nitrogen and hydrogen by corona discharge and by sparking but came to the conclusion that this method would not be the most advantageous. He decided to repeat the experiments of Sir William Ramsay and Young, who in 1884 had studied the decomposition of ammonia at about 800 degrees Celsius. They had found that a certain amount of ammonia always was left undecomposed. In other words, the reaction between ammonia and its constituent elements—nitrogen and hydrogen—had reached a state of equilibrium. Haber decided to determine the position of this equilibrium in the vicinity of 1,000 degrees Celsius. He approached the equilibrum from both sides, by reacting pure hydrogen with pure nitrogen, and by starting with pure ammonia gas and using iron filings as a catalyst. Having determined the position of the equilibrium, he next varied the catalyst and found nickel to be as effective as iron, and calcium and manganese even better. At 1,000 degrees Celsius, the rate of reaction was enough to produce practical amounts of ammonia continuously, if it were to be washed out with water from the circulating system.

Further work by Haber showed that increasing the pressure also increased the percentage of ammonia at equilibrium. For example, at 300 degrees Celsius, the percentage of ammonia at equilibrium at one atmosphere pressure was only 2 percent, but at two hundred atmospheres, the percentage of ammonia at equilibrium was 63 percent. Haber's compressor would not provide pressures above two hundred atmospheres, but with catalysts, he was able to obtain rapid combination of nitrogen and hydrogen at about 700 degrees Celsius. He found that he could obtain a yield of several grams of ammonia per hour per cubic centimeter of heated pressure chamber and believed that such a process could be a commercial success. A pilot plant was constructed and was successful enough to impress a chemical manufacturing concern, Badische Anilin- und Soda-Fabrik (BASF). They agreed to study Haber's process and to investigate different catalysts on a large scale. The company, however, substituted cheaper water-gas hydrogen for the purer electrolytic hydrogen that Haber had used and, therefore, ran into difficulties because of impurities introduced with the impure hydrogen. Once these difficulties were recognized and overcome, the process was a success.

Impact of Event

Haber's process for manufacturing ammonia from the nitrogen in the air was in such an advanced state of development when it left his laboratory that its success in industry was virtually assured. Even so, BASF did not see much promise in Haber's method at first. It was through an old friend of Haber, Carl Engler, a member of the BASF advisory council, that Haber's process was accepted.

In July, 1909, the BASF representative, Carl Bosch, came to Karlsruhe to watch a demonstration of Haber's process. (In his usual thorough manner, Haber had not only carried out the investigations of the ammonia equilibrium but also had invented a continuously circulating system whereby the mixture of nitrogen and hydrogen, under pressure, passed through new gas mixture being added as the ammonia produced was removed from the system.) Bosch was greatly impressed; within three years, he had a synthetic ammonia factory in operation. BASF soon built two huge plants.

With the beginning of World War I, the need for nitrates for use in explosives became more pressing in Germany than their need in agriculture. After the fall of Antwerp, 50,000 tons of Chilean saltpeter were discovered in the harbor and fell into German hands. Because the ammonia from Haber's process could be converted readily into nitrates, it became an important war resource. Haber's other contribution to the German war effort was his development of gas warfare. He was responsible for the chlorine gas attack on Allied troops at Ypres in 1915. He also directed research on gas masks and other protective devices.

At the end of the war, the 1918 Nobel Prize in Chemistry was awarded to Haber for his development of the process for making synthetic ammonia. Because the war was still fresh in everyone's memory, it became one of the most controversial Nobel awards ever made. A headline in *The New York Times* for January 26, 1920, stated: "French Attack Swedes for Nobel Prize Award: Chemistry Honor Given to Dr. Haber, Inventor of German Asphyxiating Gas." In a letter to the *Times* on January 28, 1920, the Swedish legation in Washington, D.C., defended the award.

Haber left Germany in 1933 under duress from the anti-Semitic policies of the Nazi authorities. He was invited to accept a position with the University of Cambridge, England, and died on a trip to Basel, Switzerland, a few months later, a great man whose spirit had been crushed by the actions of an ungrateful regime.

Bibliography

Berl, Ernst. "Fritz Haber." *Journal of Chemical Education* 19 (May, 1937): 203-207. An account of Haber's work in science by a personal friend. Includes Haber's tribute to Justus von Liebig.

Coates, J. E. "The Haber Memorial Lecture." *Journal of the Chemical Society (London)*, 1937, 1642-1672. A thirty-page biography of Haber given three years after his death. Describes his research in detail and laments the treatment Haber received from the Nazi regime in Germany.

Haber, Fritz. *Five Lectures*. Berlin: Springer-Verlag, 1924. An account by Haber of his research on ammonia synthesis. Defends his wartime work on gas warfare.

__________. *Thermodynamics of Technical Gas Reactions*. Translated by Arthur B. Lamb. London: Longmans Green, 1908. A model of critical insight into the history of thermodynamics; discusses the problem of the indeterminate constant in the free-energy equation. Following Max Planck, Haber explains the constant in terms of heat capacity and entropy. Nearly anticipates Nernst in announcing the Nernst heat theorem.

Jaenicke, Johannes. "The Gold Episode." *Die Naturwissenschaften* 23 (1935): 57. Relates Haber's unsuccessful attempt to extract gold from seawater, with the end of helping Germany pay the heavy reparations mandated by the Versailles Treaty after World War I.

Joseph A. Schufle

Cross-References

Teisserenc de Bort Discovers the Stratosphere and the Troposphere (1898), p. 26; The First Synthetic Vat Dye, Indanthrene Blue, Is Synthesized (1901), p. 98; Kipping Discovers Silicones (1901), p. 123; Willstätter Discovers the Composition of Chlorophyll (1906), p. 345; Burton Introduces Thermal Cracking for Refining Petroleum (1913), p. 573; Fabry Quantifies Ozone in the Upper Atmosphere (1913), p. 579.

HARDY AND WEINBERG PRESENT A MODEL OF POPULATION GENETICS

Category of event: Biology
Time: 1908
Locale: Stuttgart, Germany; Trinity College, Cambridge, England

Hardy and Weinberg formalized the first model for evaluating changes in gene frequency within a population and gave birth to the field of population genetics

Principal personages:

WILHELM WEINBERG (1862-1937), a German physician, geneticist, medical statistician, and early founder of population genetics, who demonstrated the importance of Mendel's laws to the genetic composition of populations

GODFREY HAROLD HARDY (1877-1947), a professor of mathematics at Trinity College and the University of Oxford, and a leading mathematician who recognized, shortly after Weinberg, the relevance of Mendel's laws of inheritance to the study of population genetics

GREGOR JOHANN MENDEL (1822-1884), an Austrian monk whose cross-breeding experiments on garden peas led to the discovery of the laws of inheritance, which eventually fueled Hardy and Weinberg's discovery

CHARLES DARWIN (1809-1882), an English naturalist whose book *On the Origin of Species* (1859) revolutionized science with its theory of natural selection, and provided the necessary context for Hardy and Weinberg's model

Summary of Event

Within a few decades following the publication of *On the Origin of Species* (1859) by Charles Darwin, the theory of natural selection had gained considerable approval by the scientific community and had revolutionized the way biologists viewed the natural world. This work provided a comprehensive explanation for both the origin and maintenance of the seemingly endless variation in nature. Stated simply, natural selection is the differential reproduction of heritable characteristics within a population, which ultimately leads to evolutionary change.

Despite almost immediate acceptance, the theory was incomplete in that it failed to account for a mechanism of heritability and an explanation for how such a mechanism could translate into the kinds of changes in populations and species that originally were predicted in Darwin's theory of natural selection.

The first of these problems was overcome in 1900, when the work by Gregor Johann Mendel was rediscovered independently by several researchers. In his quantitative assessment of breeding experiments on garden peas, Mendel demonstrated that many inherited traits in diploid organisms (those containing two sets of chromosomes) are determined by two factors, which are known now as genes (segments of chromosomes). From the frequency of traits in his populations of pea plants, Mendel reasoned that an organism receives one gene from each parent for all such heritable traits. In addition, he argued that alleles (alternate forms of the same gene) separate independently and randomly from one another when gametes (egg and sperm) form, but combine again during fertilization.

Almost immediately following the discovery of Mendel's laws of inheritance, several scientists began to realize the implications for the study of population genetics and evolutionary change. The first observations were noted by W. E. Castle in 1903. More complete analyses were presented in 1908 by the English mathematician, Godfrey Harold Hardy, and the German physician, Wilhelm Weinberg. These later works came to be known collectively as the Hardy-Weinberg law (or equilibrium) and eventually became the foundation for the study of population genetics.

The law of population genetics, credited to Hardy and Weinberg, was discovered and published independently, but almost simultaneously, by the two scientists. Weinberg was a physician with research interests in multiple births, medical statistics, and population genetics. Between 1908 and 1910, he published four critical papers relating Mendel's laws of inheritance to genetic changes in populations. For this work, he is considered by some as the father of population genetics. Hardy was a prominent mathematician who invested most of his time in the study of pure mathematics, on which he published nearly 350 papers. In late 1908, however, he deviated slightly from his main research interests when he published a short note to the editor of the journal *Science*, in which he made several observations concerning the relevance of Mendel's laws to populations, which were very similar to those pointed out by Weinberg. Together, their observations became known as the Hardy-Weinberg law.

The Hardy-Weinberg law states that, given simple patterns of Mendelian inheritance, the frequency of alleles in a population will remain constant from generation to generation, assuming certain ideal conditions are met. In other words, if these conditions hold true, allelic frequencies will not change, the genetic structure of the population will remain constant over time, and evolutionary change will not occur. These ideal conditions are as follows. First, the population must be a large, randomly breeding, or panmictic, population. In other words, all individuals in the population must have equal reproductive success. If this condition is not met, and certain individuals experience greater reproductive success than others, or if nonrandom breeding occurs as a result of small population size, then certain genes will be overrepresented in the next generation and the population's gene frequencies will change. A second necessary condition is that the population must be closed; that is, there must be no immigration or emigration of individuals in or out of the population. Third, there must be no spontaneous changes in alleles (mutations). Finally, there must be no differential success (selection) of certain alleles. Alleles must share equal probability of transmission to the next generation. For example, if some individuals possess alleles or combinations of alleles that, under certain environmental conditions, enhance their chances of survival and subsequent reproduction, then their genes will be represented more than those of others in the next generation, gene and allelic frequencies will change, and the population will adapt to environmental conditions. Such differential success of alleles is the essence of Darwin's theory of natural selection and the primary mechanism by which evolutionary changes proceed.

Given these conditions, the Hardy-Weinberg law asserts that changes in gene frequency in a population (evolution) will not occur. When any one or more of these conditions are violated, however, gene frequencies will be altered, and evolution will take place. Thus, by demonstrating the conditions necessary for evolution not to occur, Hardy and Weinberg were able to illustrate those factors that actually contribute to evolutionary change. In addition to the qualitative discussion of their argument, Hardy and Weinberg also presented their model in a purely mathematical form. To illustrate this quantitative expression of their model, it is easiest to consider a hypothetical population with a single gene and two alleles, represented by A and a. Recall also that each individual in the population possesses two alleles for this gene. Furthermore, let the frequency (numerical proportion) of the A and a in the population equal P and Q, respectively, and assume $P+Q=1$. As an example, assume P and Q equal 0.8 and 0.2; that is, the proportion of the A allele in the population is 80 percent and that of the a allele is 20 percent.

Hardy and Weinberg reasoned that if Mendel was correct and these two alleles separate randomly and independently as discrete units during formation of the egg and sperm, the frequency of allelic combinations in the next generation could be predicted mathematically. They stated that if P and Q are known, and if all conditions of the model hold true, then the frequency of allelic combinations in the next generation could be calculated easily from the equation: $P^2 + 2PQ + Q^2$, where P^2 is the frequency of individuals with two A alleles, 2PQ the frequency of individuals with one A and one a allele, and Q^2 the frequency of individuals with two a alleles. This mathematical expression of allele combinations, or genotypic frequencies, follows from a simple rule of mathematical probabilities, which states that the joint probability of two independent events is equal to the product of their individual probabilities. Thus, in the example above, the probability of acquiring two A alleles (or the proportion of individuals acquiring two A alleles) is equal to 0.8×0.8, or 0.64. Similarly, the probability of acquiring one A allele and an a allele in any order is $P \times Q \times 2$, or 0.32. In the absence of nonrandom mating, mutation, immigration, emigration, and selection, these frequencies will remain fixed at these values in all subsequent generations.

This mathematical treatment allowed biologists to predict the frequencies of allelic combinations in a population from the individual allelic frequencies. In a similar way, the frequencies of each allele could be calculated by working backward from the allelic combinations. The value of this model, however, was not in its applicability to the real world. In fact, biologists believe that natural populations rarely, if ever, meet all the ideal conditions required by the model. Instead, the

Hardy-Weinberg equation was a conceptual model that clearly illustrated how evolutionary change occurs at the population-genetic level. Although this model is limited in that the behavior of some alleles is not governed by Mendelian laws of inheritance, it is an important starting point for the study of population genetics.

Impact of Event

Hardy and Weinberg's law was a critical breakthrough in evolutionary biology that effectively linked Mendel's laws of inheritance with Darwin's theory of natural selection. It demonstrated clearly how cellular mechanisms of inheritance can translate into the microevolutionary changes that Darwin had predicted.

This synthesis was accomplished in two ways. First, by defining the conditions necessary for a population to remain genetically unchanged indefinitely, Hardy and Weinberg were able to specify those conditions that contribute directly to evolutionary change. Hence, a large, randomly breeding population in the absence of mutation, gene flow in and out of the population, and natural selection, will remain at equilibrium with respect to its allele or gene frequencies. When any one of these conditions is violated, however, gene frequencies will change in subsequent generations. Thus, Hardy and Weinberg's model helped biologists to understand how changes in gene frequencies can occur, and that such changes are, in fact, small-scale evolutionary events. In addition, the equilibrium model helped identify mechanisms of evolutionary change that were not obvious from Darwin's theory of natural selection. Although most biologists agree that differential reproduction of alleles, or natural selection, is the primary force driving evolution, it became clear from Hardy and Weinberg's work, and is still widely accepted, that other factors—such as random events and differing degrees of gene flow between populations—will contribute also to evolutionary change.

The second important observation to emerge from the Hardy-Weinberg law was that sexual reproduction alone does not result in a reduction in genetic variability or in genetic change. One of the problems faced by Darwin was explaining the maintenance of genetic variability. Darwin correctly assumed that genetic variability was the essential raw material on which natural selection acted. In the absence of an accurate mechanism of inheritance, however, he had incorrectly assumed that the inheritance process must result in a constant blending of genetic material and a loss in genetic variability as well. Because of his insistence that variability was necessary for natural selection, this posed a serious problem. From Mendel's work, it soon became clear that genetic material was transmitted in discrete units and therefore not mixed during reproduction. A short time later, Hardy and Weinberg demonstrated that, under the ideal conditions, gene frequencies and genetic variability will remain fixed indefinitely. Thus, a major problem in the theory of natural selection finally was resolved.

The synthesis of Mendel's and Darwin's work resulted in renewed interest in evolutionary biology and soon gave birth to the new field of population genetics. This field was advanced greatly during the 1920's, with the work of Sir Ronald Aylmer

Fisher, Sewall Wright, and John Burdon Sanderson Haldane; it continues to be one of the major fields of biological research.

In addition to its impact on basic research, the Hardy-Weinberg law has had several practical applications. Perhaps the most important of these is its use as a conceptual teaching model. Hardy-Weinberg equilibrium is employed in nearly every college-level biology text as a starting point for discussions on evolution, adaptation, and population genetics. In essence, it is a teaching tool that instructs beginning students in the way it taught evolutionary biologists at the beginning of the twentieth century. A second important application derived from the model concerns the manner and degree to which deleterious alleles manifest themselves within a population. The Hardy-Weinberg model shows how lethal alleles, such as those that code for fatal genetic diseases, can be maintained in a population at low frequencies. This simple conceptual model has had a tremendous impact on evolutionary biology and is certainly one of the great events in the history of science.

Bibliography

Arms, Karen, and Pamela S. Camp. *Biology: A Journey into Life*. 3d ed. Philadelphia: Saunders College Publishing, 1987. A general biology text for the layperson that provides a clear and concise description of the Hardy-Weinberg law and its implications. Well illustrated with colored photographs and graphs; includes a glossary of general biology terms.

Hanson, Earl D. *Understanding Evolution*. New York: Oxford University Press, 1981. A basic text on evolutionary biology that devotes considerable attention to population genetics and its relevance to evolutionary processes. Gives thorough coverage to Mendel's laws of inheritance and the Hardy-Weinberg law. Includes modest historical accounts of other relevant events. Well illustrated, with limited references.

Merrell, David J. *Ecological Genetics*. Minneapolis: University of Minnesota Press, 1981. A college-level text, best used as a reference. One of the first attempts to synthesize ecological processes with those of population genetics. Gives the Hardy-Weinberg law moderate coverage. Extensive references, with a few diagrams and figures.

Mettler, Lawrence E., Thomas G. Gregg, and Henry E. Schaffer. *Population Genetics and Evolution*. 2d ed. Englewood Cliffs, N.J.: Prentice-Hall, 1988. A modern text on population genetics that assumes a certain basic level of understanding. Provides a complete overview of current research efforts in the field. Well referenced, with short chapter summaries.

Raven, Peter H., and George B. Johnson. *Biology*. St. Louis, Mo.: Times Mirror/Mosby, 1989. A comprehensive introductory college text on the science of biology. Part 3, which contains five chapters, covers the basic concepts of genetics and heredity, and includes a detailed account of Mendel's laws of inheritance. Chapter 20 provides a friendly mathematical treatment of the Hardy-Weinberg law and a qualitative summary of the kinds of natural events that violate this law, and thereby alter population genetics. Colored illustrations and a detailed glossary.

Wilson, Edward O., and William H. Bossert. *A Primer of Population Biology.* Stamford, Conn.: Sinauer Associates, 1971. Intended for the beginning biologist. A small, concise handbook, perhaps one of the best introductory texts on classic population genetics. More than half the book concerns the basis of genetic change in populations. Although there is a strong mathematical orientation, problems are explained clearly. Well illustrated, with suggested readings and a short glossary of technical terms.

Michael A. Steele

Cross-References

De Vries and Associates Discover Mendel's Ignored Studies of Inheritance (1900), p. 61; Sutton States That Chromosomes Are Paired and Could Be Carriers of Hereditary Traits (1902), p. 153; Punnett's *Mendelism* Contains a Diagram for Showing Heredity (1905), p. 270; Bateson and Punnett Observe Gene Linkage (1906), p. 314; Morgan Develops the Gene-Chromosome Theory (1908), p. 407; Johannsen Coins the Terms "Gene," "Genotype," and "Phenotype" (1909), p. 433; Avery, MacLeod, and McCarty Determine That DNA Carries Hereditary Information (1943), p. 1203; Watson and Crick Develop the Double-Helix Model for DNA (1951), p. 1406.

HUGHES REVOLUTIONIZES OIL WELL DRILLING

Category of event: Applied science
Time: 1908
Locale: Houston, Texas

Oil well drillers-inventors Hughes and Sharp developed the rotary cone rock bit to accelerate rotary oil well drilling through hard rock formations

Principal personages:

HOWARD R. HUGHES (1869-1924), an American lawyer, drilling engineer, and inventor of the steel-toothed cone drill bit

WALTER B. SHARP (1860-1912), an American drilling engineer, inventor, and partner to Hughes

Summary of Event

The earliest oil well drilling had been accomplished by various adaptations of the cable tool, or the rod-and-drop percussion tool. As oil-bearing formations investigated became progressively deeper, the limits of these two techniques became increasingly apparent. Although conceived of and in limited use as early as the 1880's, the growth and spread of the hydraulic rotary drilling system was almost concurrent with the Texas Spindletop well of 1901, one of the first deep wells penetrating harder rock formations as well as softer clay and sand.

A rotary drill rig of the 1990's is basically unchanged in its essential components from its earlier versions of the 1900's. A drill bit is screwed to a line of hollow drill pipe. The latter passes through a square hole on a rotary table, which acts essentially as a horizontal gear wheel driven by an engine. The drill bit itself at the hole bottom is rotated by the connected sections of the above-lying pipe in contact with the rotating table. While being rotated, the drill bit and stem are free for feeding washwater, usually an oil and mud mixture stabilized with flocculating montmorillonite clay that increases its specific gravity, to prevent side caving and to seal off water and oil-bearing subsurface strata. The mud-laden water is pumped under high pressure down the sides of the rotary drill pipe, and jets out with great force through small holes in the rotary drill bit against the bottom of the borehole. This fluid then returns outside the drill pipe to the surface, carrying with it rock material cuttings from the subsurface. Circulated rock cuttings and fluids are regularly examined at the surface ("shale shaker"/"mud logger") to determine the precise type and age of rock formation and for signs of oil and gas. This continual washing of drill cuttings notably reduces rig downtime compared to cable and impact tools. Rotary drilling rates usually increase as additional drill-collar weights are applied to the drill bit by carefully adjusting drill pipe tension at the surface.

A key part of the total rotary drilling system is the drill bit, encompassing sharp cutting edges of some design, which make direct contact with geologic formations to be drilled. The first bits used in rotary drilling were paddlelike "fishtail" bits, fairly successful for softer formations, and tubular coring bits for harder geologies. In 1893, M. C. Baker and C. E. Baker brought a rotary water-well drill rig to Corsicana, Texas, for modification to deeper oil drilling. This rig led to the discovery of the large Corsicana-Powell oil field in Navarro County, Texas. This success also motivated its operators, the American Well and Prospecting Company, to begin the first large-scale manufacture of rotary drilling rigs for commercial sale.

In the earliest rotary drilling for oil, short fishtail bits were the tool of choice, insofar as they were at that time the best configured to "make hole" over a wide range of geologic strata without needing frequent replacement. Even so, in the course of any given oil well, many bits were required typically in Gulf coastal drilling. Especially when encountering locally harder rock units such as limestone, dolomite, or gravel beds, fishtail bits would typically either curl backward or break off in the hole, requiring the time-consuming work of pulling out all drill pipe and "fishing" to retrieve fragments and clear the hole. Because of the frequent bit wear and damage, numerous small blacksmith shops established themselves near drill rigs, dressing or sharpening bits by a hand forge and hammer. Each bit forging shop had its own particular ways of shaping bits, producing a wide variety of three- and four-winged bits of numerous long and short bit designs. Nonstandard bit designs were frequently modified further as experiments to meet the specific requests of local drillers encountering specific drilling difficulties in given rock layers.

In the early 1900's, there was no consensus as to which drill bit shape was best for a given type of subsurface rock. For example, although short, one-wing fishtails drilled holes faster and cleaner in soft shales, they offered less cutting edge and less reliability than multi-winged bits in the same material. Within a few years, several prototypes were patented for new bits variously seeking to combine short cutting bases with numerous smaller cutting teeth. In 1907 and 1908, patents wereobtained in New Jersey and Beaumont, Texas, for steel, cone-shaped drill bits incorporating a roller-type coring device with many serrated teeth. Later in 1908, both patents were bought by lawyer-driller-inventor Howard R. Hughes.

In 1906, with his partner Walter B. Sharp, Hughes successfully drilled the first oil well in overpressured shales using controlled drilling mud weights. Directly motivated by his petroleum prospecting interests, Hughes sought by trial and error to combine the best features of prior and recent drill bits into one overall design of greater reliability and flexibility. The general subsurface geologies encountered at the Corsicana and Spindletop fields often included complex sequences of argillaceous, arenaceous, and calcareous layers, in addition to thin stringer layers of hard anhydrites, gravels, and salt from diapiric intrusions. Between 1885 and 1937, the American geologist R. T. Hall (of the United States Geological Survey) worked tirelessly to establish the primary exploration stratigraphic nomenclatures for the highly variable lithofacies of the East Texas/Gulf Coastal Plain areas. The type of rock and total rock layer thickness, were the primary controls on drilling rate and efficiency. For example, although comparatively weak rocks like sands, clays, and soft shales could be drilled rapidly (at rates exceeding 30 meters per hour), in harder shales, lime-dolostones, and gravels, drill rates of 1 meter per hour or less were not uncommon. Conventional drill bits of the time had average operating lives of three to twelve hours. Economic drilling mandates increases in both bit life and drilling rate. Hughes undertook what were probably the first recorded systematic studies of drill bit performance as a function of specific rock physical properties.

Although many improvements in detail and materials have been made to the Hughes cone bit since its inception in 1908, its basic design has remained in use in rotary drilling. One of Hughes's major innovations was the much larger size of the cutters, symmetrically distributed as a large number of small individual teeth on the outer face of two or more cantilevered bearing pins. In addition, hard facing was employed to drill bit teeth to increase usable life. Hard facing is a metallurgical process basically consisting of welding a thin layer of a hard metal or alloy of special composition onto a metal surface to increase its resistance to abrasion and heat. A less noticeable but equally essential innovation, not included in other cone bit patents, was an ingeniously designed gauge surface that provided strong uniform support for all the drill teeth. The force-fed oil lubrication was another new feature also included in Hughes's patent and prototypes, reducing the power necessary to rotate the bit by 50 percent over that of prior mud or water lubricant designs.

Impact of Event

Within six years of their patents and first manufacturing, Hughes's cone drill bits were being widely used with great success in oil and water well drilling in numerous states and foreign countries. The period 1908 to 1925 was one of intensive innovation in rotary drilling and bit designs. In 1913, a patent was filed on the first cross-roller rock bit. During his sixteen remaining years as chief of manufacturing and research of rotary rock bits, Hughes developed and tested many more ideas, and obtained an additional seventy-three patents, which ensured his company of a secure leadership role for many years.

As oil wells were drilled progressively deeper and through harder formations, a new variety of the rolling cone bit, the so-called reaming roller, was required to ensure a stable and uniformly sized drill hole. Although roll-cutter rock bits did not come into wide and regular use before the development of high-powered circulating pumps specifically for oil field service in 1915, cone bits and rotary rigs gradually replaced other drilling technologies.

This trend was notably furthered when, in 1925, the first superhard facing allowed, such as stellite, were used on cone drill bits. In addition, the first so-called self-cleaning rock bits appeared from Hughes, with significant advances in roller bearings and bit tooth shape translating into increased drilling efficiency. The much larger teeth were more adaptable to drilling in a wider variety of geological formations than earlier models. In 1928, tungsten carbide was introduced as an additional bit facing hardener by Hughes metallurgists. This, together with other improvements, resulted in the Hughes "ACME" tooth form, which has been in almost continuous use since 1926.

To further efforts in reducing drill bit wear, the Hughes company established an active and continually funded research team of mechanical engineers and metallurgists to develop methods of preparing and testing new alloys and treatments for improving the all-critical bit contact face cutting edges. These required extensive ongoing tests over many years, comparatively uncommon in a boom-and-bust oil industry where research and development has traditionally been the first department to be cut. The American Petroleum Institute in 1932 recommended mandatory practices for hard facing rotary drill bits, largely derived from results of the Hughes Company. Even after Hughes's death in 1924, part of the reason for the continued success of his rotary drill bits was his company's combined efforts at improving both engineering design and metallurgical composition through research. Eventually, rotary cone rock bits have increased footage drilled per hour by more than 80 percent over original figures. Many other improvements in other drilling support technologies, such as drilling mud, mud circulation pumps, blowout detectors and preventers, and pipe properties and connectors have enhanced rotary drilling capabilities to new depths (exceeding 5 kilometers in 1990). The successful experiments by Hughes in 1908 were critical initiators of these developments.

Bibliography

Brantly, John Edward. *History of Oil Well Drilling*. Houston, Tex.: Gulf Publishing, 1971. The primary source for the history of the developments in rotary and other drilling methods. Very well referenced and illustrated.

Cernica, John N. *Principles of Rock Fragmentation*. New York: Holt, Rinehart and Winston, 1982. An advanced undergraduate reference. Gives the theory, as well as comparative laboratory field results for rock fracture under various conditions of loading and drill rotation.

Isler, C. *Well-Boring for Water, Brime, and Oil*. London: E. & F. N. Spon, 1902. Notes independent and concurrent developments in rotary drilling.

Jackson, Elaine. *Lufkin: From Sawdust to Oil*. Houston, Tex.: Gulf Publishing, 1982. Accurately recounts much of the economic and personal history of the early growth of the Texas oil industry.

Jeffrey, Walter H. *Deep Well Drilling*. 2d rev. ed. Houston, Tex.: Gulf Publishing, 1925. Chapter 5 indicates the sequences of technical improvements associated with the wider adoption of the steel-toothed cone rotary bit.

Paxson, Jeanette. *Basic Tools and Equipment for the Oil Field*. Austin, Tex.: Petroleum Extension Service, 1982. Part of a specially designed self-tutoring course on the principles and practices of rotary drilling.

Sidorov, N. A. *Drilling Oil and Gas Wells*. Chicago: Imported Publications, 1986. A technical treatment of rotary drilling. Includes discussions of some of the deepest wells drilled on land.

Gerardo G. Tango

STEINMETZ WARNS OF POLLUTION IN *THE FUTURE OF ELECTRICITY*

Category of event: Earth science
Time: 1908
Locale: New York City

Steinmetz warned of air pollution from the burning of coal and water pollution from sewage disposal into rivers, summarizing early recognition of environmental impacts from population growth and urbanization

Principal personages:

CHARLES PROTEUS STEINMETZ (1865-1923), an American electrical engineer at General Electric Co. and professor at Union College, Schenectady, New York

WILLIAM THOMPSON SEDGWICK (1855-1921), a professor of biology and lecturer on public health and sanitary science at the Massachusetts Institute of Technology in Cambridge

ALLEN HAZEN (1869-1930), a chemist and sanitary engineer who was the head of the Lawrence Experiment Station of the Massachusetts Board of Health

Summary of Event

In a 1908 lecture delivered at the New York Electrical Trade School, Charles Proteus Steinmetz presented a prophetic message on future impacts of the development of electricity. Steinmetz challenged students to reduce the cost of electricity by finding more efficient methods of distributing electrical demand evenly over twenty-four hours and 365 days. Electrical consumption would continue to grow, and it was up to the electrical engineer to make electricity economical whether it was created by steam power from coal or by water power. This need for greater efficiency of electrical use, he warned, would soon shift from a purpose of economy to one of necessity as the nation faced declining supplies of coal and posed a greater impact on free-running streams for hydroelectric power.

Steinmetz cautioned that when coal reserves run out, there would be increasing pressure to develop the nation's water courses for electrical generation. "[T]here will be no more rapid creeks and rivers," he said, "streams which furnish electric power will be slow-moving pools, connected with one another by power stations." Reserves of hard, anthracite (low sulfur) coal already were in short supply, and energy produced by the burning of soft, bituminous (high sulfur) coal created serious air pollution problems. "Probably even before the soft coal is used up" Steinmetz predicted, "we will have awakened to the viciousness of poisoning nature and ourselves with smoke and coal gas."

Smoke pollution from the burning of high sulfur coal had become a nuisance for most industrial cities by the beginning of the twentieth century. A major problem for the smoke abatement campaign that ensued as part of the Progressive Era reforms was the prevailing ethos that equated smoke with economic growth and prosperity. The Department of Public Utilities in Boston, for example, reported that prior to 1910, industries depicting factories on their letterheads "invariably represented the stack with a black plume of smoke trailing away from it to typify activity and prosperity. . . . [I]f a stack was not belching out great volumes of dense smoke it signified that the plant was shut down." Cities such as Milwaukee, Chicago, and Pittsburgh were commonly plagued by the "smoke evil," which blocked the sun, blackened the lungs, covered buildings with soot, and dirtied laundry that had been hung out to dry.

There were three principal groups behind the smoke abatement campaigns. The most ardent supporters were women's groups; they helped install smoke ordinances in many cities. Representing a wider range of interests were civic groups. Civic groups, however, were keenly aware of the economic interests of the community, and this often tempered the spirit of their activities for clean air. The third group was represented by stationary engineers—technical experts who served as smoke inspectors, argued for technological improvements, and sometimes helped draft local smoke ordinances.

World War I ended the progress made earlier by proponents of clean air because smoke was a sign of industrial output in support of the war effort. In Milwaukee, the weather bureau reported a fourfold increase in the number of smoky days from 1916 to 1918. The smoke abatement crusades that reappeared in the 1920's lacked the intensity of those of the Progressive Era. In the 1930's and 1940's, greater use of natural gas, electricity, and diesel fuel, as well as technological advances, all combined to help reduce the smoke pall in many cities.

In his lecture, Steinmetz warned of another abuse of resources. He was concerned about the steady depletion of nutrients from the nation's prime farmland. Instead of returning their wastes to agricultural land to be recycled, cities were dumping into rivers the nitrogen and phosphorus that had been taken from the soil in the form of crops, thus, not only polluting the streams but also wasting valuable fertilizer. Moreover, additional electricity would be consumed in the making of the fertilizer necessary to replenish those soils.

Rapid growth in population and industrialization, and the consequent urbanization created by this growth, had placed heavy demands on the nation's streams for both water consumption and waste disposal. As a result, residents of many cities soon found themselves drinking the sewage of their upstream neighbors. In the late nineteenth century, the growing trend of exporting pollution to downstream communities was greatly accelerated by the introduction of municipal sewerage systems. Prior to the development of the modern method of municipal wastewater collection, human waste products were either deposited in nearby cesspools (stone-lined holes) or privy vaults. As the privy vaults reached their holding capacity, either new holes were dug or the contents were emptied by private scavengers. The wastes were deposited on farmland as fertilizer, sold to companies to be processed into fertilizer, or dumped on unused land or into streams. The problems with this system of waste collection were that it created health hazards, was labor intensive, and typically became an aesthetic nuisance. As population density increased, there simply was not enough space for adequate cesspools and privy vaults. Both types of collection frequently overflowed and sometimes contaminated nearby wells. The final breakdown of this system occurred with the introduction of water supply technology.

The development of running water supply systems created a demand for the new water closet that not only effectively removed human wastes from homes but also increased water consumption. Initially, these water closets were connected to privy vaults and cesspools that were unable to handle the increased loads placed upon them. Citizens began to demand that cities construct sewer systems capable of removing these wastes. Efficiency dictated that such systems be large public works constructed only at great expense to the communities. The idea of a water carriage system of waste removal had been around for centuries, but the first practical system was constructed in the 1840's in England. In many large American cities, the debate over whether to build sewerage systems occurred at the end of the nineteenth century. Advocates stressed the sanitary and health aspects, arguing that mortality rates from typhoid fever would decline and, therefore, would justify the high cost of construction. Other proponents argued that modern sewerage systems would improve the community image, attracting industries and thus contributing to the local economy. Opponents countered that water carriage systems would waste the valuable fertilizing potential of human excreta.

Largely unforeseen by advocates arguing from the perspective of health and sanitation was that typhoid death rates actually increased for cities drawing their drinking water downstream from towns that had constructed sewerage systems that discharged raw sewage into rivers. Sanitary engineers and city authorities had traditionally assumed that contaminants were diluted in rivers to the extent that any hazards would be eliminated by the self-purification capacity of running water. Unfortunately, it took the dramatic increase in typhoid deaths for cities such as Trenton, Pittsburgh, and Atlanta during the period of sewerage construction to indicate otherwise. In 1902, William Thompson Sedgwick, a professor of biology and sanitary science at Massachusetts Institute of Technology, declared that "self-purification is only partial and absolutely unreliable."

By 1908, leading public health officials were convinced that untreated sewage should not be discharged into rivers used for drinking water. Not only was there a risk of waterborne disease but also sewage disposal in waterways limited their use for recreation and industry. Sanitary engineers, on the other hand, supported the disposal of sewage into rivers, arguing that drinking water drawn from these sources could be purified by the employment of new filtration technology, which had proven effective in disease control. Sanitary engineers such as Allen Hazen, for example, a chemist working for the Massachusetts Board of Health, said it was much cheaper to purify the water taken from rivers than to purify the sewage prior to disposal in those rivers. This argument convinced public officials; by 1920, the practice of discharging raw sewage into waterways and purification of water supplies had become well established.

Impact of Event

The Progressive Era antipollution crusades for clean air and clean water had little lasting effect. These campaigns were on the fringe of, and to some degree ran counter to, the more prominent conservation movement normally associated with Theodore Roosevelt and his chief forester, Gifford Pinchot. While the smoke abatement advocates and proponents of sewage treatment were concerned with environmental quality and protection of air and water resources, the interests of the conservation movement were directed toward the development and use of resources. Steinmetz saw the continued development of water power as an inevitable outcome of a growing public demand for electricity. This demand, he thought, posed a serious threat to the quality of life for future generations; electrical engineers should be prepared to develop methods to minimize that impact.

Partial relief from smoke in the cities came primarily through technological advances and alternative sources of energy rather than through enforcement of air quality standards. In general, dirty air was viewed as one of the costs of industrial prosperity. This perspective prevailed with respect to stream quality as well. Rivers became the common receptors of raw sewage, which received treatment prior to disposal only when deemed a nuisance. Population growth, along with increased urbanization and industrialization, combined to place heavy loads of contamination into the nation's waterways. Sanitary engineers and public officials opted for deriving the maximum utility from the capacity of large volumes of river or lake water to assimilate and dilute wastes. It was more cost-effective to purify drinking water drawn from polluted sources than to treat sewage prior to disposal.

This approach to water quality policy sometimes resulted in severe pollution of rivers, which, in turn, affected larger systems such as the Great Lakes. For example, Lake Erie—receiving wastes from Detroit, Toledo, and Cleveland, as well as fertilizer and insecticide runoff from farmland—is often cited as the classic symbol of environmental degradation. Between 1930 and 1965, Lake Erie's average content of nitrogen and phosphorus increased threefold, causing overfertilization of the lake's waters and a subsequent increase in production of phytoplankton. By the late 1950's, dense mats of algae commonly formed on the surface, drifting ashore and rotting on the beaches. The increase in phytoplankton indirectly produced dangerously low concentrations of oxygen for some valuable species of fish. Scientists believed that conditions created by excessive nutrient enrichment caused a change in the character of fish species in Lake Erie, reducing the value of the commercial fishery.

The precedent of waste dilution established in the early part of the century allowed the routine disposal of toxic compounds such as mercury and petroleum into waterways. By the 1960's, Detroit's industrial area was dumping hundreds of barrels of oil per day into the Detroit River, which then flowed into Lake Erie. The Cuyahoga River in Cleveland carried so much oil and other flammable wastes that it would sometimes catch fire. On June 22, 1969, two railroad bridges over the Cuyahoga were destroyed by river fires. Eventually, public pressure became intense enough to provoke legislative action. Lake Erie and the Cuyahoga River became national environmental icons, nagging at a renewed public consciousness that had been aroused in 1962 with the publication of Rachel Carson's *Silent Spring*. In 1972, Prime Minister Pierre Elliott Trudeau and President Richard M. Nixon signed a water quality agreement that pledged to protect and restore the Great Lakes. This agreement led to the allocation of billions of dollars of federal money to municipalities for construction of sewage treatment plants and to farmers for agricultural improvements designed to curb runoff of nutrients and insecticides. The Clean Air Acts of 1965 and 1970 provided citizens with some relief from air pollution. In the 1980's and 1990's, contamination of air and water continued to provoke debate over the standards of purity society must achieve.

Bibliography

Grinder, R. Dale. "The Battle for Clean Air: The Smoke Problem in Post-Civil War America." In *Pollution and Reform in American Cities, 1870-1930*, edited by Martin V. Melosi. Austin: University of Texas Press, 1980. A twenty-page essay replete with quotations from diverse primary sources of the period. The brevity of the article and style of reporting make it a useful introduction to further study of smoke abatement.

Hammond, John Winthrop. "Steinmetz the Prophet." In *Charles Proteus Steinmetz: A Biography*. New York: Century, 1924. An accessible book that, while focusing on the life and career of Steinmetz, includes several paragraphs from his lecture, "The Future of Electricity."

Hays, Samuel P. *Conservation and the Gospel of Efficiency: The Progressive Conservation Movement, 1890-1920*. Cambridge, Mass.: Harvard University Press, 1959. A good source for the conservation movement from the perspective of public policy decision-making. Hays explains the role of the scientist as expert and the importance of "efficiency" in Theodore Roosevelt's America. Although he does not address the antipollution campaigns per se, Hays does include detailed discussions on the politics of water power development. Includes index and bibliographic note.

Sedgwick, William T. *Principles of Sanitary Science and the Public Health with Special Reference to the Causation and Prevention of Infectious Diseases*. New York: Macmillan, 1902. A leader in water pollution control, Sedgwick provides a clear summary of the scientific principles on which sanitation rested while on the threshold of the twentieth century. Chapters 7 to 9, in particular, address sewage as a carrier of disease, pollution of public water supplies, and mechanisms of water purification.

Steinmetz, Charles Proteus. *The Future of Electricity*. New York: New York Electrical Trade School, n.d. The text of his lecture to students of the New York Electrical Trade School. Limited publication, available in few libraries. (See Hammond, cited above.)

Tarr, Joel A., James McCurley, and Terry F. Yosie. "The Development and Impact of Urban Wastewater Technology: Changing Concepts of Water Quality Control, 1850-1930." In *Pollution and Reform in American Cities, 1870-1930*, edited by Martin V. Melosi. Austin: University of Texas Press, 1980. A useful essay describing the impact of wastewater technology on human health in American cities. Supported by tables and figures.

Robert Lovely

Cross-References

Cottrell Invents the Electrostatic Precipitation Process (1906), p. 303; Insecticide Use Intensifies When Arsenic Proves Effective Against the Boll Weevil (1917), p. 640; Carson Publishes *Silent Spring* (1962), p. 1739.

MORGAN DEVELOPS THE GENE-CHROMOSOME THEORY

Category of event: Biology
Time: 1908-1915
Locale: New York City

Morgan's experiments on Drosophila *led to the discovery of the principles of the gene-chromosome theory of hereditary transmission*

Principal personages:

THOMAS HUNT MORGAN (1866-1945), a professor of experimental zoology at Columbia University who was primarily responsible for development of the gene-chromosome theory, for which he won the 1933 Nobel Prize in Physiology or Medicine

CALVIN BLACKMAN BRIDGES (1889-1938), a geneticist and one of Morgan's assistants who discovered nondisjunction

ALFRED HENRY STURTEVANT (1891-1970), a geneticist and another of Morgan's assistants who is credited with having developed the first chromosome map

HERMANN JOSEPH MULLER (1890-1967), a geneticist and Morgan's student who later won the 1946 Nobel Prize in Physiology or Medicine for demonstrating the effect of X rays on mutation rates

Summary of Event

In 1904, Thomas Hunt Morgan, a young professor of biology, was invited by E. B. Wilson to join him at Columbia University as professor of experimental zoology. The new position would allow him more time for laboratory research. That same year, through a friend, Jacques Loeb, Morgan met Hugo de Vries, the Dutch biologist who had been one of the trio of scientists who in 1900 had rediscovered the work of Gregor Johann Mendel. (Mendel's work had been unknown since he first propounded it in 1866, and it was promptly ignored and forgotten.) De Vries had a theory that new species originated through mutation. The rediscovery of Mendel and the contact with de Vries influenced Morgan to initiate experiments to try to discover mutations and to test the Mendelian laws. He began to experiment with *Drosophila melanogaster*, the fruit fly, an enormously propitious choice for a laboratory animal. It bred rapidly, ate little, and Morgan and his students found that thousands of these minuscule experimental animals could be contained in a small collection of milk bottles that they "borrowed" from the Columbia cafeteria.

In 1908, Morgan had one of his graduate students, Fernandus Payne, perform an experiment in which he bred generations of *Drosophila* in the dark, in an attempt to produce flies whose eyes would atrophy and become dysfunctional. On Morgan's advice, Payne secured his experimental animals by capturing some of the flies that were attracted to bananas he had left on the ledge of a window of the laboratory. After sixty-nine generations of fruit flies never saw light, however, nothing came of the experiment. The sixty-ninth generation could see as well as the first.

Morgan's second experiment on *Drosophila* was an attempt to induce mutations by subjecting flies to X rays and other environmental stimuli, such as wide ranges of temperature, salt, sugars, acids, and alkalies. Again, the results were negative. (Why these results were negative is unexplained, since it was later demonstrated by Hermann Joseph Muller that X rays do produce mutations.)

Morgan's experiments did not reveal mutations. Nor did they bear out the Mendelian laws but turned up enough exceptions that Morgan began to doubt the validity of the laws. One day in 1910, however, he found a single male fly with white eyes, rather than the standard red. Morgan bred the white-eyed male with a red-eyed female and the first-generation offspring were all red-eyed, suggesting that white-eyed was a Mendelian recessive factor. He then bred the first-generation flies among themselves, and the second-generation were red-eyed and white-eyed in a 3:1 ratio, appearing to confirm that white eyes were Mendelian recessive. Morgan noted that all the white-eyed flies were males. He had discovered sex-limited heredity. The Mendelian factor, or gene, that determined white eyes was located on the same chromosome as the gene that determined male sex, or as it turned out, on the male chromosome.

Chromosomes—which appear as stringlike structures within cells—had been discovered by cell researchers in the 1850's. It had been theorized, without much hard evidence, that they were involved in heredity. Mendelian theory originally had nothing to do with chromosomes (Mendel did not know about chromosomes). By the first decade of the twentieth century, when Mendelian-chromosome theories had been suggested, again without much hard evidence, Morgan had considered such theories and rejected them. Meanwhile, researchers had discovered an odd-shaped chromosome (all other chromosomes occurred in similarly shaped pairs) that seemed to be related to male sex (now called the Y chromosome). The discovery of sex-limited heredity revealed the association of Mendelian genes with chromosomes and the function of chromosomes in heredity.

Following the discovery of sex-limited heredity, Morgan saw that a concerted effort was required to expound fully the Mendelian-chromosome theory and therefore enlisted a group of exceptional students to share the work in his so-called fly room. The nucleus of the group consisted of Calvin Blackman Bridges, Alfred Henry Sturtevant, and Muller. Between 1910 and 1915, Morgan and his team developed and perfected the concepts of linkage, in which various genes are found to be located on the same chromosome and the appearance of their associated characteristics in the offspring occur together, and crossing-over, in which paired chromosomes break and rejoin during meiosis. Crossing-over produces a contrary effect from linkage. That is, characteristics that would be expected to appear in the offspring because they are associated with genes on the same chromosome will not appear because during meiosis (when the paired chromosomes split apart, whereupon only one half of the pair is conveyed to the new cell), part of the chromosome and the genes on that part were replaced by a portion of the paired chromosome. Based on an understanding of linkage and crossing-over, the team was also able to create chromosome maps, plotting the relative locations and distances of the genes on the chromosomes. That is, the occurrence of combinations and recombinations (the latter produced by crossing-over) of linked characteristics will indicate the relative locations of the genes for those characteristics on the chromosome. The frequency of recombinations should be proportional to the distance between the genes for the recombined characteristics.

The culmination of the work of Morgan's team was the publication in 1915 of *The Mechanism of Mendelian Heredity*, coauthored by Morgan, Sturtevant, Muller, and Bridges. For the next twelve years, the strictly genetics studies were performed mainly by Sturtevant and Bridges and other team members, while Morgan returned to his previous areas of interest of embryology and evolution, pursuing connections between those areas and the new discoveries in genetics. He also was occupied in publicizing the new views of heredity and their ramifications through publications and lectures.

By the late 1920's, Morgan was acknowledged as the world's leading geneticist. In 1927, Robert Andrews Millikan was reorganizing the California Institute of Technology at Pasadena to make it one of the premier scientific schools in the country, and he asked Morgan to organize the Division of Biology. Morgan accepted and joined the staff that included physicist J. Robert Oppenheimer and chemist Linus Pauling. In 1933, Morgan was awarded the Nobel Prize in Physiology or Medicine (the first Nobel Prize in the field of genetics) for his work on hereditary research.

Impact of Event

The discovery and demonstration that genes reside on chromosomes was the key to all further work in the area of genetics. Mendel, who in 1866 discovered genes, was extremely fortunate in the design of his experiments in that each of the characteristics he investigated in his pea plants happened to reside on a separate chromosome. This facilitated discovery of the Mendelian hereditary principles but made for experimental results that were rather tidy. When a larger number of characteristics is investigated, because of linkage of genes located on the same chromosome and crossing-over of chromosomes, the results will be much more complicated and will not reveal so clearly the Mendelian pattern. This is what happened in Morgan's early experiments. It was only by carefully examining—using tweezers and magnifying glass—generations upon generations of the tiny *Drosophila* that the mechanism of inheritance began to emerge.

Although Morgan did not pursue medical studies, his studies laid the groundwork for all genetically based medical research. As the presenter of the Nobel Prize to Morgan stated, without Morgan's work "modern human genetics and also human eugenics would be impractical." It was generally accepted that Mendel's and Morgan's discoveries were responsible for the investigation and understanding of hereditary diseases.

The darker side of the genetics research and discoveries was that they also laid the groundwork for Hitler's iniquitous eugenics experiments and associated racial purity fantasies. Morgan, however, was always extremely distrustful of any such experiments on the human species. In his Nobel acceptance speech, Morgan pointed out that geneticists were now able to produce populations of species of animals and plants that were free from hereditary defects by suitable breeding. He noted, however, that it was not advantageous to perform genetic experiments on humans, except to attempt to correct a hereditary defect. Morgan believed it was improper to "purify" the human race. One must look to hereditary research for new developments to ensure a healthy human race.

Bibliography

Allen, Garland E. *Thomas Hunt Morgan: The Man and His Science*. Princeton, N.J.: Princeton University Press, 1978. This scholarly, yet quite readable, volume is the most complete biography of Morgan and exposition of his work. It must serve as the standard reference for any further study of Morgan, as it has been the basic source work for the present essays. Contains an extensive bibliography of primary and secondary works.

Carlson, Elof Alex. *The Gene: A Critical History.* Philadelphia: W. B. Saunders, 1966. Rather than a standard, comprehensive review of the subject, this book is thematic in outline. The chapter entitled "The *Drosophila* Group: An Enigmatic Appraisal" provides Muller's perspective on Morgan that gives less credit to Morgan and more to his assistants for the ideas of the group after Morgan's initial discoveries.

Dunn, L. C. *A Short History of Genetics.* New York: McGraw-Hill, 1965. Dunn provides a rather technical history of the classical period of genetics from 1864 to 1939. The presentation gives a sense of the multifaceted nature of the search for the principles of genetics. More for the specialist than the general reader.

Morgan, Thomas Hunt, A. H. Sturtevant, H. J. Muller, and C. Bridges. *The Mechanism of Mendelian Heredity.* New York: Henry Holt, 1915, rev. ed. 1922. The classic text in which Morgan and his students describe their *Drosophila* research.

Shine, Ian, and Sylvia Wrobel. *Thomas Hunt Morgan: Pioneer of Genetics.* Lexington: University of Kentucky Press, 1976. This very readable biography and popular treatment of his work focuses on Morgan's likable personality and contains many personal anecdotes. It expounds Morgan's interesting family history: His uncle was a Confederate Civil War hero; and his father, as U.S. consul to Italy, discarded protocol and fought at the side of Garibaldi.

Sturtevant, A. H. *A History of Genetics.* New York: Harper & Row, 1965. A first-hand view of the development of classical genetics by a major participant. Rather technical in approach. An interesting feature for a book in the area of heredity is an appendix on the intellectual pedigrees (presenting "genealogical" charts) of the scientists and philosophers involved in the development.

Sturtevant, A. H., and G. W. Beadle. *An Introduction to Genetics.* New York: Dover,

1939, rev. ed. 1962. A textbook on genetics from 1939. The 1962 edition contained only minor revisions. The basic treatment of the subject that is also useful as a historical document. Beadle was a Nobel Prize winner in 1958.

John M. Foran

Cross-References

De Vries and Associates Discover Mendel's Ignored Studies of Inheritance (1900), p. 61; McClung Plays a Role in the Discovery of the Sex Chromosome (1902), p. 148; Punnett's *Mendelism* Contains a Diagram for Showing Heredity (1905), p. 270; Sturtevant Produces the First Chromosome Map (1911), p. 486; Avery, MacLeod, and McCarty Determine That DNA Carries Hereditary Information (1943), p. 1203; Watson and Crick Develop the Double-Helix Model for DNA (1951), p. 1406; Kornberg and Coworkers Synthesize Biologically Active DNA (1967), p. 1857; Cohen and Boyer Develop Recombinant DNA Technology (1973), p. 1987.

GEIGER AND RUTHERFORD DEVELOP THE GEIGER COUNTER

Categories of event: Applied science and physics
Time: February 11, 1908
Locale: Manchester University, England

Geiger and Rutherford developed the first electronic radiation counter able to detect atomic particles

Principal personages:

HANS GEIGER (1882-1945), a German physicist who became famous for the radiation detector named for him

ERNEST RUTHERFORD (1871-1937), a British physicist who was the main figure in the development of nuclear physics

SIR JOHN SEALY EDWARD TOWNSEND (1868-1957), an Irish physicist who studied electrical discharges in low-pressure gases

SIR WILLIAM CROOKES (1832-1919), an English physicist who studied the passage of electric currents in evacuated tubes and radioactivity

WILHELM CONRAD RÖNTGEN (1845-1923), a German physicist who discovered and studied X rays

ANTOINE-HENRI BECQUEREL (1852-1908), a French physicist who discovered the radioactivity of uranium

Summary of Event

When radioactivity was discovered and first studied, it was done with rather simple devices. In the 1870's Sir William Crookes learned how to create a very good vacuum in a glass tube. He placed electrodes in each end of the tube and studied the passage of electricity through the tube. This simple device became known as the Crookes tube. In 1895, Wilhelm Conrad Röntgen was experimenting with a Crookes tube. It was known that when electricity went through a Crookes tube, one end of the glass tube might glow. Certain mineral salts placed near the tube would also glow. In order to observe carefully the glowing salts, Röntgen had darkened the room and covered most of the Crookes tube with dark paper. Suddenly, a flash of light caught his eye. It came from a mineral sample placed some distance from the tube and in a direction shielded by the dark paper, yet when the tube was switched off, the mineral sample went dark. Experimenting further, Röntgen convinced himself that some ray from the Crookes tube caused the mineral to glow. Since light rays were blocked by the black paper, he called it an X ray, where X stands for unknown.

Antoine-Henri Becquerel heard of the discovery of X rays, and in February, 1886, set out to discover if nature might also reverse the process—that is, if glowing minerals emitted X rays. Some minerals begin to glow when activated by sunlight. If then placed in the dark, the glow fades more or less swiftly, depending upon the mineral involved. Such minerals are called phosphorescent. Becquerel's test procedure was to wrap photographic film in enough black paper that setting it in the sun all day would not expose it. He then placed various phosphorescent minerals on top of the film package and left them in the sun. He soon learned that those phosphorescent minerals containing uranium would expose the film. To make certain that the film was not being exposed by chemicals from the uranium minerals, he covered the film with glass and also placed various pieces of metal on the film. He discovered that metal seemed to protect the film from exposure. The greatest surprise, however, came during a series of cloudy days. Anxious to continue his experiments, he decided to develop film from a test that had not been exposed to sunlight. He was astonished to discover that the film was deeply exposed. Some emanation must be coming from the uranium, and it had nothing to do with sunlight. Thus, natural radioactivity was discovered with a simple piece of photographic film.

Ernest Rutherford joined the world of international physics at about the same time that radioactivity was discovered. He came from New Zealand for advanced study in physics at Cambridge in 1895. His previous work had been with radio waves, but now he plunged into a study of radioactivity. A remarkably talented man, it took only a few years before he became the dominant figure in the field. Studying the "Becquerel rays" emitted by uranium, Rutherford eventually distinguished three different types of radiation, which he named alpha, beta, and gamma from the first three letters of the Greek alphabet. He showed that alpha particles were easily stopped by a few thin metal foils. Later, he proved that an alpha particle is the nucleus of a helium atom (a group of two neutrons and two protons tightly bound together).

It took more thicknesses of metal foil to stop beta particles. It was later shown that beta particles are electrons. Gamma rays proved to be far more penetrating than either alpha or beta particles. Eventually, it was shown that gamma rays are similar to X rays, but they have higher energies.

Rutherford became director of the associated research laboratory at Manchester University in 1907. Hans Geiger became an assistant. Up to this time, Rutherford had been unable to prove his conjecture that the alpha particle carried a double positive charge, but he would soon prove it with Geiger's assistance. The obvious way to proceed would be to allow a stream of alpha particles to fall on a target and to measure the electric charge the particles brought to the target. Divide that charge by the total number of alpha particles and one would have the charge of an alpha particle. The problem lay in counting the particles and in proving that every particle had been counted.

Basing their design upon work done by Sir John Sealy Edward Townsend, a former colleague of Rutherford, Geiger and Rutherford constructed an electronic counter. It consisted of a long brass tube sealed at both ends, and from which most of the air had been pumped. A thin wire, insulated from the brass, was suspended down the middle of the tube. This wire was connected to batteries producing about thirteen hundred volts and to an electrometer, a device that could measure the voltage of the wire. This voltage could be increased until a spark jumped between the wire and the tube. If the voltage was turned back down a little from the spark over value, the tube was ready to operate. An alpha particle entering the tube would ionize (knock some electrons away from) at least a few atoms. These electrons would be accelerated by the high voltage and, in turn, would ionize more atoms which freed new electrons. These would ionize still more atoms, and so on until a veritable avalanche of electrons would strike the central wire and the electrometer would register the voltage change. Two key points became evident: First, since the tube was nearly ready to arc because of the high voltage, every alpha particle, even if it had very little energy, would initiate a discharge. Second, because of the electron avalanche, each electric discharge was several thousand times larger than the alpha particle alone could have caused and was easily measured with the electrometer.

Geiger and Rutherford completed their apparatus by connecting their alpha detector tube to a 4.5-meter-long "firing tube." The air had been evacuated from the firing tube and a small film of radium was fixed at the far end. Only a few alpha particles per minute were headed in the right direction to travel the length of the firing tube, pass through a small hole in a stopper, and pass through a thin mica window into the detector tube. There they traveled parallel with the central wire while they lost energy to ionization which in its turn caused an electron avalanche. Restricting the number of alpha particles to a few per minute was necessary because the electrometer was somewhat slow to respond and to recover.

On February 11, 1908, Geiger and Rutherford reported on the electrical method of counting alpha particles to the Manchester Literary and Philosophical Society. The most complex of the early radiation detection devices—the forerunner of the Geiger counter—had just been developed.

Impact of Event

Their first measurements showed that one gram of radium emitted 34 thousand million alpha particles per second. Soon, the number was refined to 32.8 thousand million per second. Next, Geiger and Rutherford measured the amount of charge emitted by radium each second. Dividing this number by the previous number gave them the charge on a single alpha particle. Just as Rutherford had anticipated, the charge was double that of a hydrogen ion (a proton). This proved to be the most accurate determination of the fundamental charge until Robert Andrews Millikan's classic oil drop experiment in 1911.

Another fundamental result came from a careful measurement of the volume of helium emitted by radium each second. Using that value, other properties of gases, and the number of helium nuclei emitted each second, they were able to calculate Avogadro's number more directly and accurately than had previously been possible. Avogadro's number enables one to calculate the number of atoms in a given amount of material.

One of the most important applications of the "proto" Geiger counter was proving that a different way to count alpha particles worked. In 1903, Crookes invented a device he called a spinthariscope, or "spark viewer." It consisted of a tiny spot of radium mounted several millimeters in front of a small glass screen coated with a fluorescent mineral—zinc sulfide with a trace of copper. When the screen was observed in a darkened room through a magnifying glass, tiny flashes of light, called scintillations, could be seen as the alpha particles struck the screen. Using both photographic film, and an advanced model of the spinthariscope, Geiger and Rutherford noted that a beam of alpha particles tended to spread out after penetrating a thin mica window or a metal foil. Rutherford came to see this as a key to the structure of the atom. It could be compared to that of a traveler in unfamiliar territory. Suppose this traveler suddenly comes to a thick fog bank. It is so thick that he cannot see anything that lies ahead, whether it be trees, a cliff, or a haystack. The traveler could throw pebbles into the fog to learn what lay ahead. That is, by throwing pebbles in various directions and listening to hear whether they strike something, the traveler could gain some notion of what was there. In just this fashion, Rutherford threw alpha particles at atoms. Using a zinc sulfide screen, he could observe where the alpha particles came out after interacting with the atoms of a target foil. The "proto" Geiger counter was used to prove that when appropriate precautions were taken, observers could accurately count the alpha particles that struck the zinc sulfide screen. In 1911, Rutherford announced that the atom must consist of a tiny, but relatively massive nucleus surrounded by a vast space through which the electrons moved.

The true Geiger counter evolved when Geiger replaced the central wire of the tube with a needle whose point lay just inside a thin entrance window. This counter was much more sensitive to alpha and beta particles and also to gamma rays. By 1928, with the assistance of Walther Müller, Geiger made his counter much more efficient, responsive, durable, and portable. Today, there are probably few radiation facilities in the world that do not have at least one Gieger-Müller counter, or one of its compact modern relatives.

As for the other early instruments, X-ray-sensitive film is still used today. Both the television tube and the X-ray tube are direct descendants of the Crookes tube. The zinc sulfide screen is closely related to the phosphor-coated television screen, which glows under a beam of beta particles.

Bibliography

Andrade, Edward Neville da Costa. *Rutherford and the Nature of the Atom*. Garden City, N.Y.: Doubleday, 1964. This delightful book is easily read but does not stint on science. Andrade, a physicist, joined Rutherford's group from 1913 to 1914, about the time Geiger changed the "proto" Geiger counter into the Geiger counter. Includes some interesting photographs. Highly recommended.

Berks, J. B., ed. *Rutherford at Manchester*. London: Heywood, 1962. This volume resulted from a commemorative conference held at Manchester University in 1961 to mark the fiftieth anniversary of Rutherford's discovery of the nucleus. There are several speeches by colleagues who reminisce about Rutherford and his work. Also included are several key scientific papers, including three by Rutherford and two by Geiger. Includes historical photographs, a complete bibliography of Rutherford's publications, and a bibliography of papers published by other members of his group. Recommended for the interested reader.

Keller, Alex. *The Infancy of Atomic Physics: Hercules in His Cradle*. Oxford, England: Clarendon Press, 1983. This book describes the works of Rutherford, Geiger, and many others. The major portion of the book deals with the time period from about 1880 to 1920. The author makes special efforts to convey pictures of scientists as "real" people and to place their works within their cultural and historical backgrounds. Includes a useful bibliography. Suitable for the thoughtful reader.

Mann, Wilfrid B., and S. B. Garfinkel. *Radioactivity and Its Measurement*. Princeton, N.J.: D. Van Nostrand, 1966. This book contains exceptionally complete descriptions of some of the early experiments in the discovery of radioactivity, including the work of Geiger and Rutherford. While the descriptive parts should be easily understood by the layperson, at least a good high-school-level background in physics, chemistry, and algebra is required to take full advantage of this work.

Rhodes, Richard. *The Making of the Atomic Bomb*. New York: Simon & Schuster, 1986. The descriptions of the key experiments by Geiger and Rutherford are somewhat brief, but they are well placed in historical context and include interesting personal details. Written for a wide audience.

Rowland, John. *Ernest Rutherford*. New York: Philosophical Library, 1957. This is an easily read biography of Rutherford. His work with Geiger and the advent of the Geiger tube is described.

Charles W. Rogers

Cross-References

Elster and Geitel Demonstrate Radioactivity in Rocks, Springs, and Air (1901), p. 93; Röntgen Wins the Noble Prize for the Discovery of X Rays (1901), p. 118; Becquerel Wins the Nobel Prize for the Discovery of Natural Radioactivity (1903), p. 199; Thomson Wins the Nobel Prize for the Discovery of the Electron (1906), p. 356; Rutherford Presents His Theory of the Atom (1912), p. 527; Rutherford Discovers the Proton (1914), p. 590; Chadwick Discovers the Neutron (1932), p. 973; Frédéric Joliot and Irène Joliot-Curie Develop the First Artificial Radioactive Element (1933), p. 987; Hahn Splits an Atom of Uranium (1938), p. 1135.

HALE DISCOVERS STRONG MAGNETIC FIELDS IN SUNSPOTS

Category of event: Astronomy
Time: June 26, 1908
Locale: Mount Wilson, Pasadena, California

Hale discovered that magnetic fields are associated with sunspots, giving astronomers valuable information about the formation of sunspots

Principal personages:

GEORGE ELLERY HALE (1868-1938), an American astronomer who designed instruments and observed the heavens

PIETER ZEEMAN (1865-1943), a Dutch physicist who discovered that a magnetic field affects the electromagnetic radiation emitted by a source placed in the field

HENRY AUGUSTUS ROWLAND (1848-1901), an American physicist who proved that a moving electric charge has the same magnetic effect as an electric current

Summary of Event

In the seventeenth century, Galileo first observed dark spots on the surface of the sun with a small telescope. Since then, astronomers have invented new ways to observe the sun and have used these new methods to learn more about what causes the spots and other features of the sun. In the early twentieth century, George Ellery Hale developed several solar-observing instruments and used them to determine that the dark spots on the sun exist in an intense magnetic field. This discovery fueled further research and theorized about the nature and causes of the spots and about the magnetic fields of the sun in general.

In 1908, Hale observed the solar disk with a special instrument he had developed called the spectroheliograph. This instrument worked essentially by filtering the light of the sun so that the various surface features like sunspots could be viewed in the light of one particular wavelength. This was useful for studying the processes going on in the sun, since each wavelength represents a particular atomic process in a particular type of atom. When he observed the sun in a wavelength coming from hydrogen atoms, he noted the very interesting fact that there appeared to be huge swirls of hydrogen gas resembling a terrestrial storm or tornado. These "vortices" seemed to be associated with the formation of sunspots. Hale considered what the consequences of this rotating motion might be. Research by the English physicist Joseph John Thomson had shown that hot bodies emit electrons, and as the sun is very hot, Hale thought that perhaps the sun was emitting electrons which, if caught up in this whirling motion, might create a magnetic field in the sunspots. (Henry Augustus Rowland had proven earlier that a moving electric charge acts magnetically in the same way as an electric current. Hans Christian Ørsted had discovered earlier that electric currents affect magnets, and electromagnetic theory explains that electric currents produce a magnetic field.)

Hale had a way to check this hypothesis. A star's spectrum (the pattern of bright light and dark lines that results when its light is passed through a prism or reflected from a grating) can reveal much about the temperature of the star's constituent gases, the velocity of its rotation, and other information. In 1896, Pieter Zeeman discovered that if a source of light is placed in a magnetic field, the lines in the spectrum of the light source will be split into two or more components by the magnetic field. Zeeman discovered this effect (the Zeeman effect) in the laboratory, but it proved valuable in deducing even more information from a star's light. When Hale learned of the Zeeman effect, he conducted laboratory work of his own to observe what the effect looked like in certain test cases. He carried out observations of light from an iron arc (or spark) from other sources and recorded much diversity in the splitting of the spectral lines from the different materials. Some lines split into two parts, some into three, and one even into twenty-one components. Once he had this information, he was ready to compare any observed splitting behavior in solar spectral lines to the laboratory results; if there was a magnetic field associated with sunspots, one could expect a close match between the solar line splitting and the laboratory line splitting. Each dark line in the spectrum represents a particular element undergoing a particular process; therefore, solar elements could be recognized and identified by the distinctive wavelengths of light at which their characteristic dark lines appeared in the spectrum.

Since the sun is much closer and thus much brighter to Earth than other stars, its spectrum can be photographed and examined in greater detail than can be done for other stars, which are fainter and harder to photograph. The sun emits light at many wavelengths, each associated with a different color of light. White light is seen from the sun because the colors of these many wavelengths are all blended and viewed together; a prism or finely spaced grating to spread the light back into its separate colored components must be used. With enough care and effort, it is possible to spread the light and observe detail in the dark lines. Hale expended much care and effort in building the 18-meter tower telescope on Mount Wilson. This tower had mirrors on the top to catch the sun's light and reflect it through a telescope with a 30-centimeter lens and down to a spectrograph about 9 meters underground. At the spectrograph, the light was spread out into a spectrum. Since the mirrors that collected the sunlight were so high above the ground and were fairly thick, they were not subjected to as much distortion from the sun's heat as previous solar telescopes had been, and the height of the mirrors also helped to avoid some of the warm air currents near the ground. The underground room where the final image was obtained had a nearly constant temperature, thus avoiding possible air currents caused by changes and differences in air temperature. Since moving air and warped mirrors distort and muddy the images obtained from telescopes, the net result of all these precautions was to produce an image that was unusually steady and sharp, compared to previous images from solar telescopes. Hale used this tower telescope to observe the sun's spectrum and its surface and, in particular, to search for the Zeeman effect of splitting of lines in the sun's spectrum. A doubling of certain lines in the sun's spectrum had been observed previously, but this result had been misinterpreted at the time and no one, until Hale, was able to observe the lines with sufficient precision to arrive at other ideas about why they appeared split.

Hale was an energetic and active man and spent much of the summer of 1908 running up and down the ladder of the tower in the course of his work. He was fortunate in having a period of clear skies during the second half of June, 1908, which enabled him to spend much of his time observing. One June 26, 1908, Hale observed a doubling of spectral lines that he thought was caused by the Zeeman effect. He was greatly excited and immediately compared his observations of the sun with his laboratory observations of the effect. He found a correlation between the two sets of observations. For the first time, an extraterrestrial magnetic field had been detected and related to laboratory observations. Hale wanted to be absolutely certain of his discovery and continued to make observations for two more weeks. By July 6, he was confident that what he had found was indeed the Zeeman effect, indicative of magnetic fields in sunspots. Zeeman considered Hale's results and agreed that Hale's hypothesis was the best explanation for the observed phenomena. Astronomers in general were excited and impressed by Hale's result and the possibilities for further study of the sun's magnetic properties.

Impact of Event

One important effect Hale's discovery had on astronomers is that it enabled them to extend their application of the relatively new field of electromagnetic theory to the cosmos at large. Astronomers work by understanding processes that can be observed on Earth and applying them in the heavens. With Hale's discovery, astronomers realized that they could apply their knowledge of terrestrial magnetic and electric phenomenon to the distant stars; this gave astronomers a new tool to use.

Zeeman had found—in addition to the fact that spectral lines split in the presence of a magnetic field—that the distance between the components of the split line is directly proportional to the strength of the magnetic field causing the split. Hale measured the separations in lines split in the laboratory by a magnetic field of known strength, measured the separations in split lines in the spectrum of a sunspot, and derived the strength of the sunspot's magnetic field. He found impressively large magnetic fields. Also, Hale was able to study how the strength of the field varied at different places in a sunspot; he discovered that the magnetic field is strongest at the center and weakest toward the edges. This has implications for what the structure of a sunspot might be like.

Hale then turned his attention to the question of whether the sun had an overall magnetic field in addition to the fields associated with sunspots. Although Hale worked at this question periodically for the rest of his life, it was never settled. Later astronomers not only have determined that the sun has a magnetic field overall but also have learned how this field changes over the sun's twenty-two-year sunspot cycle, how the field is affected by the sun's rotation, and the role that the field has in sunspot formation and other types of solar activity. Magnetic fields have poles (like the north and south poles of magnets), and astronomers have worked on discovering the polarity of the sun's magnetic field. Understanding the sun's magnetic properties has been crucial in understanding the sun's activity. Another question Hale considered was whether the magnetic fields associated with sunspots could be strong enough to cause magnetic storms on Earth. Hale was not in a position to answer this question, since it required records on both solar and terrestrial magnetic events over a period of time. The question, however, indicates why the study of the sun is important to astronomers. Sunspots are associated often with energetic events on the sun's surface, which can send charged particles and energetic radiation out into space to interact with Earth's atmosphere. This interaction can cause such benign phenomena as the aurora borealis, or northern lights; also, it can interfere with communications on Earth. In a long-term space colony, the radiation from solar flares could prove hazardous to the colonists. Since it has an effect on humans, astronomers are interested in studying the magnetic processes driving sunspot and flare formation.

Bibliography

Kaufmann, William J. *Discovering the Universe.* New York: W. H. Freeman, 1987. Intended as a textbook for an introductory descriptive course on astronomy, it discusses Hale's discovery and considers other solar features as well as sunspots and the role that magnetic fields play in these features. Includes photographs and drawings, as well as various study aids, such as chapter summaries, review questions, and a glossary.

Mitton, Simon, ed. *The Cambridge Encyclopædia of Astronomy.* New York: Crown, 1977. Includes a section on sunspots, which discusses Hale's participation in discovering the role that magnetic fields play in sunspots and other solar disturbances. Discusses current ideas on solar magnetism. Includes many photographs and drawings, as well as a brief physics primer at the end of the book that gives a concise overview of the magnetic and hydrodynamic facts used to explain solar activity.

Noyes, Robert W. *The Sun, Our Star.* Cambridge, Mass.: Harvard University Press, 1982. Describes Hale's discovery of the magnetic field in sunspots, as well as his work with the spectroheliograph, and discusses the sun and its magnetic properties. Includes photographs, drawings, and graphs; intended for both scientists and nonscientists.

Struve, Otto, and Velta Zebergs. *Astronomy of the Twentieth Century.* New York: Macmillan, 1962. Cowritten by Struve, who witnessed some of the astronomical events the book covers. Contains a chapter on the sun that discusses Hale's work and its implications for theories of solar magnetic fields and their effects. Discusses the growth of the solar observing facilities at Mount Wilson.

Wentzel, Donat G. *The Restless Sun.* Washington, D.C.: Smithsonian Institution Press,

1989. Although this book does not specifically mention Hale in connection with the discovery of a magnetic field associated with sunspots, it does discuss the Zeeman effect and gives a thorough presentation of current views on the sun's magnetic field. Includes diagrams, drawings, and black-and-white photographs. A good discussion of solar physics, written for the nonscientist, is presented.

Wright, Helen. *Explorer of the Universe: A Biography of George Ellery Hale.* New York: E. P. Dutton, 1966. Drawing heavily on letters and diaries written by the participants in the events described, this is an excellent source for the story of Hale's discovery of the magnetic field associated with sunspots. Includes drawings and photographs of sunspots and solar spectra, as well as bibliographies of both books and articles on Hale. Pleasant and engaging style.

Wright, Helen, Joan N. Warnow, and Charles Weiner, eds. *The Legacy of George Ellery Hale: Evolution of Astronomy and Scientific Institutions, in Pictures and Documents.* Cambridge, Mass.: MIT Press, 1972. Relying heavily on photographs and original documents, this collection of material on Hale's astronomical work includes the text of an address Hale gave in 1909, "Solar Vortices and Magnetic Fields." Also valuable for the letters, photographs, and newspaper clippings about Hale and his work.

Mary Hrovat

Cross-References

Lyot Builds the Coronagraph for Telescopically Observing the Sun's Outer Atmosphere (1930), p. 911; Parker Predicts the Existence of the Solar Wind (1958), p. 1576; A Radio Astronomy Team Sends and Receives Radar Signals to the Sun (1959), p. 1597.

EHRLICH AND METCHNIKOFF CONDUCT PIONEERING RESEARCH IN IMMUNOLOGY

Categories of event: Biology and medicine
Time: November-December, 1908
Locale: Messina, Soviet Union, and the Pasteur Institute, Paris, France (Metchnikoff); the Royal Prussian Institute for Experimental Therapy, Frankfurt-am-Main (Ehrlich)

The concept of phagocytosis as a defense mechanism of the body and the mechanism of action of antitoxins and cell receptors provided a springboard for future advances in immunity

Principal personages:
ÉLIE METCHNIKOFF (1845-1916), a Soviet biologist, microbiologist, and pathologist who advanced phagocytosis as a defense mechanism of the animal body in the inflammatory response
PAUL EHRLICH (1854-1915), a German research physician and chemist who advanced the concepts of acquired and passive immunity
CARL CLAUS (1835-1899), an Austrian zoologist, keenly interested in the phagocytosis theory

Summary of Event

In 1796, Edward Jenner introduced the first vaccination against smallpox, so named to distinguish it from syphilis, the "Great Pox," by employing cowpox pustule exudate as the inoculant. By so doing, he used a principle of animal biology with which he and others were not familiar. Jenner's success rested on a principle of immunity to prevent disease.

Immunity can be defined simply as the state of protection against disease, particularly infectious disease. It is the response of the animal body and its tissues to an assault by a variety of antigens. In the context of present-day knowledge, understanding, and research in the field of immunity, the definition is as uninformative as a definition of life. The weakness of the definition is not in what it covers, but rather what it leaves out for the sake of brevity.

In the context of late nineteenth century science, the field of immunity, aside from being young, was fraught with seemingly incongruous results. Élie Metchnikoff and Paul Ehrlich, who shared the 1908 Nobel Prize in Physiology or Medicine for their work in immunity, provided two significant insights that opened the door to the understanding and direction of research in immunity. Their work paved the way for the effective development and use of vaccination, chemotherapeutic treatment of infectious disease, and even organ transplants.

Metchnikoff correctly interpreted and advanced the concept of phagocytosis as a major mechanism by which the animal organism combats foreign particles and disease organisms invading the body. He first used the term "phagocyte," derived from the Greek, in Carl Claus's *Arbeiten* (1893; work). Metchnikoff was educated as a zoologist, but his studies led him increasingly into the field of pathology. In 1865, he made the first observation that would lead to his concept of phagocytosis as a disease-fighting mechanism. Metchnikoff examined the intracellular digestion in the round worm *Fabricia*, which compared with that of protozoans. Although this phagocytic-type process was originally discovered and noted in 1862 by Ernst Heinrich Philipp August Haeckel, it was Metchnikoff who correctly interpreted the relationship between phagocytic digestion and phagocytic defense mechanisms.

In 1882, while studying transparent starfish larvae, Metchnikoff observed mobile cells engulf foreign bodies introduced into the larva. He noted that these cells arose from the mesoderm layer (middle layer of the embryo) rather than the endoderm layer, which is associated with the digestive system. Metchnikoff examined the degeneration of the tadpole tail and observed that it occurred by the phagocytic process also. These observations led him to spend the next twenty-five years in developing and advancing his theory of phagocytosis. The need for phagocytosis in an actively diseased animal led him to study a fungus infection of the water flea, daphnia.

Metchnikoff demonstrated that human white blood cells also develop from the mesodermal layer and serve the role of attacking foreign bodies, particularly bacteria. These ideas were revolutionary because, at the time, one school of thought held that the leukocytes were responsible for nurturing and spreading bacterial infection throughout the body. Indeed, the observation of many white blood cells in the blood of patients who died of infection added resistance to Metchnikoff's phagocytosis theory. As advanced by Julius Cohnheim, the inflammatory response was believed to be operative only in higher animal life that possessed a cardiovascular system. Metchnikoff had demonstrated the principle of inflammatory response in lower life-forms devoid of such a system. It was in the study of the higher animal systems that Metchnikoff faced his most significant challenge in understanding phagocytosis and disease. His choice of infection was the anthrax bacillus. His observations appeared to conflict because phagocytosis seemed to be limited, depending upon the virulence—very virulent bacillus were not attacked while weaker bacillus were. Complicating the study was the observation that resistant animals exhibited active phagocytosis, while susceptible stock displayed no phagocytosis. Metchnikoff was up against the multifaceted complexity of immunity—the humoral versus cellular dichotomy—and the basis of his day's confusion and controversy in immunity studies.

Beginning in 1883, Metchnikoff's ideas were published in Rudolf Virchow's *Archiv für pathologische Anatomie und Physiologie, und für klinische Medizin* (archives for pathological anatomy and physiology and clinical medicine). By 1892, phagocytosis in combating disease was established and published in *The Comparative Pathology of Inflammation*. Metchnikoff wrote a comprehensive book in French in 1901, *L'Immunité dans les maladies infectieuses*, reviewing both comparative and human immunology, which proved a defense of his phagocytosis theory. The book was translated in 1905 under the title *Immunity in Infective Diseases.*

While Metchnikoff wrestled with establishing phagocytosis as a mechanism of defense in disease, Ehrlich studiously examined antitoxins. His first major accomplishment was the improvement of the effectiveness of the diphtheria antitoxin discovered and developed by Emil von Behring. He also performed fundamental experiments leading to his views on active and passive immunity. His research on antitoxin and immunity led to the development of his side-chain theory, a concept of specific cellular responses toward toxins and antitoxins. This theory led to the concept of cell receptors—the basis of the cell's chemical specificity for certain chemical substances.

Whereas Metchnikoff studied the factors associated with phagocytosis and thus the cellular aspect of immunity, Ehrlich studied the factors associated with the humoral aspects of immunity—immunity embodied in the body fluids. Ehrlich's research of toxins and antitoxins aided later studies and the development of an understanding of antigens and antibodies (immunoglobulins). Ehrlich's accomplishments were the introduction of quantification and graphical representations of the relationships existing between toxins and antitoxins. Additionally, he introduced practical and appropriate in vitro systems within which to study these complex associations, selecting erythrocytes as the simplest case in test-tube experiments. Ehrlich reproduced essentially the same effects in these test-tube experiments as observed in the animal body, particularly in the case of the plant toxin ricin. Most important, a complex series of experiments established that the lethality of a toxin and its ability to bind to antitoxin are two separate and independent properties of the toxin.

Ehrlich's work with ricin established that animals can build an immunity to such toxic substances by administration of initially minute doses, gradually increasing over time. Furthermore, he demonstrated in mice the transference of this immunity to the offspring through maternal milk. From his work on diphtheria toxin, he developed a quantitative standardization method for antitoxin dosage characterization.

Ehrlich's work in immunity arose from his study of blood, particularly staining with various dyes. These studies convinced him that the cell does bind certain dyes by distinct chemical affinities. The variously discovered blood components and the differential staining methods he developed not only prepared him for his immunity researches but also marked Ehrlich as the founder of modern hematology. Additionally, the Wassermann test for syphilis, developed by August von Wassermann in 1906, was a direct outgrowth of Ehrlich's immunological research and views.

Impact of Event

Prior to the phagocytosis doctrine advanced by Metchnikoff, the commonly held view was that resistance to bacterial infection resided in chemical properties of the blood. This view enjoyed reinforcement because antibodies in blood had been demonstrated. In 1903, the English scientists Sir Almroth Edward Wright and Stewart Douglas demonstrated the presence of substances (opsonins) in blood that seem to prepare bacteria for phagocytosis by white blood cells by binding to the bacterial surface. Thus, the phagocytosis doctrine appeared to require a precondition, a precoated bacterium that was engulfed. In the ensuing years, the role of phagocytosis and antibody formation became better defined. Phagocytosis is but one mechanism of defense offered by the host against infection. Its activation depends upon whether the infective organism is within the cell (intracellular) or outside the cell (extracellular).

Phagocytosis is most pronounced in acute (extracellular) infection, although bacteria protected by a capsular coat are not readily attacked. Phagocytosis is of limited importance in the case of intracellular infections, such as viruses. In the early stages of infection, antibody production has not yet begun. Thus, phagocytosis is the first defensive action initiated against foreign microorganisms. The administration of antibiotics slows bacterial growth and multiplication, permitting phagocytic blood cells to kill these small populations.

Present-day knowledge of defense against infection lists several types of cells involved. A division of labor exists in which some cells detect by-products of infectious organisms and release immunoglobulins to inactivate the toxic properties of these antigens. Cells of this type are B cells and usually are short-lived. Cells that kill foreign cells are known as T cells. Cytotoxic, or killer T cells, eliminate foreign cells directly. T_H cells assist B cells to differentiate and proliferate. T_A cells amplify differentiation and proliferation of the T cells. T_S cells suppress the immune response and are important in policing the body's own attack on itself.

Ehrlich's work in immunity had far-reaching consequences for general medicine. His studies provided the earliest methods of standardization of bacterial toxins and antitoxins that are still employed and are unaltered essentially from his original methods. He demonstrated further that the lethal action of toxin and its antitoxin-binding potential are actually two separate and distinct properties of the toxin. Additionally, with ricin, Ehrlich demonstrated that a lag-time exists between exposure to a toxin and the manufacture of antibodies against it. Furthermore, Ehrlich distinguished clearly between the concepts of active immunity and passive immunity during his studies of immunity transmission through milk and placenta.

Through these studies—all of which rested on Ehrlich's fundamental belief in chemical affinities between a cell and chemical substances—he went on to study ways of curing disease by chemical means. His early work and successes with trypanosome infection utilizing trypan-red and Atoxyl derivatives led to his most celebrated application of Salvarsan, an arsenical, as a cure for syphilis.

Armed with new knowledge, understanding, biotechnology, and insight into the mechanics of immunity and chemotherapy, modern medicine can fight microorganisms now on their own ground—the molecular level—and win.

Bibliography

Bender, George A., and Park, Davis & Co. *Great Moments in Medicine: A History of Medicine in Pictures.* Detroit: Northwood Institute Press, 1965. Arranged according to the great contributors to the advance of medicine. Simply and clearly written, the book takes the reader on a medical frontiers journey through time from ancient Egypt to the twentieth century and the discovery of insulin. Describes the critical events that led to the discoveries made by the great personages in medicine. Illustrated.

Glasser, Ronald J. "Our Immune System: Early Discoveries," "Our Immune System: The Protectors," "Our Immune System: The Killers," and "Our Immune System: The Mastermind." In *The Body Is the Hero.* New York: Random House, 1976. Provides in simple language the historical and scientific basis of knowledge concerning the major cellular components of the immune system. The final chapter cited discusses the lymph system as a major transportation system for the immune participants.

Marks, Marguerite. "Paul Ehrlich: The Man and His Work." *McClure's*, December, 1910, 186-200. Avoids the scientific jargon and relates in simple terms the work done by Ehrlich and its historical importance. An excellent source to review before examining more scientifically based commentaries on Ehrlich.

Marquardt, Martha. *Paul Ehrlich.* London: William Heinemann Medical Books, 1949. Marquardt, Ehrlich's secretary, provides glimpses into the man and his work. Gives a mental picture of his thinking, his mannerisms, and his temperament. Uses a minimum of technical jargon.

Metchnikoff, Élie. *Immunity in Infective Diseases.* Translated by Francis G. Binnie. Cambridge, England: Cambridge University Press, 1905. Written by Metchnikoff himself in French and translated to English. A treatise of Metchnikoff's views and findings in his research on immunity. The seventeen chapters contain few illustrations. For the more scientifically versed reader who wants an insight into the thinking of the researcher himself.

Metchnikoff, Olga. *Life of Élie Metchnikoff.* New York: Houghton Mifflin, 1921. The second wife of Élie Metchnikoff provides a record of his life from childhood to death at the Pasteur Institute. Discusses his family and his own development; depicts Metchnikoff's loves, dislikes, and weaknesses. An excellent book for those interested in the man of science more so than his studies.

Reinfeld, Fred. "The Germ Theory of Disease." In *Miracle Drugs and the New Age of Medicine.* New York: Sterling, 1957. Well illustrated and simply written. Introduces the reader to the development of the germ theory. Deals principally with the work of Pasteur and Koch and the impact their research had on medicine.

Vernon-Roberts, B. "Phagocytosis, Pinocytosis, and Vital Staining." In *The Macrophage.* Cambridge, England: Cambridge University Press, 1972. Provides a fine overview of phagocytosis, covering chemotaxis, mechanical factors, recognition of foreign substances, ingestion, the cytochemical changes of significance to ingestion, and the fate of ingested or engulfed materials.

Woglom, William H. "Phagocytosis." In *Discoverers for Medicine.* New Haven, Conn.: Yale University Press, 1949. Describes Metchnikoff's theory of phagocytosis. Begins with a brief description of his life and education. Discusses Metchnikoff's theory and how he strived to establish it. An easy-reading work; recommended to be read first before Metchnikoff's *Immunity in Infective Diseases* (cited above).

Wood, W. Barry, Jr. "White Blood Cells v. Bacteria." *Scientific American* 184 (February, 1951): 48-52. Written for a general biological science literate reader. Examines the role of white blood cells in acute infection. Particular emphasis upon Metchnikoff's theory of phagocytosis in disease states.

Eric R. Taylor

Cross-References

Rous Discovers That Some Cancers Are Caused by Viruses (1910), p. 459; Schick Introduces the Schick Test for Diphtheria (1913), p. 567; Calmette and Guérin Develop the Tuberculosis Vaccine BCG (1921), p. 705; Zinsser Develops an Immunization Against Typhus (1930), p. 921; Theiler Introduces a Vaccine Against Yellow Fever (1937), p. 1091; Isaacs and Lindenmann Discover Interferons (1957), p. 1517; A Vaccine Is Developed for German Measles (1960), p. 1654.

BOULE RECONSTRUCTS THE FIRST NEANDERTHAL SKELETON

Category of event: Anthropology
Time: December, 1908
Locale: Paris, France

Boule's reconstitution of a near-complete Neanderthal skeleton sparked controversy over the possibility of identifying the presumed "missing link" between higher apes and humans

Principal personages:

MARCELLIN BOULE (1861-1942), a French paleontologist, whose reconstitution of the first near-complete skeleton of *Homo neanderthalensis* won for him recognition as a foremost authority in paleontology

RUDOLF VIRCHOW (1821-1902), a German professor of anatomy who, nearly thirty-five years before Boule's reconstitution of the skeleton, suggested that certain simian-like features in the limited number of recovered specimens came from environmental influences

A. BOUYSSONIE, J. BOUYSSONIE, and L. BARDON, the French clerics who discovered the substantial remains of a Neanderthal skeleton at La Chapelle-aux-Saints in Southwest France

Summary of Event

Late in the fall of 1908, the French paleontologist Marcellin Boule carried out a project that would contribute to a series of reactions against a "school" of hominid (human) evolutionary thought with which he came to be identified. The event was Boule's reconstitution of a nearly complete skeleton of *Homo neanderthalensis*, drawn from fossil remains found in a cave at La Chapelle-aux-Saints in the Correze in France. His project was carried out after three priests who were recognized already for their research into prehistoric archaeology—A. Bouyssonie, J. Bouyssonie, and L. Bardon—transferred the remains they had discovered to Boule's laboratory in Paris.

The fossil bones consisted of a skull (cranium and quite massive lower jaw), twenty-one vertebras (some fragmented), twenty ribs, a clavicle (collar bone), two humeri (upper arm bones), two fragmented radii and cubitals (lower arm bones), two incomplete femurs (upper leg bones), two knee-caps, portions of two tibiae (lower leg bones), an astragalus (talus bone), a heel bone, five right metatarsals (instep bones), fragments of two left metatarsals, and one phalanx (toe bone). The layers in which the bones were found contained artifacts (dressed flint tools) that could be identified in archaeological terms as Mousterian, a time period (extending roughly from 100,000 to 40,000 B.C.) that fit well with certain paleontologists' views of how *Homo neanderthalensis* should be placed in human evolutionary terms.

The archaeologists who delivered the *Homo neanderthalensis* skeletal remains to Boule reported that the fossil body was found in a narrow trench, approximately one foot deep. It appeared, therefore, that a grave had been prepared for a burial ceremony, a fact that might have explained the nearby concentration of chipped flint implements associated with another assumed Neanderthal ceremonial ritual: the hunt. Indeed, animal remains in the same archaeological layer included bones of the woolly rhinoceros, reindeers, cave hyenas, and marmots.

When Boule set out to assemble as complete a skeleton as possible from the fossils recovered from La Chapelle-aux-Saints, he had very few firm points of reference to guide him. Earlier Neanderthal discoveries were much less extensive, and he had no comparative models to help him reconstruct the fossil. Subsequent to the work Boule carried out in 1908, a number of other discoveries would be made (notably at Le Moustier and La Ferrassie, both in Dordogne in 1909; at La Quina in French Charente district in 1911; and others in Palestine and on the Italian peninsula in the 1920's and 1930's). The book he published with fellow paleontologist Henri Vallois (*L'Homme fossile*) makes it clear that, over the years, Boule would believe he would have no trouble assimilating observable characteristics of Neanderthal fossils found later to the image and theory he originally set forward in 1908. When Boule finished piecing together the Chapelle-aux-Saints fossil, characteristics emerged that he and others considered to be evolutionary gaps between Neanderthals and *Homo sapiens.* These included a bent-over skeletal posture supporting an inordinately large-jawed skull. The latter had a prominent brow ridge and a sloping forehead, with a large nasal construct and little or no chin structure. Finally, the curvature of the leg bones, as well as a certain "pigeon toe" effect in the feet, tended to accentuate basic simian (apelike) features that have disappeared in *Homo sapiens.*

Boule knew that considerable support existed for views expressed on *Homo neanderthalensis* in the mid-nineteenth century by the German anatomist Rudolf Virchow. Briefly stated, the Virchow school had argued that Neanderthal remains should be assigned a place along with fossils of *Homo sapiens,* essentially as a contemporaneous "cousin" of what has come to be known as modern humans. If *Homo neanderthalensis* became an extinct species of humans, this could be explainable in terms of Neanderthal's relative disadvantages for survival as compared with *Homo sapiens.* It was wrong, the Virchow school maintained, to push the prime period of *Homo neanderthalensis* back, limiting it to the Mousterian age, mainly to facilitate an argument in support of the prevailing Darwinist interpretations of evolution. Had Virchow lived to confront Boule over the Chapelle-aux-Saints skeleton, he most likely would have objected to Boule's implication that *Homo neanderthalensis* was too "primitive" to be placed on a comparable evolutionary level with *Homo sapiens.*

Evolutionists, including Boule, were anxious to find a pre-sapiens hominid link between apes and *Homo sapiens.* To a significant degree, Boule's reconstruction of the Chapelle-aux-Saints skeleton seemed to respond to this need, as did his eventual separation of *Homo neanderthalensis* from the three fossil human predecessors of modern *Homo sapiens.* Regardless of whether it was Boule's intention initially to lend support to a particular school of thought concerning the place of *Homo neanderthalensis* in evolutionary theory, his work sparked a debate that—although mainly a popular one—involved a number of paleontologists of different camps, at least into the late 1950's.

Impact of Event

The role that Boule would play in the seemingly unending controversy over presumed "links" in the evolutionary chain leading to *Homo sapiens* was in some ways prepared well before the specific contribution he made in 1908. In fact, soon after discovery of the first remains of *Homo neanderthalensis* in the Neander river valley near Düsseldorf, Germany, in 1856, several schools of thought concerning the evolution of modern humans tried either to adopt or reject assumptions that would make Boule's reputation (or notoriety) as a human paleontologist.

A debate arose in the generation after the first fossils were unearthed in the Neander Valley. Several features of the first Neanderthals (and others discovered later, at Spy, Belgium, in 1886, and then in fairly substantial numbers in southwestern France at the beginning of the twentieth century) suggested that they were closer to the presumed simian ancestors of humans than any previously studied paleontological prototype. At some point in the mid-nineteenth century, for example, a sufficient number of Neanderthal sites had been identified to place them, and the tools found with them, in the Mousterian period, well before the time of *Homo sapiens* cultures. This posed a problem of linkage: if it was possible that *Homo sapiens* were biological descendants of *Homo neanderthalensis.* Boule implied that the *Homo sapiens* had replaced *Homo neanderthalensis.* The other theory was that the two species existed side by side until the superior adaptive qualities of *Homo sapiens* guaranteed their continuation while *Homo neanderthalensis* declined in numbers and eventually disappeared.

Well before Boule's time, unwillingness to accept *Homo neanderthalensis* as an evolutionary "cousin" of *Homo sapiens* had divided opinions among prominent representatives of the nineteenth century scientific world, including the geologist Thomas Huxley, who insisted upon the Charles Darwin-dominated 1860's view on the essentially "bestial" characteristics of *Homo neanderthalensis,* which he associated with later apes, not early humans.

More important as a source of future controversy over the Chapelle-aux-Saints Neanderthal reconstruction was the mid-nineteenth century contribution of the German anatomist Virchow. Unlike Huxley, Virchow refused to see *Homo neanderthalensis* as belonging to a species separate from *Homo sapiens,* thus perturbing evolutionists who needed proof of a separate, lower link from which *Homo sapiens* might have sprung. Virchow came to the conclusion that certain of Neanderthal's apelike characteristics could be explained by "ailments"—specifically the effects of rickets from malnutrition—suffered by the (basically human) individuals.

Ironically, Virchow's student, Ernst Haeckel (1834-1919), would soon take a theoretical position that seemed to be reflected in Boule's 1908 Neanderthal reconstruction and in his defensive writing some years later. Haeckel tended to seek in hominid fossils such as *Homo neanderthalensis* an evolutionary link between humans and apes. The most notable prototype associated with Haeckel's name was the so-called *Pithecanthropus alalus,* or "speechless ape-man," but he was also associated in the debate in the 1890's over the "human-like transitional form" of the Java man, *Pithecanthropus erectus,* discovered by the Dutch paleontologist Eugene Dubois.

With the passage of time and discovery of additional Neanderthal specimens, however, the scientific community would grow skeptical of Boule's insistence on placing *Homo neanderthalensis* next to *Homo sapiens.* A culmination would be reached in 1957, when W. L. Strauss and A. J. E. Cave reexamined the actual reconstructed skeleton. Combining alternative views that had preexisted Boule's 1908 contribution with studies of more recently discovered Neanderthal remains, they essentially put Boule's theory to rest. Not only was the question of naturally (namely, by disease) distorted bone structure reopened but also the accuracy of Boule's own work of skeletal reconstruction was questioned, particularly in the foot area, where angles may have been exaggerated to accentuate simian features. Strauss and Cave's refutation of Boule generally held throughout the second half of the twentieth century, with the effect of elevating *Homo sapiens neanderthalensis* to the status of a branch of *Homo sapiens.* If Neanderthal communities tended to disappear, it was the result of a loss of security in a limited competitive ecological arena in which *Homo sapiens sapiens* held an upper hand, not to their having occupied a distinctly lower level on the biological evolutionary ladder.

Bibliography

Boule, Marcellin. "L'Homme fossile de la Chapelle-aux-Saints." *Comptes rendues de l'Académie des sciences,* December 14, 1908, 1012-1017. Boule's first publication describing the Neanderthal skeleton he had reconstituted.

Boule, Marcellin, and Henri V. Vallois. *Fossil Men.* Translated by Michael Bullock. New York: Dryden Press, 1957. The English translation of Vallois and Boule's major general work, *Les Hommes fossiles.* Outlines in considerable detail Boule's views on a number of other key phases of discovery, including controversies over *Pithecanthropus* (discovered in 1891) and *Australopithecus* (discovered in 1925).

Bowden, M. *Ape-Men: Fact or Fallacy?* Bromley, England: Sovereign Publications, 1977. A semischolarly work that reviews the controversies that have surrounded a number of famous fossil finds, ranging from the clearly "invented" Piltdown man and the Java man, through the much more complex issues involved in Neanderthal paleontology.

Keith, Arthur. "The Relationship of Neanderthal Man and Pithecanthropus to Modern Man." *Nature* 89 (April, 1912): 155-156. Shows that, despite criticisms leveled at Boule's presumed bias by later critics, the scientific community that first reacted to his theory of *Homo neanderthalensis* viewed the prevailing evidence in terms that were very balanced.

Oakley, Kenneth P. *Frameworks for Dating Fossil Man*. London: Weidenfeld & Nicolson, 1964. Concerned with the development of various methods for dating human fossil finds. Because such technical skills were quite rudimentary in Boule's day, the record of subsequent advances, combining climatic stratigraphy in different geographical areas of the globe, is very relevant to the topic of Neanderthal man.

Shackley, Myra. *Neanderthal Man*. London: Gerald Duckworth, 1980. The most complete overall view of evidence on Neanderthals. Less concerned with the controversy of Boule's generation, offers chapters on tools and technology, and ritual and habitations that have been tied to Neanderthal communities in various areas of the globe.

Byron Cannon

Cross-References

Zdansky Discovers Peking Man (1923), p. 761; Dart Discovers the First Recognized Australopithecine Fossil (1924), p. 780; Weidenreich Reconstructs the Face of Peking Man (1937), p. 1096; Boyd Defines Human Races by Blood Groups (1950), p. 1373; Leakey Finds a 1.75-Million-Year-Old Hominid (1959), p. 1602; Simons Identifies a 30-Million-Year-Old Primate Skull (1966), p. 1814; Sibley and Ahlquist Discover a Close Human and Chimpanzee Genetic Relationship (1984), p. 2267; Hominid Fossils Are Gathered in the Same Place for Concentrated Study (1984), p. 2279; Scientists Date a *Homo sapiens* Fossil at Ninety-two Thousand Years (1987), p. 2341.

JOHANNSEN COINS THE TERMS "GENE," "GENOTYPE," AND "PHENOTYPE"

Category of event: Biology
Time: 1909
Locale: Copenhagen, Denmark

Johannsen helped found the science of genetics with his experimental support for and creation of the concepts of gene, genotype, and phenotype

Principal personages:

WILHELM LUDWIG JOHANNSEN (1857-1927), a Danish plant physiologist and chemist who was first to state clearly the symbolic distinction between phenotype and genotype

HERBERT SPENCER JENNINGS (1868-1947), a zoologist who tested Johannsen's pure-line work by studying the inheritance of size in the microorganism *Paramecium*

RAYMOND PEARL (1879-1940), a geneticist who conducted pure-line studies on egg production in fowls

THOMAS HUNT MORGAN (1866-1945), an embryologist and geneticist and winner of the 1933 Nobel Prize in Physiology or Medicine

Summary of Event

Natural selection is one of the mechanisms for evolutionary change and was formulated first in 1859 in *On the Origin of Species* by Charles Darwin. He described a process in which certain individual organisms are better able to survive and reproduce offspring relative to other individual organisms because of inherited differences. Over time, the inherited properties of a population gradually and continuously change, and sometimes the changes lead to a new population of organisms (speciation). Darwin also tried to account for the origins of inherited differences with his "provisional hypothesis of pangenesis," featured in a chapter in *Variation of Animals and Plants Under Domestication* (1868). He integrated a variety of ideas into his theory and believed that environment sometimes could directly influence the inherited characteristics that were passed to the offspring (soft inheritance).

As William Provine and Peter Bowler note, not even some of Darwin's strongest supporters, such as the biologists Thomas Henry Huxley and August Weismann or the biostatistician Francis Galton, accepted all the parts of his theory. Weismann was a strong supporter of natural selection but rejected soft inheritance. He believed in the continuity of the "germ plasm," which was separated from the body (soma). Galton also argued that the inherited properties, or "stirps," were passed from generation to generation with little change. Both Galton and Huxley questioned Darwin's emphasis on small, heritable differences and continuous change. They argued instead for selection of "sports," or mutations, and believed that evolution proceeded through rapid, discontinuous jumps. Galton also discovered the "law of filial regression," in which offspring tend to exhibit the average of the race or type to which they belong rather than the average of their parents for quantitative characteristics such as height. Thus, two very short parents would tend to have taller children.

Rediscovery of Gregor Johann Mendel's theory of heredity in 1900 only exacerbated the disputes. Karl Pearson and Walter Weldon, founders of the biometrical school, considered themselves to be followers of Galton but rejected Galton's belief in discontinuous evolution. They sided with Darwin in favor of natural selection of small, heritable variations and rejected Mendel's theory. Early Mendel supporters opposed the biometricians and argued that evolution proceeded by discontinuous leaps. The Mendelians believed that differences in continuous traits were too small to generate sufficient selection pressure to cause evolutionary change.

Wilhelm Ludwig Johannsen played a significant role in helping to bridge the gap between the opposing camps. Johannsen was trained as a pharmacist and attended lectures in botany at the University of Copenhagen but never obtained a degree. In 1881, he joined the chemistry department of the newly established Carlsberg laboratory. He resigned his post in 1887 but continued his research there and discovered a method of reviving dormant winter buds. The success of this discovery helped him gain a lecturer position in botany and plant physiology at the Copenhagen Agricultural College in 1892.

Johannsen was affected profoundly by Galton's work on heredity and was influenced also by Darwin's writings on natural selection. He decided to try to distinguish between Galton's ideas on selection, which had been experimentally supported by the plant physiologist, Hugo de Vries, and Darwin's theory of natural selection, as defended by Pearson and Weldon. De Vries considered selection to be ineffective whereas Pearson and Weldon believed selection was a continuous force for population change.

Unlike his predecessors, Johannsen chose a species that reproduced by self-fertilization, and "pure lines" consisted of all individuals descended from a single self-fertilized individual. In the autumn of 1900, Johannsen randomly selected five thousand princess bean (*Phaseolus vulgaris*) seeds. In the following spring, he planted one hundred seeds of average weight relative to the total of five thousand. He also planted twenty-five each of the smallest and largest seeds by weight. After the fall harvest, he discovered that, whether described in terms of the original groups (twenty-five small, one hundred average, twenty-five large) or as a combined total of 150, he got normal distributions. Yet, when he then reclassified the groups of large and small beans according to the average weight of their own offspring, he discovered that selection had resulted in larger and smaller average weights, respectively, when compared with the average for the original total population. Later, in 1902, he applied selection to the offspring of a single plant and found no change in average values. This indicated that selection on pure lines had no effect.

Johannsen presented his three years of experiments at the meeting of the Royal Danish Scientific Society on February 6, 1903, and published an article in both

Danish and German in the same year. Johannsen described his views on the evolutionary process in his book entitled *Om arvelighed og variabilitet* (on heredity and variation). An extended version of this book, now entitled *Arvelighedslaerens elementer* (the elements of heredity), was published in 1905. In 1906, he presented a paper at the Third International Conference on Genetics in which he extended his conclusions to hybridized lines on the basis of one incomplete experiment.

In 1908, Herbert Spencer Jennings at Harvard and Raymond Pearl at the Maine Agricultural Experiment Station conducted pure-line studies on the microorganism *Paramecium* and in fowls, respectively. Although their conclusions were partly erroneous, both sets of experiments were considered to provide support for Johannsen's pure-line theory.

Johannsen then wrote an even larger version of his textbook in German and published it in 1909 under the title *Elemente der exakten Erblichkeitslehre* (elements of an exact theory of heredity). He provided many statistical analyses of quantitative traits, and he discussed his experimental work. He noted how general populations represented mixtures of many pure lines, each with a different inherited constitution. Although individuals might vary for a particular continuous trait within a pure line, the inherited characteristics did not vary. He defined the term "phenotype" to mean the statistical average of the environmentally influenced variable appearances of individuals. He also recognized a need for characterizing the inherited properties of an organism. The nature of the inherited material was controversial. Physicalists such as Johannsen wanted to interpret biological processes in terms of forces. They rejected any efforts to tie Mendelian segregation patterns to the cytologist's discoveries about chromosomes and cellular division. Embryologists favored a corpuscular inherited factor. De Vries called the inherited units either factors or pangens after Darwin. Johannsen borrowed from de Vries and coined the term "gene" as a unit of calculation or accounting. He wanted a term unconnected to any of the contemporary theories about its nature, and Frederick Churchill notes that he used the term gene as if it were a chemical or physiological process rather than a thing.

Johannsen then discussed the relationship of phenotypic versus genotypic differences in terms of populations. The sense of "genotype" in this context is a reference to the inherited differences among pure lines. Churchill points out that Johannsen did not distinguish between transmission of traits and their development, nor did he apply his genotype-phenotype distinction to individuals.

Impact of Event

The geneticist Leslie Clarence Dunn notes that the *Elemente* was the first and most influential textbook of genetics on the European continent. In it, Johannsen introduced the new science and defined its key concepts. Although there were a few critics of his pure-line work, most geneticists in 1910 accepted his theory and rejected any significant role for natural selection in the evolutionary process.

From 1908 to 1911, however, while Johannsen, Jennings, and Pearl were criticizing Darwinian selection, three sets of experiments were performed that helped to harmonize the work of Darwin and Mendel. First, the geneticist William Castle experimentally demonstrated the action of selection. Second, Edward Murray East and Herman Nilsson-Ehle showed how Mendelism accounted for continuous variation and apparent blending in the offspring of parental traits. (East also noted that Johannsen's genotype-phenotype distinction was essentially a later version of August Friedrich Leopold Weismann's somatoplasm-germplasm dichotomy.) Finally, experimentation by Thomas Hunt Morgan and his students with the fruit fly *Drosophila* showed how the Mendelian factors or genes might be tiny variations. Morgan's group also showed how genes were carried on the chromosomes (color-staining bodies in the nucleus of cells). Churchill suggests that Morgan's chromosome theory hastened the acceptance of the contemporary definitions of genotype and phenotype, that is, the distinction between the expressed and outward appearance of an individual (phenotype) and the genetic information stored in the germ plasm (genotype).

In the third edition of *Elemente*, published in 1926, Johannsen used the modern definitions of genotype and phenotype and incorporated cytological discoveries and the work of Morgan's group. Yet, he still rejected a corpuscular gene. He was not alone. Disputes about the composition of the genetic material continued until Oswald Avery, Colin MacLeod, and Maclyn McCarty showed in 1944 that deoxyribonucleic acid (DNA) is the genetic material for almost all organisms, and James D. Watson and Francis Crick demonstrated in the 1950's how hereditary information is encoded in the DNA.

The science Johannsen helped found, genetics, is thriving. Since his discoveries, an ever-increasing number of behavioral, anatomical, and physiological traits have been shown to have an inherited component. Knowing how specific traits (including genetic diseases such as cystic fibrosis) are transmitted permits genetic counselors to advise parents about the risk of passing the condition on to their children. Research on the molecular biology of the gene and chromosome (the "genotype") is a fast-growing area and includes, for example, manipulation of the genetic material of food plants and animals in order to increase production or resistance to disease—identification of potent mutagens, particularly those affecting DNA, in the sperm or egg; and mapping of a specific gene's locations on a particular chromosome. In addition, studies on how traits are phenotypically expressed by the organism may provide eventually an avenue through which scientists can mediate that expression; for example, to either cure genetic diseases or prevent contemporary viruses such as the human immunodeficiency virus (HIV, responsible for AIDS—acquired immune deficiency syndrome) from exploiting a person's genetic machinery to reproduce itself.

Bibliography

Bowler, Peter J. *The Mendelian Revolution: The Emergence of Hereditarian Concepts in Modern Science and Society.* Baltimore: The Johns Hopkins University Press, 1989. Provides details about the work and life of Johannsen and the intellectual milieu. Good bibliography.

Churchill, Frederick B. "William Johannsen and the Genotype Concept." *Journal of the History of Biology* 7 (Spring, 1974): 5-30. An interesting analysis of Johannsen's central ideas and the consistency in his work during two time periods: from the pure-line studies until *Elemente* and after publication of the third edition of *Elemente*.

Dunn, L. C. *A Short History of Genetics: The Development of Some of the Main Lines of Thought, 1864-1939*. New York: McGraw-Hill, 1965. A clear, very readable description of the important personalities (including Johannsen) and discoveries contributing to the origins of genetics. Bibliography includes most of the major works by Johannsen.

Mayr, Ernst. *The Growth of Biological Thought: Diversity, Evolution, and Inheritance*. Cambridge, Mass.: Harvard University Press, 1982. A superb history of conceptual developments (including gene, genotype-phenotype) and the people responsible for them by one of the great evolutionary biologists of the twentieth century. Good bibliography.

Provine, William B., comp. *The Origins of Theoretical Population Genetics.* Chicago: University of Chicago Press, 1971. Excellent coverage of the "war" between the biometricians and the Mendelians. Details Johannsen's role in that dispute.

Roll-Hansen, Nils. "Drosophila Genetics: A Reductionist Research Program." *Journal of the History of Biology* 11 (Spring, 1978): 159-210. Describes the philosophical and scientific biases of Morgan, his students, and colleagues, including Johannsen. Interesting perspective.

Joan C. Stevenson

Cross-References

De Vries and Associates Discover Mendel's Ignored Studies of Inheritance (1900), p. 61; Sutton States That Chromosomes Are Paired and Could Be Carriers of Hereditary Traits (1902), p. 153; Morgan Develops the Gene-Chromosome Theory (1908), p. 407; Avery, MacLeod, and McCarty Determine That DNA Carries Hereditary Information (1943), p. 1203; Watson and Crick Develop the Double-Helix Model for DNA (1951), p. 1406.

THE STUDY OF MATHEMATICAL FIELDS BY STEINITZ INAUGURATES MODERN ABSTRACT ALGEBRA

Category of event: Mathematics
Time: 1909
Locale: Technical College, Breslau, Germany

Steinitz' studies on the algebraic theory of mathematics provided the basic solution methods for polynomial roots, initiating the methodology and domain of abstract algebra

Principal personages:

LEOPOLD KRONECKER (1823-1891), a German mathematician
HEINRICH WEBER (1842-1913), a German mathematician
ERNST STEINITZ (1871-1928), a German mathematician
KURT HENSEL (1861-1941), a German mathematician
JOSEPH WEDDERBURN (1882-1948), a Scottish-American mathematician
EMIL ARTIN (1898-1962), a French mathematician

Summary of Event

Before 1900, algebra and most other mathematical disciplines focused almost exclusively on solving specific algebraic equations, employing only real, and less frequently complex numbers in theoretical, as well as practical, endeavors. One result of the several movements contributing to the so-called abstract turn in twentieth century algebra was not only the much-increased technical economy through introduction of symbolic operations but also a notable increase in generality and scope. Although the axiomatic foundationalism of David Hilbert is rightly recognized as contributing the motivation and methods to this generalization by outlining how many specific algebraic operations could be reconstructed for greater applicability using new abstract definitions of elementary concepts, the other "constructivist" approaches of Henri-Léon Lebesgue, Leopold Kronecker, Heinrich Weber, and especially Ernst Steinitz had an equally concrete impact on the redevelopment and extensions of modern algebra.

Kronecker had unique convictions about how questions on the foundations of mathematics should be treated in practice. In contrast to Richard Dedekind, George Cantor, and especially Karl Weierstrass, Kronecker believed that every mathematical definition must be framed in such a way so as to be tested by mathematical constructional proofs involving a finite number of steps, whether or not the definitions or constructions could be seen to apply to any given quantity. In the older view, solving an algebraic equation more or less amounted only to determining tangibly its roots via some formula or numerical approximation. In Kronecker's view, the problem of finding an algebraic solution in general is much more problematic in principle since Évariste Galois' discoveries about (in)solvability of quartic and higher-order polynomials. For Kronecker, it required constructions of "algorithms," which will allow computation of the roots of an algebraic equation or show why this is not possible in any given case.

The question of finding algebraic roots in general had been of fundamental import since the prior work of Galois, Niels Henrick Abel, and Carl Friedrich Gauss. In particular, these efforts led Abel and Sophus Lie to formulate the first ideas of what are now known as "theory of groups." Later, Dedekind introduced the concept of "field" in the context of determining the conditions under which algebraic roots can be found. Kronecker was the first to employ the idea of fields to prove one of the basic theorems of modern algebra, which guarantees the existence of solution roots for a wider class of polynomials than previously considered. The novelty of the field approach is seen from the introduction to Weber's contemporaneous paper, "Die allgemeinen Grundlagen der Galois'chen Gleichungstheorie" (the general foundations of Galois theory). Weber first proved an important theorem stated by Kronecker, which relates the field of rational numbers to so-called cyclotomic, or Abelian, groups, a subsequently important area of the development field theory. Weber also established the notion of a "form field," being the field of all rational functions over a given base field F, as well as the crucial notion of the extension of an algebraic field. Although the main part of Weber's paper interprets the group of an algebraic equation as a group of permutations of the field of its algebraic coefficients, Weber's exposition is complicated by many elaborate and incomplete definitions, as well as a premature attempt to encompass all of algebra, instead of only polynomials. In his noted 1893 textbook on algebra, Weber calls $F(a)$ an algebraic field when a is the root of an equation with coefficients in F, equivalent to the definition given by Kronecker in terms of the "basis" set for $F(a)$ over a.

A central concern of Weber and other algebraists was that of extending the idea of absolute value, or valuation, beyond its traditional usage. For example, if F is the field of rational numbers, the ordinary absolute value $|a|$ is the valuation. The theory of general algebraic valuations was originated by Kronecker's student Kurt Hensel when he introduced the concept of p-adic numbers. In his paper "Über eine neue Begründung der algebraischen Zählen" (1899; on a new foundation of the algebraic numbers), Weierstrass' method of power-series representations for normal algebraic functions led Hensel to seek an analogous concept for the newer theory of algebraic numbers. If p is a fixed rational prime number, and a is a rational number $\neq 0$, a can be expressed uniquely in the form $a = (r/s)\, p^n$, where r and s are prime to p. If $\phi(a) = p^{-n}$, for $a \neq 0$, $\phi(a)$ is a valuation for the field of rational numbers. For every prime number p, there corresponds a number field which Hensel called the p-adic field, where every p-adic number can be represented by a sequence.

At this time, the American mathematician Joseph Wedderburn was independently considering similar problems. In 1905, he published "A Theorem on Finite Algebra," which proved effectively that every algebra with finite division is a field and that every field with a finite number of elements is commutative under multiplication, thus further explicating the close interrelations between groups and fields. Two years after Hensel's paper, Steinitz published his major report, "Algebraische Theorie der Körper" (1909; theory of algebraic fields), which took the field concepts of Kronecker, Weber, and Hensel much farther. Steinitz' paper explicitly notes that it was principally Hensel's discovery of p-adic numbers that motivated his research on algebraic fields. In the early twentieth century, Hensel's p-adic numbers were considered (by the few mathematicians aware of them) to be totally new and atypical mathematical entities, whose place and status with respect to then-existing mathematics was not known. Largely as a response to the desire for a general, axiomatic, and abstract field theory into which p-adic number fields would also fit, Steinitz developed the first steps in laying the foundations for a general theory of algebraic fields.

Steinitz constructed the roots of algebraic equations with coefficients from an arbitrary field, in much the same fashion as the rational numbers are constructable from the integers ($a X = b$), or the complex numbers from real numbers ($x^2 = -1$). In particular, Steinitz focused on the specific question of the structure of what are called inseparable extension fields, which Weber had proposed but not clarified. Many other innovative but highly technical concepts, such as perfect and imperfect fields, were also given. Perhaps most important, Steinitz' paper sought to give a constructive definition to all prior definitions of fields, therein including the first systematic study of algebraic fields solely as "models" of field axioms. Steinitz showed that an algebraically closed field can be characterized completely by two invariant quantities: its so-called characteristic number and its transcendence degree. One of the prior field concepts was also clarified.

Impact of Event

Although Steinitz announced further investigations—including applications of algebraic field theory to geometry and the theory of functions—they were never published. Nevertheless, the import and implications of Steinitz' paper were grasped quickly. It was soon realized that generalized algebraic concepts such as ring, group, and field, are not merely formally analogous to their better-known specific counterparts in traditional algebra. In particular, it can be shown that many specific problems of multiplication and division involving polynomials can be simplified greatly by what is essentially the polynomial equivalent of the unique-factorization-theorem of algebra, developed directly from field theory in subsequent studies.

In 1913, the concept of valuation was extended to include the field of complex numbers. An American algebraist, Leonard Dickson (1874-1957), further generalized these results to groups over arbitrary finite fields. Perhaps most notably, the French and German mathematicians Emil Artin and Otto Schreier in 1926 published a review paper, which in pointing out pathways in the future development of abstract algebra, proposed a program to include all of extant algebra in the abstract framework of Steinitz. In 1927, Artin introduced the notion of an ordered field, with the important if difficult conceptual result that mathematical order can be reduced operationally to mathematical computation. This paper also extended Steinitz' field theory into the area of mathematical analysis, which included the first proof for one of Hilbert's twenty-three famous problems, using the theory of real number fields.

As noted by historians of mathematics, further recognition and adoption of the growing body of work around Steinitz' original publication continued. Major texts on modern algebra, such as that by Bartel Leendert van der Waerden in 1932, already contained substantial treatment of Steinitz' key ideas. As later pointed out by the "structuralist" mathematicians of the French "Nicholas Bourbaki" group, the natural boundaries between algebra and other mathematical disciplines are not so much ones of substance or content, as of approach and method, resulting largely from the revolutionary efforts of Steinitz and others such as Emmy Noether. Thus, the theory of algebraic fields since the 1960's is most frequently presented together with the theory of rings and ideals in most textbooks.

The theory of algebraic fields is not only an abstract endeavor but also, since the late 1940's, has proven its utility in providing practical computational tools for many specific problems in geometry, number theory, the theory of codes, and data encryption and cryptology. In particular, the usefulness of algebraic field theory in the areas of polynomial factorization and combinatorics on digital computers has led directly to code-solving hardware and software such as maximal length shift registers and signature sequences, as well as error-correcting codes. Together with Noether's theory of rings and ideals, Steinitz' field theory is at once a major demarcation between traditional and modern theory of algebra and a strong link connecting diverse areas of contemporary pure and applied mathematics.

Bibliography

Artin, Emil. *Algebraic Numbers and Algebraic Functions.* New York: New York University Press, 1951. Discusses Artin's work on furthering Steinitz' field theory.

Bachman, George. *An Introduction to P-adic Numbers and Valuation Theory.* New York: Academic Press, 1964. An undergraduate treatment of the work of Kronecker, Weber, Hensel, and Steinitz.

Budden, F. J. *The Fascination of Groups.* Cambridge, England: Cambridge University Press, 1972. Represents modern efforts at elementary and intermediate-level treatments. A unique introductory treatment of groups using numerous examples for art, geometry, and music.

Dickson, Leonard E. *Algebras and Their Arithmetics.* New York: Dover, 1960. Contains some of the first simplified discussions of Steinitz' work.

McEliece, Robert. *Finite Fields for Computer Scientists and Engineers.* Boston: Kluwer Academic, 1987. Advanced text. Details the practical applications of the algebraic field theory.

Gerardo G. Tango

Cross-References

Hilbert Develops a Model for Euclidean Geometry in Arithmetic (1898), p. 31;

MILLIKAN CONDUCTS HIS OIL-DROP EXPERIMENT

Category of event: Physics
Time: January-August, 1909
Locale: Chicago, Illinois

Millikan measured electrical charges on tiny oil drops and determined that the electron is the fundamental unit of electricity

Principal personages:

ROBERT ANDREWS MILLIKAN (1868-1953), an American physicist, professor, research director, chief executive officer, and winner of the 1923 Nobel Prize in Physics

HARVEY FLETCHER (1884-1981), an American physicist and research director who worked directly with Millikan on the oil-drop experiment

Summary of Event

The first measurement of the electric charge carried by small water droplets was made in 1897 at Cambridge, England. The method timed the rate of fall of an ionized cloud of water vapor inside a closed chamber. The experiment was improved in 1903 by using a beam of X rays to produce the cloud between horizontal plates charged by a battery. The rate of descent of the top surface of the cloud between the plates was measured with an electric field switched on and off. The procedure, although an improvement, suffered from instabilities and irregularities on the top of the cloud. The cloud surface was difficult to delineate and resulted in measurements that fluctuated as much as 100 percent.

In 1909, a young graduate student, Harvey Fletcher, at the University of Chicago (then College of Chicago) went to Robert Andrews Millikan to receive suggestions for work on a doctoral thesis in physics. Millikan suggested improving upon the measurement of electronic charge previously performed at Cambridge, England. Millikan's initial plan was to use an electric field not only strong enough to increase the speed of fall of the upper surface of the ionized cloud but also powerful enough to keep the cloud surface top stationary when the electric field was reversed. This would allow the rate of evaporation to be easily observed and compensated for in the computations. This technical improvement would permit the researcher for the first time to make measurements on isolated droplets and eliminate the experimental uncertainties and assumptions involved in using the cloud method.

Millikan's improvement included the construction of a 10,000-volt small cell storage battery with enough strength to hold the top surface of the cloud suspended long enough to measure the rate of evaporation of the droplets. When the electric field was turned on, however, the result was a complete surprise to Millikan. The top of the cloud surface instantaneously dissipated and, since the experimental result assumed a rate of fall for the ionized cloud, Millikan saw this result as a complete failure. Repeated tests showed that whenever the cloud was dispersed, a few droplets would remain. By nature, however, these droplets had the proper charge to mass ratio to allow the downward force of gravity or weight of the droplet to be balanced by the upward pull of the electric field on the droplet's charge. This procedure became known as the "balanced drop method." With practice, Millikan found that he could reduce evaporation by turning off the field just prior to certain droplets in the field of view changing motion from slow downward to upward. This allowed timing of the motion for a longer period. From Stokes' law, he found the weight of the droplet. Also, by knowing the strength of the electric field, he calculated the electric charge necessary to balance its weight. It was noticed that the calculated electric charges came out to within the limits of error on his stopwatch and in multiples of whole integers (1, 2, 3, 4, and so on). The experimenters soon realized that the droplets always carried multiples of whole number charges and never fractional amounts.

The actual experimental arrangement used by Millikan and Fletcher consisted of a small box with a volume of 2 or 3 centimeters fastened to the end of a microscope. A tube was placed from the box to an expansion chamber secured by an adjustable petcock valve that allowed a rapid expansion of air to form a water vapor cloud in the box. Surrounding the box on both ends were two brass conducting plates about 20 centimeters in diameter and 4 millimeters thick. A small hole was bored into the top plate to allow the oil mist from an atomizer to enter the region between both plates which were separated by approximately 2 centimeters. A small arc light with two condensing lenses created a bright narrow beam which was in turn permitted to pass between the plates.

A physics instrument called a cathetometer was placed on the microscope that allowed it to be raised or lowered to the proper angle with the light beam for best illumination (which from practice turned out to be about 120 degrees). The plate separation allowed application of a potential difference and a resulting electric field when connected to the battery. The apparatus was operated by turning on the light; focusing the microscope, which was placed about 1 meter from the plates; and then spraying oil over the top plate, while switching on the battery. When viewed through the microscope, the oil droplets appeared like "little starlets" having the colors of the rainbow. When the electric field was first switched on, one would notice that the droplets would move at different speeds; some moved slowly upward, while the others moved downward faster. Superimposed on the droplets' downward fall was a small random back-and-forth motion (known as Brownian movement) caused by the collisions of the tiny droplets with thermally agitated air molecules within the chamber. When the electric field was reversed by changing the polarity of the battery, the same droplets that were moving downward moved upward and vice versa. The experimenters deduced that the nature of this motion indicated that part of the droplets were negatively charged, while the others carried a positive charge.

By the timely application of a polarity to the electric field, it was possible to keep selected droplets in the field of view for longer periods of time to obtain values for the calculation of electronic charges. For this condition, the electrical field interacting with the charge on the droplet created an upward force that compensated only for the weight of the droplet or the downward force. The electronic charge calculation depended upon a suitable balance between the intensity of the electrical field and the amount of electrical charge on the droplet that overcame its weight.

One major experimental problem remained, however; the water composing the droplet evaporated so fast that visibility was initially limited to only about two seconds. Discussion between Millikan and Fletcher on this led to substituting several other substances, including mercury and oil. Oil had an advantage, as it was easy to obtain and handle and its rate of evaporation was much slower than that of water. In time, the experiment was refined to obtain greater precision. The metal plates were machined more accurately, and the air between the plates was enclosed to prevent air drafts. Also, X-ray and radium sources were aimed into the chamber, producing greater ionization and more charged droplets than an atomizer.

Impact of Event

From examination of the smallest experimental values obtained, it became apparent that the charges on tiny oil droplets occur only in multiples of the smallest possible charge; no fractional amount of this basic charge was ever observed—only whole number increments. This implied that the unit charge obtained could not be subdivided into smaller charges and was independent of the droplet size. These exact values showed that the electronic charge was not merely a statistical mean, as previous experimenters believed. The experiment was, in fact, direct evidence for the existence of the electron as a finite-sized particle carrying a fundamental charge. It also made it possible to examine the attractive or repulsive properties of isolated electrons and to determine that electrical phenomena in solutions and gases are caused by electrical units that have fundamentally the same charge.

The experiment was an improvement over previous measurements in that Millikan was able to control accurately the strength of the electrical field while varying the droplet size. He also demonstrated that the oil droplet when completely discharged fell at the same rate as an uncharged droplet with the electric field on. This indicated that something fundamental, which he chose to call electricity, could be placed on or removed from the droplet only in exact amounts. Reversing the electric field to allow it to pull the droplets upward rather than downward permitted the researcher to freeze the motion of single droplets, giving more precise charge calculations than the method of trying to follow whole cloud motion, which was based only on statistical methods and could not give exact numbers.

As a result of Millikan's determination of the absolute charge on the electron and the previously known ratio of charge to mass, combined with the knowledge of the exact charges on ionized atoms from previous positive-ray analysis or electrolysis, the absolute masses of both the electron and atom could be determined with great precision. With a knowledge of the charge on the electron—a new unit of energy—the electronvolt could be defined. The kinetic energy of particles of unit charge that had moved through a potential difference now could be computed with the known mass of the particle entering the equation.

Another outcome of the experiment was the calibration of a correction factor used for Stokes' law. Millikan realized that Stokes' law, tested only for the larger spheres, would require a correction factor when used with droplets so small that their size became comparable to the mean free path of the air molecules executing Brownian movement in a gaseous state. These smallest droplets when viewed through a microscope are affected by this Brownian movement, which interferes with the droplet's rate of rise or fall and would otherwise introduce significant error into the charge computation. Thus, the measurements obtained from the oil-drop experiment served a dual purpose: as a means to determine the electronic charge and as a correction for Stokes' law.

Bibliography

Fletcher, Harvey. "My Work with Millikan on the Oil-Drop Experiment." *Physics Today* 44 (June, 1982): 43-47. In this reminiscence, Fletcher relates his experiences while working with Millikan. Photographs, a diagram, and a detailed description of how the experiment was performed will interest the lay reader.

Fraser, Charles G. "Story of Electricity and Magnetism." In *Half-Hours with Great Scientists: The Story of Physics.* Toronto: University of Toronto Press, 1948. An informative description of the great discoveries in physics organized into five chapters. The author provides the general audience with a short summary of not only experimental results but also how Millikan's work related to previous efforts.

Heathcote, Niels Hugh de Vaudrey. "Robert Andrews Millikan." In *Nobel Prize Winners in Physics, 1901-1950.* New York: Henry Schuman, 1953. The author organizes his chapter on Millikan into three sections: a biographical sketch, a description of the prizewinning work, and a summary of Millikan's contribution to science. An informative description of the experiment is given, along with diagrams of the oil-drop apparatus. Unique insights are given as Heathcote quotes Millikan often.

Millikan, Robert A. *Autobiography.* Englewood Cliffs, N.J.: Prentice-Hall, 1950. The author describes his life from early childhood and education through his later work on projects at the California Institute of Technology. An entire chapter is devoted to the oil-drop experiment with tables, procedures used, and a detailed illustration. This book is a must for the lay reader who desires more than a cursory summary.

Oldenberg, Otto. *Introduction to Atomic and Nuclear Physics.* 3d ed. New York: McGraw-Hill, 1961. Many physics texts provide a brief description of Millikan's experiment but few explain the technique and results as well as this author does. Although intended as a text, a reader with a basic mathematics background should be able to understand this book.

Romer, Alfred. "Robert A. Millikan, Physics Teacher." *The Physics Teacher* 78 (February, 1978): 78-85. A unique view into the character of Millikan from the perspective of a graduate student who knew him. Millikan's textbooks, explains Romer, were contributive in that they contained many supportive laboratory experiments. A brief description of the oil-drop experiment is given, along with an example of Millikan's insight and originality.

Michael L. Broyles

Cross-References

Thomson Wins the Nobel Prize for the Discovery of the Electron (1906), p. 356; Bohr Writes a Trilogy on Atomic and Molecular Structure (1912), p. 507; Rutherford Presents His Theory of the Atom (1912), p. 527; Rutherford Discovers the Proton (1914), p. 590; Millikan Names Cosmic Rays and Investigates Their Absorption (1920), p. 694.

BLÉRIOT MAKES THE FIRST AIRPLANE FLIGHT ACROSS THE ENGLISH CHANNEL

Category of event: Space and aviation
Time: July 25, 1909
Locale: From Calais, France, to Dover, England

Blériot's flight across the English Channel was the first international airplane flight

Principal personages:

LOUIS BLÉRIOT (1872-1936), a French inventor, businessman, aviator, and winner of the *Daily Mail* prize for successfully flying across the English Channel in a heavier-than-air flying machine

ALFRED HARMSWORTH, FIRST VISCOUNT NORTHCLIFFE (1865-1922), an outspoken proponent of the development of aviation, both civilian and military, in England

HUBERT LATHAM (1883-1912), a French-English big-game hunter, boat racer, race car driver, and aviator who failed in two attempts to pilot an airplane across the English Channel

Summary of Event

Wilbur Wright's 1908 demonstrations of powered flight in France caused an uproar throughout Europe. His ability to control his airplane astonished the European pilots and other observers. One of these was the aeronautical correspondent for the *Daily Mail*, an English newspaper owned by Alfred Harmsworth, a man with a deep interest in aviation. To spur the interest of others in his country, Harmsworth offered a prize of five hundred pounds for the first flight across the English Channel, in either direction, in a heavier-than-air device unsupported by any lifting agent; the flight was to be completed between sunrise and sunset. This was not the first time Harmsworth had attempted to interest the people and government of England in aviation. In 1906, he had offered a prize for the first flight from London to Manchester; ironically, it had been won, in 1910, by a Frenchman, Louis Paulham.

No one stepped forward in response to the *Daily Mail*'s prize, so late in 1908 the sum was increased to one thousand pounds, an amount that succeeded in bringing out contestants. The first to announce his intentions was Hubert Latham. His photograph shows a handsome, smiling young man of French and English descent, whose dark eyes seem to sparkle with some secret, or perhaps a touch of mockery. This debonair flyer immediately won the hearts of the public; even after other contestants had declared themselves, he remained the favorite. Like many early aviators, he was wealthy, which allowed him to pursue a variety of interests, from big-game hunting to racing boats and automobiles, and finally, to aviation. He had, in fact, already flown the Channel, from England to France, by balloon; he had decided to become an airplane pilot after witnessing some of Wright's demonstrations.

Latham chose to fly an *Antoinette*, a beautiful, but rather unstable, monoplane designed and built by Leon Levavasseur. The two men set up their airfield at Sangatte, not far from Calais, in the summer of 1909. The weather over the English Channel, known for its capricious and unpredictable nature, was worse than normal: High winds, rain, fog, and mist conditions precluded any attempt until mid-July. On July 19, at 6:42 A.M., Latham finally took off for his first attempt, escorted by the destroyer *Harpoon*. Seven minutes into the flight, his engine failed, and he landed in the Channel, unhurt. The *Harpoon* rescued him and attempted to bring his airplane on board, unfortunately damaging the machine beyond repair.

Once on shore, Latham ordered another plane to be sent immediately. It arrived in a short time, was quickly assembled, and he was ready for another try. In the meantime, however, word of Latham's attempt had reached the ears of his most serious competitor, Louis Blériot, spurring him to move immediately to Calais and challenge Latham.

Blériot was a heavy-set, dour-looking man, his face punctuated by a large, red, walrus mustache. His stocky, swarthy appearance, together with his no-nonsense attitude, made him less popular with the public than Latham. He had amassed a fortune by designing, manufacturing, and selling acetylene headlights for automobiles; having become infatuated with flying, he spent most of his fortune on aviation. From 1901 to 1909, he had designed and built a number of airplanes, evolving from devices with flapping wings to biplanes and, eventually, monoplanes, culminating in his *Blériot No. XI*. It is estimated that Blériot had spent about $150,000 by the time he came up with his *No. XI*, the plane that would bring him fame and more fortune.

Arriving in Calais, Blériot chose as his airfield site a farm belonging to a M. Grignon, near Les Baraques, a small village not far from Sangatte. His plane, powered by a 25-horsepower Anazini engine, crude in design and not known for its reliability, was unloaded and carted to the farm. With both aircraft standing ready for the flight, the two pilots could only fret while the weather over the Channel kept them grounded. On July 24, it appeared that there might be a change in the unfriendly weather; accordingly, that night, M. Charles Fontaine, a reporter for the Paris newspaper *Le Matin*, took the night ship to Dover. His job was to find a safe landing place for Blériot, and to signal him by waving the French flag.

At 2:30 A.M. on July 25, Blériot was awakened and given the news that the weather seemed to be improving. He dressed and ate a quick breakfast, then went to the field to prepare for departure while his wife went to Calais to alert the escort ship *Escopette* to Blériot's forthcoming departure. At 4:10 A.M., the pilot made a short test flight. By 4:35, he was ready to leave, waiting only to make sure that his takeoff would be after sunrise. Just before taking off, according to witnesses, he asked, "Au fait, où est-ce exactement, Douvres?" (By the way, where is it exactly, Dover?)

After a short time airborne, Blériot passed over his escort; now he was alone, flying through patches of mist. Suddenly, fog engulfed him, obscuring everything. Blériot, with no instruments on board his airplane, nothing to guide him, not even sure he was heading in the right direction, released the controls, letting the airplane fly itself. He flew this way for ten minutes; meanwhile, his engine began to overheat, intermittently losing and then recovering power, causing him to lose altitude. The aircraft's descent took it through a small rain shower, which seemed to resolve his problem: The engine cooled down and regained full power. A few minutes later, the mist and fog began to thin, and Blériot could see the coast of England. He realized that the winds had blown him off course: He was near St. Margaret's Bay, east of Dover. He turned and headed for the Dover lighthouse, visible through the haze, noting that the winds had increased in velocity and turbulence. Flying over the English fleet, which was anchored in Dover Harbor, he proceeded along the cliffs looking for Fontaine. Spying him standing in a depression above the cliffs, Blériot made a half circle and headed toward him. Once over the cliffs, the airplane experienced severe turbulence, which spun it around. The pilot responded by cutting the engine; the plane descended rapidly from approximately 18 meters, making a "pancake" landing, which smashed the landing gear and broke the propeller. Thirty-seven minutes after taking off from Calais, Blériot had survived his fifty-first crash, won the coveted *Daily Mail* prize, and secured his place in aviation history.

Impact of Event

Although the Wright brothers had proved that sustained, controlled flight was possible, the airplane was viewed by most people in the early 1900's as a frail, unreliable, dangerous device, a rich person's toy with no practical use. Blériot's flight across the English Channel was not particularly noteworthy for its length in either time or distance, but as the first airplane flight to traverse national boundaries and to cross a large body of water, it awakened the government's and the general public's awareness that this new invention could be something more than a passing fad; it could, in fact, have a practical use.

The worldwide fame achieved by Blériot through his successful flight, together with his investment of the prize money in his recently acquired aviation company, established him as the leading European airplane designer and manufacturer of the time and for many years thereafter. Although the controversy over the relative superiority of biplane versus monoplane would continue for a number of years, his monoplane was the most widely accepted design of that era and would become the prototype for most twentieth century airplanes.

France was quick to capitalize on the excitement and fervor surrounding its citizen's achievement. Within a month after Blériot's English Channel crossing, the champagne industry, in cooperation with the municipality of Reims, had organized the world's first international air meet, bringing together most of the leading aviators of the day to compete for prize money. The Reims air meet was soon followed by other tournaments and cross-country races, events that fed the technological development of the airplane much as early automobile races fed the development of the automobile. Within two years, these air races had become international in scope. With contestants flying from one country to another, it soon became obvious that national borders, as drawn on maps, had lost much of their effectiveness as obstacles to the movement of people and goods between countries: No longer was it necessary to stop at the border for permission to enter or pass through a country; one could even fly over one country to get to another.

While the general public was still caught up in the excitement and joy of Blériot's successful flight across the English Channel, many leaders of England and continental Europe were quick to realize its political and military significance. For hundreds of years, England's security had been guaranteed by two things: the English Channel and its Royal Navy. Blériot's flight had negated these as the country's protectors and challenged its insular safety from attack. Harmsworth was the first to predict that in the future, the airplane would play a dominant role in England's survival. Although his efforts to spur the development of aviation in Engand were supported by other farsighted leaders and reporters, as well as the general public, they were for a number of years thwarted by the Royal Navy's domination of the English military establishment. Editors on the continent, meanwhile, pointed out that the airplane could fly in both directions; hence, it was not merely England that had lost a measure of security. As a result, while England vacillated, France took the lead in nurturing the development of aviation, followed closely by Germany. By the time World War I broke out, the military establishment of both of these countries had a fledgling aviation branch, and the outstanding French fighter planes produced by Blériot's company, the Société Pour l'Aviation et ses Dérivés (better known by its acronym, SPAD) played an important part in the Allies' eventual victory. It was not until World War II, however, that Harmsworth's prediction concerning the crucial role of the airplane in England's defense against attack would come true—in 1940, when the Royal Air Force defeated the German Luftwaffe in what later became known as the Battle of Britain.

As the first international airplane flight, and the first flight over a large expanse of water, Blériot's 1909 English Channel crossing demonstrated the potential of the airplane for transporting people and goods. It may, therefore, be considered the forerunner to both military and peaceful commercial flight between countries.

Bibliography

Brennan, Dennis. "Where Is Dover?" In *Adventures in Courage: The Skymasters*. Chicago: Reilly & Lee, 1968. Presents rather romanticized stories of various important personages and events in aviation history; the chapter on Blériot purports to quote conversations between the aviator and his wife. Might interest a young reader, but relates little information to the serious student of aviation history.

Jones-Gwynn, Terry. "1909—Lord Northcliffe's Channel Challenge: The Dawn of Air Racing." In *The Air Racers: Aviation's Golden Era, 1909-1936*. London: Pelham Books, 1984. In the first chapter, Jones-Gwynn tells the story, in detail, of the two rival aviators, Latham and Blériot. Some details differ from those provided by other authors, but the overall account is in harmony with other versions. Colorized reproductions of black-and-white photographs illustrate the narrative.

Pendergast, Curtis. *The First Aviators*. Alexandria, Va.: Time-Life Books, 1980.

One of the Time-Life series, The Epic of Flight. Chapter 2, "The Great Show at Reims," includes an excellent, if brief, account of Blériot's flight. Contains a series of illustrative photographs. Like the other volumes in the series, the book is both factual and very readable.

Thomas, Lowell, and Lowell Thomas, Jr. "A Twenty-Two-Mile Flight That Startled the World." In *Famous First Flights That Changed History.* Garden City; N.Y.: Doubleday, 1968. Focuses on Blériot, while giving credit to Latham for his failed attempt to be first to fly the English Channel. A rather uninspired, although apparently historically correct, account of the event and the people involved.

Villard, Henry Serrano. *Contact! The Story of the Early Birds*. New York: Thomas Y. Crowell, 1968. Traces the first decade of powered flight, from Kitty Hawk to World War I. Chapter 4, "There Are No Islands Anymore," offers a detailed and accurate description of the first crossing of the English Channel by airplane and of the people involved; also gives insight into events that followed. An excellent source for anyone interested in the early development of aviation.

Wallace, Graham. *Flying Witness: Harry Harper and the Golden Age of Aviation*. London: Putnam, 1958. An outstanding history of early aviation, as seen by Harry Harper, who was employed by Harmsworth as the world's first aeronautical correspondent. Two chapters, "Gallant Failure" and "Across the Channel," treat the contest between Latham and Blériot to be the first to cross the English Channel by airplane.

P. John Carter

Cross-References

Zeppelin Constructs the First Dirigible That Flies (1900), p. 78; The Wright Brothers Launch the First Successful Airplane (1903), p. 203; The Fokker Aircraft Are the First Airplanes Equipped with Machine Guns (1915), p. 600; Lindbergh Makes the First Nonstop Solo Flight Across the Atlantic Ocean (1927), p. 841.

THE ELECTRIC WASHING MACHINE IS INTRODUCED

Category of event: Applied science
Time: 1910
Locale: United States

American manufacturers adapted electric power to hand-operated washing tubs and wringers, significantly reducing the labor and time involved in washing clothes

Principal personages:

O. B. WOODROW, a bank clerk who adapted electricity to a remodeled hand-operated washing machine, claiming it to be the first successful electric washer

ALVA J. FISHER (1862-1947), the founder of the Hurley Machine Company who designed the Thor electric washing machine, claiming it to be the first successful electric washer

HOWARD SNYDER, the mechanical genius of the Maytag Company who developed a washer featuring an electric motor and a swinging wringer

Summary of Event

Until the development of the electric washing machine in the twentieth century, washing clothes was a tiring and time-consuming process. With the development of the washboard, patented in the United States in 1833, dirt was loosened by rubbing. Clothes and tubs had to be carried to the water, or the water from a stream or well had to be carried to the tubs and clothes. After washing and rinsing, clothes were hand-wrung, hang-dried, and ironed with heavy heated irons.

In nineteenth century America, the laundering process became more arduous with the greater use of cotton fabrics resulting from the expansion of the textile industry. In addition, the invention and industrial application of the sewing machine resulted in the mass production of inexpensive ready-to-wear cotton clothing. With more clothing, there was more washing. One solution was hand-operated washing machines.

The first American patent for a hand-operated washing machine was issued in 1805. By 1857, more than 140 patents had been issued; by 1880, between four and five thousand patents had been granted. While most of these were never produced, they are evidence of the desire to find a mechanical means to relieve the burden of washing clothes. Nearly all of the early types prior to the Civil War were based upon the rubbing principle used by women when they rubbed clothes on a washboard. One of these patented in 1846 employed a swinging curved, inverted "T" that passed over clothes placed on a curved bed of rollers and must have worked fairly well. The basic model can be found advertised in an early twentieth century Sears, Roebuck catalog as its "Quick and Easy Washer" and as "Our Famous Old Faithful" in a 1927 Montgomery Ward catalog.

Washing machines based upon the rubbing principle had two limitations: They washed only one item at a time, and the constant rubbing was hard on clothes. The major conceptual breakthrough was to move away from rubbing and to design machines that would clean by forcing water through a number of clothes at the same time. Electric washing machines would eventually use one of five designs (suction, oscillating tub, dolly, horizontal rotary cylinder, or underwater agitator) to create the washing action needed to force water through fabric. Each of these designs was developed in the nineteenth century for hand-operated machines.

An early suction machine utilized a plunger fastened to a fulcrum attached to a washing tub. When a person raised one end of the handle, a plunger at the other end dropped into the tub, forcing water through the clothes. Later, electric machines would have between two to four suction cups, similar to plungers, attached to arms that went up and down and rotated on a vertical shaft. The cups pushed the water through the clothes on the down stroke and then sucked the water through the clothes on the up stroke. Another hand-operated washing machine used oscillating action by rocking a tub on a frame. The rocking action threw water through the clothes and then the clothes through the water. An electric motor was later substituted for the hand lever that rocked the tub. A third hand-operated washing machine was the dolly type. The dolly, which looked like an inverted three-legged milking stool, was attached to the inside of the tub cover and turned by a two-handled lever on the top of the enclosed tub. Clothes were washed by being pulled through the hot, soapy water in a tub with corrugated sides that increased agitation. The dolly type was the most popular of the manually operated machines and the first to be adapted for electric power.

The hand-operated machines that would later dominate the market as electric machines were the horizontal rotary cylinder and the underwater agitator types. In 1851, James King patented a machine that utilized two concentric half-full cylinders. Water in the outer cylinder was heated by a fire beneath it; a hand crank turned the perforated inner cylinder that contained clothing and soap. The inner-ribbed design of the rotating cylinder raised the clothes as the cylinder turned. Once the clothes reached the top of the cylinder, they dropped down into the soapy water. The rotary motion and dropping action created the agitation by which clothing was washed. An important advance took place in 1863, when Hamilton Smith patented a belt-driven reciprocating revolving drum, thereby demonstrating the importance of reversible action in washing machines.

The first underwater agitator-type machine was patented in 1869. In this machine, four blades at the bottom of the tub were attached to a central vertical shaft which was turned by a hand crank on the outside. The agitation created by the blades washed the clothes by driving the water through the fabric. Of the five hand-operated washing machines that were modified for electric power, the underwater agitator type was the last to be successfully marketed as an electric machine. It was not until 1922, when Howard Snyder of the Maytag Company developed an underwater agitator with reversible motion, that this type of machine was able to compete with the other electric machines. Reversible action was central to the machine's function. If the blades in this machine had turned in only one direction, clothes would soon be wrapped around the blades and cleansing agitation would be significantly reduced.

Until the nineteenth century, getting water out of clothing required hand wringing. The development of hand-operated wringing machines, similar to hand-operated washing machines, provided the basic designs later used in electric machines. The term "wringer" was first used with an 1847 device wherein wet clothes, placed in a slit sack suspended between two posts, were twisted by a hand crank. Later, the term was used in reference to the two-roller innovation marketed in 1861 that became the basic model for motor-powered machines. This innovation featured two adjustable parallel rollers, one above the other. Water was extracted as the clothing was pressed between the rollers, one of which was turned by hand crank while the other turned freely. In the twentieth century, motors underneath washing tubs were connected by belts to the wringer apparatus. The basic design was completed in 1910, when a patent was issued for a reversible, swinging wringer. Reversible rollers allowed the operator to correct improper loading and to withdraw clothes that had wrapped around one of the rollers. The swinging wringer allowed the operator to move the wringer so that it could be over the wash tub, over the rinse tub, or out of the way when either was being loaded. In the twentieth century, motor-powered machines would also extract water by the centrifugal force of rapid spinning. Like other developments in washing machines, this one preceded the application of electricity. A machine patented in 1873 featured an inner clothing basket which was made to spin very rapidly by the use of a hand crank. Unfortunately, turning the crank to get the rapid spinning action required much manual labor.

It was not until about 1900 that hand-operated washing machines began to replace washboards in American homes. By 1905, washing machines were being advertised that were operated by other than hand power. Small gasoline engines—used at this time for a variety of purposes—were now used to power washing machines. They were popular in homes that did not have electricity. Washing machines powered by water motors were advertised for homes with sufficient water pressure. For those homes wired for electricity, an electric washing machine was soon available.

Priority for the first electric washing machine was claimed by O. B. Woodrow, who founded the Automatic Electric Washer Company, and Alva J. Fisher, who developed the Thor electric washing machine for the Hurley Machine Corporation. Both Woodrow and Fisher made their innovations in 1907 by adapting electric power to modified hand-operated, dolly-type washing machines. Since only 8 percent of American homes were wired for electricity in 1907, the early machines were advertised as adaptable to electric or gasoline power and could be hand operated if the power source failed. Soon, electric power was being applied to rotary cylinder, oscillating, and suction-type machines. Separate belts were attached to wringers so that they could be operated by electric power. In 1910, a number of companies, including Woodrow's Automatic Electric Washer Company, introduced washing machines with attached wringers that could be operated by electricity. Maytag's 1911 electric washing machine, developed by Snyder, featured the Maytag swinging wringer.

Impact of Event

By 1907 (the year electricity was adapted to washing machines), electric power was already being used to operate fans, ranges, coffee percolators, flatirons, and sewing machines. As more and more homes were wired for electricity, more and more families bought electric appliances, including washing machines. By 1920, nearly 35 percent of American residences were wired for electricity; by 1941, nearly 80 percent were wired. By 1941, a majority of American homes had electric washing machines; by 1958, this had risen to an estimated 90 percent.

The growth of electric appliances, especially washing machines, is directly related to the decline in the number of domestic servants in America. In 1910, there were 1,830,000 domestic servants, of whom 520,000 were laundresses. With immigration down as a result of World War I and new employment opportunities opening for women, by 1920, the number of domestic servants had declined to 1,400,000, of whom 385,000 were laundresses. This indicates that the development of the electric washing machine in this decade was, in part, a response to a decline in servants, especially laundresses. Although conditions changed in the 1920's, and by 1930 the number of domestic servants increased to 2 million, the number of laundresses actually declined. Rather than easing the work of laundresses with technology, American families replaced their laundresses with washing machines.

Commercial laundries were also affected by the growth of electric washing machines. Commercial laundries used steam and pioneered the development of rotary washing and spin drying. At the end of the nineteenth century, they were in every major city and utilized by people of all incomes. Observers noted that as spinning, weaving, and baking had once been done in the home but now were done in commercial establishments, laundry work had now begun its move out of the home. Commercial laundry business grew impressively, doubling total receipts each decade, until the depression of the 1930's. Many customers, responding to intense advertising by manufacturers, purchased electric washing machines in the belief that the overall cost of owning a machine was less than commercial laundry service. Although commercial laundries grew after World War II, their business centered more and more on institutional laundry, rather than residential laundry, which they had lost to the home washing machine.

The return of residential laundry to the home is the only example of a household task that began to move out of the home and then returned. This occurred in part because of technological developments. Each advance—electric machines over hand-powered machines, spin drying and rotary driers over wringers—meant that the operator no longer had to give constant attention to each segment of the laundering process. Bendix's introduction of automatic washers in 1937 meant that washing machines could change phases without the action of the operator.

Some scholars have argued that the return of laundry to the home was also the product of marketing strategies that developed the image of the American woman as one who is home operating her appliances. On the other hand, it probably resulted from the fact that the electric washing machine greatly reduced the time and labor involved in laundry, that American women, still primarily responsible for the family laundry, were able to pursue careers outside the home.

Bibliography

Allen, Edith. *Mechanical Devices in the Home*. Peoria, Ill.: Manual Arts Press, 1922. Although dated, the book is valuable in showing what kinds of electric washing machines were expected to be industry leaders at that time. Chapters on laundry equipment, wringers, and water motors are good in describing how these technologies actually worked.

Buehr, Walter. *Home Sweet Home in the Nineteenth Century.* New York: Thomas Y. Crowell, 1965. One of the few books devoted to household technologies. Although it is limited to the nineteenth century, the book contains interesting diagrams of early hand-powered washing machines.

Cowan, Ruth S. *More Work for Mother: The Ironies of Household Technology from the Open Hearth to the Microwave*. New York: Basic Books, 1983. A very important study of the relationship of household technologies and housework. Shows how labor-saving devices may actually reorganize the work process and increase the amount of labor involved. Describes how advertisements in women's magazines set new standards for housework. Examines why alternative approaches to housework—commercial laundries and cooperative kitchens—failed.

Davidson, Caroline. *A Woman's Work Is Never Done: A History of Housework in the British Isles, 1650-1950*. London: Chatto & Windus, 1983. In a chapter on laundry, Davidson offers an excellent survey of the history of clothes washing in England since 1650. The chapter is particularly good at explaining why England lagged behind the United States in adopting electric washing machines.

De Armond, Fred. *The Laundry Industry.* New York: Harper & Brothers, 1950. Written by one sympathetic to the problems of the commercial laundry. Describes the history of the commercial laundry from the middle of the nineteenth century to 1948. Good description of changing attitudes of commercial laundries toward residential washing and the competition between commercial laundries and home electric washing machines.

Du Vall, Nell. *Domestic Technology: A Chronology of Developments*. Boston: G. K. Hall, 1988. A survey of developments in domestic technologies. Helpful in dating some, but not all, of the major innovations in washing machines, irons, and laundry aids. Unfortunately, the descriptions in most cases are too short to be helpful in understanding how the technology worked or how the innovations were related.

Giedion, Sigfried. *Mechanization Takes Command*. New York: Oxford University Press, 1948. A fascinating book that places the development of the washing machine within the overall development of mechanization in the nineteenth century. Although there are only about ten pages on the washing machine, they contain some of the most insightful comments written. Giedion clearly shows how agitator and cylinder washers of the twentieth century were based upon concepts developed in the nineteenth century.

Katzman, David M. *Seven Days a Week: Woman and Domestic Service in Industrializing America*. New York: Oxford University Press, 1978. An impressive survey of domestic service in the United States between 1880 and 1930. The book is excellent in describing changing patterns of service in various parts of the United States during this period when the electric washing machine was being introduced. Also includes an extensive list of tables charting changes in domestic service.

Strasser, Susan. *Never Done: A History of American Housework*. New York: Pantheon Books, 1982. The chapter "Blue Monday" is superb in describing the arduous nature of laundry work in the nineteenth century and how American women sought relief through the use of laundresses, commercial and cooperative laundries, and electric washing machines. Also contains some excellent pictures that capture the demanding nature of laundry work.

Swisher, Jacob. "The Evolution of Wash Day." *Iowa Journal of History and Politics* 38 (January, 1940): 3-49. A historical survey of innovations in washing machines, wringers, irons, and commercial laundries. While the article concentrates on developments in Iowa, the fact that many washing machine companies were from Iowa indicates that most major innovations are covered.

Thomas W. Judd

Cross-References

Booth Invents the Vacuum Cleaner (1901), p. 88; Corning Glass Works Trademarks Pyrex and Offers Pyrex Cookware for Commercial Sale (1915), p. 605; Birdseye Develops Freezing as a Way of Preserving Foods (1917), p. 635; Flourescent Lighting Is Introduced (1936), p. 1080.

ROUS DISCOVERS THAT SOME CANCERS ARE CAUSED BY VIRUSES

Category of event: Medicine
Time: 1910
Locale: Rockefeller Institute for Medical Research, New York City

Rous discovered that a chicken liver sarcoma (connective tissue cancer) is caused by a virus

Principal personages:

PEYTON ROUS (1879-1970), an American pathologist who studied cancer in various species and cowinner of the 1966 Nobel Prize in Physiology or Medicine

DMITRY IOSIFOVICH IVANOWSKY (1864-1920), a Russian microbiologist who discovered that a virus smaller than bacteria caused tobacco mosaic disease

FRIEDRICH AUGUST JOHANNES LÖFFLER (1852-1915), a German microbiologist who discovered that a virus causes foot-and-mouth disease

DAVID BALTIMORE (1938-   ), an American virologist and cowinner of the 1975 Nobel Prize in Physiology or Medicine for his discovery that retroviruses copy DNA from RNA using the enzyme reverse transcriptase

HOWARD M. TEMIN (1934-   ), an American virologist and cowinner of the 1975 Nobel Prize in Physiology or Medicine for his discovery that Rous sarcoma virus copies DNA from RNA when it infects cells

Summary of Event

During the last half of the 1800's, tremendous progress was made in a new area of biology called microbiology, the study of extremely small organisms. Primary focus was upon microbial pathogens, those microorganisms that cause infectious diseases in humans, animals, and plants. Two giants of nineteenth century microbiology, Louis Pasteur and Robert Koch, firmly established the germ theory of disease, which maintains that some microorganisms are responsible for human, animal, and plant infectious diseases. Koch was the first microbiologist to identify the bacterial pathogen (disease-producer) *Bacillus anthracis* as the causative agent of anthrax in humans and cattle.

The microbial world consists of many species ranging from protozoa to fungi to bacteria. The nineteenth century work of Matthias Jakob Schleiden and Theodor Schwann produced the cell theory, which maintains that all living organisms are composed of cells. A cell, the basic unit of life, is a membrane-bound compartment containing all of the necessary chemical ingredients for life. The human body contains approximately one hundred trillion cells. Bacteria, however, exist as single cells approximately 0.01 millimeter in diameter. Experiments aimed at disproving the spontaneous generation of life (that is, abiogenesis) and at sterilizing liquid food used special filters to remove bacteria. The liquid food containing bacterial cells was forced through a series of filter papers containing microscopic pores small enough to block cells but not liquid.

In 1892, the Russian microbiologist Dmitry Iosifovich Ivanowsky was attempting to discover the cause of tobacco mosaic disease, an infection that was decimating Russian tobacco crops. Filtering infected tobacco juice failed to remove the agent responsible for the disease. Something smaller than a bacterium was the culprit. Using filters containing smaller and smaller pores, Ivanowsky demonstrated that the causative agent was approximately one hundred to one thousand times smaller than bacteria. This new organism, called a virus, was approximately 0.001 to 0.01 micrometer long and had a unique pattern of infection. The Dutch botanist Martinus Willem Beijerinck discovered viruses independently at about the same time. In 1898, Friedrich August Johannes Löffler showed that a virus causes foot-and-mouth disease.

In the early twentieth century, viruses were shown to be noncellular in structure, consisting only of nucleic acid—that is, deoxyribonucleic acid (DNA) and ribonucleic acid (RNA)—wrapped within a protective protein covering. Viruses are immobile and inactive outside cells. They can function only within a host cell, and then only to reproduce and destroy the host cell. They are obligate intracellular parasites, always invading cells, raping cellular resources, reproducing, and destroying. Because of the noncellular structure and unusual nature of viruses, there is considerable debate over their classification as a life form.

Once a virus is carried by air or fluid to the cells of a given host species, it may be only by chance that it physically contacts a cell. Once physical contact is made, a rapid series of chemical reactions between the virus protein covering and the cell membrane triggers the injection of the viral DNA or RNA into the cell. Once it is inside the host cell, the viral nucleic acid can follow two possible infection routes, depending upon cellular conditions and certain enzymes encoded by the viral nucleic acid: the lysogenic cycle and the lytic cycle. In the lysogenic cycle, the viral nucleic acid encodes a repressor enzyme which prevents viral reproduction, followed by the viral DNA inserting itself into the host cell DNA and lying dormant indefinitely. During cellular stress, the dormant virus can enter the lytic cycle. In the lytic cycle, the viral nucleic acid commandeers the cell's resources, which are directed to synthesize up to several thousand new viruses. The end of the lytic cycle is cell rupture with the release of thousands of new viruses, each of which can infect new cells.

In 1909, Peyton Rous began research in pathology at the Rockefeller Institute for Medical Research (now Rockefeller University) in New York City. Rous was interested in the physiology of cancer within mammals and birds. He discovered a type of connective tissue cancer in chickens, called Rous sarcoma, which causes gross hypertrophy (enlargement) of certain organs, particularly the liver and gallbladder.

Rous sarcoma eventually is fatal. In his experiments, Rous grafted sarcoma tumor cells from diseased hens to healthy hens; the healthy hens contracted the disease. He then cultivated hen tumor cells, extracted a fluid not containing cells, and injected this fluid into healthy hens. Again, the healthy hens contracted the disease. His results pointed toward one possible conclusion: Some noncellular component of the tumor extract was capable of producing cancer in healthy hens. The active agent was not bacterial, protozoan, or fungal, because the tumor extract contained no cells. The most plausible explanation was a virus. Further filtration and infection experiments yielded identical results. Rous hypothesized that a Rous sarcoma virus caused this chicken sarcoma. Nevertheless, his work was derided by his peers who unsuccessfully repeated his experiments with other species. The failure of many to accept his conclusion reflected a considerable lack of understanding of both viruses and cancer by the medical and scientific community of that time. Despite the negative reactions, Rous continued his studies of liver and gallbladder physiology.

With greater understanding of viruses during subsequent decades of the twentieth century, Rous's viral theory of cancer began to be recognized. From his studies of Rous sarcoma virus, his theory maintained that some cancers could be caused by viruses. The discovery of more tumor-causing viruses during the 1950's resulted in Rous receiving the 1966 Nobel Prize in Physiology or Medicine, which he shared with Charles Brenton Huggins.

Since Rous's initial discovery, additional oncogenic (tumor-causing) viruses have been identified, including the RNA tumor viruses feline leukemia virus (cat leukemia), T-cell lymphotropic virus (human leukemia), and mouse mammary tumor virus. The Rous sarcoma virus also was shown to be an RNA retrovirus whose nucleic acid is RNA. DNA oncogenic viruses include hepatitis B (serum hepatitis and liver cancer), papilloma virus (warts), the Epstein-Barr virus (mononucleosis and Burkitt's lymphoma), and herpes simplex virus II (genital herpes and cervical cancer).

Oncogenic viruses are capable of cellular transformation, converting normal cells to abnormal growth patterns. The abnormal cell growth may proceed with rapid cellular divisions, gain or loss of chromosomes, and unusual production of certain proteins. If the tumor begins to invade neighboring healthy tissues and to enter the bloodstream for transport to other body regions (metastasis), then the tumor has become a cancer with life-threatening potential. Oncogenic viruses are transmitted like any virus, principally by air, liquid, direct contact, or especially the transfer of bodily fluids. Once an oncogenic virus contacts a target cell, it proceeds either into the lytic or lysogenic cycles. If the virus follows a lytic pathway, it will release proteins capable of cellular transformation, thus causing the host cell to become cancerous and to multiply out of control. As the tumor cell multiplies, it produces more viruses which bud off from the tumor cell membrane to infect neighboring cells. If the virus follows a lysogenic pathway, it may be dormant within the host cell DNA for years before emerging, entering the lytic cycle, and transforming the cell into a tumor.

The Rous sarcoma virus resurfaced in the 1960's with Rous's Nobel Prize and with the work of two molecular virologists, Howard M. Temin and David Baltimore. Temin discovered that the Rous sarcoma virus can copy a DNA polynucleotide chain from the viral RNA originally injected into the host cell. Soon thereafter, Temin and Baltimore independently discovered the enzyme reverse transcriptase, which RNA retroviruses use to encode DNA from RNA. This discovery overturned the molecular biological view that DNA encoded RNA exclusively. While their results initially were seriously questioned, mounting evidence led to Temin and Baltimore receiving the 1975 Nobel Prize in Physiology or Medicine.

Impact of Event

The discovery that Rous sarcoma is caused by a virus not only helped to reaffirm Ivanowsky's discovery of viruses as pathogens but also revealed that some cancers are infectious, being spread from individual to individual via viruses. Rous's viral theory of cancer placed cancer into a totally different perspective.

The Rous sarcoma virus was the first of more than twenty-four oncogenic viruses discovered during the twentieth century. Several of these viruses cause other diseases in addition to cancer. For example, the Epstein-Barr virus causes infectious mononucleosis and may be responsible for certain cases of chronic fatigue. This same virus, however, can cause a rare type of lymph node cancer called Burkitt's lymphoma. Similarly, the liver disease hepatitis afflicts approximately two hundred million people worldwide each year. Hepatitis is caused by the hepatitis A, B, and C viruses and is transmitted person-to-person by contaminated food and by contaminated drinking water. The hepatitis B virus also can cause liver cancer. A person may contract hepatitis B, recover, and then contract liver cancer many years later. The hepatitis B virus can lie dormant within host liver cell DNA for long periods of time. Health-care workers have approximately a 40 percent chance of encountering the hepatitis virus during their careers; however, antihepatitis vaccines are available.

Rous's discovery also paved the way for a better understanding of the origin of viruses. Viruses most likely evolved from cells because viruses are noncellular, because they must reproduce inside cells, and because they have the same genetic code as living cells. It is possible that, more than one billion years ago, a small group of genes capable only of reproduction and of manufacturing a protective protein covering escaped from a cell and temporarily existed outside cells in an inactive, dormant state. Viruses could be intercellular messengers whose functions went awry.

In the 1960's, Temin and Baltimore demonstrated that RNA retroviruses, such as Rous sarcoma virus, could encode DNA from RNA using a special viral enzyme called reverse transcriptase. This phenomenon went against established scientific dogma, which maintained that DNA encodes RNA. The list of such RNA retroviruses includes the notorious human immunodeficiency virus (HIV), the causative agent of acquired immune deficiency syndrome (AIDS) in humans. While HIV causes AIDS, it does not cause cancer; instead, it destroys an individual's immune system cells such that a person's body is unable to defend itself from secondary infections, such as pneumonia, and spontaneous cancers.

The knowledge that viruses can induce tumors in normal cells has been valuable for manipulating viruses as cloning vectors. A molecular biologist can clone a particular gene of interest, package the gene within a virus' DNA, and then infect a desired target cell with the virus. If the virus DNA enters the lysogenic cycle once it is inside the target cell, then the scientist effectively will have cloned a specific gene into a host cell's DNA using the virus as a transport mechanism.

Finally, the study of viruses as disease- and cancer-causing agents has led to the discovery of even smaller, bizarre life-forms which also cause disease. A prion is a little-understood mass of protein, containing no nucleic acid, that has been implicated in a number of mammalian central nervous system diseases, including scrapie, kuru, and Creutzfeld-Jacob disease.

Bibliography

Bishop, J. Michael. "Oncogenes." *Scientific American* 246 (March, 1982): 80-92. This review article, written by a major oncogenic virus researcher, is a discussion of how oncogenic viruses infect cells and initiate cancer. The Rous sarcoma virus is described, along with the work of Rous, Temin, and Baltimore. The article has superb diagrams and illustrations.

Gardner, Eldon J., and D. Peter Snustad. *Principles of Genetics.* 7th ed. New York: John Wiley & Sons, 1984. This outstanding introductory genetics textbook for undergraduate biology majors is a comprehensive survey of the field. It is clearly written with numerous diagrams and illustrations. Chapter 11, "Regulation of Gene Expression and Development," includes a good discussion of oncogenes and RNA tumor viruses, including the Rous sarcoma virus.

Lechevalier, Hubert A., and Morris Solotorovsky. *Three Centuries of Microbiology.* New York: McGraw-Hill, 1965. This outstanding book is a detailed history of microbiological research from the 1600's to the present, including the many famous experiments of Pasteur, Koch, Ivanowsky, Rous, and others. Chapter 8, "Viruses and Rickettsiae," describes Rous' famous experiment with quotations from his 1911 paper in volume 56 of the *Journal of the American Medical Association.*

Lewin, Benjamin. *Genes.* 2d ed. New York: John Wiley & Sons, 1985. This outstanding textbook of molecular biology for advanced undergraduate and graduate students is a thorough, detailed survey of the science behind biotechnology. It contains excellent illustrations and up-to-date references. Chapter 36, "Mobile Elements in Eukaryotes," includes a discussion of retroviruses, their possible origins, and how they infect cells.

Raven, Peter H., and George B. Johnson. *Biology.* 2d ed. St. Louis: Times Mirror/Mosby, 1989. This outstanding introductory biology textbook for undergraduate biology majors is clearly written with beautiful photographs and illustrations. Chapter 29, "Viruses," provides a very good description of the lytic and lysogenic cycles, classification of viruses, and tumor-causing viruses.

Wistreich, George A., and Max D. Lechtman. *Microbiology.* New York: Macmillan,

1988. This microbiology book is both an outstanding reference work and an excellent introductory textbook for undergraduate biology students. It covers every aspect of microbiology in extensive detail, supported by a plethora of charts, diagrams, tables, and photographs. It provides disease case histories and an extensive coverage of viruses. Great book for the serious microbiologist.

David Wason Hollar, Jr.

Cross-References

Sabin Develops an Oral Polio Vaccine (1957), p. 1522; A Vaccine Is Developed for German Measles (1960), p. 1654; Horsfall Announces That Cancer Results from Alterations in the DNA of Cells (1961), p. 1681; The U.S. Centers for Disease Control Recognize AIDS for the First Time (1981), p. 2149.

RUSSELL AND WHITEHEAD'S *PRINCIPIA MATHEMATICA* DEVELOPS THE LOGISTIC MOVEMENT IN MATHEMATICS

Category of event: Mathematics
Time: 1910
Locale: Trinity College, Cambridge, England

Russell and Whitehead's attempt to deduce mathematics from logic in Principia Mathematica *gave the logistic movement in mathematics its definitive expression*

Principal personages:

BERTRAND RUSSELL (1872-1970), an English philosopher, mathematician, and social reformer who was the winner of the 1950 Nobel Prize in Literature

ALFRED NORTH WHITEHEAD (1861-1947), an English philosopher and mathematician who made one of the outstanding attempts in his generation to produce a comprehensive metaphysical system that included scientific cosmology

GOTTLOB FREGE (1848-1925), a German mathematician and philosopher who was the founder of modern mathematical logic

GIUSEPPE PEANO (1858-1932), an Italian mathematician and logician who was a professor of mathematics at Turin

Summary of Event

At the end of the nineteenth century, several new approaches to the foundations of mathematics were developing as a response to a growing number of issues that challenged the stability of the previously accepted foundations of mathematics. By the first decade of the twentieth century, these new approaches divided many mathematicians into opposing schools of thought and provided the grounds for disagreement as to the proper foundations of mathematics. These new approaches provided the grounds for the three principal contemporary philosophies of mathematics: the logistic school, of which Bertrand Russell and Alfred North Whitehead are the chief expositors; the intuitionist school, led by the Dutch mathematician L. E. J. Brouwer; and the formalist school, developed principally by the German mathematician and logician David Hilbert. While there are contemporary philosophies of mathematics other than these three, none of them has been as widely followed or has developed such a large body of associated literature.

The basic thesis of the logistic school is that mathematics is derivable from logic. The logistic school maintains that mathematics is a branch of logic, rather than merely a tool of mathematics. With the development of the logistic school, logic became the forefather of mathematics in that all mathematical concepts were to be developed as theorems of logic, and the distinction between mathematics and logic became merely one of practical convenience. The logicists argued that because the laws of logic are accepted as a body of truths (at least in the early 1900's), mathematics must be accepted also as a body of truths. Furthermore, they contended that since truth is consistent, so is logic and mathematics.

The logicist thesis that mathematics is derivable from logic can be traced back to the German philosopher Gottfried Wilhelm Leibniz. Leibniz made a distinction between necessary truths and contingent truths; a truth is called "necessary" when its opposite implies a contradiction (for example, all right angles are equal) and "contingent" when it is not necessary, and there are bodies in nature that possess an angle of exactly 90 degrees. Hence, Leibniz considered all mathematical truths to be necessary, and as such, derivable from logic whose principles are also necessary and hold true in all possible worlds. Nevertheless, Leibniz did not go on to derive mathematics from logic, nor did anyone else with similar beliefs for almost two hundred years.

It was not until the late nineteenth century that the German mathematician Gottlob Frege undertook to develop the logistic thesis. Frege thought that laws of mathematics say no more than what is implicit in the principles of logic, which are a priori truths. In his *Die Grundlagen der Arithmetik* (1884; *The Foundations of Arithmetic*, 1950) and in his two-volume *Grundgesetze der Arithmetik* (1893, 1903; the basic laws of arithmetic), Frege proceeded to derive the concepts of arithmetic and the definitions and laws of number from logical premises. From the laws of number, it is possible to deduce algebra, analysis, and even geometry because analytic geometry expresses the concepts and properties of geometry in algebraic terms. Unfortunately, Frege's symbolism was complex and strange to mathematicians. He therefore had little influence on his contemporaries.

Another important forerunner of the logicist school was the Italian mathematician and logician Giuseppe Peano, who, between 1889 and 1908, had undertaken to state the theorems of mathematics by means of logical symbolization. Russell was greatly influenced by Peano's work and met him at the Second International Congress of Philosophy in Paris in 1900. Russell carefully studied Peano's work and adopted his notation as an instrument of analysis.

Although Russell was to conceive of the same program as Frege, he did it without any knowledge of Frege's program. It was only while he was developing his own program that he ran across Frege's work. In the early 1900's, Russell believed with Frege that since logic was a body of truths, if the fundamental laws of mathematics could be derived from logic, then these laws would also be truths and consistent. In *The Principles of Mathematics* (1903), Russell writes: "The fact that all Mathematics is Symbolic Logic is one of the greatest discoveries of our age; and when this fact has been established, the remainder of the principles of mathematics consists in the analysis of Symbolic Logic itself." The successful completion of the logistic school's program would leave the foundations of mathematics beyond doubt.

The logistic school received its definitive expression in the monumental *Principia Mathematica* (henceforth referred to as *Principia*), written by Russell in collaboration with his former teacher, the English philosopher Whitehead. Russell and Whitehead spent 1900 to 1911 developing what ultimately became *Principia*—the definitive version of the logistic school's position. This complex work purports to be a detailed reduction of the whole of mathematics to logic. Even though the contents of *Principia* elude summary—for to do so would be as difficult as summarizing a dictionary—a brief sketch of the contents is in order. *Principia* begins with the development of logic itself. Axioms of logic (for example, if q is true, then p or q is true) are carefully stated, from which theorems are deduced to be used in subsequent reasoning. The development starts with primitive (or undefined) ideas and propositions. These primitive ideas and propositions are taken as descriptions and hypotheses concerning the real world, and although they are explained, the explanations are not part of the logical development. The aim of *Principia* is to develop mathematical concepts and theorems from these primitive ideas and propositions, starting with a calculus of propositions, proceeding up through the theory of sets and relations to the establishment of the natural number system, and then to all mathematics derivable from the natural system. In this development, the natural numbers emerge with the unique meanings ordinarily assigned to them.

After having built up the logic of propositions, Russell and Whitehead proceeded to propositional functions. These, in effect, represent sets (or classes), for instead of naming the members of a set, a propositional function describes them by a property. For example, the propositional function "x is red" denotes the set of all red objects. This method of defining a set enables one to define infinite sets as readily as finite sets of objects. Russell and Whitehead wanted to avoid the paradoxes that arise when a collection of objects is defined that contains itself as a member. Although most sets are not members of themselves, some are. For example, while the set of all frogs is not a frog, the set of all comprehensible things is itself a comprehensible thing. Thus, to avoid the paradoxes, they require that no set is a member of itself and introduce the theory of types to carry out this restriction. The basic idea of their theory of types is that a set is on a higher level than its members; the set of which this set is a member is on a still higher level, and so on. Hence, individuals, such as a particular frog, are type 0. An assertion about a property of individuals is of type 1, and so on.

Russell and Whitehead then go on to address the theory of relations, stating that relations are expressed by means of propositional functions of two or more variables ("x loves y" expresses a relation). Next was an explicit theory of sets defined in terms of propositional functions. On this basis, Russell and Whitehead were prepared to introduce the notion of natural numbers. Given the natural numbers, it is possible to build up the real and complex number systems, functions, and all of analysis. To accomplish their objective, however, Russell and Whitehead had to introduce two more axioms: the axiom that infinite sets exist and the axiom of choice. These axioms were to become the focus of much criticism.

Impact of Event

In 1959, Russell wrote that he "used to know of only six people who had read the later parts of the book." Yet, despite Russell's reservations, *Principia* has attracted much careful study and has received extensive critical attention. In fact, Russell and Whitehead's logical investigations in the foundations of mathematics opened up a world of possibilities not only among mathematicians but also among logicians and philosophers, and the study of the philosophy of logic became a central concern for philosophy itself.

Nevertheless, the great achievement of *Principia* was not enough to shield the logistic approach to mathematics from a barrage of criticism. One point of attack has been directed toward the axioms of reducibility, choice, and infinity that were used in *Principia*. Controversy and discussion regarding these axioms has focused on the purity of the logic used in *Principia*. For example, the axiom of reducibility has been said to be arbitrary and lacking evidence. Criticism has gone as far as to question whether it is an axiom of logic.

Even Russell and Whitehead were uneasy about this axiom in the first edition of *Principia*; in the second edition, they attempted to rephrase the axiom, but in doing so only created new difficulties. Although they thought that the axiom was justified on the pragmatic basis of leading to the desired conclusion, it was not a justification with which they could rest content. They had to determine how essential this axiom was to the logistic program. All efforts to reduce mathemetics to logic without the axiom have failed so far.

The use of the axioms of reducibility, infinity, and choice has challenged the entire logistic program and has raised questions as to where the line between logic and mathematics is to be drawn. On the one hand, if logic actually contains these three controversial axioms—as proponents of the logistic program maintain—then the logic of *Principia* is pure. On the other hand, if the logic of *Principia* is not pure, opponents of the logistic program deny that mathematics, or even any important branch of mathematics, had yet been reduced to logic. Others, while arguing for the impurity of the logic of *Principia*, are willing to extend the meaning of the term logic so that it includes these axioms.

Another criticism charged that *Principia* only reduced arithmetic, algebra, and analysis to logic, but did not reduce the nonarithmetical parts of mathematics (such as geometry, topology, and abstract algebra) to logic. Others, who hold that all mathematics can be reduced to logic, claim that it is possible to reduce geometry, topology, and abstract algebra to logic.

Although both *Principia*, in particular, and the logistic program, in general, have had a long, complicated development and have been criticized on various grounds, many of which have not been mentioned, whether or not the logistic thesis has been established seems to be a matter of opinion. Although some accept the program as satisfactory, even though they might be critical of its present state, others have found its shortcomings insurmountable, charging that it produces conclusions formed in advance from unwarranted assumptions.

Bibliography

Bell, E. T. *The Development of Mathematics.* 2d ed. New York: McGraw-Hill, 1945. A broad account of the general development of mathematics, with particular emphasis on main concepts and methods. It is highly accessible to the general reader. Chapter 23, entitled "Uncertainties and Probabilities," is particularly helpful for gaining an understanding of the state of mathematics around the time of *Principia*.

Jager, Ronald. *The Development of Bertrand Russell's Philosophy.* London: George Allen & Unwin, 1972. A fine survey of the development of Russell's philosophy. One part of the book is devoted to Russell's philosophy of mathematics, and two sections of this part are particularly relevant: section A, entitled "The Poetry and Essence of Mathematics," and section B, entitled "Logicism." Accessible to the diligent general reader.

Kneebone, G. T. *Mathematical Logic and the Foundations of Mathematics: An Introductory Survey.* New York: D. Van Nostrand, 1963. Even though this book is based on a series of lectures given in the University of London to advanced undergraduates and graduate students, it has much material for the general reader interested in an introduction to mathematical logic and the philosophy of logic. Difficult sections are marked by an asterisk so that they may be easily avoided. Generous amounts of space are given to nontechnical discussions of *Principia*.

Russell, Bertrand. *Introduction to Mathematical Philosophy.* New York: Macmillan, 1919. Russell presents results "hitherto only available to those who have mastered logical symbolism, in a form offering the minimum of difficulty to the beginner." An excellent nontechnical introduction to Russell's work on mathematics and logic, and a highly recommended source for this subject.

____________. *My Philosophical Development.* New York: Simon & Schuster, 1959. This is the place to read Russell's thoughts on the mathematical and philosophical aspects of *Principia*. Russell's discussion is addressed to a general audience and is informative and enjoyable reading. Cites influences on his work and provides a good general account of *Principia*. Particularly relevant are chapters 7 and 8.

Schoenman, Ralph, ed. *Bertrand Russell: Philosopher of the Century.* London: George Allen & Unwin, 1967. An excellent selection of essays. Section 4, entitled "Mathematician and Logician," is especially useful. Hilary Putnam's article in section 4 entitled "The Thesis That Mathematics Is Logic" is particularly relevant.

Whitehead, Alfred North, and Bertrand Russell. *Principia Mathematica.* 3 vols. Cambridge, England: Cambridge University Press, 1910-1913. A monumental work, the culmination of ten years of development, and the founding work of the logistic school. It is virtually incomprehensible to those without a knowledge of formal logic, although the introduction is somewhat accessible to the diligent general reader.

Jeffrey R. DiLeo

Cross-References

Hilbert Develops a Model for Euclidean Geometry in Arithmetic (1898), p. 31; Levi Recognizes the Axiom of Choice in Set Theory (1902), p. 143; Russell Discovers the "Great Paradox" Concerning the Set of All Sets (1902), p. 184; Brouwer Develops Intuitionist Foundations of Mathematics (1904), p. 228; Zermelo Undertakes the First Comprehensive Axiomatization of Set Theory (1904), p. 233; Gödel Proves Incompleteness-Inconsistency for Formal Systems, Including Arithmetic (1929), p. 900.

GREAT EVENTS FROM HISTORY II

CHRONOLOGICAL LIST OF EVENTS

VOLUME I

VOLUME II

VOLUME III

VOLUME IV

VOLUME V